Silicon Anode Systems for Lithium-Ion Batteries

Silicon Anode Systems for Lithium-Ion Batteries

Edited by

Prashant N. Kumta
Edward R. Weidlein Endowed Chair Professor, Swanson School of Engineering, University of Pittsburgh, Pittsburgh, PA, United States

Aloysius F. Hepp
Chief Technologist, Nanotech Innovations LLC, Oberlin, OH, United States

Moni K. Datta
Assistant Professor, Bioengineering Department, University of Pittsburgh, Pittsburgh, PA, United States

Oleg I. Velikokhatnyi
Assistant Professor, Bioengineering Department, University of Pittsburgh, Pittsburgh, PA, United States

ELSEVIER

Elsevier
Radarweg 29, PO Box 211, 1000 AE Amsterdam, Netherlands
The Boulevard, Langford Lane, Kidlington, Oxford OX5 1GB, United Kingdom
50 Hampshire Street, 5th Floor, Cambridge, MA 02139, United States

Copyright © 2022 Elsevier Inc. All rights reserved.

No part of this publication may be reproduced or transmitted in any form or by any means, electronic or mechanical, including photocopying, recording, or any information storage and retrieval system, without permission in writing from the publisher. Details on how to seek permission, further information about the Publisher's permissions policies and our arrangements with organizations such as the Copyright Clearance Center and the Copyright Licensing Agency, can be found at our website: www.elsevier.com/permissions.

This book and the individual contributions contained in it are protected under copyright by the Publisher (other than as may be noted herein).

Notices
Knowledge and best practice in this field are constantly changing. As new research and experience broaden our understanding, changes in research methods, professional practices, or medical treatment may become necessary.

Practitioners and researchers must always rely on their own experience and knowledge in evaluating and using any information, methods, compounds, or experiments described herein. In using such information or methods they should be mindful of their own safety and the safety of others, including parties for whom they have a professional responsibility.

To the fullest extent of the law, neither the Publisher nor the authors, contributors, or editors, assume any liability for any injury and/or damage to persons or property as a matter of products liability, negligence or otherwise, or from any use or operation of any methods, products, instructions, or ideas contained in the material herein.

Library of Congress Cataloging-in-Publication Data
A catalog record for this book is available from the Library of Congress

British Library Cataloguing-in-Publication Data
A catalogue record for this book is available from the British Library

ISBN: 978-0-12-819660-1

For information on all Elsevier publications
visit our website at https://www.elsevier.com/books-and-journals

Publisher: Matthew Deans
Acquisitions Editor: Christina Gifford
Editorial Project Manager: Rachel Pomery
Production Project Manager: Sojan P. Pazhayattil
Cover Designer: Victoria Pearson

Typeset by STRAIVE, India

		5.2.1 Energy balance of the Li ion half-cell discharge process	162
	5.3	Computational method	164
	5.4	Problem description	164
	5.5	Results and discussion	165
		5.5.1 Effect of buffer layer stiffness on the *a*-Si pattern anode stability	166
		5.5.2 Effect of interface properties on the *a*-Si pattern anode stability	172
	5.6	Discussion	175
	5.7	Conclusions	176
		Appendix	176
		Acknowledgments	178
		References	178

Part III
Electrolytes and surface electroyle interphase (SEI) issues

6. SEI layer and impact on Si-anodes for Li-ion batteries
Partha Saha, Tandra Rani Mohanta, and Abhishek Kumar

6.1	Introduction	183
6.2	Energetics of SEI	189
6.3	SEI compositions via electrolyte and salt decomposition	191
6.4	Electrolyte additives and their roles during SEI formation	195
	6.4.1 Role of vinylene carbonate	202
	6.4.2 Role of fluoroethylene carbonate	203
6.5	SEI layer properties	208
6.6	SEI formation on Si/SiO$_x$ electrodes	208
6.7	SEI growth model	212
6.8	Mechanical deformation and associated strain of SEI layer	212
6.9	Computational work on SEI formation	217
6.10	Coating strategies and core-shell/yolk-shell morphology for stable SEI formation	219
6.11	Binders	222
	6.11.1 Problem with conventional binders	222
	6.11.2 Polymeric binders with supramolecular chemistry	224
	6.11.3 Self-healing polymeric binders	245
	6.11.4 Polymer binders: Impact of various architectures	246
6.12	Conclusions and future outlook	247
	References	248
	Further reading	263

7. Active/inactive phases, binders, and impact of electrolyte
Chen Fang and Gao Liu

7.1	Active and inactive phases of Si-based materials	265
7.2	Silicon electrode binders	272
	7.2.1 Adhesion	272
	7.2.2 Ionic conductivity	275
	7.2.3 Electrical conductivity	276
	7.2.4 A holistic approach	277
7.3	Electrolytes and additives	277
	7.3.1 Carbonate electrolytes and lithium salts	278
	7.3.2 Electrolyte additives	281
	7.3.3 Polymer electrolytes	283
	7.3.4 Ionic liquid electrolytes	286
	7.3.5 Analytical methods of surface electrolyte interphases	287
7.4	Conclusions	289
	References	290

Part IV
Achieving high performance

8. Performance degradation modeling in silicon anodes
Partha P. Mukherjee and Ankit Verma

8.1	Introduction	299
8.2	Morphology performance interactions in silicon anodes	300
	8.2.1 Electrochemistry transport dynamics in silicon anodes	301
	8.2.2 Representative highlights	306
8.3	Degradation phenomena in silicon anodes	312
	8.3.1 Mechano-electrochemical stochastics in silicon anodes	312
	8.3.2 Representative highlights	316
8.4	Summary and outlook	326
	Acknowledgments	327
	References	327

9. Nanostructured 3D (three dimensional) electrode architectures of silicon for high-performance Li-ion batteries
Surendra K. Martha, Liju Elias, and Sourav Ghosh

9.1	Introduction	331
9.2	Structure and electrochemistry of silicon	332
9.3	The failure mechanism of silicon anodes	334

9.4	Mitigation strategies			335
9.5	Nanostructured 3D electrode architectures			340
	9.5.1	3D porous silicon-carbon composite		340
	9.5.2	3D porous metallic current collector as a scaffold		346
	9.5.3	3D carbon-based scaffold		352
	9.5.4	3D porous conductive polymer framework		360
9.6	Conclusions			363
	References			364

10. Processing and properties of silicon anode materials
Raj N. Singh

10.1	Introduction		373
10.2	Experimental methods		375
	10.2.1	Single crystal silicon anode preparation and electrochemical testing for residual stress	375
	10.2.2	Processing of etched anode from Si powder	376
	10.2.3	Processing of anode from Si powder (unetched)	378
10.3	Results and discussion		379
	10.3.1	Evolution of residual stress in single crystal Si and fracture	379
	10.3.2	Processing and properties of hierarchical Si anode from etched silicon powder	385
	10.3.3	Processing and properties of Si anode from unetched-silicon powder	394
10.4	Conclusions		401
	Acknowledgments		402
	References		403

Part V
Applications and future directions

11. Advanced silicon-based electrodes for high-energy lithium-ion batteries
Dominic Leblanc, Abdelbast Guerfi, Myunghun Cho, Andrea Paolella, Yuesheng Wang, Alain Mauger, Christian Julien, and Karim Zaghib

11.1	Introduction		411
11.2	Silicon as a low-cost anode material		414
	11.2.1	Metallurgical-grade silicon produced by the carbothermal process	414
	11.2.2	High-purity silicon produced via the Siemens process	414
11.3	Mechanically milled silicon nanopowder		415
	11.3.1	Fracture mechanics: Critical defect size	416
	11.3.2	Decrease in the mechanical size of silicon	417

		11.3.3	Size effects on electrochemical properties	418
		11.3.4	Wet grinding optimization	420
	11.4	\multicolumn{2}{l	}{Silicon nanospheres using induced plasma}	422
		11.4.1	Synthesis of nano-silicon using the physical process	423
		11.4.2	Synthesis of nano-silicon using the chemical process	423
		11.4.3	In situ TEM lithiation of silicon nanospheres	424
		11.4.4	Si thin films	425
		11.4.5	Si nanowires	426
		11.4.6	Porous silicon	426
		11.4.7	Nano-Si/pyrolysis carbon microparticles	429
		11.4.8	Nano-Si/graphite microparticles	430
		11.4.9	Nano-Si/graphene composite	432
	11.5	\multicolumn{2}{l	}{Nano-structured SiO_x and its use as a lithium-ion battery anode}	434
		11.5.1	SiO_x nanowire synthesis	435
		11.5.2	Electrochemical evaluation of SiO_x nanowires	436
	11.6	\multicolumn{2}{l	}{Binder selection}	437
	11.7	\multicolumn{2}{l	}{Reducing surface reactivity of nano-silicon}	440
		11.7.1	Protection against water	440
		11.7.2	Stabilization of the SEI	440
	11.8	\multicolumn{2}{l	}{Conclusion}	444
		\multicolumn{2}{l	}{References}	445

12. Batteries for integrated power and CubeSats: Recent developments and future prospects

Aloysius F. Hepp, Prashant N. Kumta,
Oleg I. Velikokhatnyi, and Ryne P. Raffaelle

	12.1	Introduction		457
	12.2	\multicolumn{2}{l	}{Integrated micropower source in space: Starshine 3}	458
		12.2.1	Integrated power technologies: Multi-junction devices and motivation	458
		12.2.2	Starshine 3 flight opportunity: Integrated micro power system design	460
		12.2.3	Starshine 3 integrated power system flight experiment and results	462
	12.3	\multicolumn{2}{l	}{Integrated power technologies — 2005–2020}	464
		12.3.1	Next steps after Starshine 3: Solar charging of electric vehicle batteries	464
		12.3.2	Monolithic integration and further examples of integrated power technologies	467
		12.3.3	Integrated power technologies: Applications, challenges, and practical concerns	475
	12.4	\multicolumn{2}{l	}{Energy storage options for CubeSats}	479
		12.4.1	CubeSats: Background, present issues, and future missions	481

	12.4.2	Potential failure modes for commercial Li-ion batteries in low-earth orbit	485
	12.4.3	Integrating batteries into structural elements to enhance performance	487
	12.4.4	Use of commercial batteries on a successful inter-planetary CubeSat mission	490
12.5	**Conclusions**		497
	Acknowledgments		498
	References		498

Index   509

Contributors

Numbers in parentheses indicate the pages on which the author's contributions begin.

Myunghun Cho (411), Center of Excellence in Transportation Electrification and Energy Storage (CETEES), Hydro-Québec, Varennes, QC, Canada

Sameer S. Damle (157), Department of Chemical and Petroleum Engineering, University of Pittsburgh, Pittsburgh, PA, United States

Moni K. Datta (47), Department of Bioengineering; Center for Complex Engineered Multifunctional Materials, University of Pittsburgh, Pittsburgh, PA, United States

Liju Elias (331), Department of Chemistry, Indian Institute of Technology Hyderabad, Sangareddy, Telangana, India

Chen Fang (265), Energy Storage and Distributed Resources Division, Lawrence Berkeley National Laboratory, Berkeley, CA, United States

Sourav Ghosh (47,331), Department of Chemistry, Indian Institute of Technology Hyderabad, Sangareddy, Telangana, India

Abdelbast Guerfi (411), Center of Excellence in Transportation Electrification and Energy Storage (CETEES), Hydro-Québec, Varennes, QC, Canada

Ivana Hasa (3), WMG, The University of Warwick, Coventry, United Kingdom

Aloysius F. Hepp (47,157), Nanotech Innovations, LLC, Oberlin, OH, United States

Christian Julien (411), Institut de Minéralogie, de Physique des Matériaux et de Cosmochimie (IMPMC), Sorbonne Université, UMR-CNRS 7590, Paris, France

Abhishek Kumar (183), Department of Ceramic Engineering, National Institute of Technology, Rourkela, Odisha, India

Prashant N. Kumta (47,157,457), Department of Chemical and Petroleum Engineering; Department of Bioengineering; Mechanical Engineering and Materials Science; Center for Complex Engineered Multifunctional Materials, University of Pittsburgh, Pittsburgh, PA, United States

Dominic Leblanc (411), Center of Excellence in Transportation Electrification and Energy Storage (CETEES), Hydro-Québec, Varennes, QC, Canada

P.-K. Lee (119), School of Energy and Environment, City University of Hong Kong, Kowloon, Hong Kong

Gao Liu (265), Energy Storage and Distributed Resources Division, Lawrence Berkeley National Laboratory, Berkeley, CA, United States

Spandan Maiti (157), Department of Chemical and Petroleum Engineering; Department of Bioengineering; Mechanical Engineering and Materials Science, University of Pittsburgh, Pittsburgh, PA, United States

Surendra K. Martha (47,331), Department of Chemistry, Indian Institute of Technology Hyderabad, Sangareddy, Telangana, India

Alain Mauger (411), Institut de Minéralogie, de Physique des Matériaux et de Cosmochimie (IMPMC), Sorbonne Université, UMR-CNRS 7590, Paris, France

Tandra Rani Mohanta (183), Department of Ceramic Engineering, National Institute of Technology, Rourkela, Odisha, India

Partha P. Mukherjee (299), School of Mechanical Engineering, Purdue University, West Lafayette, IN, United States

Siladitya Pal (157), Department of Bioengineering, University of Pittsburgh, Pittsburgh, PA, United States

Andrea Paolella (411), Center of Excellence in Transportation Electrification and Energy Storage (CETEES), Hydro-Québec, Varennes, QC, Canada

Stefano Passerini (3), Helmholtz Institute Ulm (HIU), Ulm; Karlsruhe Institute of Technology (KIT), Karlsruhe, Germany

Ryne P. Raffaelle (457), Department of Physics, Rochester Institute of Technology, Rochester, NY, United States

Partha Saha (183), Department of Ceramic Engineering; Centre for Nanomaterials, National Institute of Technology, Rourkela, Odisha, India

Raj N. Singh (373), School of Materials Science and Engineering, Oklahoma State University, Tulsa, OK, United States

T. Tan (119), School of Energy and Environment, City University of Hong Kong, Kowloon, Hong Kong

Oleg I. Velikokhatnyi (457), Department of Bioengineering, University of Pittsburgh, Pittsburgh, PA, United States

Ankit Verma (299), School of Mechanical Engineering, Purdue University, West Lafayette, IN, United States

S. Wang (119), School of Energy and Environment, City University of Hong Kong, Kowloon, Hong Kong

Yuesheng Wang (411), Center of Excellence in Transportation Electrification and Energy Storage (CETEES), Hydro-Québec, Varennes, QC, Canada

Haimin Yao (95), Department of Mechanical Engineering, The Hong Kong Polytechnic University, Kowloon, Hong Kong SAR, PR China

Qifang Yin (95), Department of Mechanical Engineering, The Hong Kong Polytechnic University, Kowloon, Hong Kong SAR; Department of Mechanical Engineering, School of Civil Engineering, Wuhan University, Wuhan, PR China

D.Y.W. Yu (119), School of Energy and Environment, City University of Hong Kong, Kowloon, Hong Kong

Karim Zaghib (411), Center of Excellence in Transportation Electrification and Energy Storage (CETEES), Hydro-Québec, Varennes, QC, Canada

Preface

Rechargeable Li-ion batteries (LIBs) have emerged as the flagship energy storage technology amongst various electrochemical energy storage (EES) systems explored since Sony commercialized the first LIB in 1991. The technology is projected to, in the near future, deliver a high usable specific energy ($\geq$500 Wh/kg) at the high operating ($\sim$3–5 V) voltage with continued advances forthcoming in the anode, cathode, and electrolyte systems. Current state-of-the-art (SOA) LIB cells still use graphite as the active anode material with Li-nickel manganese cobalt oxide (NMC) alongside SOA $LiCoO_2$ (LCO) as the cathode giving a usable energy density (100–200 Wh/kg). Despite considerable advancements in the graphite/Li-NMC-based LIBs, high energy density battery systems meeting the incessantly growing higher energy demand of electric vehicle (EV) applications as well as the sustained operation in low temperature low earth orbit (LEO) and deep space conditions are still elusive, due to the limited energy and power densities of current LIBs, thus warranting further research in advanced materials systems. To achieve the desired energy density of LIBs rendering them amenable for higher energy application ($\geq$350 Wh/kg, $\geq$750 Wh/L), silicon (Si)-based anode materials, a much anticipated next-generation LIB anode, were identified following the initial work of scientists at Fujifilm introducing tin (Sn)-based composite oxide (TCO) as a potential alternative to carbon anodes in 1997 presenting the use of an in-situ-generated lithium oxide (Li_2O) serving as a compliant Li-ion conducting channel for realizing the high capacity of Sn.

Silicon-based anodes falling in the same group of the periodic table (IVA or 14) as Sn are extremely attractive, exhibiting low operational voltage with a theoretical storage capacity of >3500 mAh/g; an order of magnitude greater than the theoretical capacity of traditionally used and extensively studied graphite (372 mAh/g) anodes. Unfortunately, the applicability and commercial implementation of universally pure Si or Si-based composite LIB anodes utilizing practically desired $\geq$30 wt.% Si composition (in the electrode) are hindered due to the colossal volume expansion/contraction ($\sim$300%) of Si occurring during alloying/dealloying with Li causing electrode decrepitation resulting in loss of particle contact eventually leading to failure of the battery with poor cycle life. Furthermore, irreversible chemical reactions between an organic electrolyte system and the reducing environment arising from the low potential of the ensuing Li-Si intermetallic alloy cause an unstable irreversible solid surface electrolyte interphase (SEI) at the electrode surface enhancing the charge transfer resistance further degrading the battery performance giving rise to unacceptable calendar life for automotive applications. Improvements in the structural stability/integrity with formation of stable SEI at the Si surface commensurate with achieving long-term cyclability of Si-based anodes have been recorded by researchers over the past two decades exploiting nanotechnology, primarily generation of nanoscale architectures of Si (e.g., nanotube, nanowire, nanodroplet) combined with nanocomposites of Si and C (e.g., Si/C nanocomposite). These systems harness the advantage of having a relatively inactive but compliant second phase such as carbon similar to Li_2O in TCO to endure the colossal volume expansion-related stresses and ensuing electrode cracking.

Thus the last two decades have witnessed numerous scientific reports demonstrating improvement in the cyclability of Si/C-based active materials by implementing proper design and effectively developing composite electrode (active materials + binder + conducting carbon) exploiting the mechanical properties of the binder combined with efficient interactions of the binder functional groups with the native silicon oxide layers inherent on the Si surface. A quick Scopus search of "silicon (Si) anodes" and "lithium-ion (Li-ion) batteries" produced 2447 results, approximately half (1201) of them since 2017 indicate the importance of the field. More than 85% of all such publications appeared within the last decade. There has been a nine-fold increase in publications in this area comparing the last (2110) and previous decades (235). Recognizing the importance of this topic, the editors of *Silicon Anode Systems for Lithium-Ion Batteries* selected and organized 12 (and coauthored three) chapters authored by experts in the fields of battery materials science, application, and energy storage. The main objective is to present the SOA and potential future developments in Si anodes for LIBs, including fundamentals, recent advances, mechanical and surface electrolyte interface challenges, approaches to achieve high(er) performance, and prospects for current and future energy storage-related applications of LIBs. The book is divided into five parts, each devoted to an important aspect of Si anodes for LIBs.

Part I: Introduction and background

The initial section includes two chapters that provide a background and introduce the main topic(s) of the book: LIBs and Si(containing) anode materials. Chapter 1, coauthored by Dr. I. Hasa from the University of Warwick (UK) and Prof. Stefano Passerini, Editor-in-Chief of Elsevier's *Journal of Power Sources*, from Helmholtz Institute Ulm and Karlsruhe Institute of Technology, provides an overview of Si electrodes for LIBs; it is presented from a material and an electrode perspective, highlighting the material's fundamental properties and electrode behavior. Strategies associated with Si nanostructuring and production of composite materials, aimed at improving the electrode cyclability, are also discussed as well as the unique nature of the SEI. Chapter 2 is a collaboration between several coeditors and Prof. S. Martha from the Indian Institute of Technology, Hyderabad, India; it covers three aspects of recent advances in Si materials for Li-ion batteries: novel processing, alternative starting materials, and more practical considerations. The authors discuss the progress achieved in making hybrid materials with Si as composite anodes, specifically carbon-silicon hybrid, oxide-silicon hybrid, silicon-metal hybrid, and silicides for LIBs. Finally, this chapter covers several topics that pertain to practical considerations such as alternative and low-cost processing methods.

Part II: Mechanical properties

The three chapters in the second part discuss various approaches to enhance understanding of the impact of the mechanical properties on the performance of LIBs with Si anodes. Chapter 3, a collaborative work authored by Dr Q. Yin from Wuhan University and Prof. H. Yao from Hong Kong Polytechnic University, is devoted to the investigation of the effects of the mechanical constraint on the performance(s) of Si nanosheets as the active material for LIBs using customized techniques for simulating lithiation and delithiation using molecular dynamics (MD) approaches. As mechanical constraints would have a significant impact on the capacity and lithiation rate of Si nanosheets, strategies for improving the capacity and lithiation rate of the constrained Si nanosheets are proposed. Chapter 4, authored by Prof. D. Yu from the City University of Hong Kong, concerns the mechanical properties of silicon-based electrodes. In this chapter, mechanical issues facing Si anodes during charge and discharge cycling are discussed including some of the recent developments to alleviate the problem. Characterization of single particles and nanowires provides visual observations of expansion and contraction mechanisms. As Si materials comprise anodes for practical applications, thickness change and its reversibility become important factors that govern the battery stability; other developments to sustain the cycling performance of Si electrodes are also introduced. Chapter 5, the final chapter of the section, coauthored by Prof. S. Maiti and colleagues at the University of Pittsburgh, Pittsburgh, PA, USA, is a theoretical study exploring the effect of insertion of an elastic buffer layer on the stability of an amorphous silicon (a-Si) thin-film patterned Li-ion anode serving to validate the previously published experimental evidence. In this study an innovative custom-developed multiphysics finite element approach revealed that the modulus mismatch between the Si thin film and the elastic layer sandwiched between this amorphous Si layer and the current collector is a key design parameter critical for improving the mechanical integrity of the silicon thin film as evinced by the published experimental studies. The study also showed that the adhesion between the different anode components is important in prolonging the silicon anode cycle life.

Part III: Electrolytes and surface electrolyte interphase issues

The third section consists of two chapters. In Chapter 6, Prof. P. Saha with coauthors from the National Institute of Technology, Rourkela, India, provides an overview of the SEI layer and its impact on Si-anodes for Li-ion batteries. The formation of SEI on crystalline and amorphous Si, at the Si/SiO$_x$ interface or surface is a complicated and complex process. This chapter provides a detailed understanding of SEI evolution and its formation on silicon anodes in liquid organic electrolytes and explains the stability and decomposition products of lithium salts, carbonate-based solvents, along with the formation of radical species including various inorganic and organic products during electrochemical cycling of Li$^+$ ions with Si/SiO$_x$ anodes. In Chapter 7, Drs. C. Fang and G. Liu from the Lawrence Berkeley National Laboratory, USA, present an in-depth discussion in three areas of advancements in Si-based electrode materials and electrolyte for advanced Li-ion rechargeable battery applications. First, it delves into the effects of active and inactive phases of Si and SiO$_2$ for battery performance, and the effects of need for carbon coating and composites with Si materials. Second, it discusses the use of electrode binders as an enabling technology for Si-based electrodes. Finally, the important role of electrolyte and additives for Si materials is discussed in detail.

Part IV: Achieving high(er) performance: Modeling and experimental perspectives

This section of the book includes three chapters, with a theoretical and experimental approaches. In Chapter 8, Prof. P. Mukherjee from Purdue University, USA, reports an overview of performance-degradation modeling in silicon anodes for Li-ion batteries. The chapter focuses on strategies used to model the performance-degradation interactions in silicon anodes as a result of their large volume expansion. The need for coupling of multiscale modeling paradigms for achieving a deep understanding of the interfacial and bulk phenomena in silicon anodes via multiphysics simulations is also discussed. In Chapter 9, Prof. S. Martha with coauthors outlines nanostructured 3D architectures of Si for future high-performance LIBs. This chapter presents the nanoengineering efforts undertaken to achieve a rational design of nanostructured silicon electrodes with excellent electrochemical performance and stability by giving a special emphasis on nanostructured 3D electrode architectures of Si. Chapter 10, authored by Prof. R. Singh at Oklahoma State University, covers SOA descriptions of the stress evolution, failure upon lithiation of silicon anodes, and different ways of processing silicon anodes to overcome the ensuing capacity fade. Several innovative approaches to understand failure mechanisms and alleviate battery capacity fade based upon research at Oklahoma State are described in some detail. The final section (Part V) of the book goes into greater depth covering potential future and practical applications.

Part V: Future directions: Novel devices and space applications

The two chapters in the final section discuss novel devices and structures as well as advanced applications, including small satellites. Chapter 11 is a collaborative work of researchers from Hydro-Québec and Sorbonne Université, introduces LIBs and processing of key anode materials. The purpose of this chapter is to review the progress on Si-based anodes with different morphologies in recent years. The performance of Si-based anodes largely depends on their size, geometry, and structure/microstructure. In practice, combining high-rate capability and high energy density in a single battery is difficult, both micro- and nanostructures have potential commercial opportunities, depending on which parameter should be favored according to the use envisioned for the cells. Finally, Chapter 12 is a collaboration between several coeditors and Dr. Ryne Raffaelle, Vice-Provost for Research at the Rochester Institute of Technology, providing an overview of batteries for integrated power and CubeSats. This chapter covers a number of topics that pertain to practical considerations for integrated power devices (and systems) and microsatellites (i.e., CubeSats), including their applicability to future space exploration. The chapter begins with a review of a successful integrated power device flight experiment demonstration on a small satellite (Starshine 3) in the early 2000s. Integrated power technologies that involve energy storage and energy conversion address several concerns about limitations imposed upon space flight hardware and systems related to power, space, and mass limitations. Four basic types of integrated power devices (or systems) are discussed, as well as potential applications; the authors also discuss advantages, challenges, and practical considerations. A successful demonstration of a CubeSat mission to Mars using commercial LIBs is outlined and finally, factors contributing to this successful mission are examined.

The success of this edited book project is clearly the result of the full commitment of each contributing author. Without their availability and cooperation to share their valuable knowledge and provide a critical in-depth review of the respective topics, this book could not be published meeting the high publication standards of intellectual information. Our heartfelt thanks also go to Rachel Pomery, our project manager, for her timely support and assistance. Finally, Christina Gifford, the acquisitions editor, is acknowledged for her constant support given throughout the entire publication process. Despite the extenuating circumstances surrounding the Coronavirus pandemic, the editors and the publication team are especially thankful to all the authors for providing their contributions in time resulting in this compilation. We sincerely hope that this collection of high quality work from accomplished and renowned researchers in a very germane area of LIBs will be received very favorably by the scientific community at large.

Prashant N. Kumta
University of Pittsburgh

Aloysius F. Hepp
Nanotech Innovations

Moni K. Datta
University of Pittsburgh

Oleg I. Velikokhatnyi
University of Pittsburgh

Part I

Introduction and background

Chapter 1

Silicon anode systems for lithium-ion batteries

Ivana Hasa[a] and Stefano Passerini[b,c]
[a]WMG, The University of Warwick, Coventry, United Kingdom, [b]Helmholtz Institute Ulm (HIU), Ulm, Germany, [c]Karlsruhe Institute of Technology (KIT), Karlsruhe, Germany

1.1 Introduction

The Nobel Prize 2019 in Chemistry was awarded jointly to John B. Goodenough, M. Stanley Whittingham and Akira Yoshino for the development of LIBs [1]. It took almost thirty years after the first commercialization of LIBs, to recognize the importance of the work conducted by the three Nobel laureates as well as the many other pioneering scientists around the world that contributed to this field. In 1991 Sony Energytec Inc. began to produce commercial lithium-ion cells based on the Asahi patents [2] and the world was on the verge of changing, fundamentally altering the way society communicates, works, travels, and lives. But the journey that led to this success was much longer and like in every successful story it started decades before.

It was in 1913, when pioneering research was first conducted by G.N. Lewis on the electrochemistry of lithium (Li) metal [3], but it was not until the 1960's that the electrochemistry of nonaqueous solvents gained attention in the academic community [4]. In 1958, William S. Harris described in his Ph.D. thesis the work conducted at the University of California on the electroplating of different metals in nonaqueous electrolytes (cyclic ester solvents) [5]. The study triggered interest in the use of propylene carbonate (PC) used in combination with alkali metals for electrochemical application. Interestingly, this solvent as well as other carbonates remain the standard electrolyte components for Li-ion cells today. It took another twenty years to formalize the conceptualization of the working principle of LIBs and to commence developing their practical use [6,7].

From 1970 until the mid-1980s several ground-breaking discoveries paved the way for LIBs commercialization. Among the research efforts conducted in that period, Stanley Whittingham stands out as a pioneer. At the time, Whittingham discovered the ability of titanium disulfide layered compounds (TiS$_2$) to reversibly host lithium ions through an intercalation mechanism. By exploiting this type of mechanism, the first commercial lithium metal batteries employing a liquid based electrolyte were developed and manufactured in USA by the Exxon Company using a TiS$_2$ cathode, and in Canada by Moli Energy utilizing a MoS$_2$ cathode [7].

However, several concerns affected the widespread use of the Li/TiS$_2$ Exxon cell, including safety concerns associated with the dangerous nature of lithium metal, namely, the presence of shock sensitive lithium perchlorate (LiClO$_4$) salt in the electrolyte and the problematic handling of the TiS$_2$ cathode in ambient atmosphere due to the spontaneous release of toxic H$_2$S gas, upon contact with moisture. The problems associated with the use of lithium metal were experienced first-hand by Moli Energy which had to issue a recall of their Li/MoS$_2$ batteries due to reports of battery fires [8].

The breakthrough came in the 1980's. During this time, not only did J. Goodenough file his patent on lithium cobalt oxide (LiCoO$_2$) as a lithium intercalation cathode material [9], but researchers had also demonstrated the practical use of graphite as an alternative anode material to lithium metal. H. Ikeda at Sanyo, was the first to patent the use of graphite as an intercalation material in an organic solvent [10]. In 1981, Yoshino started basic research on a nonaqueous secondary battery using polyacetylene as a negative electrode [6]. Several works conducted by many excellent scientific figures have finally led to the golden combination of the components of the lithium-ion cell including a LiCoO$_2$ cathode, a carbonaceous anode, and a carbonate-based electrolyte solution enabling lithium ions transport between the two host electrode materials.

In the years following commercialization, the performance of LIBs was strongly enhanced. While the cathode remained mostly unchanged, the electrolyte and the anode were modified and optimized in order to achieve higher energy density, higher power capability and longer cycle life with increased safety content. The second generation of lithium-ion cells, which included a hard carbon anode, exhibited a 10% increase in volumetric energy density compared to the first generation and was rated at $\sim$130 Wh kg^{-1} [8, 11].

On the other hand, the use of hard carbon at the anode resulted in a huge first cycle irreversible capacity, which consumed significant amounts of Li from the cathode and required extra cathode capacity to compensate, lowering the overall energy density [8]. The exceptional flat voltage plateau of graphite was a highly desired characteristic compared to the sloppy voltage profile of hard carbon. Though, the use of graphite was hindered by the unstable SEI formed in PC-based electrolyte due to the co-intercalation processes occurring upon cycling. The world had to wait until 1995, for the first commercially available Li-ion cell including graphite at the anode, after the discovery that a mixture of ethylene carbonate (EC) and PC at the electrolyte was the solution to avoid the undesired co-intercalation mechanism [11, 12]. Since then, graphite has been considered the anode of choice for LIBs, and currently in most of the commercially available cells, graphite is the predominant anode material [13].

There have been few other successful examples of anode materials. For instance, the spinel phase of $Li_4Ti_5O_{12}$ (LTO), has seen large use in high power applications [14]. Its incorporation in lithium-ion cells has an important advantage compared to graphite anodes. In spite of the lower energy density, LTO ensures high safety standards due to its relatively higher operation voltage, which prevents possible lithium plating and the detrimental associated phenomena [15, 16].

In the last decade, the Li-ion technology based on intercalation chemistry has seen tremendous technological improvements approaching the theoretical limit. Material improvement as well as enhanced compatibility of the cell components, both active, and inactive, together with advanced engineered cell manufacturing processes has translated into a continuous improvement of the battery performance. Cells are smaller and lighter, with increased energy density and power [17]. In addition, these cells are cost-effective when compared to the first generation of lithium-ion cells. The steep decrease in price of LIBs observed during the last years has been driven by the technological improvements and the wide competition among manufacturers [18].

In a continuous effort to achieve high performance rechargeable batteries, electrode materials operating with alternative mechanisms have been investigated. A discovery in contrast with the standard insertion-type materials could be achieved in principle by exploiting the lithium alloying reaction [19]. Lithium alloys have had some success in commercial application. In 2005, Sony developed a new high-capacity Sn-Co-C [20] composite material. The Nexelion anode was used in commercial cells exploiting the alloying mechanism between lithium and amorphous tin [21]. Among the elements of the periodic table able to alloy with lithium, silicon has without a doubt attracted copious attention especially in view of its environmentally friendly nature, its low cost when compared to tin or germanium and its extremely high theoretical specific capacity which combined with its relatively low operating voltage, guarantees (in principle) high gravimetric energy density [22, 23].

Currently, an increasing number of environmental policies set by the governments have rapidly accelerated the need for efficient, economical and high-performance batteries for the electrified transport sector, where an even more stringent requirement than just the gravimetric capacity is represented by the volumetric capacity [24, 25]. In this context, silicon represents the most appealing, valid alternative to graphite. Silicon has a theoretical specific capacity of $4200\,mAh\,g^{-1}$ and a volumetric capacity of about $9800\,mAh\,cm^{-3}$, which is ten times higher than that of graphite. A comparison of the properties of the most promising anode materials for LIBs is reported in Table 1.1 while Fig. 1.1 reports a comparison of the theoretical and experimental gravimetric and volumetric energy density (at a materials level) of thirty different anodes against a Ni rich cathode material as calculated by Andre et al. [24].

TABLE 1.1 Comparison of the electrochemical properties of graphite, LTO, and pure silicon.

	Lithiated phase	Gravimetric capacity/ mAh g^{-1}	Volumetric capacity/ mAh cm^{-3}	Average potential/V vs Li$^+$/Li	Volume change/%	Density
Lithium	Li	3862	2047	0	100	0.53
Graphite	LiC$_6$	372	837	0.05	12	2.25
Li$_4$Ti$_5$O$_{12}$	Li$_4$Ti$_5$O$_{12}$	175	613	1.6	1	3.5
Silicon	Li$_{4.4}$Si	4200	9786	0.4	420	2.3
Tin	Li$_{4.4}$Sn	994	7246	0.6	260	7.3

Silicon anode systems for lithium-ion batteries **Chapter | 1** 7

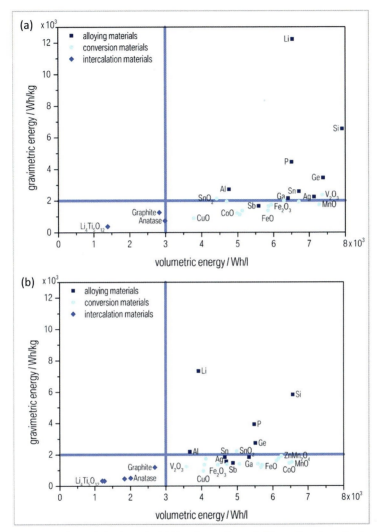

Fig. 1.1 Comparison of the theoretical (A) and experimental (B) gravimetric and volumetric energy densities on material level of thirty anode materials against a Ni-rich cathode with the derived energy targets (blue lines) based on the lithiated state. *(Adapted and reproduced from D. Andre, H. Hain, P. Lamp, F. Maglia, B. Stiaszny, Future high-energy density anode materials from an automotive application perspective, J. Mater. Chem. A 5 (2017) 17174–17198 with permission from © The Royal Society of Chemistry, 2017.)*

In the next section, an overview of the Si—Li alloy properties will be presented adopting a material perspective. An outline of the differences observed upon lithiation between crystalline and amorphous Si will be given as well as some considerations on the presence of the naturally occurring silicon oxide surface layer and its effects.

1.2 The Si—Li alloy: A material perspective

W. Weppner and R. A. Hugging first presented the thermodynamic treatment of binary systems including one electroactive species into a solid alloy electrode under the assumption of being at complete equilibrium [26]. In these conditions, the Gibbs phase rule could be applied to evaluate the thermodynamic properties of alloy systems and detect the variation of the potential upon charge and discharge. Under the assumption of a complete equilibrium, the electrochemical potentials vary according to the composition in a single-phase region, while it is composition-dependent in the two-phase regions if temperature and pressure are kept constant [26].

Several studies were conducted on a number of lithium alloy systems, especially at 400°C, because of the potential use of these alloys as components in molten salt electrolyte lithium batteries operating at high temperature. Among them, the Si—Li system was one of the most studied. The reaction of Li with elemental Si has been known for decades [26, 27]. C. J. Wen and R. A. Hugging determined the equilibrium open-circuit voltage in the Li—Si system as a function of composition by Coulometric titration [27]. Their study was conducted at 415°C and consisted of the electrochemical lithiation of a silicon wafer (crystalline silicon). Four intermediate Li_xSi phases were observed. The nominal composition of these phases was $Li_{12}Si_7$, Li_7Si_3, $Li_{13}Si_4$, and $Li_{22}Si_5$, with the last phase being stable between 44 and 2 mV with respect to pure lithium.

Other phases have also been reported such as a richer $Li_{17}Si_4$ [28] or a revised $Li_{21}Si_5$ phase [29]. Interestingly, none of these phases is obtained by electrochemical lithiation at room temperature. The typical voltage profile of crystalline Si upon lithiation exhibits a relatively flat voltage plateau around 0.1 V, instead of the stepped voltage profile obtained at high temperature. This suggests a two-phase lithiation process [30].

Crystalline silicon is lithiated following a two-phase mechanism. Silicon is consumed to produce an amorphous silicide phase first observed through X-ray diffraction (XRD) and high resolution transmission electron microscopy (HR-TEM) experiments at room temperature by Limthongkul et al. [31]. The study revealed that the crystallization of equilibrium Li—Si intermetallic compounds is inhibited during lithiation, and that the formation of highly lithiated glass occurs. The study was later confirmed by in situ TEM [32], revealing that as the lithiation proceeds, the two phases are separated by a sharp nanometer thick reaction front between the crystalline silicon and an amorphous Li_xSi alloy.

The sharp interface formation is explained considering the large activation energy required to break Si—Si bonds. A high concentration of Li atoms near the reaction front is required to weaken the Si—Si bonds, leading to enhanced lithiation kinetics and to the two-phase reaction. The mechanism has been confirmed by Key et al. [33]. They observed through in situ nuclear magnetic resonance (NMR) studies that the first lithiation occurs through the formation of isolated Si atoms and Si—Si clusters embedded in a Li matrix.

Overall, this explains the differences observed in the voltage profile of crystalline silicon at high and room temperature. The room temperature lithiation is a solid-state amorphization reaction involving formation of metastable amorphous Li_xSi phases, while at high temperatures, equilibrium intermetallic compounds are formed. In the amorphous lithiated phase, Li_xSi, x ranges between $0 < x < 3.75$ with the most highly lithiated phase ($x = 3.75$) corresponding to $Li_{15}Si_4$. The highly lithiated phase corresponds to a crystalline metastable phase formed after a second two-phase reaction observed at high lithiation depth (50 mV).

Ding et al. [34] reported on the formation of a crystalline highly lithiated $Li_{15}Si_4$ phase observed only in the discharge process. However, before the formation of the highly lithiated crystalline phase, they also observed two plateaus associated to the formation of intermediate a-Li_7Si_3 and a-$Li_{13}Si_4$, suggesting that the change of x in Li_xSi is not linear and smooth. The formation of a second crystalline phase at a high degree of lithiation corresponding to about 3579 mA h g^{-1} ($Li_{15}Si_4$), has been detected by XRD [35] and in situ TEM [36] analysis.

These phenomena at high and low temperature are quite different. The formation of the highly lithiated phase observed at 415°C, $Li_{22}Si_5$ (theoretical capacity of 4200 m Ah g^{-1}), has never been observed at room temperature; only the $Li_{15}Si_4$ phase has been detected. This highly lithiated phase is observed only when the potential drops below 50 mV, and it is a very unstable and reactive crystalline nonequilibrium (metastable) phase. Nonetheless, it is worth mentioning that by holding the cell potential at 0 V after lithiation for 24 h, diffraction peaks belonging to the $Li_{22}Si_5$ phase have been detected by XRD study, suggesting that the metastable (kinetically favored) phase can transform into the thermodynamically stable phase if sufficient time is given [37].

Upon de-lithiation, the crystalline $Li_{15}Si_4$ phase transforms back into amorphous Li_xSi phase through a two-phase reaction mechanism. In situ XRD studies [38] revealed that upon de-lithiation the small diffraction peaks observed as a consequence of the crystalline $Li_{15}Si_4$ phase disappear again. Also, during the second lithiation, the Li_xSi phase crystallizes once more when the electrode potential drops below 50 mV.

It has been reported that the formation of the metastable $Li_{15}Si_4$ phase is deleterious for the cycling behavior of silicon anodes. Indeed, this metastable phase undergoes self-discharge by losing lithium, which in turn reacts with the electrolyte resulting in the lithium loss from the electrode when kept at open circuit voltage. By avoiding the crystallization of the lithiated silicon at low voltages, de-lithiation occurs through a single phase process and an important source of capacity loss is avoided [30, 33, 39].

Besides crystalline forms of silicon, also amorphous silicon-based anodes have been investigated. Amorphous silicon typically presents a sloping voltage profile when compared to crystalline silicon, suggesting a single-phase reaction mechanism. Though, further studies have evidenced a two-phase reaction [30].

Moreover, during lithiation a slight decrease in voltage is observed at about 50% lithiation. The effect is associated with the local environment surrounding inserted Li atoms [40]. Key et al. [41] have proposed that the Si lattice is broken up into Si clusters by the end of the observed higher voltage process, and further lithiation breaks up these clusters into isolated Si atoms.

The Li atoms that are inserted after this step in the lithiation process are primarily surrounded by other Li atoms, which results in less charge transfer to Si atoms and therefore a slightly lower electrochemical potential. At full lithiation, the metastable crystalline $Li_{15}Si_4$ phase has been observed as in the case of crystalline Si. The two-phase mechanism has been observed by in situ TEM [42, 43] and in situ attenuated total reflection Fourier transform infrared spectroscopy (ATR-FTIR) [44] studies. On the contrary other studies conducted through time-of- flight secondary ion mass spectrometry (ToF-SIMS) suggest a diffusive process [45].

This distinction is important since the two different reaction mechanisms lead to different stress evolution and fracture behavior upon cycling. Assuming that both amorphous and crystalline silicon, undergo a two-phase lithiation process, the two reactions are still different as highlighted by the different voltage profiles. The main divergence arises from the difference in structure between the two materials. While for crystalline silicon anisotropic expansion is observed as a result of the interface limited reaction with differing reaction rates at different crystallographic planes, no anisotropy is found either in the interphase evolution, or in the volume expansion of the material in the case of amorphous silicon [30].

It has been reported that for amorphous silicon nano-spheres an isotropic volume expansion occurs upon lithiation [42]. Another beneficial effect of amorphous silicon compared to crystalline is that mechanical failure and cracking of the particles occurs at larger particle size. However, while the use of amorphous compounds as stabilizers for the mechanical failure mechanism of alloying anodes has already been demonstrated by the Nexelion anode, it is worth noting that amorphous silicon also has a lower electronic conductivity and lithium ion diffusivity when compared to crystalline silicon [39]. Moreover, the chemical-physical properties of amorphous silicon strongly depend on the preparation method adopted. Most of the investigated amorphous silicon anodes come in the form of silicon thin films, with also nano-sized particles generally of an amorphous nature [30, 39]. Typically, silicon comes with a native oxide layer (SiO_2 or SiO_{2-x}) formed spontaneously. In ambient conditions, this layer is found to be typically amorphous [46]. Great attention has been devoted to the understanding of the effect of the silicon oxide layer on the lithiation process and the effect on the electrochemical properties of silicon based anodes [47–49].

In 1999, Huang et al. [50] suggested SiO_2 to be electrochemically inert toward lithiation, due to the strength of the Si—O bond which cannot be electrochemically broken under the normal operating conditions of anode materials cycled down to 0.1 V vs Li$^+$/Li. Several years later, Ariel et al. [51] suggested that the thickness of the SiO$_2$ is a crucial parameter affecting lithium mobility, reporting that with a 9 nm SiO$_2$ layer, lithium diffusion is possible, however without the electrochemical reactivity of SiO$_2$ with Li due to the limited electronic conductivity of the SiO$_2$ layer. Larcher and co-workers [52] determined from thermodynamic calculations that the reaction conversion of SiO$_2$ with Li into Li$_2$O and Si should proceed at a potential value of 0.19 V vs Li$^+$/Li. Notwithstanding, based on the evidence that conversion reactions generally occur at 1 V above the equilibrium potentials, it was suggested that the likelihood for the conversion reaction to progress was limited. On the contrary, a thermodynamically favored reaction of SiO$_2$ with Li has been proposed by Graetz et al. [53], suggesting SiO$_2$ as the origin of the higher first cycle irreversible capacity generated in the investigated high surface area Si nanocrystals.

Guo et al. reported on the formation of Li$_2$O and Li$_4$SiO$_4$ as reaction products of the lithiation of nano-SiO$_2$ in hard carbon anode [54]. In addition, Sun et al. [55] reported on the lithiation mechanism of sputtered SiO$_2$ thin films suggesting the occurrence of the Li—Si alloying process in concomitance with the reversible conversion reaction of SiO$_2$ into Li$_2$Si$_2$O$_5$. Despite the great efforts toward a comprehensive understanding of the oxide role, the reversibility of the lithiation process, the interaction of the formed species with the electrolyte and the influence on the SEI stability is still largely disputed [56–58]. The observed discrepancies may be attributed to the different nature (amorphous or crystalline) [59], chemical composition (SiO$_2$ or nonstoichiometric SiO$_x$), thickness [51, 60] and density [61] of the oxide layer.

1.3 The Si—Li alloy: An electrode perspective

Besides the promising properties in terms of volumetric and gravimetric capacities, alloying anodes, and specifically silicon, presents several issues which still need to be addressed for a successful commercialization [62]. The alloying reaction involves a huge volume variation upon cycling. The fully lithiated phase leads to a volume expansion of about 300% [19, 35]. So far, three key failure mechanisms, reported in Fig. 1.2, have been identified as originating from the large volume variation [63].

The first challenge results from the pulverization of the active material as a consequence of particle cracking. The second failure mechanism, deriving from the first one, is associated with the delamination and contact loss between particles and the current collector, leading to loss of active material upon cycling and consequent cell failure. While these issues have been addressed by downsizing of the particle size and exploiting nano-structuration by efficiently employing mechanically compliant carbon matrixes buffering the volume expansion, smart electrode design and implementation of advanced binders [63–67], the third failure mechanism is the most critical one. The instability of the SEI layer represents the key phenomena to be tackled for the successful

12 PART | I Introduction and background

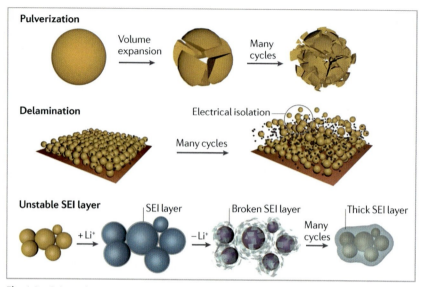

Fig. 1.2 Schematic representation of the main degradation mechanisms of Si anodes originating from the large volume expansion of Si upon the lithiation/de-lithiation process. *(Adapted and reproduced from J.W. Choi, D. Aurbach, Promise and reality of postlithium-ion batteries with high energy densities, Nat. Rev. Mater. 1 (2016) 16013 with permission from © Macmillan Publishers Limited, 2016.)*

development of the next generation high energy density silicon anodes. The SEI instability is certainly triggered by the volume expansion and ensuing cracking which silicon undergoes exposing new silicon with each cycle, however, even with optimized morphologies and electrode design, low coulombic efficiencies and lithium inventory loss are generally observed, undermining the achievement of high performance Si anode containing LIBs [68–72]. Each of these failure mechanisms and associated proposed strategies to mitigate the undesired effects is discussed in the following sub-sections.

1.3.1.1 Volume expansion and material pulverization: The importance of size and nano-structuring

The performance degradation observed for silicon anodes during the first cycles is a consequence of the large volume variations occurring upon the charging/discharging process. The high stress experienced by the silicon particles leads to cracking, pulverization, delamination hence capacity fading and low coulombic efficiency translated into reduced cyclability [63, 73]. In addition, Sethuraman et al. [74] detected that another consequence of the volume expansion can be observed as a few hundred mV to the voltage hysteresis, which detrimentally affects the energy efficiency. Several techniques have been adopted to inspect and elucidate the formation and propagation of cracks, among which atomic force microscopy [75], in situ scanning [76] and TEM appear prominent [77]. Acoustic emission experiments [78] have also shown that fractures mainly occur during the initial lithium alloying induced crystalline-to-amorphous phase transition.

Whether the cracking is the consequence of the lithiation or de-lithiation process is still debated by the battery community. Stress measurement performed on a 35 nm thick amorphous silicon thin film by optical cantilever method [79] suggested that cracks appear at the end of lithiation. It was evident that volume expansion upon lithium uptake induced a compressive stress in the electrode, especially elevated upon lithiation below 0.1 V vs Li^+/Li, when microcracking appeared. Limiting the lower cut-off voltage to 0.1 V significantly improved cycle life, probably due to the prevention of microcracking upon lithiation. While the conclusions of the study are supported by the experimental results, many other studies did not report cracking for 35 nm thick silicon thin films. In addition, later studies confirmed that silicon thin-film electrodes with a thickness < 100 nm do not crack upon cycling [80]. Yet other studies performed by using different silicon nanostructures such as nanowires and nanospheres have reported on cracks appearing after lithiation [81, 82].

On the other hand, several other studies have reported on crack formation only during the de-lithiation cycle [75, 80]. Dahn et al. [75] investigated by in situ atomic force and optical microscopy the volume changes occurring as lithium is electrochemically added and removed from lithium alloy films. It was found that when lithium is first added to alloy films on rigid substrates, the films expand perpendicular to the substrate. On the other side, when lithium is removed, the films shrink both perpendicular and parallel to the substrates, leading to crack patterns similar to those found in dried mud cracking. The study also revealed the reversible formation and closing of fractures in a 1 μm SiSn film during the first de-lithiation and subsequent lithiation/de-lithiation cycles [75].

Thin films and nanoparticles with different shapes can exhibit considerably different electrochemical interactions with lithium ions. For silicon thin films, volume expansion can only occur through an increase of film thickness, with a probable rise of the internal pressure in the directions parallel to the substrate. In this case, no crack would be expected during lithiation, if the films thickness does not exceed 100 nm. On the other hand, three dimensional nanoparticles can have a uniform expansion leading to a favorable condition for crack formation, in turn releasing the stress from the system. In all of the above-mentioned cases, size is a crucial parameter to tune as well as control the stress hence, reducing the cracking phenomenon in silicon-based anodes. For instance the critical particle size for crystalline silicon before any crack occurs is about 150 nm, with Si nano-powders (78 nm) exhibiting an improved cycling stability when compared to Si powder (250 mesh) [83]. However for amorphous silicon, the threshold is about 6 times larger [39].

Nanostructured electrodes are able to easily absorb the strains associated with the change of volume, and avoid cracking [84]. Furthermore, the observed improved electrochemical performance is also attributed to the shortening of the lithium ions' diffusion distance, enhancing the electroactivity toward Li uptake and release. The large surface area displayed by nanoparticles however, leads to a pronounced reactivity with the electrolyte, which can affect the formation of a stable SEI and can be detrimental to the electrode performance. For a given geometry (wires or spheres), experimental evidence has established the beneficial effect of the size reduction [85]. Specifically, it has been reported that the decrease of the diameter of the particles (volume shrinkage) upon de-lithiation is proportional to the size of the particles, suggesting that smaller particles can keep contact with the matrix while allowing full extraction of the lithium, thus limiting the irreversible capacity loss and enhancing capacity retention. Apart from the size, the electrochemical performance is also strongly affected by the geometry of the silicon particles.

Zero-dimensional (0D) nest-like Si nanospheres prepared by solvothermal method have been reported to show a large specific capacity of 3052 mAh g^{-1} at 2000 mA g^{-1} (i.e., *ca.* 0.5C rate) by using 1 M LiPF$_6$ in a 3:1:1 mixture of ethylene carbonate (EC), propylene carbonate (PC) and diethyl carbonate (DEC) [86]. Compared to solid structures, hollow structures can provide empty interior space, which offers a buffer for the volume expansion. An optimized 0D silicon electrode based on interconnected Si hollow nanospheres has been prepared by Cui and co-workers via chemical vapor deposition (CVD) of Si on silica particles followed by HF etching of SiO$_2$ [87]. The materials exhibited a very high initial discharge capacity of 2725 mAh g^{-1} at 0.1C rate in carbonate-based electrolyte. After 700 cycles, the electrode still retained 1420 mAh g^{-1} at 0.5C rate (see Fig. 1.3 panel (I)). Silicon thin films and nanoparticles however, tend to display satisfactory cycling behavior only when low electrode loadings are employed, which is one of the main drawbacks of nanostructured anodes.

On the other hand, one dimensional nano-structures (1D) such as silicon nanowires (NWs) have been reported to exhibit extraordinary electrochemical performance. Si NWs can offer empty space able to accommodate the large volume variation upon lithium uptake and release. In addition, the Si NWs can be electrically connected to the current collector making conductive carbon additives and polymer binders not strictly necessary [23]. The most common method adopted to obtain Si NWs is the vapor-solid-liquid (VLS) growth performed in a CVD reactor by using a flowing Si-bearing gas and a metal catalyst at temperatures ranging from 300°C to 1000°C [88]. The Si NWs obtained by VLS-CVD usually have a crystalline Si core and an amorphous shell with a total diameter ranging from several nm to hundreds of microns. However, in order to avoid any cracking effect upon lithium uptake and release, their diameters should not exceed 360 nm for fully crystalline Si NWs and about 900 nm for amorphous Si NWs [85].

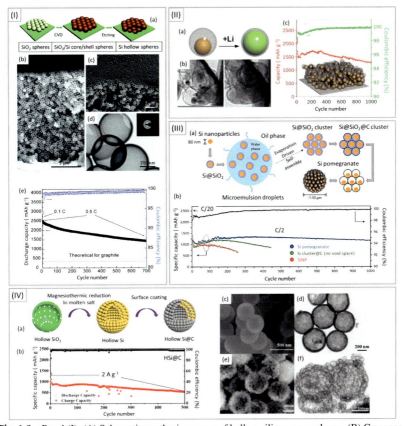

Fig. 1.3 Panel (I): (A) Schematic synthesis process of hollow silicon nanospheres. (B) Cross sectional SEM image of a 12 μm thick silicon hollow nanosphere based electrode and (C) corresponding 45° tilted side view image. (D) TEM image of the interconnected hollow silicon nanosphere highlighting two interconnected spheres. (E) Cycling behavior in terms of discharge capacity and coulombic efficiency of the interconnected silicon hollow nanospheres during 700 cycles. Panel (II): (A) Schematic of an individual Si/void/C particle showing that the silicon nanoparticles expand without breaking the carbon coating or disrupting the SEI layer on the outer surface. (B) In situ TEM images captured during lithiation process. (C) Corresponding long-term cycling test with de-lithiation capacity and coulombic efficiency over 1000 cycles at C/10 for the first cycle, C/3 for following 10 cycles, and 1C for the later cycles. Panel (III): (A) Schematic of the fabrication process of pomegranate-inspired silicon nanoparticles encapsulated in a conductive carbon layer. (B) Corresponding electrochemical behavior in terms of de-lithiation capacity over 1000 cycles compared with other silicon-based electrodes tested under the same conditions. Panel (IV): (A) Schematic illustration of the preparation of raspberry-like hollow silicon nanospheres with highly conductive carbon shells (HSi@C) nanocomposite. (B) Long-term cycling performance of HSi@C composite at 2 A g^{-1} over 500 cycles. (C, D) SEM and TEM images of SiO$_2$ hollow spheres. (E, F) SEM and TEM images of HSi@C. *(Panel (I) is adapted and reproduced from Y. Yao, et al., Interconnected silicon hollow nanospheres for lithium-ion battery anodes with long cycle life, Nano Lett. 11 (2011) 2949–2954 with permission from © American Chemical Society, 2011; Panel (II) is adapted and reproduced from N. Liu, et al., A yolk-shell design for stabilized and scalable Li-ion battery alloy anodes, Nano Lett. 12 (2012) 3315–3321 with permission from © American Chemical Society, 2012, Panel (III) is adapted and reproduced from N. Liu, et al., A pomegranate-inspired nanoscale design for large-volume-change lithium battery anodes, Nat. Nanotechnol. 9 (2014) 187–192 with permission from © Nature Publishing Group, 2014; Panel (IV) is adapted and reproduced from S. Fang, Z. Tong, P. Nie, G. Liu, X. Zhang, Raspberry-like nanostructured silicon composite anode for high-performance lithium-ion batteries, ACS Appl. Mater. Interfaces 9 (2017) 18766–18773 with permission from © American Chemical Society, 2017.)*

A low cost and scalable synthesis method for Si NWs production has been proposed. The method consists of a variation of the metal assisted wet chemical etching process that uses c-Si powders (instead of c-Si wafers), hence NWs can also be etched [89]. This process is far more scalable and cost efficient than the widely employed etching method which is feasible only for lab scale studies [90]. By combining a metal deposition and metal-assisted chemical etching process, B. M. Bang et al. [89] produced nano-porous SiNWs with a length of about 5–8 µm length and with a pore size of 10 nm, directly formed in the bulk silicon particle. The multidimensional structure of the developed materials showed the ability of the electrode to accommodate the volume variation upon cycling. A reversible capacity of 2400 mA h g^{-1} was observed as well as a stable cycling behavior and an initial coulombic efficiency of 91%.

Among the 2D nanostructures, Si thin films represent in principle a promising strategy to overcome volume expansion issues [91]. Due to their inherent structure, thin films can efficiently buffer volume variation (up to a certain thickness) and maintain the electrode structural integrity. A specific capacity of about 3500 mA h g^{-1} has been shown by 50 nm-thick Si thin films cycled over 200 cycles at 2C rate. By increasing the thickness up to 150 nm, the thin film electrodes delivered 2000 mA h g^{-1} for over 200 cycles at 1C rate [92]. Nevertheless, despite the promising results in terms of cycle life and rate capability, a real implementation of silicon thin films is stalled by the low active material content and the high costs of the preparation methods which certainly hampers their practical application.

Three-dimensional bulk porous Si architectures have also been reported by Kim et al. [93]. The developed highly interconnected porous structure could accommodate the large volume changes without pulverization. The electrode preserved a charge capacity of about 2800 mA h g^{-1} at 1C rate (2000 mA g^{-1}) even after 100 cycles. Void engineering as well as the incorporation of carbon composites have led to the acquisition of 3D nano-porous structure with very promising properties. Among them are materials obtained by implementing yolk-shell- [94], pomegranate- [95] and raspberry- [96] silicon morphologies exhibiting much higher capacities than graphite-based anodes (at least in half-cell configuration). Morphologies and corresponding electrochemical behavior are reported in Fig. 1.3 (panels II, III, and IV).

Combining nanoscale silicon with various morphologies and structures has been proven to be an efficient approach to improve the LIB performance. Although nanosized silicon can circumvent the pulverization, the volume expansion of silicon still exists, which means that the SEI films will be destroyed and continuously produced with every cycle exposing a new Si surface for the SEI to form. This will consume a large quantity of lithium ions, causing irreversible loss, seriously reducing the coulombic efficiency and consequently, the reversible specific capacity of batteries. In addition, the large surface area of nanosized structures, results in an increased reactivity toward the electrolyte, irreversibly consuming Li, thus further limiting the efficiency.

Moreover, the conductivity of nano-silicon is still an issue which is directly translated in limited rate performance. To date, one of the most successful strategies toward the commercialization of silicon anodes is the development of silicon composites with carbon or other metals/metal oxides [97].

Carbon composites are among the most promising solution to address the aforementioned issues. Due to the outstanding electrical conductivity and mechanical strength of carbon, silicon/carbon composites benefit from an improved Li ion diffusion and increased electrode integrity conferred by the buffering effect of the carbon matrix toward volume expansion. Conventional Si/C composites have been commonly produced by mechanical mixing methods, leading to inhomogeneous dispersion and poor connectivity of each component. Some graphite/silicon composite materials have been obtained by mechanical ball milling. By using this preparation method, Si nanoparticles were embedded into a relatively dense carbon matrix [98–100]. The as-prepared Si/C nanocomposites showed to some extent an improved cycle stability and capacity, however because of the dense carbon matrix nature, volume changes were only hindered to a limited degree.

On the other hand, carbon uniformly coated on silicon with different structures has been reported to be a more efficacious approach toward improving electrochemical performance [101, 102]. Nie et al. [103] reported on the encapsulation of mesoporous silicon microspheres into a carbon coating obtained by using polybenzimidazole as a novel carbon sources. The composite material exhibited an initial charge/discharge capacity of $1713\,mAh\,g^{-1}$ when cycled at $1\,A\,g^{-1}$. After 200 cycles a capacity of $1128\,mAh\,g^{-1}$ could be retained with coulombic efficiency values approaching 100%. The promising electrochemical performance is attributed to the multiple synergistic effects of the polymeric carbon coating. The carbon layer prevents direct contact between electrolyte and silicon particles, thus limiting the reactivity toward the electrolyte and preventing uncontrolled growth of the SEI. In addition, the nitrogen content within the carbon coating layer greatly enhanced the electronic conductivity of the materials as well the capacity delivered upon cycling due to the presence of functional groups suitable for lithium ions storage [103].

Additionally, the cycle life of carbon coated silicon-based anodes could be modified by efficiently engineering the composite. Conformal carbon coating on Si would still rupture upon volume expansion exposing silicon to the electrolyte and exacerbating electrolyte decomposition. Introduction of voids between silicon and carbon has been reported to be a strategy of greater efficiency toward volume expansion mitigation also preventing silicon aggregation. The advantage of these structures relies on two main aspects. The void between carbon and silicon particles allows for silicon expansion without breaking the shell, while the electrical conductivity of the shell improves the lithiation kinetic, preventing electrolyte reactivity with silicon nanoparticles. Luo et al. [104] reported on sub-micrometric capsules made of Si nanoparticles wrapped by crumpled graphene shells. The composite material, obtained by one-step capillary-driven assembly route in aerosol droplets, exhibited greatly improved performance in terms of capacity, cycling stability, and coulombic efficiency when compared to bare silicon nanoparticles [104]. Guo et al. [105] instead designed porous silicon nanoparticles loaded in controllable void carbon spheres with yolk-shell structure. The optimized material showed excellent performance, delivering a capacity of 1400 mAh g^{-1} after 100 cycles at 0.2 A g^{-1}. Notable results have also been obtained by Liu et al. [94] with a similar yolk-shell structure Si/carbon composite material. Si nanoparticles were coated with a SiO$_2$ layer and a polydopamine layer, which was subsequently carbonized to form a nitrogen-doped carbon coating. The yolk-shell Si/Void/C structure was obtained after selectively removing the SiO$_2$ layer by hydrofluoric acid (HF) treatment. The material showed an excellent capacity of 2833 mAh g^{-1} when cycled at 0.1C rate, with a capacity retention of 74%, after 1000 cycles and a coulombic efficiency of 99.84%.

Extraordinary results in terms of cycling stability have also been reported by adopting porous Si/C composite electrodes. Magasinski et al. [106] reported a large-scale hierarchical bottom-up fabrication approach for the formation of a porous Si/C architecture (see Fig. 1.4A). The material was procured by anchoring silicon nanoparticles into annealed carbon-black dendritic particles consequently assembled into rigid spheres with open interconnected internal channels during further carbon deposition. The designed electrode architecture enabled fast Li ions diffusion while the internal particles porosity could accommodate the Si volume changes upon cycling, facilitating a stable cycling behavior and reversible specific capacities of about 1950 mAh g^{-1}.

Another scalable synthetic approach for porous Si/C composites was developed by An et al. [107] reporting on a thermal nitridation of a Mg—Si alloy and acid etching method to produce ant-nest-like porous silicon. The designed material comprised of 3D interconnected silicon nano-ligaments and bi-continuous nanopores that could prevent pulverization and enable volume expansion during cycling. The material manifested extraordinary capacity retention (about 90%) after 1000 cycles while delivering 1271 mAh g^{-1} and a very limited swelling of 17.8% at a high areal capacity (5.1 mAh cm^{-2}). The composite electrode showed propitious electrochemical performance also in full cell configuration. By using a prelithiated anode and a Li(Ni$_{1/3}$Co$_{1/3}$Mn$_{1/3}$)O$_2$ cathode an energy density of 502 Wh kg^{-1} could be achieved with 84% capacity retention after 400 cycles.

A 3D porous Si/conductive polymer hydrogel composite electrode produced by in situ polymerization of conducting hydrogel to conformally coat silicon nanoparticles has been reported by Wu et al. [66]. The electrode exhibited stable reversible capacity of 1600 mAh g^{-1} after 1000 deep cycles and a cycle life of 5000 cycles with over 90% capacity retention at a current density of 6 A g^{-1} (see Fig. 1.4B). This performance is attributed to the fast electronic and ionic transfer channels provided by the conductive polymer 3D network, as well as the limited volume expansion accommodated by the porous space.

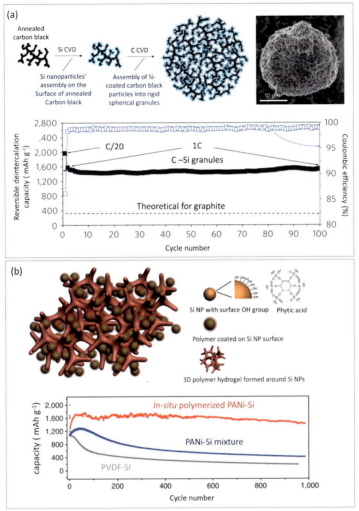

Fig. 1.4 (A) Schematic illustration of the hierarchical bottom-up assembly of a Si—C nanocomposite granule including annealing of carbon-black dendritic particles coated by Si nanoparticles and then assembled into rigid spheres with open interconnected internal channels during carbon deposition. Associated SEM micrograph of the Si—C nanocomposite granule after electrochemical cycling with long term cycling test in comparison with the theoretical capacity of graphite. (B) Schematic illustration of a 3D porous silicon nanoparticles (SiNPs)/conductive polymer hydrogel composite electrodes. Each SiNP is conformally coated by a conductive polymer surface obtained through interactions between -OH surface groups and the phosphonic acids in the cross-linker phytic acid (right column), or through the electrostatic interaction between negatively charged -OH groups and positively charge polyaniline (PANi) due to phytic acid doping. The coated Si NPs are further connected to the highly porous hydrogel framework. The associated electrochemical cycling performance of the in-situ polymerized SiNP-PANi composite electrodes is reported at a current density of $1 A g^{-1}$ for 1000 cycles (red line) in comparison with SiNP-PANi without in-situ polymerization (blue line) and PVDF-SiNP electrode (green line). *Part figure (A) adapted and reproduced from A. Magasinski et al., High-performance lithium-ion anodes using a hierarchical bottom-up approach, Nat. Mater. 9 (2010) 353–358 with permission from © Nature Publishing Group, 2010; Part figure (B) adapted and reproduced from H. Wu, et al., Stable Li-ion battery anodes by in-situ polymerization of conducting hydrogel to conformally coat silicon nanoparticles, Nat. Commun. 4 (2013) 1943–1946 with permission from ©Nature Publishing Group, 2013.*

Other than carbon composites, advancements on the electrochemical performance of silicon anodes especially at high rates have also been obtained by compounding silicon with other metals. Doping of silicon with inactive metals provides an efficient strategy to improve the poor electronic conductivity of silicon, thus promoting charge transfer reactions. In addition, the partial substitution of silicon with inactive metals enables a reduced volume expansion (less silicon in the composite materials) where the inactive metal acts as a buffering matrix for silicon expansion strongly enhancing electrode integrity.

Yu et al. [108] proposed a synergistic combination of a porous 3D nanostructure combined with noble metal coating to improve the rate capability of silicon anodes. A silver-coated 3D macro-porous silicon anode was obtained by magnesiothermic reduction reaction under Ar/H_2 atmosphere in an open crucible followed by Ag nanoparticles incorporation using an Ag-mirror reaction. Ag nanoparticles only act as an electronic additive and do not introduce any electrochemical reactivity. The nanostructured electrode was able to provide electron pathways from the current collector to the whole surface area of the 3D porous Si particles showing a reversible capacity of 1163 mAh g^{-1} at 0.2C rate after 100 cycles, and an enhanced rate capability of 1930 mAh g^{-1} when cycled at 1C rate outperforming bare silicon nanoparticles and also 3D macro-porous silicon electrodes.

In search for high-rate and high-capacity durable silicon-based anodes, Song et al. [109] proposed a novel alloy-forming approach to convert amorphous silicon-coated copper oxide core shell nanowires into hollow interconnected Si—Cu alloy nanotubes. By using a template-free synthesis, coating of amorphous silicon on CuO nanowires was achieved through plasma enhanced CVD. The final hollow Si—Cu nanotubes were obtained by an in situ H_2 annealing. The electrode displayed a stable cycling over 100 cycles (88% capacity retention) delivering 780 mAh g^{-1} at 20 A g^{-1}. By tuning the operating voltage range, the conversion reaction of CuO was avoided inducing electrochemical inactivity for the Cu^{2+}/CuO reaction.

Besides inactive metals, Silicon compounding with Li-storage active metals has been put forth as an attractive strategy. It has been observed that when Li is inserted into one active component, the other will act as a buffer and alleviate the volume expansion maintaining electrode integrity as in the case of inactive metals. Three-dimensional nanoporous SiGe alloy has been fabricated by means of a de-alloying method employing a ternary AlSiGe ribbon as the precursor [110]. By tuning the aluminium content in the precursor different morphology and porosity of the alloy could be obtained. Specifically, by using a precursor with 80% Al content, the 3D nanoporous SiGe alloy presented a coral-like structure with hierarchical micropores and mesopores, which led to a high reversible capacity of 1158 mAh g^{-1} after 150 cycles at a current of 1 A g^{-1} with excellent rate capacity. The electrochemical activity of both Si and Ge was detected at approximately 0.4 V. By comparing the developed SiGe alloy with bare Si, a clear improved cycling stability was achieved demonstrating the beneficial effect of Ge incorporation with its favorable electronic conductivity and rapid Li$^+$ mobility as well as buffering effect [110].

Apart from alloys, Si/metal oxide composites with a core-shell structure have also been suggested as alternatives to Si/C composites. The advantage of these structures is that the metal oxide coating provides protection from the electrolyte to the silicon particles, while the void engineered structure contains the volume expansion. Yang et al. [111] proposed a sol-gel strategy to synthesize amorphous TiO_2 coated core-shell Si particles. The amorphous TiO_2 shell had a thickness of about 3 nm and was able to maintain elasticity upon cycling, preserving the structural integrity of the electrode and avoiding direct electrolyte contact of the Si core. It was determined that a crystalline TiO_2 coating of the Si core was not as effective when compared to the amorphous phase. The material displayed impressive cycling stability over 200 cycles, with a specific capacity of about 1720 mA h g^{-1}. More importantly, the electrode also presented a promising first coulombic efficiency of 86%, considering the high silicon loading in the composite (about 89%) [111]. Much improvement in terms of areal loading (0.621 mg cm^{-2}) for a practical application are required, though the designed material offers a valuable strategy toward high silicon content electrodes.

To adequately address the low initial coulombic efficiency of silicon electrode (50%–80% vs 90%–94% for graphite anodes), Wang et al. proposed a lithiated silicon anode [112]. Li$_x$Si alloys have the potential for an easy assembly of full lithium-ion cells but also for full cells including the next generation lithium free conversion type cathodes, such as sulfur or oxygen. The process to improve upon the concept first described Yang et al. [111] revolved around lithiated-TiO_2 protected Li$_x$Si nanoparticles prepared via a thermal alloy lithiation approach [112]. One of the greatest advantages of this method is that the dense TiO_2 coating layer enables protection of the inner Li$_x$Si alloy from ambient corrosion, leading to high dry-air stability. Even by exposing the material to air with 10% relative humidity for 30 days, 87% of the capacity was retained. The unexposed material showed a capacity of about 1300 mA h g^{-1} with a capacity retention of over 500 cycles of about 77%.

1.3.1.2 Pulverization and delamination: The importance of polymer composites and binders

Beside Si/C and Si/metal and metal oxide composites, large attention has been focused on Si-conductive polymer composites. Conducting polymers exhibiting enhanced chemical stability, superior electronic conductivity and structural flexibility have been employed in silicon anodes to advance the reaction kinetics and to allow an improved containment of the volume expansion. A properly engineered bonding at the interface between the conducting polymer and silicon particles enables reduced risk of particle contact loss, maintaining a high electronic conductive network in the electrode [113].

Conductive polymers such as polyaniline (PANI) and polypyrrole (PPy) have shown unique morphologies which can be tailored by tuning pH and temperature of the binding reaction [114]. This aspect represents a great advantage for application in silicon based anodes, since the morphologies can be designed according to the silicon particles [115, 116]. Du et al. [115] reported on the beneficial effect of PPy coating of porous hollow Si nanospheres obtained by using a magnesiothermic reaction of mesoporous silica hollow nanospheres (MHSiO$_2$) and in situ chemical polymerization of PPy on the porous Si hollow spheres (SiPHSi) surface. By comparing the coated and uncoated Si nanostructured electrodes, it was observed that the coated material was exceeding the electrochemical performance of the uncoated. Meanwhile the coated material retained 88% of its initial discharge capacity (2500 mA h g^{-1}) over 250 cycles. The uncoated one manifests a lower capacity retention of about 44% over 100 cycles despite its initially higher capacity (3000 mA h g^{-1}).

Another approach has been proposed by Wu et al. [66]. By incorporating a conductive polymer hydrogel into a Si-based anode through in situ polymerization, a well-connected 3D network of Si nanoparticles conformally coated by the conductive polymer was acquired. The result is a porous hydrogel structure with Si nanoparticles embedded into a matrix wrapped with PANI and interconnected by interparticle polyaniline bridge. The fast synthetic approach developed required only 2 min and the high viscosity of the obtained hydrogel enabled direct pasting of the slurry into the current collector without the use of additional binder. The electrode showed a cycle life of 5000 cycles with 90% capacity retention at a current density of 6 A g^{-1}. The impressive electrochemical behavior was attributed to an improved buffering effect of the volume expansion and a stable SEI formation and evolution upon cycling. However, the electrodes had a very low loading of around 0.3 mg cm^{-2}. Further improvements in terms of areal loadings are a necessity, however, the fast preparation method, the absence of the binder, and the excellent electrochemical performance, demonstrate the validity of the approach.

Beside polymer-coated Si nanostructures, polymeric binders play a vital role in the electrochemical stability and cycle life of silicon-based anodes. One of the main challenges for Si nanostructures is the electrical connection between the particles and the current collector. A strong adhesion is necessary to avoid delamination and active material loss with consequent cell failure. The role of the binder in an electrode is to act as a glue to adhere all components of the composite electrode layer together and to the current collector. The binder should also be able to alleviate the stress induced during the electrode drying process, maintaining the mechanical integrity of the electrode. Moreover, the binder should have minimal swelling in the electrolyte, be flexible, display adequate Li ion conductivity and remain stable in the operational voltage window upon cycling [73].

Poly (vinylidene fluoride) (PVDF) has been used as a binder in graphite anodes. However, its mechanical stability was not able to sufficiently withstand the large volume expansion that silicon-based anodes undergo [117]. PVDF has been known to swell considerably in organic solvents, gaining up to 130% in electrolyte uptake [73] and losing to some extent its adhesion ability, facilitating detachment of Si. In addition, given its symmetrical fluorine-bonded carbon backbone structure, PVDF does not offer polar functional groups able to closely surround the silicon particles. On the contrary it only offers weak van-der Waals interactions. Alternative binders such as sodium carboxymethyl cellulose (CMC), styrene butadiene rubber (SBR) with CMC, and poly acrylic acid (PAA) are alternatives that present strong dipole or charge in their chemical structure which can interact strongly with hydroxyl groups on the surface of Si particles [73, 118, 119]. Moreover, the use of aqueous binders such as CMC, is seen as a step forward for sustainable and greener batteries due to its solubility in water, which enables water processing instead of the use of hazardous solvents such as N-methyl-2-pyrrolidone (NMP) (as required for PVDF) [120].

Among the vast variety of polymer binders, one would expect the most efficient types to possess high elasticity and mechanical resistance. However, it was found that SBR which is a highly stretchable polymer does not sufficiently improve the electrochemical performance of the silicon anode [121]. On the other hand, CMC which is rather brittle, can suitably accommodate the volume expansion of silicon particles [67]. This suggests that elasticity is not the key parameter, but rather the nature of the bonding with the active silicon particles [121]. Bridel et al. suggested that CMC, rather than forming an ester-like covalent bond with the Si surface, tends to generate hydrogen bonding which is strong enough to create the primary texture, but weak enough to be broken when mechanical stress is applied according to a mechanism referred to as self-healing [121]. As a result, electrode integrity is maintained as demonstrated by the cycling stability of the Si/C composite electrode including the CMC binder that manifests a coulombic efficiency of 99.9% per cycle over 100 cycles [121]. The Si-CMC interaction is strongly affected by the pH of the slurry. Mazouzi et al. confirmed that the esterification reaction, instead of hydrogen bonding, is preferentially occurring at $pH = 3$, and that the strong covalent bonding between the CMC and Si improved the cycling stability of the electrode remarkably [122].

Polyacrylate binders (e.g., PAA) have also been reported to improve the electrochemical performance of Silicon anodes. The large amount of carboxyl groups in polyacrylates, exceeding those in CMC, enables a higher degree of grafting between the binder and silicon particles through hydrogen bonds [123]. Komaba et al. [124] proved that during the drying process of a SiO-composite electrode, a uniform covering polymeric network with improved mechanical properties is formed when using PAA. The PAA chains bridge and cross-link together via the formation of hydrogen-bonds and carboxylic anhydride groups generating the polymeric network.

The SiO-PAA composite electrode delivered about 700 mA h g^{-1} reversible when cycled at 100 mA g^{-1}. The electrochemical activity of the SiO composite electrode was noticeably improved by using the PAA binder in comparison to the conventional PVDF binder. The beneficial effect of the PAA binder is most likely related to the uniform coating layer of PAA on the silicon surface, which lessened the collapse of the SiO composite electrode upon cycling and effectively reduced the direct contact between silicon and electrolyte acting as an artificial SEI layer [124].

Polyacrylic acid can also be used in its neutralized form. In fact, PAA possesses carboxylic functional groups, which can be dissociated in water by adding an alkaline salt such as LiOH to generate Li-PAA. The reaction leads to the formation of water while the pH of the solution changes from 2.7 to 9.7 [125]. Polyacrylic acid exhibits an agglomerated conformation in water due to the intramolecular hydrogen bonds between the carboxylic groups, while Li-PAA offers more linear chains due to the electrostatic repulsion between the carboxylate groups.

Porcher et al. [126] reported on the effect of PAA and Li-PAA on the coulombic efficiency for silicon anodes with practical loading of about 2.5 mA h cm^{-2}. It was shown that both PAA and Li-PAA outperformed CMC and polyether imide (PEI) containing electrodes especially in terms of coulombic efficiency. The study revealed that the improved electrode behavior is associated to a specific spring-like conformation that the polymer adopted as a consequence of the hydrogen bonding interactions with the particles. This particular conformation, not observed for CMC electrodes, enabled a strong formation and stabilization of the electronic percolation network in the electrode responsible for the improved efficiencies [126]. The study, further confirmed that despite being a very brittle polymer, PAA retains the electrode integrity, highlighting the importance of the silicon particle/binder interaction rather than the mechanical properties of the binder [121, 126].

Augmented electrochemical performance has been achieved by Kovalenko et al. [127], when using alginate as a binder material rather than PVDF and CMC. Alginate, also known as alginic acid and commonly found in its Na salt form is a high-modulus natural polysaccharide extracted from brown algae containing carboxylic groups in each of the polymer's monomeric units. The extraction of alginates from algae is performed by heating algae in a hot Na$_2$CO$_3$ solution, which results in the dissolution of its alginate component. From a chemical standpoint, alginate is a copolymer of 1 → 4 linked β-D- mannuronic acid (M) and α-L-guluronic acid (G) residues. Several different compositions can be obtained by changing the ratio of M and G blocks which also influences the mechanical properties of the binder.

In the study by Kovalenko et al. [127] it was found that the alginate binder presents similar mechanical properties to CMC and PVDF binders, however, with alginate offering a lower swelling capacity leading to a lower binder/electrolyte interaction crucial for an improved electrode cyclability. The improved electrochemical performance of alginate containing electrodes is most likely attributed to the chemical structure of alginate which contains a high concentration of uniformly distributed carboxylic groups enhancing Li^+ ions transport while covering the silicon surface uniformly permitting a more stable SEI. Furthermore, the increased polarity of the alginate could upgrade the adhesion to the Cu current collector thus fine-tuning electrode stability and mitigating delamination effects [127].

An important parameter to consider when selecting binders is their electrochemical reactivity with the electrolyte components. It has been reported that binders can react with lithium salts, e.g., $LiPF_6$, and accelerate electrolyte degradation [117]. The presence of hydroxyl groups, both present in CMC and alginate, can strongly interact with decomposition products of $LiPF_6$ such as PF_5 leading to the formation of $O=PF_3$. This inevitably leads to the decomposition of carbonate solvents in turn leading to electrolyte degradation and reduced cycling performance [117]. Studies in this area have been explored by Delpuech et al. [128], indicating that removing the hydroxyl group on the SiO_2 surface through the esterification reaction of CMC, reduces the $LiPF_6$ degradation in a similar manner. In order to reduce free hydroxyl groups in CMC a *co*-polymerization approach has been proposed. Following this approach, Koo et al. [129] cross linked PAA with CMC, enabling high mechanical stability and reduced electrolyte reactivity translated in an improved capacity retention and coulombic efficiency of the resulting Si composite electrode outperforming those made with the single polymers.

The molecular structure of polysaccharide binders determines the mechanical properties and electrochemical performances of silicon anodes. Yoon et al. [130] revealed that glycosidic linkages (α and β) and side chains substituents (-COOH and -OH) are the main factors affecting the binder's behavior. They investigated three different single-component polysaccharides. Pectin was compared to CMC for the understanding of the effect of α-linkages versus β-linkages. In addition, pectin as a COOH-containing polymer was compared to amylose as its non-COOH counterpart.

Pectin was remarkably superior to CMC and amylose in cyclability and rate capability of battery cells based on silicon anodes. The α-glycosidic linkage, present in pectin and amylose, induces enhanced elasticity when compared to the β counterpart encountered in CMC and a lower resistance to elongation as confirmed with elastic modulus measurements. The study also revealed the role of side groups, indicating that amylose (with only hydroxyl groups) presents an elastic modulus similar to the more rigid, β-linked CMC. The elastic nature of the pectin binder allowed volume expansion of silicon electrodes while maintaining the physical integrity of the electrodes during repeated lithiation/de-lithiation, seemingly due to the elastic nature caused by the chair-to-boat conformation in the α-linkages of its backbone [130]. On the contrary, the CMC-based electrode revealed cracks developing as a consequence of silicon volume expansion. Additionally, the covalent bonds formed between carboxylic side chains of pectin and silicon surface oxide allowed for mechanical stability preventing silicon loss and electrical disconnection. In contrast, the hydrogen bonds between hydroxyl side chains of amylose and silicon surface oxide were not strong enough to maintain the electrode integrity [130].

Inspired by biological systems, Ryou et al. [131] proposed the introduction of catechol functionalities into the polymer backbones of PAA and alginate to increase the binders adhesion ability. Catechol is naturally present in mussels and plays a decisive role in the exceptional wetness-resistant adhesion ability which resembles the operative conditions of batteries with components in contact with each other in a liquid environment. The unique properties of the developed binders containing catechol groups allowed substantial improvement of the cyclability, both in terms of specific capacity and cycle life of silicon nanoparticle electrodes, upholding structural integrity upon cycling. The improved mechanical stability of the above-mentioned tailored polymeric binders has resulted in conventional PVDF being replaced by cross-linked chains, self-healing polymeric matrices [65] (see Fig. 1.5) and electronically conducting polymers as mentioned in this section which effectively preserve the electrode structure upon repeated volume change of Si, mainly via 3D interchain interactions [63].

Lessons learned in the research and development for functional silicon binders can be summarized as follows. Optimal binders for silicon anodes should be able to provide self-healing properties in order to be able to regenerate the interaction with the silicon particles surface upon volume variation as demonstrated by Wang et al. [65] (see Fig. 1.5). The polymer structure should offer the formation of bridges between silicon particles and current collectors to prevent delamination. Elongation and adhesion are important properties which have to be tuned considering the backbone structure of the polymers, carboxylic groups as well as β-glycosidic bonds have been proven to induce ion-dipole interactions allowing for improved mechanical stability. It is essential that the binder permits Li^+ ion conduction while limiting its wettability toward the electrolyte thus inhibiting electrolyte degradation and SEI instability.

Future efforts devoted to the development of binders for silicon anodes should be focused on improved compatibility of the polymeric binders with Si/C composites, given that the industrial demands are shifting toward the implementation of Si/C rather than Si-based anodes [73, 97]. This can be attained through the rational design of hybrid polymers with improved binding abilities toward silicon, but also carbonaceous matrixes (e.g., graphite). In addition, limiting the quantity of inactive materials is very desirable while developing high energy density systems. As a result, reducing the binder content to less than 5% is an important industrial target [63]. The coulombic efficiency upon cycling and the first irreversible capacity of silicon anodes needs to be upgraded

Silicon anode systems for lithium-ion batteries Chapter | 1 **27**

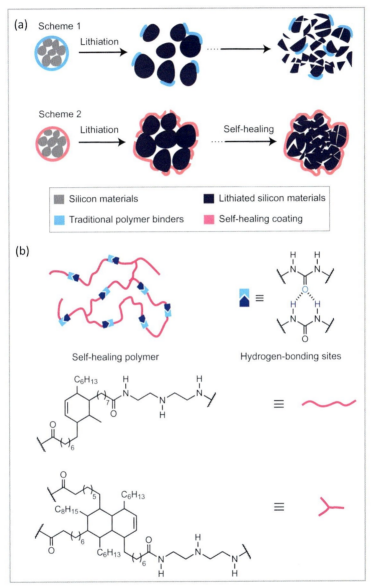

Fig. 1.5 (A) Schematic illustration of the lithiation behavior of silicon electrodes employing conventional (Scheme 1) and self-healing (Scheme 2) binder components. Particle cracking and electrode integrity loss can be mitigated by using stretchable self-healing binder. (B) Typical structure of a self-healing polymer enabling hydrogen bonds formation. The red lines represent the polymer backbones while the light-blue and dark-blue boxes represent hydrogen-bonding sites. *(Reproduced from C. Wang, et al., Self-healing chemistry enables the stable operation of silicon microparticle anodes for high-energy lithium-ion batteries, Nat. Chem. 5 (2013) 1042–1048 with permission Copyright © Nature Publishing Group, 2013.)*

for an efficient implementation in full lithium-ion cells, thus the development of binders allowing high Li$^+$ ions conductivity and limiting Li$^+$ ions trapping is crucial, artificial polymeric SEI layer are a valid solution for the stabilization of the SEI and the limitation of electrolyte degradation whilst sustaining fast Li$^+$ ions kinetics of diffusion.

1.3.1.3 The silicon/electrolyte interphase

In 2009, M. Winter published a review article claiming that the "*SEI is the most important and the least understood solid electrolyte in rechargeable Li batteries*" [132]. Indeed, the SEI represents perhaps one of the most crucial components of every alkali ion rechargeable battery. Even though the first model was initially introduced in 1979 by Peled et al. [133], many processes occurring at the electrode/electrolyte interface are yet to be comprehensively understood.

Nonaqueous LIBs are thermodynamically unstable systems, which can operate under a delicate kinetic stability thanks to the formation of the SEI layer. This layer is responsible for the passivation of the anode surface and limits the electrolyte decomposition while guaranteeing Li$^+$ ions conduction [133–135]. It is well known that the SEI layer formation and evolution strongly depend upon the electrolyte system as well as on the electrode surface. Metallic lithium, LTO, graphite and silicon anodes manifest diverse surface characteristics especially when comparing the stability upon cycling of the SEI layer and the cycling efficiencies of the anodes.

Lithium metal is quite reactive toward the electrolyte upon contact, forming a spontaneous passivation layer. However, this layer is not stable and Li corrosion along with poor passivation properties of the lithium/electrolyte interphase added to the pernicious dendrite formation characteristics have limited to some extent, the implementation of metallic lithium as anode in LIBs [132]. On the contrary, in layered carbon materials, such as graphite, the SEI formation and the charge storage mechanism ensues in a completely different way. The SEI formation in graphite electrodes mostly occurs during the first cycle and it remains reasonably stable upon cycling. The uniqueness of the SEI stability in graphite arises from its structural properties. In an electrochemical cell, two main graphite surfaces are exposed to the electrolyte, namely the edge and basal planes. While the latter ones are smooth surfaces mainly consisting of carbon atoms, the edges often present oxygen-containing surface groups. Therefore, the SEI properties and functions of these two surfaces differ profoundly [132, 136]. Li ion transport and electrolyte reduction take place predominantly at the edge, or prismatic sites, so the charge loss transpiring upon SEI formation is strongly affected by the basal to prismatic sites ratio. Accordingly, only the prismatic sites contribute to processes such as formation, dissolution and aging of the SEI [137]. The structure and stability of the SEI in graphite is strongly influenced by the composition of the electrolyte [12, 135]. In spite of the SEI formation in graphite electrodes, this contributes to almost

5%–10% of the capacity loss; a properly engineered LIB can still provide an acceptable cycle life [138].

Contrary to graphite, Si anodes never fully passivate, with consequent continuous electrolyte decomposition [58, 68] responsible for the charge/discharge related coulombic inefficiency and consequent rapid cell degradation during long-term cycling of Li-ion cells [139]. It has been confirmed in numerous studies that the SEI formation is strongly influenced by the electrolyte composition and the use of additives [12], yet recently it has been stated that the electrochemical reduction of the electrolyte is an electrocatalytic process strongly affected by the surface of the electrodes, with kinetics strongly dependent on the electrode material [140, 141]. Silicon presents an intrinsic nonpassivating activity in the most commonly employed carbonate based electrolytes further aggravated by the severe volume expansion of the Si particles upon lithiation/de-lithiation [139]. The unique electrochemical reactivity of silicon toward the electrolyte leads to large irreversible capacity loss, poor passivating behavior with consequent gradual electrolyte consumption, lithium inventory shift in full cell configuration and cell impedance increase due to surface film accumulation upon cycling [142].

A cyclic variation (breathing effect) of the SEI thickness has been observed upon lithiation and de-lithiation, indicating the instability of this passivating layer and its continuous change as a function of charge state [39]. Veith et al. [72] revealed through neutron reflectometry that the breathing performance of the SEI is associated with a change in the chemistry of the compounds constituting the SEI (see Fig. 1.6A). The SEI formation is strongly dependent on the electrolyte composition and the use of additives [12]. By using a fluoroethylene carbonate (FEC)-added electrolyte, the SEI thickens upon lithiation and becomes more organic-like, while upon de-lithiation it becomes thinner and richer in inorganic compounds [72]. The direction of the changes observed in thickness is the opposite when using an FEC-free electrolyte [143] (see Fig. 1.6B). Toney et al. [144] observed through X-ray reflectivity measurements (Fig. 1.6C) and high precision coulombic efficiency measurements that the SEI grows with increased time spent at low voltage values and that the growing rate is described by a parabolic function.

Recently, Hasa et al. [145] linked the SEI layer "breathing effect" to the formation of lithium ethylene dicarbonate (solvent decomposition product) and P—F and P-O-F containing compounds (salt decomposition products) and their disappearance upon de-lithiation. The study revealed that $LiPF_6$ decomposition products undergo a noncomplete dissolution during the first de-lithiation suggesting an accumulation of these species on the surface of the electrode. LiF instead, forms early on the silicon surface and it resides there permanently, forming a stable component of the SEI layer (see Fig. 1.6D). The observed "breathing effect" of the SEI layer on silicon is a clear manifestation of the intrinsic interfacial instability and poor passivating behavior of the silicon anode in organic carbonate electrolyte and is explained by considering a slow

30 PART | I Introduction and background

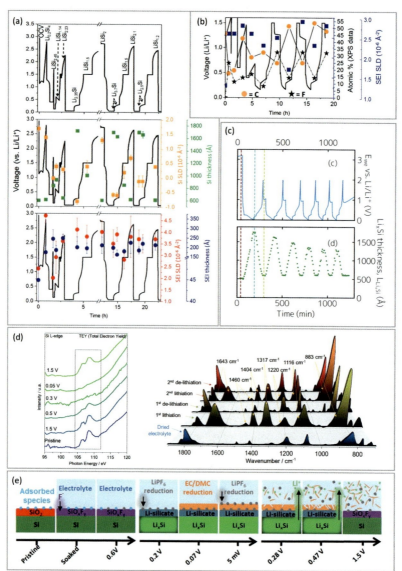

Fig. 1.6 Breathing effect of the silicon/electrolyte interphase observed with different analytical techniques. (A) Neutron reflectometry measurements evidence the SEI thickness variation upon cycling of a silicon thin film electrode cycled in FEC-free electrolyte. (Top) Voltage profile of the silicon thin film electrode as a function of time with the corresponding Li:Si ratios determined from Coulometric measurements. (Middle) Refined Si thickness (green), and Si scattering length density (SLD) values (orange). (Bottom) Refined SEI thickness (blue), and SEI SLD values (red). (B) Neutron reflectometry measurement performed on a silicon thin film electrode in an FEC-added electrolyte evidence SEI thickness variation with an opposite direction of changes when compared to the FEC-free systems. (C) Evolution of the silicon voltage profile as a function of time

dissolution of the species in the electrolyte (see Fig. 1.6E) [57, 58, 144, 146]. The precise identification of chemical species, their exact distribution and function in the electrode/electrolyte interphase with regard to its passivating properties have been investigated thoroughly. Nevertheless, barriers remain for the development and implementation of practical silicon-based anodes for high-energy Li-ion cells, highlighting the importance of the need for further studies.

Most of the above-mentioned studies have been performed by using model silicon electrodes, such as Si thin films with limited thickness (<100 nm). Model electrodes provide a very useful platform for the investigation of the silicon/electrolyte interphase enabling a separate analysis on the evolution of the SEI and its dependency upon cracking effects and upon the pure catalytic activity of silicon. While the fundamental understanding of these processes has the potential to pave the way toward the development of degradation-proof strategies, the transition remains nevertheless still very challenging to migrate the acquired fundamental knowledge gained and apply into practical electrodes in which inevitably several degradation processes (induced by the catalytic activity [140] of silicon and its expansion/cracking upon cycling [63]) occur simultaneously. A major concern with the SEI in silicon anodes is related to the large volume changes of the material upon cycling. This leads to SEI rupture and exposure of fresh silicon as mentioned above to the electrolyte with continuous electrolyte degradation and accumulation of film on the silicon surface with increased cell resistance accompanied by low efficiencies eventually leading to cell failure.

In order to form a stable and robust SEI layer, one can act at the electrolyte or at the electrode level. A commonly employed strategy toward the stabilization

---

Fig. 1.6, Cont'd compared with the Li_xSi thickness variation calculated from X-ray reflectivity measurements. (D) Direct observation of the silicon/electrolyte interphase breathing effect upon cycling by X-ray absorption spectroscopy and ATR-FTIR measurements on a silicon thin film electrode. (E) Schematic representation the proposed reaction mechanisms occurring during the early cycling of an amorphous Si thin film in liquid electrolyte based on X-ray photoelectron spectroscopy measurements. *(Part figure (A) reproduced from G.M. Veith et al., Direct determination of solid-electrolyte interphase thickness and composition as a function of state of charge on a silicon anode, J. Phys. Chem. C 119 (2015) 20339–20349 with permission from © American Chemical Society, 2015; Part figure (B) reproduced from G.M. Veith et al., Determination of the solid electrolyte interphase structure grown on a silicon electrode using a fluoroethylene carbonate additive, Sci. Rep. 7 (2017) 1–15 distributed under Creative Commons CC BY license; Part figure (C) reproduced from H.G. Steinrück, C. Cao, G.M. Veith, M.F. Toney, Toward quantifying capacity losses due to solid electrolyte interphase evolution in silicon thin film batteries, J. Chem. Phys. 152 (2020) with permission from AIP Publishing; Part figure (D) reproduced from I. Hasa et al., Electrochemical reactivity and passivation of silicon thin-film electrodes in organic carbonate electrolytes, ACS Appl. Mater. Interfaces 12 (2020) 40879–40890 with permission from © American Chemical Society, 2020; Part figure (E) reproduced from G. Ferraresi, L. Czornomaz, C. Villevieille, P. Novák, M. El Kazzi, Elucidating the surface reactions of an amorphous Si thin film as a model electrode for Li-ion batteries, ACS Appl. Mater. Interfaces 8 (2016) 29791–29798 with permission from © American Chemical Society, 2016.)*

of the SEI is the use of electrolyte additives. FEC and vinylene carbonate (VC) have been widely studied as effective additives for improving cyclability of Si anodes [147]. Both VC and FEC are considered SEI forming additives and are indispensable for silicon anodes. The positive role of FEC on the SEI formation and stability was demonstrated by Choi et al. [148] on the electrochemical performance of a lithium/silicon thin-film using a 1.3 M LiPF$_6$ solution in EC/DEC with and without 3 wt% FEC. The additive provided markedly improved capacity retention and coulombic efficiency while generating a smooth and uniform SEI layer.

Schroder et al. [149] investigated the SEI structure and chemical composition by X-ray photoelectron spectroscopy (XPS) and ToF-SIMS depth profiling. It was determined that the main decomposition product is LiF, consistently with previously reported studies. They proposed that the effectiveness of FEC at improving the coulombic efficiency and capacity retention of silicon anodes is due to fluoride ion formation from reduction of the solvent, leading to the chemical reactivity of the silicon-oxide surface passivation layer and the formation of a kinetically stable SEI constituted predominantly by LiF and Li$_2$O [149].

Shkrob et al. [150] also stated that the reduction of FEC can lead to the formation of a vinoxyl radical that can abstract an H atom from another FEC molecule, causing FEC decomposition and radical polymerization. The resulting polymer can further de-fluorinate yielding to radicals that migrate and recombine to produce a highly cross-linked network which may exhibit elastomeric properties, explaining its cohesion during expansion and contraction of silicon particles upon Li alloying/de-alloying. Despite a comprehensive understanding, the knowledge of the role played by FEC is still incomplete [151]; it is generally accepted that FEC decomposes and forms an SEI layer rich in poly(FEC), alkyl-carbonates, lithium fluoride and carbonate [72, 152–155]. It has been concluded that a high concentration of LiF is commonly reported together with FEC, but the role of LiF is not fully understood. From the studies on graphite, LiF appears to have a negative impact on reversibility, nonetheless a direct transfer of knowledge from graphite to silicon is not straightforward with many aspects that need to be considered [145]. For instance, Xu et al. emphasized that it is not only the chemical properties of LiF that determines its role in the SEI, but also its distribution [156].

In contrast to FEC, the role of VC is commonly accepted [157, 158]. VC was employed in combination with Si thin films by Xie et al. [158]; stable cycling behavior up to 200 cycles was reported. The improved electrochemical performance was attributed to the favorable properties of the SEI layer uniformly passivating the silicon surface with polycarbonates reduction products and a limited content of LiF [157, 159, 160]. Even though VC offers some benefits such as its high thermal stability and its inactivity toward the cathode electrode [147], FEC containing electrolytes present a smaller impedance when compared to VC-added electrolytes. Aurbach et al. [151] suggested that the smaller impedance and improved capacity retention of silicon anodes cycled in FEC-containing electrolytes can be attributed to the nature of specific decomposition products, i.e., polyenes, which may serve as a key component for the formation of an effectively passivating surface layer.

Jaumann et al. investigated the effect of FEC on silicon/carbon electrode and attributed the enhanced reversibility of the lithiation/de-lithiation process to the formation of a very thin polymer layer, presumably a vinyl polymer [161]. In addition, a high concentration of Li_2O, a decomposition product arising from the interaction of the native SiO_2 layer with the electrolyte, was found for electrodes cycled in the FEC-containing electrolyte suggesting that the diffusion of HF through the SEI is prevented. On the other hand, the FEC-free electrolyte caused a high electrolyte decomposition, showing no ability to prevent Li_2O dissolution.

Blending of additives has also been proved to be a good strategy for an efficient passivation of silicon anodes. By mixing FEC and di (2,2,2-trifluoroethyl) carbonate (DFDEC), Song et al. [162] observed the formation of a robust, thick SEI layer resulting in a structural stabilization of the Si/graphite composite anode when compared with the FEC only added electrolyte. The SEI was formed specifically by FEC decomposition products such as lithium carbonate anhydride, as well as from carboxylates, alkyl carbonates, LiF, and P—F and/or P-O-F containing compounds (e.g., $OPF_{3-y}(OR)_y/Li_xPF_yO_z$) that mainly arise from DFDEC decomposition [162].

Siloxanes and silanes have also been used as effective additives for Si-based electrodes by forming a strong SEI layer [163]. When used in combination with silicon thin film electrodes, alkoxysilane-based additives were found to suppress the mass accumulation of SEI layer improving the electrochemical stability and cycle life. The effective Si passivation was mainly accredited to the reactivity of the alkoxysilane functional groups in the additives with hydroxyl groups on the Si electrode surface, forming organosilicon compounds effective in recruiting interfacial resistance and preventing Si cracking [163]. Several additives of various chemical compositions have been successfully examined as film-forming additives such as methylene ethylene carbonate [164], lactic acid O-carboxyanhydride [165], succinic anhydride [166], or pentafluorophenyl isocyanate [167]. The main effect is observed by the formation of a stable SEI layer, effective in passivating the silicon surface and reducing continuous electrolyte consumption as well as the ability to mitigate the silicon electrode damage from the large volume changes during cycling [147].

Another strategy to optimize the interphase stability at the silicon anode is to design and tune the surface properties of the silicon particles. For instance, Dalla Corte et al. [168] reported that grafting a molecular monolayer of carboxydecyl moieties (acid grafting) or poly (oxoethylene) (PEG) chains on a hydrogen terminated silicon anode decreases the irreversible capacity and stabilizes the SEI. In contrast, Cui's group [69] designed a well-engineered active silicon nanotube surrounded by an ion-permeable silicon oxide shell cycling over 6000 times in half cells while retaining more than 85% of its initial capacity. The anode consisted of a double-walled Si-SiO$_x$ nanotube-based anode, with Si as the main constituent of the inner wall and SiO$_x$ confined in the outer wall. The well-designed oxide layer enabled Li$^+$ ion diffusion, while avoiding silicon direct contact with the electrolyte. In addition, while the outer surface of the Si nanotube was prevented from expansion by the oxide shell, the expanding inner surface was not exposed to the electrolyte, thus strongly stabilizing the SEI. The double wall silicon nanotube-based anode exhibited capacities about eight times larger than conventional carbon anodes sustaining charging rates up to 20C.

A few years later the same group, inspired by the pomegranate structure, designed a robust hierarchical structured silicon anode able to tackle the main challenges faced by silicon anodes including: silicon structural degradation, silicon/electrolyte interphase instability due to volume expansion, its chemical reactivity with the electrolyte and the occurrence of side reactions as well as the typical low volumetric capacity of Si-based anode arising from the necessary use of nanometer scale material size [95]. The well-designed material consisted of the encapsulation of single silicon nanoparticles into a carbon coating layer with enough space between the silicon and carbon layer to allow for silicon expansion and contraction upon cycling. The electrodes exhibited extraordinary cyclability with a capacity retention of 97% over 1000 cycles. What's more, given the hierarchical arrangements of the material, the limited interaction of silicon with the electrolyte and the stable SEI layer formed (not subjected to stressful volume variations) permitted the achievements of coulombic efficiency values of 99.87% and an initial value for the first cycle of 82% for a 9% carbon content. The coulombic efficiency decreased to 75% when increasing the thickness of the carbon layer to about 23% carbon content. Future efforts must tackle this by potentially developing efficient prelithiation methods or by replacing the amorphous carbon coating with carbon material less prone to lithium ions trapping. Other than the electrochemical performance, further imperative challenges in the development of well-performing silicon-based anodes relies on the definition of a low cost, easy to implement scalable preparation method and on the achievement of high areal electrode loadings. Interestingly, the electrode material proposed by Cui et al. [95] showed a volumetric capacity of about 1270 mA h cm^{-3} with impressive cycling stability also when increasing the areal capacity to commercial LIBs value, i.e., 3.7 mA h cm^{-2}. Also, authors claimed that the pomegranate inspired materials could be easily prepared in the powder form without involving specialized equipment or processes. Contrary to other nanostructures, it could be processed as other common electrode materials during the slurry preparation. The authors also highlighted the importance of reducing the cost of starting materials such as the silicon nanoparticles in order to be able to meet the requirement for practical market application.

1.4 Conclusions: Summary and perspective

Silicon appears to be the most promising anode materials for the next generation LIBs, owing to its desirable properties such as the high volumetric and gravimetric capacity, the relatively low operating voltage and its natural abundance. However, several hurdles impede the practical implementation of silicon-based anodes in commercial LIBs. Among them the inherent catalytic activity of silicon toward the electrolyte and the poor passivation properties leading to instability of the SEI along with the occurrence of side reactions responsible for the poor coulombic efficiency, combined with the large volume expansion leading to electrode integrity limitation, active material loss, and consequent rupture of the SEI, lead to unsatisfactory electrochemical performance.

The various strategies so far proposed to successfully overcome the limitations for silicon-anodes implementation have been discussed above. Overall, tremendous progresses have been achieved in the rational design of functional silicon-based anodes. The lessons learned can be summarized as follow:

- Downsizing of the silicon particles helps mitigating silicon particles fracture upon lithiation/de-lithiation. The shape and size of silicon particles greatly affect the deriving electrode performance.
- While nanoscaling improves the structural stability of the silicon particles, from an electrode perspective this is not sufficient, thus nano-structuration and design of composites electrode including use of mechanically compliant carbon, metals and metal oxides composites can improve electrode integrity.
- The adoption of porous silicon nanostructures constitutes a further improvement for the obtainment of high-performance anode able to accommodate the huge volume expansion.
- The synergetic effect of combining void-engineered structures containing carbon coated silicon porous nanostructures greatly improves the electrochemical performance by improving the electronic conductivity of silicon (fast electron conducting pathways), stabilizing the silicon/electrolyte interface by forming a stable, elastic SEI and well-containing the silicon expansion during the lithiation/de-lithiation process.
- Third components such as the binder and the electrolyte play a vital role in the cycling performance of silicon anodes, thus the strategic approaches toward the implementation of silicon-based anodes should consider not only the silicon properties but also the interaction with the other cell components.

Despite the promising results achieved by implementing the above-mentioned strategies at a lab scale, the implementation of high silicon content anodes in commercial LIBs still remains a challenge. Currently, coating or embedding silicon with carbon layers/matrixes appears to be the most effective way toward practical implementation of silicon in LIBs. Three main aspects warrant much attention of the R&D community for the implementation of silicon anodes produced at industrial scale:

- Improvement of the silicon's tap density (0.4–1.0 mg cm^{-3} vs 1.4–1.8 mg cm^{-3} in graphite) and areal loading of silicon anodes are necessary to meet the volumetric energy density demands for practical applications.
- Development of cost-effective synthesis processes for the preparation of nanosized materials. Cost is a crucial parameter and plays a fundamental role for the widespread of a technology in the market.
- Sustainable and environmentally friendly preparation methods for silicon-based anodes have been largely proposed at lab scale, however feasibility studies for the up-scaling to large quantities are still at an infancy stage.

To overcome the low tap density and large surface area responsible for the low volumetric capacities and efficiency (increased electrolyte reactivity of silicon anodes), recently the focus has been devoted toward the development of microstructures assembled from nanostructures subunits [17]. An interesting example are SiO_x materials consisting of Si and SiO_2 nano-domains and silicon sub-oxides at their interfaces [169]. The combined use of graphite and Si or SiO_x in mixtures and/or composites is also seen as a very effective strategy to provide the improved volumetric energy density and cycling stability.

Industry has indeed adopted silicon monoxide (SiO_x, $x \approx 1$) for the first Si-based commercial anode material. One of the main advantages of these materials is their production method. Indeed, they can be produced in large scale at relatively low cost. In 2007 Fukuoka et al. [170] patented a gas-phase production method for SiO_x ($x < 1$) to be employed for use as anode material for LIBs. Solution-based [171] approaches have also been employed successfully, ensuring large-scale prices are still high but at competitive costs when compared to graphite (US$100 per kg versus $10–20 per kg of graphite) [63].

Other advantages of SiO_x compounds rely on the absence of phase transitions leading to the formation of c-$Li_{15}Si_4$-like structure [172]. Moreover, the presence of oxide species allows for a limited volume expansion upon cycling most likely associated to the formation of silicates. It is also worth mentioning though that the effect of the oxide and the role of silicates species is not fully understood and its effect on the cycling behavior is still discussed [56]. Besides, when used in combination with graphite in blends (5% SiO_x/graphite) promising performance has been achieved. Unfortunately, by increasing the loading of SiO_x in the blend composite, very low initial coulombic efficiencies are observed, severely hindering its implementation in LIB cells, which would require very high cathode loadings thus limiting the energy density of the final cell.

Aurbach et al. [63] identified a main challenge in the design of effective silicon anode structures consisting in the simultaneous improvement of cyclability and irreversible coulombic efficiency. Indeed, while porous structure and carbon matrices are seen as effective strategies to mitigate the volume expansion effects of silicon anodes, on the other hand, they are very detrimental for the efficiency of the lithiation/de-lithiation process by irreversibly trapping Li ions. Efficient prelithiation strategies could mitigate these issues [173–175].

Currently, Si/C composites represent the most promising commercial solution, however, severe issues still need to be addressed and real breakthroughs are required to achieve the key performance indicators targets set for a full transition to electromobility [176]. In the Strategic Energy Technology (SET) plan [177], Europe has declared its renewable energy ambitions providing a framework to accelerate the development and deployment of cost-effective low carbon technologies. The implementation of the SET Plan is followed up with a number of targeted ambitious key performance indicators specific for different battery chemistries. A categorization of the different classes of Li-based battery chemistry was first introduced by the German National Electromobility platform in 2016 [178], followed by the European Commission's Joint Research Centre (JRC) in 2017 [179] and the Energy Materials Industrial Research Initiative (EMIRI) roadmap in 2019 [180]. This categorization has further been adopted in 2020 by the Strategic Research and Innovation Agenda [181] formulated by European Technology & Innovation Platform (ETIP) Batteries Europe envisioning Si-based anodes as a crucial near-future battery chemistry.

The therein defined Generation 3a cells, including high nickel content layered oxides cathodes (NMC-811) and anodes containing graphite with 5 up to 10% silicon, were forecasted for market deployment in 2020. To the best of our knowledge, current EVs are still required to depend on the relatively mature battery technologies including NMC, NCA, and LMO-NMC blended chemistries coupled with carbon anodes. Some exceptions from Tesla are worth mentioning. Silicon-containing anodes such as Si—C, SiO-C in combination with NCA chemistry beyond the pure graphitic anode have been reported [182].

However, implementation of chemistries with high specific energy and lower cost such as Ni-rich NMC (e.g., NMC-811) and Si is still foreseen in the short term [17]. In 2025, silicon/carbon composite with higher silicon content are envisioned in combination with high energy density lithium-rich or spinel type cathodes. Clear signals from industry, academia and governmental agencies indicate optimism in the implementation of silicon-based anodes, thus further accelerating their development and implementation in the market. Smart electrode optimization and cell design are essential for postgeneration 3a lithium cells. Synergetic strategies should be adopted to achieve full implementation of silicon anodes, while keeping in mind the importance of cost and sustainability as crucial indicators for the successful intake of the developed technology in the market.

References

[1] https://www.nobelprize.org/prizes/chemistry/2019/summary/. (Accessed 22 June 2021).
[2] Y. Nishi, Past, present and future of lithium-ion batteries, in: Lithium-Ion Batteries, Elsevier, 2014, pp. 21–39, https://doi.org/10.1016/B978-0-444-59513-3.00002-9.
[3] G.N. Lewis, F.G. Keyes, The potential of the lithium electrode, J. Am. Chem. Soc. 35 (1913) 340–344.

[4] C.A. Vincent, Lithium batteries: a 50-year perspective, 1959–2009, Solid State Ion. 134 (2000) 159–167.
[5] W.S. Harris, Electrochemical Studies in Cyclic Esters, University of California, Berkeley, 1958.
[6] A. Yoshino, The birth of the lithium-ion battery, Angew. Chem. Int. Ed. 51 (2012) 5798–5800.
[7] B. Scrosati, History of lithium batteries, J. Solid State Electrochem. 15 (2011) 1623–1630.
[8] M. Li, J. Lu, Z. Chen, K. Amine, 30 years of lithium-ion batteries, Adv. Mater. 30 (2018) 1800561.
[9] J.B. Goodenough, U.S. Patent 4,302,518 (issued 1980/3/31), 1980.
[10] H. Ikeda, K. Narukawa, H. Nakashim, Japanese Patent 1769661 (issued 1981/6/18), 1981.
[11] Y. Nishi, The development of lithium ion secondary batteries, Chem. Rec. 1 (2001) 406–413.
[12] K. Xu, Nonaqueous liquid electrolytes for lithium-based rechargeable batteries, Chem. Rev. 104 (2004) 4303–4418.
[13] J. Lu, Z. Chen, F. Pan, Y. Cui, K. Amine, High-performance anode materials for rechargeable lithium-ion batteries, Electrochem. Energy Rev. 1 (2018) 35–53.
[14] S. Chauque, et al., Lithium titanate as anode material for lithium ion batteries: synthesis, post-treatment and its electrochemical response, J. Electroanal. Chem. 799 (2017) 142–155.
[15] C.P. Sandhya, B. John, C. Gouri, Lithium titanate as anode material for lithium-ion cells: a review, Ionics (Kiel) 20 (2014) 601–620.
[16] S.K. Martha, et al., $Li_4Ti_5O_{12}/LiMnPO_4$ Lithium-ion battery systems for load leveling application, J. Electrochem. Soc. 158 (2011) A790–A797.
[17] M. Armand, et al., Lithium-ion batteries—current state of the art and anticipated developments, J. Power Sources 479 (2020) 228708.
[18] C. Curry, Lithium-ion Battery Costs and Market, BNEF, 2017. 5 July 2017.
[19] M.N. Obrovac, V.L. Chevrier, Alloy negative electrodes for Li-ion batteries, Chem. Rev. 114 (2014) 11444–11502.
[20] https://www.sony.net/SonyInfo/News/Press/200502/05-006E/. (Accessed 22 June 2021).
[21] Z. Lin, X. Lan, X. Xiong, R. Hu, Recent development of Sn-Fe-based materials as a substitution for Sn-Co-C anode in Li-ion batteries: a review, Mater. Chem. Front. (2020), https://doi.org/10.1039/d0qm00582g.
[22] M.N. Obrovac, L. Christensen, D.B. Le, J.R. Dahn, Alloy design for lithium-ion battery anodes, J. Electrochem. Soc. 154 (2007) A849–A855.
[23] H. Wu, Y. Cui, Designing nanostructured Si anodes for high energy lithium ion batteries, Nano Today 7 (2012) 414–429.
[24] D. Andre, H. Hain, P. Lamp, F. Maglia, B. Stiaszny, Future high-energy density anode materials from an automotive application perspective, J. Mater. Chem. A 5 (2017) 17174–17198.
[25] Z. Du, J. Li, C. Daniel, D.L. Wood, Si alloy/graphite coating design as anode for Li-ion batteries with high volumetric energy density, Electrochim. Acta 254 (2017) 123–129.
[26] R.A. Huggins, Alloy negative electrodes for lithium batteries formed in-situ from oxides, Ionics (Kiel) 3 (1997) 245–255.
[27] C.J. Wen, R.A. Huggins, Chemical diffusion in intermediate phases in the lithium-silicon system, J. Solid State Chem. 37 (1981) 271–278.
[28] M. Zeilinger, D. Benson, U. Häussermann, T.F. Fässler, Single crystal growth and thermodynamic stability of $Li_{17}Si_4$, Chem. Mater. 25 (2013) 1960–1967.
[29] R. Nesper, H.G. von Schnering, $Li_{21}Si_5$, a Zintl phase as well as a Hume-Rothery phase, J. Solid State Chem. 70 (1987) 48–57.

[30] M.T. McDowell, S.W. Lee, W.D. Nix, Y. Cui, 25th anniversary article: understanding the lithiation of silicon and other alloying anodes for lithium-ion batteries, Adv. Mater. 25 (2013) 4966–4985.

[31] P. Limthongkul, Y.I. Jang, N.J. Dudney, Y.M. Chiang, Electrochemically-driven solid-state amorphization in lithium-metal anodes, J. Power Sources 119–121 (2003) 604–609.

[32] X.H. Liu, et al., In situ atomic-scale imaging of electrochemical lithiation in silicon, Nat. Nanotechnol. 7 (2012) 749–756.

[33] B. Key, et al., Real-time NMR investigations of structural changes in silicon electrodes for lithium-ion batteries, J. Am. Chem. Soc. 131 (2009) 9239–9249.

[34] N. Ding, et al., Determination of the diffusion coefficient of lithium ions in nano-Si, Solid State Ion. 180 (2009) 222–225.

[35] M.N. Obrovac, L. Christensen, Structural changes in silicon anodes during lithium insertion/extraction, Electrochem. Solid St. 7 (2004) A93–A96.

[36] X.H. Liu, et al., Anisotropic swelling and fracture of silicon nanowires during lithiation, Nano Lett. 11 (2011) 3312–3318.

[37] J.Y. Kwon, J.H. Ryu, S.M. Oh, Performance of electrochemically generated Li21Si5 phase for lithium-ion batteries, Electrochim. Acta 55 (2010) 8051–8055.

[38] J. Li, J.R. Dahn, An in situ X-ray diffraction study of the reaction of Li with crystalline Si, J. Electrochem. Soc. 154 (2007) A156.

[39] F. Ozanam, M. Rosso, Silicon as anode material for Li-ion batteries, Mater. Sci. Eng., B 213 (2016) 2–11.

[40] V.L. Chevrier, J.R. Dahn, First principles studies of disordered lithiated silicon, J. Electrochem. Soc. 157 (2010) A392.

[41] B. Key, M. Morcrette, J.M. Tarascon, C.P. Grey, Pair distribution function analysis and solid state NMR studies of silicon electrodes for lithium ion batteries: understanding the (de)lithiation mechanisms, J. Am. Chem. Soc. 133 (2011) 503–512.

[42] M.T. McDowell, et al., In situ TEM of two-phase lithiation of amorphous silicon nanospheres, Nano Lett. 13 (2013) 758–764.

[43] J.W. Wang, et al., Two-phase electrochemical lithiation in amorphous silicon, Nano Lett. 13 (2013) 709–715.

[44] D. Alves Dalla Corte, et al., Spectroscopic insight into Li-ion batteries during operation: an alternative infrared approach, Adv. Energy Mater. 6 (2016) 1–11.

[45] A. Bordes, et al., Investigation of lithium insertion mechanisms of a thin-film Si electrode by coupling time-of-flight secondary-ion mass spectrometry, X-ray photoelectron spectroscopy, and focused-ion-beam/SEM, ACS Appl. Mater. Interfaces 7 (2015) 27853–27862.

[46] N.V. Rumak, V.V. Khatko, Structure and properties of silicon dioxide thermal films. II. 110 nm thick SiO2 films, Phys. Status Solidi 86 (1984) 477–484.

[47] E. Radvanyi, E. De Vito, W. Porcher, S. Jouanneau Si Larbi, An XPS/AES comparative study of the surface behaviour of nano-silicon anodes for Li-ion batteries, J. Anal. At. Spectrom 29 (2014) 1120–1131.

[48] E. Radvanyi, et al., Failure mechanisms of nano-silicon anodes upon cycling: an electrode porosity evolution model, Phys. Chem. Chem. Phys. 16 (2014) 17142–17153.

[49] K.W. Schroder, A.G. Dylla, S.J. Harris, L.J. Webb, K.J. Stevenson, Role of surface oxides in the formation of solid-electrolyte interphases at silicon electrodes for lithium-ion batteries, ACS Appl. Mater. Interfaces 6 (2014) 21510–21524.

[50] H. Huang, E.M. Kelder, L. Chen, J. Schoonman, Electrochemical characteristics of $Sn_{1-x}Si_xO_2$ as anode for lithium-ion batteries, J. Power Sources 81–82 (1999) 362–367.

[51] N. Ariel, G. Ceder, D.R. Sadoway, E.A. Fitzgerald, Electrochemically controlled transport of lithium through ultrathin SiO2, J. Appl. Phys. 98 (2005) 023516.

[52] J. Saint, et al., Towards a fundamental understanding of the improved electrochemical performance of silicon-carbon composites, Adv. Funct. Mater. 17 (2007) 1765–1774.
[53] J. Graetz, C.C. Ahn, R. Yazami, B. Fultz, Highly reversible lithium storage in nanostructured silicon, Electrochem. Solid St. 6 (2003) A194.
[54] B. Guo, et al., Electrochemical reduction of nano-SiO2 in hard carbon as anode material for lithium ion batteries, Electrochem. Commun. 10 (2008) 1876–1878.
[55] Q. Sun, B. Zhang, Z.W. Fu, Lithium electrochemistry of SiO2 thin film electrode for lithium-ion batteries, Appl. Surf. Sci. 254 (2008) 3774–3779.
[56] E. Sivonxay, M. Aykol, K.A. Persson, The lithiation process and Li diffusion in amorphous SiO_2 and Si from first-principles, Electrochim. Acta 331 (2020) 135344.
[57] G. Ferraresi, L. Czornomaz, C. Villevieille, P. Novák, M. El Kazzi, Elucidating the surface reactions of an amorphous Si thin film as a model electrode for Li-ion batteries, ACS Appl. Mater. Interfaces 8 (2016) 29791–29798.
[58] B. Philippe, et al., Nanosilicon electrodes for lithium-ion batteries: interfacial mechanisms studied by hard and soft X-ray photoelectron spectroscopy, Chem. Mater. 24 (2012) 1107–1115.
[59] I.I. Abate, C.J. Jia, B. Moritz, T.P. Devereaux, Ab initio molecular dynamics study of SiO_2 lithiation, Chem. Phys. Lett. 739 (2020) 136933.
[60] S. Xun, et al., The effects of native oxide surface layer on the electrochemical performance of Si nanoparticle-based electrodes, J. Electrochem. Soc. 158 (2011) A1260.
[61] O. Renner, J. Zemek, Density of amorphous silicon films, Czech. J. Phys. B 23 (1973) 1273–1276.
[62] S. Chae, M. Ko, K. Kim, K. Ahn, J. Cho, Confronting issues of the practical implementation of Si anode in high-energy lithium-ion batteries, Joule 1 (2017) 47–60.
[63] J.W. Choi, D. Aurbach, Promise and reality of post-lithium-ion batteries with high energy densities, Nat. Rev. Mater. 1 (2016) 16013.
[64] I.H. Son, et al., Silicon carbide-free graphene growth on silicon for lithium-ion battery with high volumetric energy density, Nat. Commun. 6 (2015) 7393.
[65] C. Wang, et al., Self-healing chemistry enables the stable operation of silicon microparticle anodes for high-energy lithium-ion batteries, Nat. Chem. 5 (2013) 1042–1048.
[66] H. Wu, et al., Stable Li-ion battery anodes by in-situ polymerization of conducting hydrogel to conformally coat silicon nanoparticles, Nat. Commun. 4 (2013) 1943–1946.
[67] J. Li, R.B. Lewis, J.R. Dahn, Sodium carboxymethyl cellulose a potential binder for Si negative electrodes for Li-ion batteries, Electrochem. Solid St. 10 (2007) A17–A20.
[68] A.L. Michan, et al., Solid electrolyte interphase growth and capacity loss in silicon electrodes, J. Am. Chem. Soc. 138 (2016) 7918–7931.
[69] H. Wu, et al., Stable cycling of double-walled silicon nanotube battery anodes through solid-electrolyte interphase control, Nat. Nanotechnol. 7 (2012) 310–315.
[70] J. Yang, N. Solomatin, A. Kraytsberg, Y. Ein-Eli, In-situ spectro–electrochemical insight revealing distinctive silicon anode solid electrolyte interphase formation in a Lithium–ion battery, ChemistrySelect 1 (2016) 572–576.
[71] K.W. Schroder, H. Celio, L.J. Webb, K.J. Stevenson, Examining solid electrolyte interphase formation on crystalline silicon electrodes: influence of electrochemical preparation and ambient exposure conditions, J. Phys. Chem. C 116 (2012) 19737–19747.
[72] G.M. Veith, et al., Determination of the solid electrolyte interphase structure grown on a silicon electrode using a fluoroethylene carbonate additive, Sci. Rep. 7 (2017) 1–15.
[73] K. Feng, et al., Silicon-based anodes for lithium-ion batteries: from fundamentals to practical applications, Small 14 (2018).

[74] V.A. Sethuraman, V. Srinivasan, A.F. Bower, P.R. Guduru, In situ measurements of stress-potential coupling in lithiated silicon, J. Electrochem. Soc. 157 (2010) A1253.
[75] L.Y. Beaulieu, K.W. Eberman, R.L. Turner, L.J. Krause, J.R. Dahna, Colossal reversible volume changes in lithium alloys, Electrochem. Solid St. 4 (2001) 7–10.
[76] P. Hovington, et al., In situ scanning electron microscope study and microstructural evolution of nano silicon anode for high energy Li-ion batteries, J. Power Sources 248 (2014) 457–464.
[77] M.T. McDowell, et al., Studying the kinetics of crystalline silicon nanoparticle lithiation with in situ transmission electron microscopy, Adv. Mater. 24 (2012) 6034–6041.
[78] K. Rhodes, N. Dudney, E. Lara-Curzio, C. Daniel, Understanding the degradation of silicon electrodes for lithium-ion batteries using acoustic emission, J. Electrochem. Soc. 157 (2010) A1354.
[79] S.J. Lee, et al., Stress effect on cycle properties of the silicon thin-film anode, J. Power Sources 97–98 (2001) 191–193.
[80] J. Li, A.K. Dozier, Y. Li, F. Yang, Y.-T. Cheng, Crack pattern formation in thin film lithium-ion battery electrodes, J. Electrochem. Soc. 158 (2011), A689.
[81] S. Huang, F. Fan, J. Li, S. Zhang, T. Zhu, Stress generation during lithiation of high-capacity electrode particles in lithium ion batteries, Acta Mater. 61 (2013) 4354–4364.
[82] I. Ryu, J.W. Choi, Y. Cui, W.D. Nix, Size-dependent fracture of Si nanowire battery anodes, J. Mech. Phys. Solids 59 (2011) 1717–1730.
[83] H. Li, X. Huang, L. Chen, Z. Wu, Y. Liang, High capacity nano-Si composite anode material for lithium rechargeable batteries, Electrochem. Solid St. 2 (1999) 547–549.
[84] Y.X. Yin, L.J. Wan, Y.G. Guo, Silicon-based nanomaterials for lithium-ion batteries, Chin. Sci. Bull. 57 (2012) 4104–4110.
[85] C. Julien, A. Mauger, A. Vijh, K. Zaghib, Lithium Batteries: Science and Technology, 2015, https://doi.org/10.1007/978-3-319-19108-9.
[86] H. Ma, et al., Nest-like silicon nanospheres for high-capacity lithium storage, Adv. Mater. 19 (2007) 4067–4070.
[87] Y. Yao, et al., Interconnected silicon hollow nanospheres for lithium-ion battery anodes with long cycle life, Nano Lett. 11 (2011) 2949–2954.
[88] V. Schmidt, J.V. Wittemann, U. Gösele, Growth, thermodynamics, and electrical properties of silicon nanowires, Chem. Rev. 110 (2010) 361–388.
[89] B.M. Bang, H. Kim, H.K. Song, J. Cho, S. Park, Scalable approach to multi-dimensional bulk Si anodes via metal-assisted chemical etching, Energ. Environ. Sci. 4 (2011) 5013–5019.
[90] M.R. Zamfir, H.T. Nguyen, E. Moyen, Y.H. Lee, D. Pribat, Silicon nanowires for Li-based battery anodes: a review, J. Mater. Chem. A 1 (2013) 9566–9586.
[91] A. Mukanova, A. Jetybayeva, S.T. Myung, S.S. Kim, Z. Bakenov, A mini-review on the development of Si-based thin film anodes for Li-ion batteries, Mater. Today Energy 9 (2018) 49–66.
[92] S. Ohara, J. Suzuki, K. Sekine, T. Takamura, A thin film silicon anode for Li-ion batteries having a very large specific capacity and long cycle life, J. Power Sources 136 (2004) 303–306.
[93] H. Kim, B. Han, J. Choo, J. Cho, Three-dimensional porous silicon particles for use in high-performance lithium secondary batteries, Angew. Chem. Int. Ed. 47 (2008) 10151–10154.
[94] N. Liu, et al., A yolk-shell design for stabilized and scalable Li-ion battery alloy anodes, Nano Lett. 12 (2012) 3315–3321.
[95] N. Liu, et al., A pomegranate-inspired nanoscale design for large-volume-change lithium battery anodes, Nat. Nanotechnol. 9 (2014) 187–192.

[96] S. Fang, Z. Tong, P. Nie, G. Liu, X. Zhang, Raspberry-like nanostructured silicon composite anode for high-performance lithium-ion batteries, ACS Appl. Mater. Interfaces 9 (2017) 18766–18773.

[97] X. Chen, H. Li, Z. Yan, F. Cheng, J. Chen, Structure design and mechanism analysis of silicon anode for lithium-ion batteries, Sci. China Mater. 62 (2019) 1515–1536.

[98] B. Fuchsbichler, C. Stangl, H. Kren, F. Uhlig, S. Koller, High capacity graphite-silicon composite anode material for lithium-ion batteries, J. Power Sources 196 (2011) 2889–2892.

[99] N. Dimov, S. Kugino, M. Yoshio, Mixed silicon-graphite composites as anode material for lithium ion batteries—influence of preparation conditions on the properties of the material, J. Power Sources 136 (2004) 108–114.

[100] M. Yoshio, T. Tsumura, N. Dimov, Electrochemical behaviors of silicon based anode material, J. Power Sources 146 (2005) 10–14.

[101] I.H. Son, et al., Graphene balls for lithium rechargeable batteries with fast charging and high volumetric energy densities, Nat. Commun. 8 (2017) 1–10.

[102] M. Ko, et al., Scalable synthesis of silicon-nanolayer-embedded graphite for high-energy lithium-ion batteries, Nat. Energy 1 (2016) 1–8.

[103] P. Nie, et al., Mesoporous silicon anodes by using polybenzimidazole derived pyrrolic N-enriched carbon toward high-energy Li-ion batteries, ACS Energy Lett. 2 (2017) 1279–1287.

[104] J. Luo, et al., Crumpled graphene-encapsulated Si nanoparticles for lithium ion battery anodes, J. Phys. Chem. Lett. 3 (2012) 1824–1829.

[105] S. Guo, X. Hu, Y. Hou, Z. Wen, Tunable synthesis of yolk-shell porous silicon@carbon for optimizing Si/C-based anode of lithium-ion batteries, ACS Appl. Mater. Interfaces 9 (2017) 42084–42092.

[106] A. Magasinski, et al., High-performance lithium-ion anodes using a hierarchical bottom-up approach, Nat. Mater. 9 (2010) 353–358.

[107] W. An, et al., Scalable synthesis of ant-nest-like bulk porous silicon for high-performance lithium-ion battery anodes, Nat. Commun. 10 (2019) 1–11.

[108] Y. Yu, et al., Reversible storage of lithium in silver-coated three-dimensional macroporous silicon, Adv. Mater. 22 (2010) 2247–2250.

[109] H. Song, et al., Highly connected silicon-copper alloy mixture nanotubes as high-rate and durable anode materials for lithium-ion batteries, Adv. Funct. Mater. 26 (2016) 524–531.

[110] Y. Yang, et al., Morphology- and porosity-tunable synthesis of 3D nanoporous SiGe alloy as a high-performance Lithium-ion battery anode, ACS Nano 12 (2018) 2900–2908.

[111] J. Yang, et al., Amorphous TiO2 shells: a vital elastic buffering layer on silicon nanoparticles for high-performance and safe Lithium storage, Adv. Mater. 29 (2017) 1–7.

[112] C. Wang, et al., Thermal Lithiated-TiO2: a robust and Electron-conducting protection layer for Li-Si alloy anode, ACS Appl. Mater. Interfaces 10 (2018) 12750–12758.

[113] H.Y. Lin, C.H. Li, D.Y. Wang, C.C. Chen, Chemical doping of a core-shell silicon nanoparticles@polyaniline nanocomposite for the performance enhancement of a lithium ion battery anode, Nanoscale 8 (2016) 1280–1287.

[114] P. Sengodu, A.D. Deshmukh, Conducting polymers and their inorganic composites for advanced Li-ion batteries: a review, RSC Adv. 5 (2015) 42109–42130.

[115] F.H. Du, et al., Surface binding of polypyrrole on porous silicon hollow nanospheres for li-ion battery anodes with high structure stability, Adv. Mater. 26 (2014) 6145–6150.

[116] Z.P. Guo, J.Z. Wang, H.K. Liu, S.X. Dou, Study of silicon/polypyrrole composite as anode materials for Li-ion batteries, J. Power Sources 146 (2005) 448–451.

[117] X. Gao, Y. Guo, D. Wei, Y. Luo, R. Su, Recent progress on binders for silicon-based anodes in lithium-ion batteries, Huagong Xuebao/CIESC J. 69 (2018) 4605–4613.

[118] J. Song, et al., Interpenetrated gel polymer binder for high-performance silicon anodes in lithium-ion batteries, Adv. Funct. Mater. 24 (2014) 5904–5910.

[119] Y.K. Jeong, et al., Millipede-inspired structural design principle for high performance polysaccharide binders in silicon anodes, Energ. Environ. Sci. 8 (2015) 1224–1230.

[120] D. Bresser, D. Buchholz, A. Moretti, A. Varzi, S. Passerini, Alternative binders for sustainable electrochemical energy storage-the transition to aqueous electrode processing and bio-derived polymers, Energ. Environ. Sci. 11 (2018) 3096–3127.

[121] J.S. Bridel, T. Azaïs, M. Morcrette, J.M. Tarascon, D. Larcher, Key parameters governing the reversibility of Si/carbon/CMC electrodes for Li-ion batteries, Chem. Mater. 22 (2010) 1229–1241.

[122] D. Mazouzi, B. Lestriez, L. Roué, D. Guyomard, Silicon composite electrode with high capacity and long cycle life, Electrochem. Solid St. 12 (2009) A215–A218.

[123] C. Erk, T. Brezesinski, H. Sommer, R. Schneider, J. Janek, Toward silicon anodes for next-generation lithium ion batteries: a comparative performance study of various polymer binders and silicon nanopowders, ACS Appl. Mater. Interfaces 5 (2013) 7299–7307.

[124] S. Komaba, et al., Study on polymer binders for high-capacity SiO negative electrode of Li-ion batteries, J. Phys. Chem. C 115 (2011) 13487–13495.

[125] Z.J. Han, et al., Electrochemical lithiation performance and characterization of silicon-graphite composites with lithium, sodium, potassium, and ammonium polyacrylate binders, Phys. Chem. Chem. Phys. 17 (2015) 3783–3795.

[126] W. Porcher, et al., Understanding polyacrylic acid and Lithium polyacrylate binder behavior in silicon based electrodes for Li-ion batteries, J. Electrochem. Soc. 164 (2017) A3633–A3640.

[127] J.A. Lawton, et al., A major constituent of brown algae for use in high-capacity li-ion batteries, Science 334 (2011) 75–79.

[128] N. Delpuech, et al., Critical role of silicon nanoparticles surface on lithium cell electrochemical performance analyzed by FTIR, Raman, EELS, XPS, NMR, and BDS spectroscopies, J. Phys. Chem. C 118 (2014) 17318–17331.

[129] B. Koo, et al., A highly cross-linked polymeric binder for high-performance silicon negative electrodes in lithium ion batteries, Angew. Chem. Int. Ed. 51 (2012) 8762–8767.

[130] D.E. Yoon, et al., Dependency of electrochemical performances of silicon Lithium-ion batteries on Glycosidic linkages of polysaccharide binders, ACS Appl. Mater. Interfaces 8 (2016) 4042–4047.

[131] M.H. Ryou, et al., Mussel-inspired adhesive binders for high-performance silicon nanoparticle anodes in lithium-ion batteries, Adv. Mater. 25 (2013) 1571–1576.

[132] M. Winter, The solid electrolyte interphase—the Most important and the least understood solid electrolyte in rechargeable Li batteries, Z. Physiol. Chem. 223 (2009) 1395–1406.

[133] E. Peled, The electrochemical behavior of alkali and alkaline earth metals in nonaqueous battery systems—the solid electrolyte interphase model, J. Electrochem. Soc. 126 (1979) 2047–2051.

[134] E. Peled, D. Golodnitsky, J. Penciner, The anode/electrolyte interface, Handb. Batter. Mater. Second Ed. (2011) 479–523.

[135] E. Peled, S. Menkin, Review-SEI: past, present and future, J. Electrochem. Soc. 164 (2017) A1703–A1719.

[136] J.P. Olivier, M. Winter, Determination of the absolute and relative extents of basal plane surface area and 'non-basal plane surface' area of graphites and their impact on anode performance in lithium ion batteries, J. Power Sources 97–8 (2001) 151–155.

[137] M. Winter, Graphites for lithium-ion cells: the correlation of the first-cycle charge loss with the Brunauer-Emmett-Teller surface area, J. Electrochem. Soc. 145 (1998) 428.
[138] A. Wang, S. Kadam, H. Li, S. Shi, Y. Qi, Review on modeling of the anode solid electrolyte interphase (SEI) for lithium-ion batteries, NPJ Comput. Mater. 4 (2018) 1–15.
[139] F. Shi, P.N. Ross, G.A. Somorjai, K. Komvopoulos, The chemistry of electrolyte reduction on silicon electrodes revealed by in situ ATR-FTIR spectroscopy, J. Phys. Chem. C 121 (2017) 14476–14483.
[140] P.N. Ross, Catalysis and interfacial chemistry in lithium batteries: a surface science approach, Catal. Lett. 144 (2014) 1370–1376.
[141] F. Shi, et al., A catalytic path for electrolyte reduction in lithium-ion cells revealed by in situ attenuated total reflection-fourier transform infrared spectroscopy, J. Am. Chem. Soc. 137 (2015) 3181–3184.
[142] Y. Jin, B. Zhu, Z. Lu, N. Liu, J. Zhu, Challenges and recent progress in the development of Si anodes for lithium-ion battery, Adv. Energy Mater. (2017) 1700715.
[143] G.M. Veith, et al., Direct determination of solid-electrolyte interphase thickness and composition as a function of state of charge on a silicon anode, J. Phys. Chem. C 119 (2015) 20339–20349.
[144] H.G. Steinrück, C. Cao, G.M. Veith, M.F. Toney, Toward quantifying capacity losses due to solid electrolyte interphase evolution in silicon thin film batteries, J. Chem. Phys. 152 (2020) 084702.
[145] I. Hasa, et al., Electrochemical reactivity and passivation of silicon thin-film electrodes in organic carbonate electrolytes, ACS Appl. Mater. Interfaces 12 (2020) 40879–40890.
[146] B. Philippe, et al., Role of the LiPF6 salt for the long-term stability of silicon electrodes in Li-ion batteries—a photoelectron spectroscopy study, Chem. Mater. 25 (2013) 394–404.
[147] Z. Xu, J. Yang, H. Li, Y. Nuli, J. Wang, Electrolytes for advanced lithium ion batteries using silicon-based anodes, J. Mater. Chem. A 7 (2019) 9432–9446.
[148] N.S. Choi, et al., Effect of fluoroethylene carbonate additive on interfacial properties of silicon thin-film electrode, J. Power Sources 161 (2006) 1254–1259.
[149] K. Schroder, et al., The effect of fluoroethylene carbonate as an additive on the solid electrolyte interphase on silicon Lithium-ion electrodes, Chem. Mater. 27 (2015) 5531–5542.
[150] I.A. Shkrob, J.F. Wishart, D.P. Abraham, What makes Fluoroethylene carbonate different? J. Phys. Chem. C 119 (2015) 14954–14964.
[151] E. Markevich, G. Salitra, D. Aurbach, Fluoroethylene carbonate as an important component for the formation of an effective solid electrolyte interphase on anodes and cathodes for advanced Li-ion batteries, ACS Energy Lett. 2 (2017) 1337–1345.
[152] Y. Jin, et al., Identifying the structural basis for the increased stability of the solid electrolyte interphase formed on silicon with the additive fluoroethylene carbonate, J. Am. Chem. Soc. 139 (2017) 14992–15004.
[153] M. Nie, D.P. Abraham, Y. Chen, A. Bose, B.L. Lucht, Silicon solid electrolyte interphase (SEI) of lithium ion battery characterized by microscopy and spectroscopy silicon solid electrolyte interphase (SEI) of lithium ion battery characterized by microscopy and spectroscopy, J. Phys. Chem. C 117 (2013) 13403–13412.
[154] H. Nakai, T. Kubota, A. Kita, A. Kawashima, Investigation of the solid electrolyte interphase formed by fluoroethylene carbonate on Si electrodes, J. Electrochem. Soc. 158 (2011) A798.
[155] V. Etacheri, et al., Effect of fluoroethylene carbonate (FEC) on the performance and surface chemistry of Si-nanowire li-ion battery anodes, Langmuir 28 (2012) 965–976.
[156] K. Xu, Electrolytes and interphases in Li-ion batteries and beyond, Chem. Rev. 114 (2014) 11503–11618.

[157] T. Jaumann, et al., Lifetime vs. rate capability: understanding the role of FEC and VC in high-energy Li-ion batteries with nano-silicon anodes, Energy Storage Mater. 6 (2017) 26–35.

[158] L. Chen, K. Wang, X. Xie, J. Xie, Effect of vinylene carbonate (VC) as electrolyte additive on electrochemical performance of Si film anode for lithium ion batteries, J. Power Sources 174 (2007) 538–543.

[159] L. Chen, K. Wang, X. Xie, J. Xie, Enhancing electrochemical performance of silicon film anode by vinylene carbonate electrolyte additive, Electrochem. Solid St. 9 (2006).

[160] D. Aurbach, et al., An analysis of rechargeable lithium-ion batteries after prolonged cycling, Electrochim. Acta 47 (2002) 1899–1911.

[161] T. Jaumann, et al., SEI-component formation on sub 5 nm sized silicon nanoparticles in Li-ion batteries: the role of electrode preparation, FEC addition and binders, Phys. Chem. Chem. Phys. 17 (2015) 24956–24967.

[162] H. Jo, et al., Stabilizing the solid electrolyte interphase layer and cycling performance of silicon-graphite battery anode by using a binary additive of fluorinated carbonates, J. Phys. Chem. C 120 (2016) 22466–22475.

[163] J. Wang, L. Zhang, H. Zhang, Effects of electrolyte additive on the electrochemical performance of Si/C anode for lithium-ion batteries, Ionics (Kiel) 24 (2018) 3691–3698.

[164] C.C. Nguyen, B.L. Lucht, Improved cycling performance of Si nanoparticle anodes via incorporation of methylene ethylene carbonate, Electrochem. Commun. 66 (2016) 71–74.

[165] R. Nölle, J.P. Schmiegel, M. Winter, T. Placke, Tailoring electrolyte additives with synergistic functional moieties for silicon negative electrode-based lithium ion batteries: a case study on lactic acid O-carboxyanhydride, Chem. Mater. 32 (2020) 173–185.

[166] Y. Li, et al., Improvement of cyclability of silicon-containing carbon nanofiber anodes for lithium-ion batteries by employing succinic anhydride as an electrolyte additive, J. Solid State Electrochem. 17 (2013) 1393–1399.

[167] R. Nölle, A.J. Achazi, P. Kaghazchi, M. Winter, T. Placke, Pentafluorophenyl isocyanate as an effective electrolyte additive for improved performance of silicon-based Lithium-ion full cells, ACS Appl. Mater. Interfaces 10 (2018) 28187–28198.

[168] D.A. Dalla Corte, et al., Molecular grafting on silicon anodes: artificial solid-electrolyte interphase and surface stabilization, Electrochim. Acta 201 (2016) 70–77.

[169] A. Hirata, et al., Atomic-scale disproportionation in amorphous silicon monoxide, Nat. Commun. 7 (2016) 1–7.

[170] H. Fukuoka, M. Aramata, S. Miyawaki, Patent Application Publication (Nov. 1, 2007) US 2007/0254102 A1. Method for producing a SiO$_x$, ($x < 1$), 2007.

[171] E. Park, et al., Dual-size silicon nanocrystal-embedded SiOx nanocomposite as a high-capacity Lithium storage material, ACS Nano 9 (2015) 7690–7696.

[172] K. Kitada, et al., Unraveling the reaction mechanisms of SiO anodes for Li-ion batteries by combining in situ 7Li and ex situ 7Li/29Si solid-state NMR spectroscopy, J. Am. Chem. Soc. 141 (2019) 7014–7027.

[173] N. Liu, L. Hu, M.T. McDowell, A. Jackson, Y. Cui, Prelithiated silicon nanowires as an anode for lithium ion batteries, ACS Nano 5 (2011) 6487–6493.

[174] Y. Yang, et al., New nanostructured Li2S/silicon rechargeable battery with high specific energy, Nano Lett. 10 (2010) 1486–1491.

[175] H.J. Kim, et al., Controlled Prelithiation of silicon monoxide for high performance Lithium-ion rechargeable full cells, Nano Lett. 16 (2016) 282–288.

[176] M. Marinaro, et al., Bringing forward the development of battery cells for automotive applications: Perspective of R&D activities in China, Japan, the EU and the USA, J. Power Sources 459 (2020) 228073.

[177] Integrated SET-Plan Action 7, Implementation Plan, Become competitive in the global battery sector to drive e-mobility and stationary storage forward, 2016.
[178] A Roadmap for an integrated cell and battery production in Germany (Roadmap integrierte Zell-und Batterieproduktion Deutschland), Nationale Plattform Elektromobilität, 2016.
[179] M. Steen, N. Lebadeva, F. Di Persio, L. Boon-Brett, EU Competitiveness in Advanced Li-ion Batteries for E-Mobility and Stationary Storage Applications—Opportunities and Actions, vol. 44, Publications Office of the European Union, 2017.
[180] M. Ierides, et al., Advanced Materials for Clean and Sustainable Energy and Mobility, Emiri, 2019.
[181] Batteries Europe, Strategic Research Agenda for Batteries, 2020.
[182] Y. Ding, Z.P. Cano, A. Yu, J. Lu, Z. Chen, Automotive Li-ion batteries: current status and future perspectives, Electrochem. Energy Rev. 2 (2019) 1–28.

Chapter 2

Recent advances in silicon materials for Li-ion batteries: Novel processing, alternative raw materials, and practical considerations

Sourav Ghosh[a], Aloysius F. Hepp[b], Prashant N. Kumta[c,d,e,f], Moni K. Datta[d,f], and Surendra K. Martha[a]

[a]*Department of Chemistry, Indian Institute of Technology Hyderabad, Sangareddy, Telangana, India,* [b]*Nanotech Innovations, LLC, Oberlin, OH, United States,* [c]*Department of Chemical and Petroleum Engineering, University of Pittsburgh, Pittsburgh, PA, United States,* [d]*Department of Bioengineering, University of Pittsburgh, Pittsburgh, PA, United States,* [e]*Mechanical Engineering and Materials Science, University of Pittsburgh, Pittsburgh, PA, United States,* [f]*Center for Complex Engineered Multifunctional Materials, University of Pittsburgh, Pittsburgh, PA, United States*

2.1 Introduction

Li-ion battery (LIB) technology continues to be the most successful electrochemical energy storage system due to its high specific energy density, leading to its extensive applications to power cellular phones, portable computers, camcorders, power tools, and even in hybrid electric vehicles and electric vehicles [1–6]. Graphite store lithium in the form of LiC_6 has been known to be a potential anode material for LIBs over the last three decades [7]. The theoretical specific capacity of graphite anode is limited to 372 mAh g^{-1}. Thus, several compounds such as carbon-based compounds like hard carbon, graphene, carbon nanotubes, etc., metal oxides such as lithium titanate (LTO), SnO_2, Mn_3O_4, Co_3O_4, etc., metal and metal alloys such as lithium, Si, Sn, Sb, Bi, In, P, As, Ge, etc. have been well studied in the literature [3, 8–17].

Nanostructured carbon materials, such as reduced graphene oxide, carbon nanotubes could potentially deliver a reversible capacity of ∼1000 mAh g^{-1} [18]. But the issue related to undesirable side reaction with carbonate-based electrolyte during the initial lithiation process results in a large first-cycle irreversible (FIR) capacity and low coulombic efficiency (CE) in subsequent cycles, which hinder its applications towards full cell prototypes [19, 20]. Also, the high surface area of these nanostructured carbons not only enhance the side reaction but also reduce the tap density of the material and the electrode, leading to lowered volumetric energy density. Similarly, conversion-type metal oxides and their composites with carbon/conducting polymer(s) demonstrate capacities of ~800–1200 mAh g^{-1}, when used as anodes in LIBs [12, 21–23]. Furthermore, various alloy-based anode materials, especially Sn, Sb, and Si have gained high interest due to their high theoretical capacities (~660–4200 mAh g^{-1}) [3, 8, 10, 12, 14, 15, 24–26]. However, both of these conversion-type and alloy-based materials suffer from ease of fracturing and accompanied decrepitation of the electrode during repeated charge-discharge cycling [27]. Besides, the average lithiation-delithiation potential is relatively higher, which affects the operating potential of the full-cell when assembled with various layered cathode.

On the contrary, Si-based alloy materials have received tremendous attention due to their relatively low average lithiation/delithiation potential, extremely high gravimetric and volumetric capacity, abundance, low cost, chemical stability, and nontoxicity. Silicon (Si), storing Li with the ability to form the alloy of composition Li$_{15}$Si$_4$, exhibits a superior theoretical charge storage capacity, nearly an order of magnitude greater at 3579 mAh g^{-1} contrasted with that of graphite of 372 mAh g^{-1} [28, 29]. Even though silicon is the second most abundant element in the Earth's crust, nontoxic, low cost, and a high capacity anode material, its application is limited in commercial LIBs. Silicon undergoes colossal volume change (~400%) during charge/discharge cycling with lithium due to several zintl phase-related transformations. As a result, Si particles are subjected to significant stress/strain causing fracture and loss of contact of the active material with the current collector. The volume induced stresses coupled with unstable solid electrolyte interphase (SEI) formation at the Si electrode-electrolyte interface, eventually leads to capacity fade and eventual failure thus reducing the battery life [30, 31].

In the second section of this chapter, we outline key findings from recent reports that cover topics such as hybrid silicon-carbon anodes [8, 10, 12, 32–35] and comparing approaches and perspectives on their processing [36–38], silicide and Si-alloy anodes [39–41], metal oxide anodes [42–44]. The third section addresses alternative low-cost including natural silicon sources [45–47], advanced (molten metal-based) processing [48–50] including bio-inspired approaches [51, 52], and performance issues as well as practical concerns for processing anode materials [53–57]. Throughout this chapter, we will include perspectives, developed through research activities at the University of Pittsburgh and several collaborators [3, 8, 10, 12, 14, 26, 30, 55, 58, 59], on key recent advances in Si-containing anode materials and novel applications.

2.2 Hybrid and alloy-based silicon-containing materials

Despite the high theoretical capacities of Si (4200 mAh g^{-1}, Li$_{4.4}$Si), the ensuing complications (such as volume changes resulting in continuous SEI formation and electrode pulverizations) associated during the lithiation/delithiation process restricts the use of pristine Si as an anode in LIBs [59, 60]. Thus, various strategies including the downsizing of Si particles to the nanoscale (nanoparticles [3, 8, 61–64], nanoflakes [65], nanowires [29, 66–69], nanorods [70, 71], nanotubes [72, 73], hollow nanotubes [74, 75], and nanosheets [76, 77]), and making composites of silicon with carbonaceous/metal oxide materials have been widely investigated to enhance the electrochemical performance of Si-based anode by mitigating the issue of pulverization (providing enough space to accommodate the volume change during lithiation) and facilitating fast Li-ion diffusion [78, 79]. This section discusses various methodologies, which have been adopted to make a hybrid material such as silicon composite anodes, specifically carbon-silicon hybrid, oxide-silicon hybrid, silicon-metal hybrid, and silicides for LIBs.

2.2.1 Carbon-silicon hybrid materials

In general, carbon-based hybrid materials have advantages such as (a) forming a stable SEI, (b) minimal volume change during cycling preventing the volume-induced stress-related cracking and ensuing decripitation, and (c) enhances the electrical conductivity of the overall matrix. It has been observed that when silicon is embedded in a continuous matrix of carbonaceous materials like graphite, graphene, carbon nanotube/nanofiber, the electrochemical performance is significantly improved [80–90]. Graphite could be used as an effective carbon-based precursor making a hybrid material with silicon due to its significant role in providing sufficient electrical conductivity and mechanical strength to the electrode materials [10, 14, 30, 91–93]. Kumta et al. reported the facile scalable generation of composites comprising silicon (Si), graphite (C) and polyacrylonitrile-based disordered carbon (PAN-C) [10, 14, 30, 94], synthesized by thermal treatment of mechanically milled silicon, graphite, and use of polyacrylonitrile (PAN) powder serving as a diffusion barrier to formation of undesired *SiC* exhibiting a reversible capacity of ~660–1000 mAh g^{-1} with an excellent capacity retention displaying almost no fade in capacity when cycled at a rate of ~C/4.

Li et al. reported a novel core-shell structure of graphite/Si-porous carbon composite synthesized using spray-drying/pyrolysis method, as shown in Fig. 2.1A [81]. Herein, the graphite and Si mainly acts as an electrochemical active site for the Li-ion storage, whereas the major role of porous carbon is to provide a buffer space for the volume expansion of the Si, suppress the agglomeration of Si-nanoparticles, and facilitating the electrolyte accessibility to the bulk of the electrode materials. To examine the formation of the Si-C

50 PART | I Introduction and background

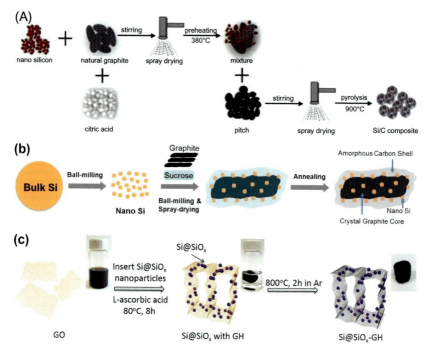

FIG. 2.1 (A) Schematic of the preparation procedure for the novel graphite/Si-porous carbon composite. (B) Schematic illustration of the procedure to synthesize Si@SiO$_x$/GH composites, with insert showing the digital photos. (C) Schematic of the synthesis process of the SiGC composite. *(Panel A reproduced with permission from: M. Li, X. Hou, Y. Sha, J. Wang, S. Hu, X. Liu, Z. Shao, Facile spray-drying/pyrolysis synthesis of core-shell structure graphite/ silicon-porous carbon composite as a superior anode for Li-ion batteries, J. Power Sources 248 (2014) 721–728. Copyright Elsevier. Panel B reproduced with permission from: X. Bai, Y. Yu, H.H. Kung, B. Wang, J. Jiang, Si@SiO$_x$/graphene hydrogel composite anode for lithium-ion battery, J. Power Sources 306 (2016) 42–46. Copyright Elsevier. Panel C reproduced with permission from: D. Sui, Y. Xi, W. Zhao, H. Zhang, Y. Zhou, X. Qin, Y. Ma, Y. Yang, Y. Chen, A high-performance ternary Si composite anode material with crystal graphite core and amorphous carbon shell, J. Power Sources 384 (2018) 328–333, Copyright Elsevier.)*

core-shell structure, the morphology of the materials is studied microscopically via SEM and TEM [81].

Fig. 2.2 shows that the raw silicon particles form 200 nm-sized secondary particles by agglomerating 30–50 nm-sized primary particles (Fig. 2.2A inset: TEM image), suggesting higher surface energy of the nanoparticles. The graphite cores are approximately 10 μm in size with an irregular morphology (Fig. 2.2B). Fig. 2.2C–E shows the SEM of the composite material. The porous structure of the composite is clearly observed from Fig. 2.2D. Further, cross-section image of the Si/C composite particle demonstrates a porous shell coated on the dense core of graphite material (Fig. 2.2E). The hybrid material contains

Recent advances in silicon materials for Li-ion batteries **Chapter | 2** 51

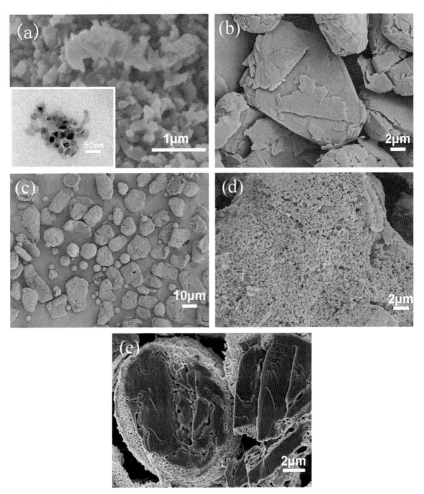

FIG. 2.2 (A) Typical SEM and TEM (the inset one) profiles of silicon raw materials for the preparation of Si/C composite; (B) typical SEM image of natural graphite; (C)–(E) SEM image, magnified surface image, and cross profile of the novel core–shell porous Si/C composite, respectively. *(Reproduced with permission from: M. Li, X. Hou, Y. Sha, J. Wang, S. Hu, X. Liu, Z. Shao, Facile spray-drying/pyrolysis synthesis of core-shell structure graphite/silicon-porous carbon composite as a superior anode for Li-ion batteries, J. Power Sources 248 (February 2014) 721–728. Copyright Elsevier (February 2014).)*

8 wt% of Si and could deliver a reversible capacity of 724 mAh g^{-1} during the initial cycles, with capacity retention of 593 mAh g^{-1} at the end of 100 cycles.

Kumta et al. [74] have developed a simple and scalable approach to hollow silicon nanotubes further modified [75] by developing a core–shell C@Si@C hollow nanotubular configuration with optimal Si thickness (~60 nm) showing no microstructural damage during lithiation and delithiation processes serving as a stable anode for LIBs with low FIR loss of ~13% and high areal capacity (~3 mAh cm^{-2}). The hollow Si nanotubes (h-SiNTs) have been generated via high-throughput, scalable and recyclable, sacrificial MgO wire template fabrication approach [74]. Generation of Si films of varying thickness by low-pressure thermal chemical vapor deposition (LPCVD) with subsequent etching yields h-SiNTs [74]. Modification/optimization of the h-SiNT physical characteristics exhibit improved performance in LIBs. Carbon coating of optimized h-SiNTs further, yields core–shell h-SiNTs exhibiting not only low FIR loss of ~13%, but also yields a specific capacity of ~1000 mAh g^{-1} at large discharge/charge currents of ~1 A g^{-1} for over 120 cycles, with also a low fade rate of ~0.072% loss per cycle. Graphene is another widely examined buffer matrix for Si materials due to its mechanical strength, flexibility, electrical conductivity, and high surface area [95–100].

Bai et al. described a solution-based self-assembly process to synthesize Si@SiO$_x$ hydrogel composite for excellent Li-ion storage capacity and C-rate performance [84]. The overall schematic illustration of the synthesis procedure is represented by Fig. 2.1C. The electrode was proven to deliver 1640 mAh g^{-1} after the 140th cycle at 100 mA g^{-1} with 80% capacity retention with reference to the 10th cycle. The unique hierarchical architecture of graphene hydrogel, as a freestanding-porous-3D carbon network could assist to accommodate high volume changes, fast ionic and electronic conduction, and electrolyte penetration within the microstructure, resulting in superior electrochemical performance.

Sui et al. developed a low-cost, scalable strategy to prepare ternary Si/graphite/pyrolytic carbon (SiGC) composite anode by ball-milling and spray drying process, as shown in Fig. 2.1B [86]. The electrochemical performance of SiGC anode is displayed in Fig. 2.3. The typical discharge–charge profile of SiGC anode shown in Fig. 2.3 shows an initial reversible capacity of 818 mAh g^{-1}. The C-rate studies depicted in Fig. 2.3B suggest that the material could deliver a capacity of about 756, and 458 mAh g^{-1} at 500 and 2000 mA g^{-1}, respectively. The capacity retention of 83.6% (610 mAh g^{-1}) at 500 mA g^{-1} is observed after 300 cycles, as shown in Fig. 2.3C–D. The superior electrochemical performance of SiGC is attributed to the novel core shell structure which support the accommodation of volume changes, as well as an amorphous carbon shell on the top of the core could enhance the overall integrity of the electrode.

Other examples of the solid Si@C core-shell structure approach have been reported [101, 102]; in fact, a new class of Si/C multiphase nanocomposites with yolk-shell structure is widely reported in the literature [103–105]. The yolk-shell structure has the advantages of additional internal void spaces between the internal core and upper shell [64]. Thus, the voids inside the yolk-shell structure could provide enough space to buffer the volume changes during lithiation/delithiation of Si particles and the external thin layer of carbon could facilitate the Li-ion and electron transfer. Hu et al. designed a scalable and feasible method to form a Si/C yolk-shell structure [35]. The overall

Recent advances in silicon materials for Li-ion batteries **Chapter | 2** 53

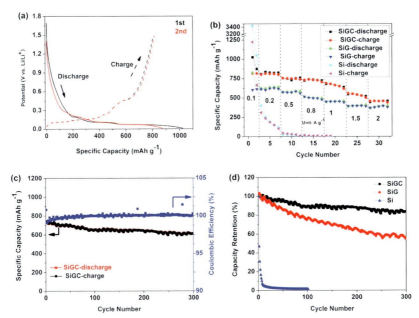

FIG. 2.3 (A) The galvanostatic charge/discharge profiles of SiGC for the first two cycles at a current density of 0.1 A g^{-1}; (B) rate capabilities of ball-milled Si, SiG and SiGC; (C) cycling performance of SiGC for 300 cycles at 0.5 A g^{-1}; (D) capacity retention of ball-milled Si, SiG and SiGC at 0.5 A g^{-1}. *(Reproduced with permission from: D. Sui, Y. Xi, W. Zhao, H. Zhang, Y. Zhou, X. Qin, Y. Ma, Y. Yang, Y. Chen, A high-performance ternary Si composite anode material with crystal graphite core and amorphous carbon shell, J. Power Sources 384 (April 2018) 328–333. Copyright Elsevier (April 2018).)*

methodology of the synthetic process is illustrated in Fig. 2.4 [35]. Herein, resorcinol-formaldehyde (RF) resin and polydopamine are used as a carbon precursor to create a double carbon coating. Si/C yolk-shell electrode could deliver an initial discharge capacity of 2108 mAh g^{-1} at 100 mA g^{-1}. Further, the advantages of the double carbon layers are depicted pictorially in Fig. 2.5 a-c. The presence of a single carbon layer could result in collapsing of the structure during deep cycling, whereas double carbon coating could provide robustness to the overall framework and maintain the structural integrity of Si/C composite electrode.

He et al. constructed a novel folded-hand Si/C binder-free 3D network using chemical vapor deposition (CVD) coupled with ultrasonic atomization technique [32]. Two different kinds of samples namely spiciform Si/C bunches and folded-hand Si/C 3D network is synthesized under similar condition, except the fabrication of transition carbon layer using squaric acid in the latter case. The SEM image shown in Fig. 2.6A illustrates a rough and porous surface of Cu-foil, which can help for the deposition and growth of carbon. As shown in Fig. 2.6B-C, spiciform Si/C bunches are naturally distributed on Cu-foil

54 PART | I Introduction and background

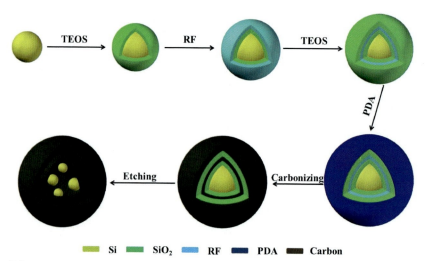

FIG. 2.4 Schematic illustration of the preparation process of yolk-shell structure of Si/C composite. *(Reproduced with permission from L. Hu, B. Luo, C. Wu, P. Hu, L. Wang, H. Zhang, Yolk-shell Si/C composites with multiple Si nanoparticles encapsulated into double carbon shells as lithium-ion battery anodes, J. Energy Chem. 32 (May 2019) 124–130. Copyright Elsevier (May 2019).)*

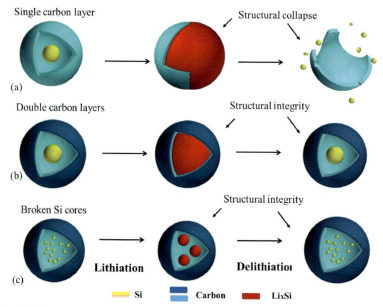

FIG. 2.5 Schematic representation of the initial Li$^+$ insertion/delithiation modes of the electrodes: (A) Si/C-HF with single carbon layer, (B) TSC-PDA with double carbon layers, and (C) TSC-PDA-B with broken Si cores and double carbon layers. *(Reproduced with permission from: L. Hu, B. Luo, C. Wu, P. Hu, L. Wang, H. Zhang, Yolk-shell Si/C composites with multiple Si nanoparticles encapsulated into double carbon shells as lithium-ion battery anodes, J. Energy Chem. 32 (May 2019) 124–130. Copyright Elsevier (May 2019).)*

FIG. 2.6 SEM images of Cu foil (A), the spiciform Si/C bunches (B, C), digital photographs of rice (D); carbon film on the Cu foil (E), the folded-hand Si/C 3D networks with different magnifications (F–H), and digital photographs of folded-hands (i). *(Reproduced with permission from: Y. He, K. Xiang, W. Zhou, Y. Zhu, X. Chen, H. Chen, Folded-hand silicon/carbon three-dimensional networks as a binder-free advanced anode for high-performance lithium-ion batteries, Chem. Eng. J. 353 (December 2018) 666–678. Copyright Elsevier (December 2018).)*

and consisting of ordered and oriented stems (alike digital photograph is represented in Fig. 2.6D). The presence of numerous interconnected carbon nanowires and nanodots, deposited from the squaric acid, are evident in Fig. 2.6E. The folded-hand Si/C binder-free 3D network is constructed by short rods and uniform grains, as shown in Fig. 2.6F–H. It is obvious to note that the formation of a folded-hand Si/C 3D network depends on the presence of a transition carbon layer.

The electrochemical performances of both the samples were evaluated in the potential window between 0 and 2 V, as shown in Fig. 2.7 [32]. Fig. 2.7A shows

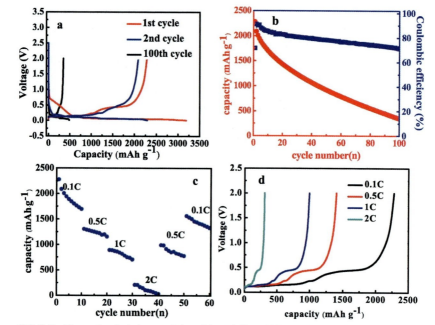

FIG. 2.7 Electrochemical characteristics of the spiciform Si/C bunches- (A) 1st, 2nd, and 100th galvanostatic charge/discharge cycles, the rate was 0.1 C, (B) Cycling performance and Coulombic efficiency tested at 0.1 C; electrochemical characteristics of the folded-hand Si/C 3D network, (C) galvanostatic charge-discharge curves, (D) corresponding cycling performance and coulombic efficiency. *(Reproduced with permission from: Y. He, K. Xiang, W. Zhou, Y. Zhu, X. Chen, H. Chen, Folded-hand silicon/carbon three-dimensional networks as a binder-free advanced anode for high-performance lithium-ion batteries, Chem. Eng. J. 353 (December 2018) 666–678. Copyright Elsevier (December 2018).)*

a long plateau below 0.15 V during lithiation of Si, followed by a plateau region at 0.25–0.5 V during the delithiation process in the subsequent charging process. Spiciform Si/C bunches could deliver a reversible capacity of 2280, 2274, and 346 mAh g^{-1} after 1st, 2nd, and 100th cycles, respectively (Fig. 2.7A–B). The continuous capacity fading could be a result of pulverization of Si nanoparticles from Cu current collector during repeated charge-discharge process. However, folded-hand Si/C 3D network could deliver an initial reversible capacity of 2277 mAh g^{-1}, with retention of 2167 mAh g^{-1} after 100 cycles at 0.1C rate (Fig. 2.7C, D). The superior electrochemical performance of the folded-hand Si/C 3D network is the result of the multidimensional channel, shortened diffusion length for Li-ions, and the presence of amorphous carbon which could improve electronic conductivity and relive the expansion for Si nanoparticles. A comparison of the electrochemical performances of core-shell structured Si/C composites of several studies [35, 64, 82, 89, 90, 103, 106–111] is presented in Table 2.1.

TABLE 2.1 The comparison of the electrochemical performances of core-shell structured Si/C composites of several studies.

Samples	Current density (mAg^{-1})	Cycle number (#)	Capacity (mAhg^{-1}) after cycle number	Initial CE %	References
Nanosilica/carbon spheres	100	300	620	~65	[89]
Hollow core-shell Si/C composites	100	100	768	–	[90]
Core-shell Si/C composites	50	150	1625	92	[103]
Hollow Si/C heterostructures	50	70	520	48	[64]
Si@void@C microparticles	250	100	600	~80	[106]
Si@void@C composites	1000	50	977	~76	[82]
Carbon@silicon-silica nanostructure	100	100	920	–	[107]
MWCNT/Si composites	400	70	520	~50	[108]
Hollow Si/SiO$_x$@C spheres	1000	100	940	65	[109]
Si/C/graphene	1000	200	650	~97	[110]
Si/C-HF	100	100	167	58	
TSC-PDA	100	200	807	56	[35]
TSC-PDA-B	100	200	1113	79	
Si@DC-yolk	50	80	944	55	[111]

Adapted from L. Hu, B. Luo, C. Wu, P. Hu, L. Wang, H. Zhang, Yolk-shell Si/C composites with multiple Si nanoparticles encapsulated into double carbon shells as lithium-ion battery anodes, J. Energy Chem. 32 (May 2019) 124–130.

2.2.2 Processing hybrid anodes: Fundamental *vs.* practical considerations

The Industrial-practical application of Si anode is mostly hindered by its low electronic conductivity, huge volume changes during lithiation/delithiation, and unstable nature of SEI [37, 38]. Even though significant progress has been achieved in Si-based hybrid materials from the viewpoint of fundamental research, the commercial viability of Si-based anode demands low irreversible capacity, better capacity retention, good tap density, high Si content, simple manufacturing process, and most importantly low manufacturing cost. Thus, from the industrial point of view, Si-graphite/carbon composite could be a promising anode material with its merits like Si-based component has high capacity and carbon buffer could provide high ionic/electronic conductivity, good capacity retention, and helps to reduce pulverization of the overall matrix [112]. Additionally, an efficient, simple, low-cost, easily scalable synthesis methodology is desired which can assure an excellent electrochemical performance of Si-carbon composite. A comparison of different approaches for preparing Si/carbon/graphite composites [81, 86, 113–139], including their advantages and disadvantages are summarized in Table 2.2. The fabrication process not only affects the structure and yield of the final material but also determines the scope of scalability. Further, the recent advancements on the Si-based anode materials from the industrial and fundamental science perspectives are illustrated in Fig. 2.8 [38]. The fundamental research on Si nanostructures will undoubtedly facilitate the further development of Si-based anodes at the industrial level.

2.2.3 Silicon-metal alloy anodes

In addition to using carbon-based composites, the research on applying porous nanostructure based on Si-metal composites is similarly conducted with other anode alloys to mitigate the negative effects of pristine Si anodes [41, 140–142]. Further, the introduction of inactive elements like Ni, Cu, Fe, Zn, etc. into the active Si matrix through alloying can effectively alleviate the mechanical stress induced by the active phase volume change during lithiation/delithiation [143–147]. This type of alloying can promote the ionic and electrical conductivity of the overall matrix.

Hao et al. reported an Si/Sn composite consisting of three dimensional hierarchical macroporous Si network, fabricated from SiSnAl alloy via a facile dealloying process [41]. To synthesize hierarchical macroporous (HMP) structure consisting of bimodal pore channels, high content Al alloy (such as $Si_{8.5}Sn_{1.5}Al_{90}$) is used as precursors so that removal of Al could result in a highly porous structure. As shown in Fig. 2.9, the significant amount of Al has started etched out of SiSnAl alloy foil to form a defective HMP after dissolving in dilute NaOH solution for 5h. With the subsequent increase in reaction time

TABLE 2.2 A comparison of different approaches for preparing Si/carbon/graphite composites.

Fabrication methods	Advantages	Disadvantages	Reference
Ball milling/mechanical milling/high energy ball milling	Simple, low-cost, scalable, high-yield	Relatively nonuniform distribution. Easily exfoliates graphite	[86, 113–117]
Spray drying	Spherical particle, uniform distribution. Simple, scalable, high-yield	Typically followed by ball milling or thermal annealing. Relatively high-cost	[81, 118–123]
Chemical vapor deposition (CVD)	Highly uniform growth of Si-based composites. Uniform surface coating	Energy-consuming. Highly toxic and expensive raw materials and synthesis process	[124–127]
Wet-processing	Special particle morphology. Better dispersing of precursors	Typically followed by hydrothermal, sol-gel, carbonization, or thermal annealing. Relatively high-cost	[128–135]
Self-assembly assisted annealing	Uniform particle dispersion. Special particle morphology	Complex reaction steps	[136, 137]
Acid or electrochemical etching	Design porous structure	High-cost complex reaction steps	[138, 139]

Adapted from: P. Li, G. Zhao, X. Zheng, X. Xu, C. Yao, W. Sun, S.X. Dou, Recent progress on silicon-based anode materials for practical lithium-ion battery applications, Energy Storage Mater. 15 (November 2018) 422–446.

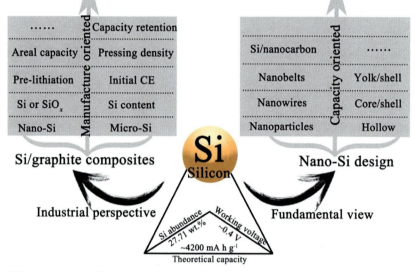

FIG. 2.8 Schematic illustration of the recent advancement on Si-based anode materials from two perspectives: industrial perspective and fundamental science point of view. *(Reproduced with permission from: P. Li, G. Zhao, X. Zheng, X. Xu, C. Yao, W. Sun, S.X. Dou, Recent progress on silicon-based anode materials for practical lithium-ion battery applications, Energy Storage Mater. 15 (November 2018) 422–446. Copyright Elsevier (December 2018).)*

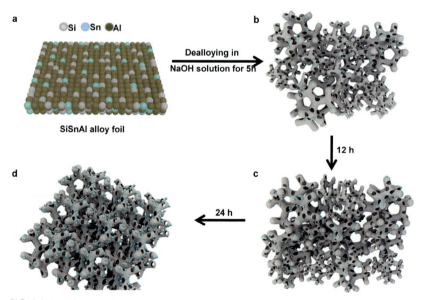

FIG. 2.9 A schematic diagram for the fabrication of Si/Sn composites. *(Reproduced with permission from: Q. Hao, J. Hou, J. Ye, H. Yang, J. Du, C. Xu, Hierarchical macroporous Si/Sn composite: Easy preparation and optimized performances towards lithium storage, Electrochim. Acta 306 (May 2019) 427–436. Copyright Elsevier (May 2019).)*

(after 24 h) the formation of regular HMP structure with uniform distribution of Sn nanoparticles on the surface of the pores. Additionally, HMP $Si_{80}Sn_{20}$, and Si samples are fabricated to compare the electrochemical performance using a similar method from the $Si_8Sn_2Al_{90}$, and Si_5Al_{95} alloy, respectively. The electrochemical Li-ion storage behavior of all the three samples are investigated in the potential window of 0.01–1.5 V vs Li/Li$^+$. Even though the HMP Si could deliver an initial capacity of 1777 mAh g^{-1}, only 9.5% (169 mAh g^{-1}) of the capacity was retained after 70 cycles at 200 mA g^{-1}. Whereas $Si_{85}Sn_{15}$ and $Si_{80}Sn_{20}$ samples could deliver an initial capacity of 2466, and 1856 mAh g^{-1}, with a capacity retention of 62.7, and 56% after 70 cycles, respectively.

Fig. 2.10A shows the cycling performance of all three samples at a higher current density of 2 A g^{-1} for 200 cycles. Although there is an initial decay of capacity for the initial 50 cycles for HMP $Si_{85}Sn_{15}$, the capacity remained around 600 mAh g^{-1} during the remaining 150 cycles. It is evident that HMP $Si_{85}Sn_{15}$, $Si_{80}Sn_{20}$, and Si retained 44%, 13%, and 4% capacity after 200 cycles, respectively. The C-rate performance of HMP $Si_{85}Sn_{15}$ is superior in comparison to the other two samples, as shown in Fig. 2.10B; impedance measurements of HMP Si, $Si_{85}Sn_{15}$, and $Si_{80}Sn_{20}$ electrodes are shown in Fig. 2.10C. The excellent Li-ion storage of HMP $Si_{85}Sn_{15}$ is the result of unique microstructure, which could shorten the Li-ion diffusion and reduce the agglomeration tendency. The porous network originating from the bimodal microstructure could be beneficial for the free expansion of Si during lithiation/delithiation. Further, higher conductivity and faster diffusion rate of Li into Sn could act as a conductive buffer and effectively reduce diffusion-induced stress, respectively. The comparison of the electrochemical performances of the related Si-metal/silicide composite reported in the literature [41, 145, 148–155] is summarized in Table 2.3.

2.2.4 Oxide-containing anodes

Various nanostructured metal oxides with different Li-ion storage mechanisms have been reported in the literature due to their high theoretical capacity (~900 mAh g^{-1}), safety, stability, and cost-effectiveness [11, 12]. Nanostructured titanium dioxide (TiO_2) is extensively used as a multifunctional material because of its stability, robustness, and physicochemical properties with the advantage of its moderate cost [11]. The advantages of excellent structural stability, environmental benignity with volume expansion of about 3% during lithiation depicts the extensive studies on TiO_2 as anode for LIBs [42, 156].

Wei et al. reported the synthesis of novel Si/TiO_2 nanowire array (TNA) using solvothermal process followed by magnetron sputtering of Si [42]. To optimize the best ratio of Si to TiO_2, the different mass percentages of Si (0%, 25%, 50%, 75%, and 100%) are prepared (Fig. 2.11A–G) and the Li-ion storage behavior is carried out in the potential window between 2.7 and 0 V vs Li/Li$^+$. The cycling stability of the TNA sample (0% Si) demonstrates

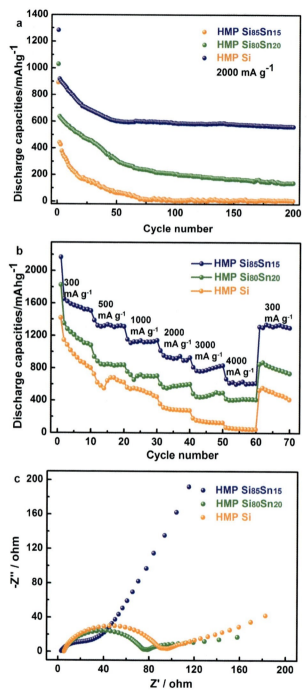

FIG. 2.10 (A) The cycling performances at 2000 mA g^{-1} for 200 cycles, (B) rate performances, and (C) impedance measurements of the HMP Si, Si$_{85}$Sn$_{15}$, and Si$_{80}$Sn$_{20}$ electrodes. *(Reproduced with permission from: Q. Hao, J. Hou, J. Ye, H. Yang, J. Du, C. Xu, Hierarchical macroporous Si/Sn composite: easy preparation and optimized performances towards lithium storage, Electrochim. Acta 306 (May 2019) 427–436. Copyright Elsevier (May 2019).)*

TABLE 2.3 Electrochemical performance(s) of the silicon-metal/silicide composite reported in the literature.

Samples description	Initial capacity (mAh g^{-1})	Residual capacity (mAh g^{-1})	Test period (cycle)	Current density (mA g^{-1})	References
Sn$_{0.9}$Si$_{0.1}$/carbon nanoparticles	964	~770	50	360	[148]
Si$_{92}$Sn$_8$ thin film	~2400	~100	30	0.1C (<400)	[149]
	~1800	~900	—	—	
Si/Sn nanowires	3192	1874	100	~350	[150]
Si/Sn ribbon	~1400	~1100	50	400	[151]
Si-Ni-Sn nanospheres	1065	402	50	300	[152]
Si/Sn@ carbon-graphite composite	833	613	100	100	[153]
Hierachical macroporous	2466	1545	70	200	[41]
Si$_{85}$Sn$_{15}$ composite	1642	748	100	1000	
Fe-Cu-Si ternary composite	460	410	50	0.5C (1C = 0.42 A g^{-1})	[154]
Silicon/iron silicide/three-dimensional (3D) carbon nanocomposite	1483 (NaCl-0)	655	100	140 mA g^{-1}	[145]
	1524 (NaCl-26.7)	823	100		
	1508 (NaCl-45.9)	1061	100		
Titanium silicide coated porous silicon nanospheres	1780	1180	50	400	[155]

Adapted from: Q. Hao, J. Hou, J. Ye, H. Yang, J. Du, C. Xu, Hierarchical macroporous Si/Sn composite: Easy preparation and optimized performances towards lithium storage, Electrochim. Acta 306 (2019) 427–436.

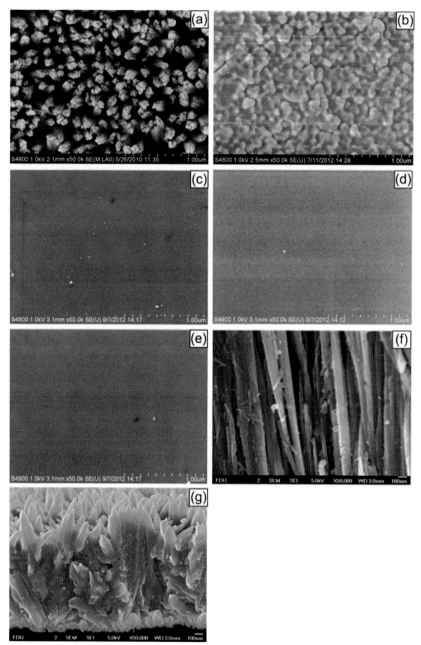

FIG. 2.11 FE-SEM images of the Si/TiO$_2$ nanowire array (TNA) composites with different Si percentages after 200 cycles: (A) 100% TNA, (B) 25% Si, (C) 50% Si, (D) 75% Si, (E) 100% Si; (F) sectional view of TNA, (G) sectional view of Si (75%)/TNA composite. *(Reproduced with permission from: Z. Wei, R. Li, T. Huang, A. Yu, Fabrication and electrochemical properties of Si/TiO$_2$ nanowire array composites as lithium-ion battery anodes, J. Power Sources 238 (September 2013) 165–172. Copyright Elsevier (September 2013).)*

a reversible capacity retention of 104 mAh g^{-1} after 200 cycles; pure Si could deliver a reversible capacity of 440 mAh g^{-1} after the third cycle and remains at 300 mAh g^{-1} after 200 cycles [42]. The composites containing 25, 50, and 75 wt % Si could deliver a reversible capacity of approximately 550, 600, and 800 mAh g^{-1}, respectively, after 200 cycles [42]. To understand the electrochemical behavior of these composites, the SEM images with various percentages of Si were obtained and analyzed. Even though the nanowire structure of TiO$_2$ seems to be disappeared, the interspacing between the TNA ends up in a porous structure (Fig. 2.11A and F). Similarly, pure Si films end up in huge structural disintegration, which further leads to the formation of isolated islands, as shown in Fig. 2.11E. However, the composite of Si-TNA (i.e. 25%–75% Si) suggests no severe mechanical disintegration in the electrode and the porous structure of 75% sample (Fig. 2.11D and G) still remains, which could be the reason behind the stable cycles life of this anode in LIBs.

Although the natural abundance of SiO$_2$ is quite high (and hence inexpensive), the electrochemical inactivity of bulk SiO$_2$ toward Li-storage requires extensive processing to produce the appropriate morphology and crystallinity for nano-silica to be electrochemically active [129, 157, 158]. Wang et al. demonstrated a methodology to form hollow porous SiO$_2$ nanobelts using CuO as a sacrificial template [43]. The process involves green, scalable, and very simple methodologies. The overall schematic representation of the process involved is shown in Fig. 2.12 [43]. To understand the morphological aspects of the development of porous SiO$_2$ nanobelts, SEM, and TEM images of CuO nanobelt, CuO@SiO$_2$ nanobelts, and hollow porous SiO$_2$ nanobelts are shown in Fig. 2.13. The CuO templates evident a typical belt-like morphology, which is single crystalline in nature (Fig. 2.13A and B). Further, the core-shell structure of CuO@SiO$_2$ intermediates (Fig. 2.13C and D) is transformed into a

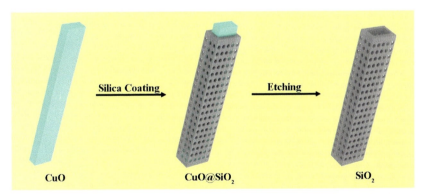

FIG. 2.12 Schematic illustration for the synthesis of the hollow porous SiO$_2$ nanobelts. *(Reproduced from: H. Wang, P. Wu, H. Shi, W. Tang, Y. Tang, Y. Zhou, P. She, T. Lu, Hollow porous silicon oxide nanobelts for high-performance lithium storage, J. Power Sources 274 (January 2015) 951–956. Copyright Elsevier (January 2015).)*

FIG. 2.13 SEM and TEM images of the CuO nanobelts (A, B), CuO@SiO$_2$ nanobelts (C, D), and hollow porous SiO$_2$ nanobelts (E, F). *(Reproduced with permission from: H. Wang, P. Wu, H. Shi, W. Tang, Y. Tang, Y. Zhou, P. She, T. Lu, Hollow porous silicon oxide nanobelts for high-performance lithium storage, J. Power Sources 274 (January 2015) 951–956. Copyright Elsevier (January 2015).)*

SiO$_2$ hollow structure (Fig. 2.13E and F) by removing the CuO template with 1 M HCl solution under stirring condition.

An example of the use of carbon nanobelts is illustrated by the work of Zeng et al. reporting disproportionate-SiO/C nano-spheres coated with a network of carbon nanobelts (d-SiO/C@NCNBs) to alleviate the issue of volume change

Recent advances in silicon materials for Li-ion batteries **Chapter | 2** **67**

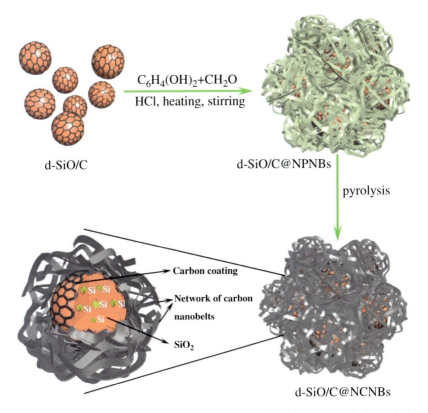

FIG. 2.14 Schematic illustration of the preparation process of the d-SiO/C@NCNBs. *(Reproduced with permission from: S.-Z. Zeng, Y. Niu, J. Zou, X. Zeng, H. Zhu, J. Huang, L. Wang, L.B. Kong, P. Han, Green, and scalable preparation of disproportionated SiO anode materials with cocoon-like buffer layer, J. Power Sources 466 (August 2020) 228234. Copyright Elsevier (August 2020).)*

and capacity degradation [44]. To synthesize d-SiO/C@NCNBs, initially, carbon-coated disproportionate SiO (d-SiO/C) was prepared using the chemical vapor deposition (CVD) method. Then, d-SiO/C powder was dispersed in a solution constituting of formaldehyde, hydroquinone, and hydrochloric acid with constant stirring for 6h at 80°C. Finally, to obtain d-SiO/C@NCNBs anode materials, the resulting sample from the last step is carbonized at 850°C for 2h under Ar atmosphere; the overall process is schematically illustrated in Fig. 2.14 [44].

The corresponding SEM images are shown in Fig. 2.15A–D where the porous structure of d-SiO/C@NCNBs and carbon nanobelts are intimately attached to the d-SiO/C particles, which enable good ionic/electronic transport and provide space for volume changes during discharge/charge process. The cycling stability and C-rate performance of both d-SiO/C and d-SiO/C@NCNBs anodes are studied in LIBs. The initial discharge/charge capacities of d-SiO/C and d-SiO/C@NCNBs are 1882/1374 and 1375/1004 mAh g^{-1}, respectively. Even

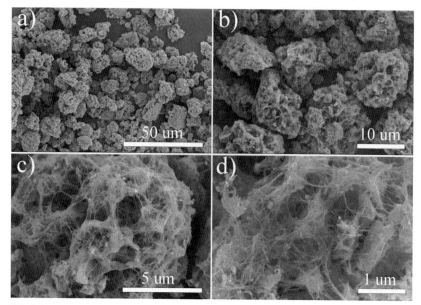

FIG. 2.15 (A–D) SEM images of the d-SiO/C@NCNBs sample at different magnifications. *(Reproduced with permission from: S.-Z. Zeng, Y. Niu, J. Zou, X. Zeng, H. Zhu, J. Huang, L. Wang, L.B. Kong, P. Han, Green and scalable preparation of disproportionate SiO anode materials with cocoon-like buffer layer, J. Power Sources 466 (August 2020) 228234. Copyright Elsevier (August 2020).)*

though the initial capacity of d-SiO/C sample is higher in comparison to d-SiO/C@NCNBs, the reversible capacity of 490 and 1043 mAh g^{-1} was remained after 100 cycles at 0.1C (1C = 1500 mA g^{-1}), respectively. Further, the C-rate study suggests a superior electrochemical performance of d-SiO/C@NCNBs even at higher current rate of 1C in comparison to d-SiO-C.

2.3 Alternative raw materials and novel processing methods

It is very important to have the availability of suitable raw materials for the production of Si-based anode, especially when approaching the industrial facility. The successful implementation of alternative raw materials is very essential for the conservation of natural resources towards sustainable development. The increased use of alternative raw materials in the production of Si-based electrode materials could be one of the core elements on the way to climate strategy. In this section, we will discuss various alternative raw materials and novel processing methodologies, which have been explored in literature towards the development of Si-based anode for LIB applications. Thus, the use of Si-containing biomass, clays, waste materials, and by-products from other industries as valuable raw materials from a material viewpoint is investigated. Further, different processing methodologies such as magnesiothermic/aluminothermic reduction, metallic melt are discussed.

2.3.1 Recycling of silicon-containing industrial sources

Even though various works based on designing nanostructured Si-based materials for stable electrochemical performance are reported in the literature, most of these synthesis methodologies are complex and use expensive raw materials. Thus, recycling waste generating from the industry into Si-based anode materials is a sustainable strategy, wherein the waste is reduced and being converted to the functional anode for Li-ion storage [45, 46, 159, 160]. Li et al. converted glass bottles into high purity and interconnected Si networks [45]. As shown in Fig. 2.16A, quartz powder derived from the glass bottle could be directly used for the reduction of SiO_2 to Si without using any other side reaction/process. This process is much more feasible in terms of environmentally-benign, energy-saving, and efficiency in the synthesis process. A beverage glass bottle (Fig. 2.16B) is crushed into glass followed by mechanical milling to reduce the particle size of silica (Fig. 2.16C—milled glass).

Eventually, a uniform mixture constituting SiO_2-NaCl-Mg is prepared and transferred into a small reactor to carry out the reaction at 700°C for 6h. Finally, glass derived Si (gSi) is obtained by washing the powder with DI water and concentrated HCl to remove NaCl, MgO, and Mg_2Si. The overall synthesis process of Si from SiO_2 is pictorially illustrated in Fig. 2.16D. Further, a thin layer of carbon is coated onto gSi using acetylene gas by the CVD method. The gSi and gSi@C delivered a stable reversible capacity of 796 and 1420 mAh g^{-1} with a capacity retention of 20, and 72% after 400 cycles when tested as an anode in a potential range of 0.01–1.5 V (vs. Li/Li$^+$) at C/2 rate, respectively.

The semiconducting nature of crystalline Si has led to its intensive application towards renewable energy conversion, i.e., Si-based solar cells consist of 90% of the photovoltaic market share. During the manufacturing of solar cells, diamond wire cutting technology is employed to cut Si wafers, which led to the generation of a huge amount of Si powder as waste. Due to the addition of organic-based solvents to the cutting fluid, the wasted Si powders are contaminated during diamond wire cutting. Wang et al. reported a series of chemical treatments to recycle the impure Si-based waste generating from diamond wire cutting [159]. As illustrated in Fig. 2.17, typically 0.8–18.9 micrometer sized Si powders (represented as S0) are obtained by ultrasonic cleaning of the slurry generated during Si wafer cutting. A series of chemical treatments comprising consecutive applications of HF, ammonia, H_2O_2, and HCl-H_2O_2-DI are used to clean Si-powder. Further, Si/C composites are synthesized using petroleum asphalt powder as carbon precursors and studied as an anode in LIBs.

2.3.2 Silicon sourced from biomass and clays

Recently, biomass-derived electrode materials especially carbon-based have attracted enormous attention as anodes for LIB [161–164]. Biomasses, for

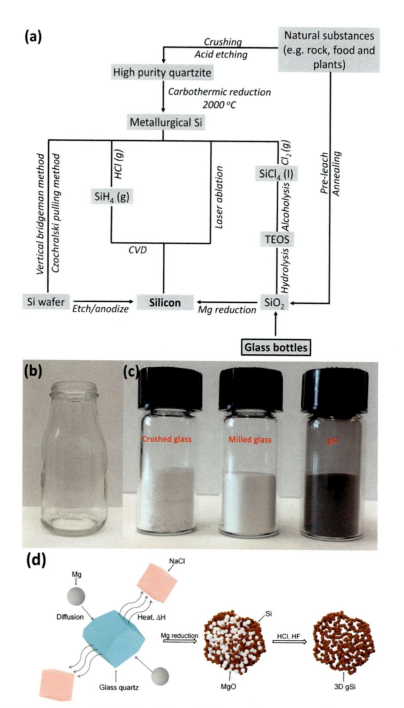

FIG. 2.16 (A) Flow chart showing existing synthesis routes for nano-Si, including the synthesis method from glass bottles. (B) A collected beverage glass bottle. (C) (from left to right) vials of crushed glass, milled glass, and gSi powder. (D) Schematic of the Mg reduction process using NaCl as a heat scavenger. *(Reproduced from: C. Li, C. Liu, W. Wang, Z. Mutlu, J. Bell, K. Ahmed, R. Ye, M. Ozkan, C.S. Ozkan, Silicon derived from glass bottles as anode materials for lithium-ion full cell batteries, Sci. Rep. 7 (April 2017) 917. (Creative Commons 4.0 CC BY).)*

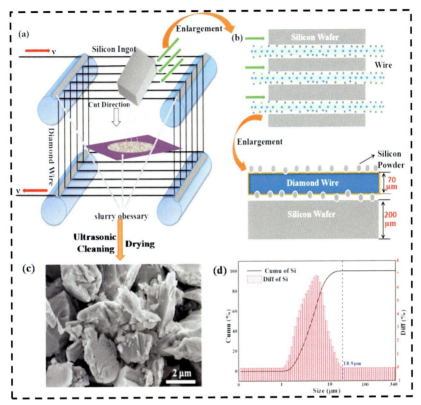

FIG. 2.17 The schematic illustration of (A) Si powders from diamond wire cutting; (B) Enlargement of cutting; (C) SEM of S0 powders; (D) size distribution of S0 powders. *(Reprinted with permission from: K. Wang, B. Xue, Y. Tan, J. Sun, Q. Li, S. Shi, P. Li, Recycling of micron-sized Si powder waste from diamond wire cutting and its application in Li-ion battery anodes, J. Cleaner Production 239 (December 2019) 117997. Copyright Elsevier (December 2019).)*

instance, rice husk, corn cob, cellulose, wood, rapeseed shuck, fruit peel, etc. are organic-inorganic based materials derived from plants and animals and it could be a source of energy [161–166]. Typically, these are dumbed as waste or burnt in the air, which leads to serious pollution in the environment. Thus, it is worth converting these biomasses into functional materials. Further, it serves as a cheap raw material for the production of materials. For instance, rice (*Oryza sativa*) is one of the most staple and primary food source for billions of people all over the world. Rice husk is the outer coating on the grain of rice, and the major components are cellulose (25%–30%), lignin (26%–30%), silica (15%–20%), and organic compounds [167].

Zhang et al. reported the fabrication of Si/N-doped carbon/CNT (SNCC) nano/micro hierarchical structured spheres using the electrospray approach, wherein Si nanoparticles are synthesized from rice husk (RH) [168]. The RH was initially treated with HCl to remove the metallic impurities, and followed by heat treatment at 700°C in an atmospheric condition which eliminates all the organics and carbon-based components. The reaction is resulting in the formation of nano-SiO$_2$, which is further reduced to nano-Si using magneisothermic reduction. Later on, RH derived Si nanoparticles along with CNTs are dispersed in DMF solution containing PAN to obtain SNCC nano/microspheres via electrospray method. The schematic of the synthesis of SNCC nano/microstructured spheres is illustrated in Fig. 2.18 [168].

Similarly, clays, which are natural soil material bearing silicate-based compounds, could be another cost-effective precursor for the synthesis of Si-based materials [50, 169]. Similar to magnesiothermic reduction, aluminothermic reduction processing is also a very attractive approach for the synthesis of porous Si due to its low environmental impact and improved energy/cost-efficiency. Gao et al. reported a cost-efficient approach for the synthesis of porous Si using aluminothermic reduction of attapulgite (clay soil) at 200–300°C for 2h [170]. To mitigate the issue of poor electronic/ionic conductivity of pristine Si, the composite Si/graphite @carbon was prepared by ball-milling followed by carbonization at 800°C. The overall synthesis and electrochemical performance of the electrode materials derived is represented in Fig. 2.19 [170].

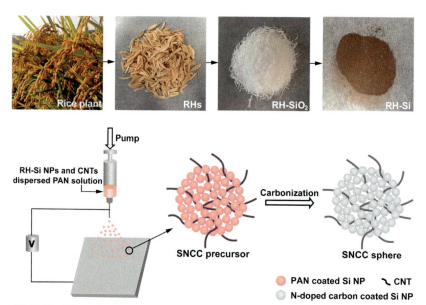

FIG. 2.18 Schematic illustration of the preparation process for SNCC nano/microstructured spheres. *(Reproduced with permission from: Y.-C. Zhang, Y. You, S. Xin, Y.-X. Yin, J. Zhang, P. Wang, X.-s. Zheng, F.-F. Cao, Y.-G. Guo, Rice husk-derived hierarchical silicon/nitrogen-doped carbon/carbon nanotube spheres as low-cost and high-capacity anodes for lithium-ion batteries, Nano Energy 25 (July 2016) 120–126. Copyright Elsevier (July 2016).)*

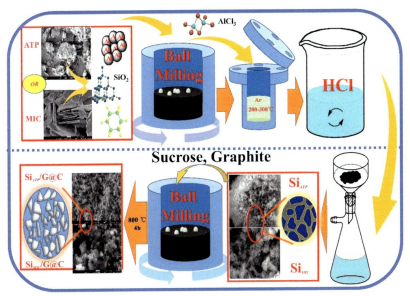

FIG. 2.19 Schematic illustration for the synthesis processes of porous Si and corresponding composites. *(Reproduced with permission from: S. Gao, D. Yang, Y. Pan, L. Geng, S. Li, X. Li, P.-F. Cao, H. Yang, From natural material to high-performance silicon-based anode: Towards cost-efficient silicon-based electrodes in high-performance Li-ion batteries, Electrochim. Acta 327 (December 2019) 135058. Copyright Elsevier (December 2019).)*

2.3.3 Magnesiothermic and metallic melt processing

The process of magnesiothermic reduction of silicon dioxide into silicon has gained wide attention due to its advantages of maintaining the original micro/nano-structures even after conversion to the final product [48, 49, 91, 171, 172]. Additionally, this process typically takes place at an approximate temperature range of 600–700°C, which is relatively lower. The advantages of this processing method have been taken forward for the development of various kinds (in terms of morphology) of Si-based materials as the anode in LIBs. Liu et al. reported a magneisothermic reduction of one-dimensional nanostructured SiO_2 into Si at a lower temperature of 500°C [171]. Initially, linear polyamine, polyethyleneimine (LPEI) is used as a template for the synthesis of one-dimensional SiO_2, as shown in Fig. 2.20A. After that, LPEI components are removed from the LPEI@SiO_2 samples, and the as-obtained nanostructured SiO_2 is mixed with Mg and undergo magnesiothermic reduction as picturized in Fig. 2.20B [171].

Similarly, metallic melt processing is another effective way to process nanostructured materials, by utilizing the selective dissolution of the elements into a melt [173]. The following example illustrates this phenomenon: element A is

74 PART | I Introduction and background

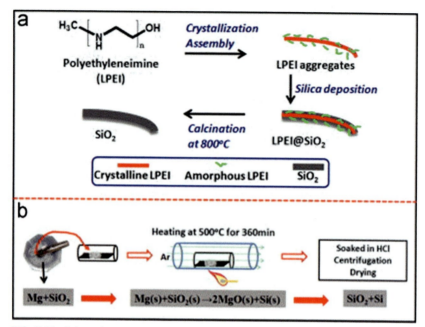

FIG. 2.20 Schematic representation of the experimental procedure. (A) Crystallization and assembly of LPEI followed by silica mineralization on LPEI to produce LPEI@SiO₂ hybrids that are converted into SiO₂ via calcination. (B) Low-temperature magnesiothermic reduction of SiO₂ to Si. *(Reprinted with permission from: X. Liu, Y. Gao, R. Jin, H. Luo, P. Peng, Y. Liu, Scalable synthesis of Si nanostructures by low-temperature magnesiothermic reduction of silica for application in lithium-ion batteries, Nano Energy 4 (March 2014) 31–38. Copyright Elsevier (March 2014).)*

miscible in X, whereas B is immiscible in X, if an alloy of A-B is immersed into X, only B will remain as a solid with nanoporous structure (A will dissolve in X). Thus, this is selective leaching (or dealloying) of an element into the metallic melt. Wada et al. reported a three-dimensional nanoporous interconnected Si material synthesized by immersing Mg-Si alloy compound precursor into Bi melt [173]. During these processes, Mg atoms are selectively dissolved into Bi melt and the remaining Si atoms get re-organized to form porous nanostructure materials, as shown in Fig. 2.21 [173]. Different nanoporous Si samples were prepared by varying the temperature and studied the cyclic behavior of these materials in LIBs, as shown in Fig. 2.21 (Right). As discussed in the previous section, manesiothermic reduction is an affordable and effective method for the synthesis of nanostructured porous Si. Table 2.4 details the cycling performance of a variety of Si-based anodes synthesized via magnesiothermic reduction of both natural and industrial sources [50, 76, 91, 165, 166, 169, 174, 175].

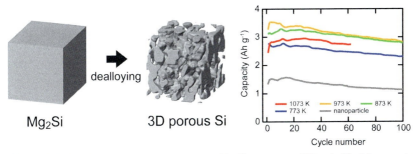

FIG. 2.21 Left: Schematic of dealloying using an Mg-Si precursor and Bi melt. The Mg atoms in the precursor selectively dissolve into Bi, and the remaining Si atoms self-organize into a nanoporous structure with characteristic length ranging from several ten to hundred nanometers. Right: Cyclic properties of NP–Si electrodes prepared at temperatures between 773 and 1073 K at a constant current mode of 1.0C. *(Reproduced with permission from: T. Wada, J. Yamada, H. Kato, Preparation of three-dimensional nanoporous Si using dealloying by metallic melt and application as a lithium-ion rechargeable battery negative electrode, J. Power Sources 306 (February 2016) 8–16. Copyright Elsevier (February 2016).)*

2.3.4 Nano-silicon derived from diatomite and inspired by nature

Improvements in designing hierarchical electrode materials for battery applications could be achieved by taking inspiration from nature. For instance, a pomegranate-inspired hierarchical structured nanoscale design has been used to tackle the issues related to the large volume change of silicon anode and achieve high areal capacity by improving the tap density of the materials [176]. Diatomite is an environmentally benign, highly abundant, low-cost siliceous rocks, mainly composed of SiO_2 [51, 52]. The unique hierarchical structure with extremely high porosity of diatomite signifies its applicability in designing Si-based anode. Further, the unique hierarchical structure of diatomite could be inherited into the as-obtained silicon framework, using magnesiothermic reduction. Cambell et al. reported the use of diatomite-derived nano silicon as an anode in LIBs, which has been synthesized by using an efficient magnesiothermic reduction process. The electrode could deliver a reversible capacity of 1102 mAh g^{-1} at C/5 rate after 50 cycles [177]. Later on, Zheng et al. demonstrated the fabrication of mesoporous Si/SiC from diatomite [178]. The overall schematic, illustrating the various steps involved during the conversion process is depicted in Fig. 2.22.

2.3.5 Other novel processing methods

In addition to the above-mentioned processing methodology, several other approaches have been explored to open the pathways for scalable, low-cost, and feasible fabrication of Si-based anode materials for practical applications [55–57,179,180]. Kumta et al. have demonstrated [55] a novel electrode fabrication technique involving a manual scribing action of vertically aligned silicon

TABLE 2.4 Cycling performance of Si-based anodes synthesized through magnesiothermic reduction of natural and industrial silicon sources.

Si-based anodes	Si source	Initial charge capacity (mAh g^{-1})	Capacity retention (# cycles)	Current density (mA g^{-1})	Reference
Si nanoparticles	Rice husks	2790	86% (300)	700	[165]
BL-derived Si nanoparticles	Bamboo leaves	~2800	64% (100)	840	[166]
Hollow vesica-like Si	Fumed SiO$_2$	1846	39% (200)	360	[174][a]
Monodisperse porous Si-CNT	TEOS	3105	71% (50)	210	[175][a]
Si/NanoGs composite	fumed SiO$_2$	1703	57% (100)	100	[91][a]
Annealed Si nanosheet	montmorillonite	~1400	31% (200)	400	[169]
H-Si nanoparticles	halloysite clay	2723	82% (100)	700	[50]
Si nanosheet	sand	2431	–	500	[76]

[a] It indicates that these are industrial or man-made vs. natural sources.

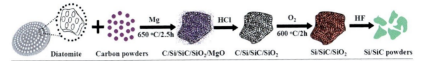

FIG. 2.22 Schematic illustrating the synthesis process of Si/SiC powders from diatomite starting materials. *(Reproduced with permission from: Y. Zheng, H. Fang, F. Wang, H. Huang, J. Yang, G. Zuo, Fabrication and characterization of mesoporous Si/SiC derived from diatomite via magnesiothermic reduction, J.Solid State Chem. 277 (September 2019) 654–657. Copyright Elsevier (September 2019).)*

coated multiwall carbon nanotubes on a copper foil as a viable approach to Li-ion battery electrodes. The scribed electrodes were prepared without the use of any conductive additives and binders, and they were directly assembled in a coin cell. These "binder-less" scribable Si-CNT electrodes exhibited a very high discharge capacity in excess of 3000 mAh g^{-1} and a low FIR loss (19%). In addition, the electrodes also showed good cyclability with capacity retention of 76% at the end of 50 cycles corresponding to a fade rate of 0.48% loss per cycle rendering the technique attractive for suitable Li-ion applications.

Chang et al. also demonstrated a novel binder-free Si-based anode for LIB [56]. Three different kinds of carbon-based materials (reduced graphene oxide, CNT, and polymer-derived amorphous carbon) are used to stabilize the overall electrochemical performance during cycling. Electron micrographs of materials produced during the fabrication of Si-based novel architecture are provided in Fig. 2.23. Initially, CNTs are grown on the Ni-foam via a CVD method, under continuous C_2H_2 flow at 700 °C for 20 min. The CNT-embedded Ni-foam was immersed into a homogenized mixture of Si nanoparticles and graphene oxide (GO) sheets in poly(methyl methacrylate) (PMMA) solutions, and dried thereafter; the novel Si-based architecture is finally obtained by reducing GO and carbonizing PMMA at 700°C for 4h, under Ar-H_2 atmosphere (Fig. 2.23A) [56].

The top view of the electrodes shows the Si NPs are almost covered by the rGO layers (Fig. 2.23B). The rGO networks can be easily identified due to folds in the rGO; these nanostructures improve the electronic conductivity of Si NPs and also provide void space (Fig. 2.23C) to accommodate the volume change of Si NPs during lithiation, thereby enhancing cycling performance in LIB applications. After carbonizing the PMMA, cellular carbon was fully coated onto Si NP surfaces (Fig. 2.23D); cross-sectional SEM images (Fig. 2.23E and F) show CNTs interspersed with rGO. The detailed nanostructure of the carbon-coated Si NPs was further investigated using TEM, indicating the Si NPs were uniformly coated with carbon (Fig. 2.23G); the amorphous carbon thickness was observed to be approximately 10 nm (Fig. 2.23H).

The electrochemical characterization of the as-fabricated binder-free Si/rGO/CNT (SiGC) nanostructured electrode is studied in LIBs [56]. The electrode could deliver an initial alloying/dealloying capacities of 3717, and 2735 mAh g^{-1} at 0.05C rate (1C = 2600 mA g^{-1}), as shown in Fig. 2.24A.

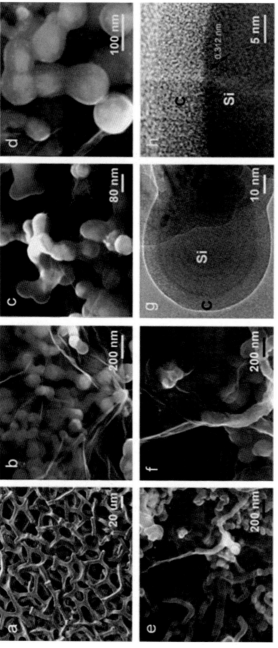

FIG. 2.23 A binder-free Si/rGO/CNT (SiGC) nanostructured electrode. (A) CNT/3D porous Ni foam electrode coated with carbon-coated Si/rGO nanostructures. Top view: (B) carbon-coated Si NPs in rGO networks; (C) carbon-coated Si NPs; (D) high-magnification SEM image of carbon-coated Si NPs. Cross-sectional view: (E) SEM image and (F) higher magnification image of the penetration of carbon-coated Si NPs/rGO nanostructures by CNTs. TEM images of (G) carbon-coated Si NPs and (H) HRTEM image of the carbon-coated Si NPs. *(Reproduced with permission from: J. Chang, X. Huang, G. Zhou, S. Cui, S. Mao, J. Chen, Three-dimensional carbon-coated Si/rGO nanostructures anchored by nickel foam with carbon nanotubes for Li-ion battery applications, Nano Energy 15 (July 2015) 679–687. Copyright Elsevier (July 2015).)*

Recent advances in silicon materials for Li-ion batteries **Chapter | 2** 79

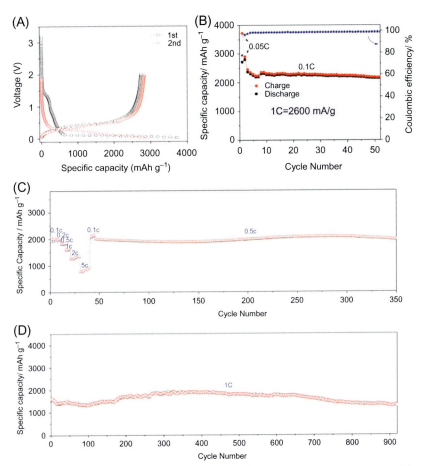

FIG. 2.24 Electrochemical characteristics of the SiGC electrode. (A) Galvanostatic charge/discharge profiles of the first two cycles at 0.05C-rate, (B) capacity retention and Coulombic efficiency at 0.05C-rate for the first two cycles, and then 0.1C for 50 cycles. (C) Rate capability and (D) long-term cyclic performance at 1C. *(Reproduced with permission from: J. Chang, X. Huang, G. Zhou, S. Cui, S. Mao, J. Chen, Three-dimensional carbon-coated Si/rGO nanostructures anchored by nickel foam with carbon nanotubes for Li-ion battery applications, Nano Energy 15 (July 2015) 679–687. Copyright Elsevier (July 2015).)*

The cycling stability study of SiGC electrode at 0.1C shows a stable capacity of about 2292 mAh g^{-1} with 93% retention after the 50th cycle (Fig. 2.24B). Further, the C-rate performance of SiGC anodes at the various current rate of 0.1, 0.2, 0.5, 1, 2, and 5C delivers reversible capacities of 2010, 2000, 1835, 1616, 1321, and 830 mAh g^{-1}, respectively, as shown in Fig. 2.24C. Besides, the electrode could deliver a reversible capacity of 1311 mAh g^{-1} at 1C, after 900 cycles (Fig. 2.24D). The excellent electrochemical performance of the designed SiGC electrodes could be attributed to several factors (Fig. 2.23): (i) nanosized Si particles, which can reduce the diffusion pathways for Li-ions; (ii) the interconnected rGO could stabilize the overall electronic transport by providing a 2D network; (iii) CNTs provide strong interfacial connection between active material Ni foam, which maintain active material stabilizations during cycling; (iv) PMMA leads to generate a dense cellular carbon on the overall electrode, enhancing the overall integrity of the electrode [56].

Li et al. proposed an efficient roll-to-roll production method for Si-based anodes via electrochemical reduction of SiO_2 (ERSiO$_2$), as shown in Fig. 2.25 [57]. Initially, SiO_2 particles are anchored within the porous carbon electrode, which provides the framework for the electrolytic reduction of SiO_2 particles. Further, the as-fabricated electrodes undergo chemical treatment, resulting in removal of contamination from molten salt as well as unreduced SiO_2.

Similarly, 3D carbon-based scaffold electrode architecture has been widely demonstrated in the literature to tackle the volume expansion and structural disintegration during cycling, which is discussed broadly in Chapter 9 of this book. Martha et al. detailed how a 3D electrode architecture of Si-NPs on carbon fibers provides a continuous conducting framework and flexible network of carbon which accommodates the volume change of silicon during charge and discharge [180]. Most of the works presented in 3D carbon fiber-based electrode architecture are limited mostly to button type cells. For tabbing large format pouch cells, carbon felt can be metalized by using sputtering) or metal coating followed by spot welding with metal foil or by using conducting epoxy with metal foil.

2.4 Conclusions

Silicon is a promising anode material for rechargeable LIBs and can replace the conventional carbon-based anodes due to its high theoretical specific capacity. However, the drawback of significant volume changes during charge-discharge cycling urge an appropriate materials engineering and electrode design which can meet the industrial requirements for practical applications. In this chapter, the systematic development of Si-containing materials in terms of its novel processing methods, and raw materials are expressed. Si-carbon-based hybrid materials, for instance, Si-graphite, Si-graphene, Si-CNT/CNF, hollow Si nanotubes, etc. are widely studied due to the advantageous of mitigating the issue related to electronic conductivity, stable SEI formation, and the stress developed on the electrode during charge-discharge cycling process.

Further, various alternative raw materials and novel processing methodologies are explored by the researchers towards the development of Si-based anode for LIB applications. Si-containing biomass (for instance, rice husk, bamboo leaf, etc.), clays, sand, waste materials and by-products from other industries (such as glass bottles, Si-powder from solar cell) are used as, a precursors to

FIG. 2.25 Schematic of a roll-to-roll nano-MG-Si battery material production process based on electrochemical reduction of SiO_2 (ERSiO_2). *(Reproduced with permission from: X. Li, R.B. Wehrspohn, Nanometallurgical silicon for energy application, Joule 3 (May 2019) 1172–1179, Copyright Elsevier (May 2019).)*

synthesize the Si-based anode. Further, different processing methodologies such as, magneisothermic/aluminothermic reduction and metallic melt are used to fabricate Si-based anode materials by reduction of oxide-based/alloy-based precursors.

Fundamental research in Si-based hybrid materials is primarily focused on achieving desirable gravimetric capacity and stable cycle life. Specifically, when hierarchical nanostructured Si is fabricated, the tap density of the materials and electrode is very poor, consequently resulting in low volumetric capacities and limiting the energy density of the whole cell. Nevertheless, the commercial viability of Si-based anode demands low first cycle irreversible capacity, good coulombic efficiency, better capacity retention, good tap density, high Si content, simple and scalable manufacturing process, and most importantly low manufacturing cost. Thus, an efficient, simple, low-cost, easily scalable synthesis methodology is desired which can assure an excellent electrochemical performance of Si-based anode materials. The successful implementation of suitable raw materials and economical processing methods are very essential for the production of Si-based anode, especially when approaching the industrial facility.

References

[1] M. Armand, J.-M. Tarascon, Building better batteries, Nature 451 (2008) 652–657.
[2] V. Etacheri, R. Marom, R. Elazari, G. Salitra, D. Aurbach, Challenges in the development of advanced Li-ion batteries: a review, Energ. Environ. Sci. 4 (2011) 3243–3262.
[3] R. Teki, M.K. Datta, R. Krishnan, T.C. Parker, T.M. Lu, P.N. Kumta, N. Koratkar, Nanostructured silicon anodes for lithium ion rechargeable batteries, Small 5 (2020) 2236–2242.
[4] S. Megahed, B. Scrosati, Lithium-ion rechargeable batteries, J. Power Sources 51 (1994) 79–104.
[5] J.R. Owen, Rechargeable lithium batteries, Chem. Soc. Rev. 26 (1997) 259–267.
[6] J.-M. Tarascon, M. Armand, Issues and challenges facing rechargeable lithium batteries, Nature 414 (2001) 359–367.
[7] R. Yazami, P. Touzain, A reversible graphite-lithium negative electrode for electrochemical generators, J. Power Sources 9 (1983) 365–371.
[8] R. Epur, M.K. Datta, P.N. Kumta, Nanoscale engineered electrochemically active silicon–CNT heterostructures-novel anodes for Li-ion application, Electrochim. Acta 85 (2012) 680–684.
[9] N. Nitta, F. Wu, J.T. Lee, G. Yushin, Li-ion battery materials: present and future, Mater. Today 18 (2015) 252–264.
[10] M.K. Datta, P.N. Kumta, Silicon and carbon based composite anodes for lithium ion batteries, J. Power Sources 158 (2006) 557–563.
[11] Z. Yang, D. Choi, S. Kerisit, K.M. Rosso, D. Wang, J. Zhang, G. Graff, J. Liu, Nanostructures and lithium electrochemical reactivity of lithium titanites and titanium oxides: a review, J. Power Sources 192 (2009) 588–598.
[12] W. Wang, P.N. Kumta, Nanostructured hybrid silicon/carbon nanotube heterostructures: reversible high-capacity lithium-ion anodes, ACS Nano 4 (2010) 2233–2241.

[13] C.-M. Park, J.-H. Kim, H. Kim, H.-J. Sohn, Li-alloy based anode materials for Li secondary batteries, Chem. Soc. Rev. 39 (2010) 3115–3141.
[14] N.L. Rock, P.N. Kumta, Synthesis and characterization of electrochemically active graphite–silicon–tin composite anodes for Li-ion applications, J. Power Sources 164 (2007) 829–838.
[15] J. He, Y. Wei, T. Zhai, H. Li, Antimony-based materials as promising anodes for rechargeable lithium-ion and sodium-ion batteries, Mater Chem. Front. 2 (2018) 437–455.
[16] Z. Hu, S. Zhang, C. Zhang, G. Cui, High performance germanium-based anode materials, Coord. Chem. Rev. 326 (2016) 34–85.
[17] M. Obrovac, V. Chevrier, Alloy negative electrodes for Li-ion batteries, Chem. Rev. 114 (2014) 11444–11502.
[18] Y. Chen, X. Li, K. Park, J. Song, J. Hong, L. Zhou, Y.-W. Mai, H. Huang, J.B. Goodenough, Hollow carbon-nanotube/carbon-nanofiber hybrid anodes for Li-ion batteries, J. Am. Chem. Soc. 135 (2013) 16280–16283.
[19] M. Winter, J.O. Besenhard, M.E. Spahr, P. Novak, Insertion electrode materials for rechargeable lithium batteries, Adv. Mater. 10 (1998) 725–763.
[20] J.R. Dahn, T. Zheng, Y. Liu, J. Xue, Mechanisms for lithium insertion in carbonaceous materials, Science 270 (1995) 590–593.
[21] Y. Lu, L. Yu, X.W.D. Lou, Nanostructured conversion-type anode materials for advanced lithium-ion batteries, Chem 4 (2018) 972–996.
[22] S.-H. Yu, X. Feng, N. Zhang, J. Seok, H.C.D. Abruna, Understanding conversion-type electrodes for lithium rechargeable batteries, Acc. Chem. Res. 51 (2018) 273–281.
[23] X. Gu, J. Yue, L. Li, H. Xue, J. Yang, X. Zhao, General synthesis of MnOx (MnO$_2$, Mn$_2$O$_3$, Mn$_3$O$_4$, MnO) hierarchical microspheres as lithium-ion battery anodes, Electrochim. Acta 184 (2015) 250–256.
[24] J.H. Sung, C.-M. Park, Amorphized Sb-based composite for high-performance Li-ion battery anodes, J. Electroanal. Chem. 700 (2013) 12–16.
[25] M. Obrovac, Si-alloy negative electrodes for Li-ion batteries, Curr. Opin. Electrochem. 9 (2018) 8–17.
[26] J.P. Maranchi, A.F. Hepp, P.N. Kumta, High capacity, reversible silicon thin-film anodes for lithium-ion batteries, Electrochem. Solid State Lett. 6 (2003) A198–A201.
[27] J. Wang, Y.C.K. Chen-Wiegart, J. Wang, In situ three-dimensional synchrotron X-ray nano-tomography of the (De) lithiation processes in tin anodes, Angew. Chem. Int. Ed. 53 (2014) 4460–4464.
[28] M. Gu, Z. Wang, J.G. Connell, D.E. Perea, L.J. Lauhon, F. Gao, C. Wang, Electronic origin for the phase transition from amorphous Li$_x$Si to crystalline Li$_{15}$Si$_4$, ACS Nano 7 (2013) 6303–6309.
[29] M.R. Zamfir, H.T. Nguyen, E. Moyen, Y.H. Lee, D. Pribat, Silicon nanowires for Li-based battery anodes: a review, J. Mater. Chem. A 1 (2013) 9566–9586.
[30] M.K. Datta, P.N. Kumta, In situ electrochemical synthesis of lithiated silicon–carbon based composites anode materials for lithium ion batteries, J. Power Sources 194 (2009) 1043–1052.
[31] M. Gu, Y. He, J. Zheng, C. Wang, Nanoscale silicon as anode for Li-ion batteries: the fundamentals, promises, and challenges, Nano Energy 17 (2015) 366–383.
[32] Y. He, K. Xiang, W. Zhou, Y. Zhu, X. Chen, H. Chen, Folded-hand silicon/carbon three-dimensional networks as a binder-free advanced anode for high-performance lithium-ion batteries, Chem. Eng. J. 353 (2018) 666–678.

[33] J. Nzabahimana, S. Guo, X. Hu, Facile synthesis of Si@ void@ C nanocomposites from low-cost microsized Si as anode materials for lithium-ion batteries, Appl. Surf. Sci. 479 (2019) 287–295.

[34] H. Chen, S. He, X. Hou, S. Wang, F. Chen, H. Qin, Y. Xia, G. Zhou, Nano-Si/C microsphere with hollow double spherical interlayer and submicron porous structure to enhance performance for lithium-ion battery anode, Electrochim. Acta 312 (2019) 242–250.

[35] L. Hu, B. Luo, C. Wu, P. Hu, L. Wang, H. Zhang, Yolk-shell Si/C composites with multiple Si nanoparticles encapsulated into double carbon shells as lithium-ion battery anodes, J. Energy Chem. 32 (2019) 124–130.

[36] A. Casimir, H. Zhang, O. Ogoke, J.C. Amine, J. Lu, G. Wu, Silicon-based anodes for lithium-ion batteries: effectiveness of materials synthesis and electrode preparation, Nano Energy 27 (2016) 359–376.

[37] Y. Sun, K. Liu, Y. Zhu, Recent progress in synthesis and application of low-dimensional silicon based anode material for lithium ion battery, J. Nanomater. 2017 (2017).

[38] P. Li, G. Zhao, X. Zheng, X. Xu, C. Yao, W. Sun, S.X. Dou, Recent progress on silicon-based anode materials for practical lithium-ion battery applications, Energy Storage Mater 15 (2018) 422–446.

[39] I.N. Lund, J.H. Lee, H. Efstathiadis, P. Haldar, R.E. Geer, Influence of catalyst layer thickness on the growth of nickel silicide nanowires and its application for Li-ion batteries, J. Power Sources 246 (2014) 117–123.

[40] K. Mishra, X.-C. Liu, M. Geppert, J.J. Wu, J.-T. Li, L. Huang, S.-G. Sun, X.-D. Zhou, F.-S. Ke, Submicro-sized Si-Ge solid solutions with high capacity and long cyclability for lithium-ion batteries, J. Mater. Res. 33 (2018) 1553.

[41] Q. Hao, J. Hou, J. Ye, H. Yang, J. Du, C. Xu, Hierarchical macroporous Si/Sn composite: easy preparation and optimized performances towards lithium storage, Electrochim. Acta 306 (2019) 427–436.

[42] Z. Wei, R. Li, T. Huang, A. Yu, Fabrication and electrochemical properties of Si/TiO$_2$ nanowire array composites as lithium ion battery anodes, J. Power Sources 238 (2013) 165–172.

[43] H. Wang, P. Wu, H. Shi, W. Tang, Y. Tang, Y. Zhou, P. She, T. Lu, Hollow porous silicon oxide nanobelts for high-performance lithium storage, J. Power Sources 274 (2015) 951–956.

[44] S.-Z. Zeng, Y. Niu, J. Zou, X. Zeng, H. Zhu, J. Huang, L. Wang, L.B. Kong, P. Han, Green and scalable preparation of disproportionated SiO anode materials with cocoon-like buffer layer, J. Power Sources 466 (2020) 228234.

[45] C. Li, C. Liu, W. Wang, Z. Mutlu, J. Bell, K. Ahmed, R. Ye, M. Ozkan, C.S. Ozkan, Silicon derived from glass bottles as anode materials for lithium ion full cell batteries, Sci. Rep. 7 (2017) 1–11.

[46] T. Matsumoto, K. Kimura, H. Nishihara, T. Kasukabe, T. Kyotani, H. Kobayashi, Fabrication of Si nanopowder from Si swarf and application to high-capacity and low cost Li-ion batteries, J. Alloys Compd. 720 (2017) 529–540.

[47] M. Yuan, X. Guo, Y. Liu, H. Pang, Si-based materials derived from biomass: synthesis and applications in electrochemical energy storage, J. Mater. Chem. A 7 (2019) 22123–22147.

[48] W. Luo, X. Wang, C. Meyers, N. Wannenmacher, W. Sirisaksoontorn, M.M. Lerner, X. Ji, Efficient fabrication of nanoporous Si and Si/Ge enabled by a heat scavenger in magnesiothermic reactions, Sci. Rep. 3 (2013) 2222.

[49] X. Zuo, X. Wang, Y. Xia, S. Yin, Q. Ji, Z. Yang, M. Wang, X. Zheng, B. Qiu, Z. Liu, Silicon/carbon lithium-ion battery anode with 3D hierarchical macro-/mesoporous silicon network: self-templating synthesis via magnesiothermic reduction of silica/carbon composite, J. Power Sources 412 (2019) 93–104.

[50] X. Zhou, L. Wu, J. Yang, J. Tang, L. Xi, B. Wang, Synthesis of nano-sized silicon from natural halloysite clay and its high performance as anode for lithium-ion batteries, J. Power Sources 324 (2016) 33–40.
[51] Z. Bao, M.R. Weatherspoon, S. Shian, Y. Cai, P.D. Graham, S.M. Allan, G. Ahmad, M.B. Dickerson, B.C. Church, Z. Kang, Chemical reduction of three-dimensional silica microassemblies into microporous silicon replicas, Nature 446 (2007) 172–175.
[52] R. Gordon, D. Losic, M.A. Tiffany, S.S. Nagy, F.A. Sterrenburg, The glass menagerie: diatoms for novel applications in nanotechnology, Trends Biotechnol. 27 (2009) 116–127.
[53] G. Bucci, S.P. Nadimpalli, V.A. Sethuraman, A.F. Bower, P.R. Guduru, Measurement and modeling of the mechanical and electrochemical response of amorphous Si thin film electrodes during cyclic lithiation, J. Mech. Phys. Solids 62 (2014) 276–294.
[54] J. Yang, L. Zhang, T. Zhang, X. Wang, Y. Gao, Q. Fang, Self-healing strategy for Si nanoparticles towards practical application as anode materials for Li-ion batteries, Electrochem. Commun. 87 (2018) 22–26.
[55] R. Epur, M. Ramanathan, M.K. Datta, D.H. Hong, P.H. Jampani, B. Gattu, P.N. Kumta, Scribable multi-walled carbon nanotube-silicon nanocomposites: a viable lithium-ion battery system, Nanoscale 7 (2015) 3504–3510.
[56] J. Chang, X. Huang, G. Zhou, S. Cui, S. Mao, J. Chen, Three-dimensional carbon-coated Si/rGO nanostructures anchored by nickel foam with carbon nanotubes for Li-ion battery applications, Nano Energy 15 (2015) 679–687.
[57] X. Li, R.B. Wehrspohn, Nanometallurgical silicon for energy application, Joule 3 (2019) 1172–1175.
[58] I.S. Kim, P.N. Kumta, G.E. Blomgren, Si/TiN nanocomposites novel anode materials for Li-ion batteries, Electrochem. Solid St. 3 (2000) 493.
[59] J. Maranchi, A. Hepp, A. Evans, N. Nuhfer, P. Kumta, Interfacial properties of the a-Si/Cu: active–inactive thin-film anode system for lithium-ion batteries, J. Electrochem. Soc. 153 (2006) A1246.
[60] M. Ashuri, Q. He, L.L. Shaw, Silicon as a potential anode material for Li-ion batteries: where size, geometry and structure matter, Nanoscale 8 (2016) 74–103.
[61] H. Kim, M. Seo, M.H. Park, J. Cho, A critical size of silicon nano-anodes for lithium rechargeable batteries, Angew. Chem. Int. Ed. 49 (2010) 2146–2149.
[62] J.K. Lee, K.B. Smith, C.M. Hayner, H.H. Kung, Silicon nanoparticles–graphene paper composites for Li ion battery anodes, Chem. Commun. 46 (2010) 2025–2027.
[63] R.D. Cakan, M.-M. Titirici, M. Antonietti, G. Cui, J. Maier, Y.-S. Hu, Hydrothermal carbon spheres containing silicon nanoparticles: synthesis and lithium storage performance, Chem. Commun. (2008) 3759–3761.
[64] X.-Y. Zhou, J.-J. Tang, J. Yang, J. Xie, L.-l. Ma, Silicon@ carbon hollow core–shell heterostructures novel anode materials for lithium ion batteries, Electrochim. Acta 87 (2013) 663–668.
[65] B. Gattu, P.H. Jampani, M.K. Datta, R. Kuruba, P.N. Kumta, Water-soluble-template-derived nanoscale silicon nanoflake and nano-rod morphologies: stable architectures for lithium-ion battery anodes, Nano Res. 10 (2017) 4284–4297.
[66] C.K. Chan, R. Ruffo, S.S. Hong, R.A. Huggins, Y. Cui, Structural and electrochemical study of the reaction of lithium with silicon nanowires, J. Power Sources 189 (2009) 34–39.
[67] H.T. Nguyen, F. Yao, M.R. Zamfir, C. Biswas, K.P. So, Y.H. Lee, S.M. Kim, S.N. Cha, J.M. Kim, D. Pribat, Highly interconnected Si nanowires for improved stability Li-ion battery anodes, Adv. Energy Mater. 1 (2011) 1154–1161.

[68] X.H. Liu, L.Q. Zhang, L. Zhong, Y. Liu, H. Zheng, J.W. Wang, J.-H. Cho, S.A. Dayeh, S.T. Picraux, J.P. Sullivan, Ultrafast electrochemical lithiation of individual Si nanowire anodes, Nano Lett. 11 (2011) 2251–2258.

[69] K.-Q. Peng, X. Wang, L. Li, Y. Hu, S.-T. Lee, Silicon nanowires for advanced energy conversion and storage, Nano Today 8 (2013) 75–97.

[70] S.H. Nguyen, J.C. Lim, J.K. Lee, Electrochemical characteristics of bundle-type silicon nanorods as an anode material for lithium ion batteries, Electrochim. Acta 74 (2012) 53–58.

[71] Y. Zhou, X. Jiang, L. Chen, J. Yue, H. Xu, J. Yang, Y. Qian, Novel mesoporous silicon nanorod as an anode material for lithium ion batteries, Electrochim. Acta 127 (2014) 252–258.

[72] J.K. Yoo, J. Kim, Y.S. Jung, K. Kang, Scalable fabrication of silicon nanotubes and their application to energy storage, Adv. Mater. 24 (2012) 5452–5456.

[73] M.-H. Park, M.G. Kim, J. Joo, K. Kim, J. Kim, S. Ahn, Y. Cui, J. Cho, Silicon nanotube battery anodes, Nano Lett. 9 (2009) 3844–3847.

[74] R. Epur, P.H. Jampani, M.K. Datta, D.-H. Hong, B. Gattu, P.N. Kumta, A simple and scalable approach to hollow silicon nanotube (h-SiNT) anode architectures of superior electrochemical stability and reversible capacity, J. Mater. Chem. A 3 (2015) 11117–11129.

[75] B. Gattu, R. Epur, P.H. Jampani, R. Kuruba, M.K. Datta, P.N. Kumta, Silicon–carbon core–shell hollow nanotubular configuration high-performance lithium-ion anodes, J. Phys. Chem. C 121 (2017) 9662–9671.

[76] W.-S. Kim, Y. Hwa, J.-H. Shin, M. Yang, H.-J. Sohn, S.-H. Hong, Scalable synthesis of silicon nanosheets from sand as an anode for Li-ion batteries, Nanoscale 6 (2014) 4297–4302.

[77] J. Tang, Q. Yin, Q. Wang, Q. Li, H. Wang, Z. Xu, H. Yao, J. Yang, X. Zhou, J.-K. Kim, Two-dimensional porous silicon nanosheets as anode materials for high performance lithium-ion batteries, Nanoscale 11 (2019) 10984–10991.

[78] F.-H. Du, K.-X. Wang, J.-S. Chen, Strategies to succeed in improving the lithium-ion storage properties of silicon nanomaterials, J. Mater. Chem. A 4 (2016) 32–50.

[79] L. Liu, J. Lyu, T. Li, T. Zhao, Well-constructed silicon-based materials as high-performance lithium-ion battery anodes, Nanoscale 8 (2016) 701–722.

[80] J.Y. Howe, D.J. Burton, Y. Qi, H.M. Meyer III, M. Nazri, G.A. Nazri, A.C. Palmer, P.D. Lake, Improving microstructure of silicon/carbon nanofiber composites as a Li battery anode, J. Power Sources 221 (2013) 455–461.

[81] M. Li, X. Hou, Y. Sha, J. Wang, S. Hu, X. Liu, Z. Shao, Facile spray-drying/pyrolysis synthesis of core–shell structure graphite/silicon-porous carbon composite as a superior anode for Li-ion batteries, J. Power Sources 248 (2014) 721–728.

[82] L. Pan, H. Wang, D. Gao, S. Chen, L. Tan, L. Li, Facile synthesis of yolk–shell structured Si–C nanocomposites as anodes for lithium-ion batteries, Chem. Commun. 50 (2014) 5878–5880.

[83] H. Wang, P. Wu, H. Shi, F. Lou, Y. Tang, T. Zhou, Y. Zhou, T. Lu, Porous Si spheres encapsulated in carbon shells with enhanced anodic performance in lithium-ion batteries, Mater. Res. Bull. 55 (2014) 71–77.

[84] X. Bai, Y. Yu, H.H. Kung, B. Wang, J. Jiang, Si@ SiOx/graphene hydrogel composite anode for lithium-ion battery, J. Power Sources 306 (2016) 42–48.

[85] H. Wang, X. Li, M. Baker-Fales, P.B. Amama, 3D graphene-based anode materials for Li-ion batteries, Curr. Opin. Chem. Eng. 13 (2016) 124–132.

[86] D. Sui, Y. Xie, W. Zhao, H. Zhang, Y. Zhou, X. Qin, Y. Ma, Y. Yang, Y. Chen, A high-performance ternary Si composite anode material with crystal graphite core and amorphous carbon shell, J. Power Sources 384 (2018) 328–333.

[87] H. Akbulut, D. Nalci, A. Guler, S. Duman, M. Guler, Carbon-silicon composite anode electrodes modified with MWCNT for high energy battery applications, Appl. Surf. Sci. 446 (2018) 222–229.
[88] Z. Yi, N. Lin, Y. Zhao, W. Wang, Y. Qian, Y. Zhu, Y. Qian, A flexible micro/nanostructured Si microsphere cross-linked by highly-elastic carbon nanotubes toward enhanced lithium ion battery anodes, Energy Storage Mater. 17 (2019) 93–100.
[89] M. Li, Y. Yu, J. Li, B. Chen, X. Wu, Y. Tian, P. Chen, Nanosilica/carbon composite spheres as anodes in Li-ion batteries with excellent cycle stability, J. Mater. Chem. A 3 (2015) 1476–1482.
[90] H. Tao, L.-Z. Fan, W.-L. Song, M. Wu, X. He, X. Qu, Hollow core–shell structured Si/C nanocomposites as high-performance anode materials for lithium-ion batteries, Nanoscale 6 (2014) 3138–3142.
[91] Y. Zhang, Y. Jiang, Y. Li, B. Li, Z. Li, C. Niu, Preparation of nanographite sheets supported Si nanoparticles by in situ reduction of fumed SiO_2 with magnesium for lithium ion battery, J. Power Sources 281 (2015) 425–431.
[92] B.-C. Kim, H. Uono, T. Sato, T. Fuse, T. Ishihara, M. Senna, Li-ion battery anode properties of Si-carbon nanocomposites fabricated by high energy multiring-type mill, Solid State Ion. 172 (2004) 33–37.
[93] H. Wang, J. Xie, S. Zhang, G. Cao, X. Zhao, Scalable preparation of silicon@ graphite/carbon microspheres as high-performance lithium-ion battery anode materials, RSC Adv. 6 (2016) 69882–69888.
[94] R. Kuruba, M.K. Datta, K. Damodaran, P.H. Jampani, B. Gattu, P.P. Patel, P.M. Shanthi, S. Damle, P.N. Kumta, Guar gum: structural and electrochemical characterization of natural polymer based binder for silicon–carbon composite rechargeable Li-ion battery anodes, J. Power Sources 298 (2015) 331–340.
[95] M. Ko, S. Chae, S. Jeong, P. Oh, J. Cho, Elastic a-silicon nanoparticle backboned graphene hybrid as a self-compacting anode for high-rate lithium ion batteries, ACS Nano 8 (2014) 8591–8599.
[96] J. Luo, X. Zhao, J. Wu, H.D. Jang, H.H. Kung, J. Huang, Crumpled graphene-encapsulated Si nanoparticles for lithium ion battery anodes, J. Phys. Chem. Lett. 3 (2012) 1824–1829.
[97] X. Zhou, Y.X. Yin, L.J. Wan, Y.G. Guo, Self-assembled nanocomposite of silicon nanoparticles encapsulated in graphene through electrostatic attraction for lithium-ion batteries, Adv. Energy Mater. 2 (2012) 1086–1090.
[98] T. Mori, C.-J. Chen, T.-F. Hung, S.G. Mohamed, Y.-Q. Lin, H.-Z. Lin, J.C. Sung, S.-F. Hu, R.-S. Liu, High specific capacity retention of graphene/silicon nanosized sandwich structure fabricated by continuous electron beam evaporation as anode for lithium-ion batteries, Electrochim. Acta 165 (2015) 166–172.
[99] R. Yi, J. Zai, F. Dai, M.L. Gordin, D. Wang, Dual conductive network-enabled graphene/Si–C composite anode with high areal capacity for lithium-ion batteries, Nano Energy 6 (2014) 211–218.
[100] M. Zhou, X. Li, B. Wang, Y. Zhang, J. Ning, Z. Xiao, X. Zhang, Y. Chang, L. Zhi, High-performance silicon battery anodes enabled by engineering graphene assemblies, Nano Lett. 15 (2015) 6222–6228.
[101] D. Shao, D. Tang, Y. Mai, L. Zhang, Nanostructured silicon/porous carbon spherical composite as a high capacity anode for Li-ion batteries, J. Mater. Chem. A 1 (2013) 15068–15075.
[102] H.-C. Tao, X.-L. Yang, L.-L. Zhang, S.-B. Ni, Double-walled core-shell structured Si@ SiO2@ C nanocomposite as anode for lithium-ion batteries, Ionics 20 (2014) 1547–1552.

[103] S. Chen, M.L. Gordin, R. Yi, G. Howlett, H. Sohn, D. Wang, Silicon core–hollow carbon shell nanocomposites with tunable buffer voids for high capacity anodes of lithium-ion batteries, Phys. Chem. Chem. Phys. 14 (2012) 12741–12745.

[104] J. Yang, Y.-X. Wang, S.-L. Chou, R. Zhang, Y. Xu, J. Fan, W.-x. Zhang, H.K. Liu, D. Zhao, S.X. Dou, Yolk-shell silicon-mesoporous carbon anode with compact solid electrolyte interphase film for superior lithium-ion batteries, Nano Energy 18 (2015) 133–142.

[105] J. Xie, L. Tong, L. Su, Y. Xu, L. Wang, Y. Wang, Core-shell yolk-shell Si@ C@ void@ C nanohybrids as advanced lithium ion battery anodes with good electronic conductivity and corrosion resistance, J. Power Sources 342 (2017) 529–536.

[106] C. Pang, H. Song, N. Li, C. Wang, A strategy for suitable mass production of a hollow Si@ C nanostructured anode for lithium ion batteries, RSC Adv. 5 (2015) 6782–6789.

[107] Q. He, C. Xu, J. Luo, W. Wu, J. Shi, A novel mesoporous carbon@ silicon–silica nanostructure for high-performance Li-ion battery anodes, Chem. Commun. 50 (2014) 13944–13947.

[108] Y. Chen, N. Du, H. Zhang, D. Yang, Facile synthesis of uniform MWCNT@ Si nanocomposites as high-performance anode materials for lithium-ion batteries, J. Alloys Compd. 622 (2015) 966–972.

[109] W. Li, Z. Li, W. Kang, Y. Tang, Z. Zhang, X. Yang, H. Xue, C.-S. Lee, Hollow nanospheres of loosely packed Si/SiO x nanoparticles encapsulated in carbon shells with enhanced performance as lithium ion battery anodes, J. Mater. Chem. A 2 (2014) 12289–12295.

[110] F. Zhang, X. Yang, Y. Xie, N. Yi, Y. Huang, Y. Chen, Pyrolytic carbon-coated Si nanoparticles on elastic graphene framework as anode materials for high-performance lithium-ion batteries, Carbon 82 (2015) 161–167.

[111] Z. Sun, S. Tao, X. Song, P. Zhang, L. Gao, A silicon/double-shelled carbon yolk-like nanostructure as high-performance anode materials for lithium-ion battery, J. Electrochem. Soc. 162 (2015) A1530.

[112] A. Guerfi, P. Charest, M. Dontigny, J. Trottier, M. Lagacé, P. Hovington, A. Vijh, K. Zaghib, SiOx–graphite as negative for high energy Li-ion batteries, J. Power Sources 196 (2011) 5667–5673.

[113] W. Zhou, S. Upreti, M.S. Whittingham, Electrochemical performance of Al–Si–graphite composite as anode for lithium–ion batteries, Electrochem. Commun. 13 (2011) 158–161.

[114] S.-O. Kim, A. Manthiram, A facile, low-cost synthesis of high-performance silicon-based composite anodes with high tap density for lithium-ion batteries, J. Mater. Chem. A 3 (2015) 2399–2406.

[115] J. Li, J. Wang, J. Yang, X. Ma, S. Lu, Scalable synthesis of a novel structured graphite/silicon/pyrolyzed-carbon composite as anode material for high-performance lithium-ion batteries, J. Alloys Compd. 688 (2016) 1072–1079.

[116] L. Qian, J.-L. Lan, M. Xue, Y. Yu, X. Yang, Two-step ball-milling synthesis of a Si/SiO x/C composite electrode for lithium ion batteries with excellent long-term cycling stability, RSC Adv. 7 (2017) 36697–36704.

[117] S. Huang, L.-Z. Cheong, D. Wang, C. Shen, Nanostructured phosphorus doped silicon/graphite composite as anode for high-performance lithium-ion batteries, ACS Appl. Mater. Interfaces 9 (2017) 23672–23678.

[118] J. Lai, H. Guo, Z. Wang, X. Li, X. Zhang, F. Wu, P. Yue, Preparation and characterization of flake graphite/silicon/carbon spherical composite as anode materials for lithium-ion batteries, J. Alloys Compd. 530 (2012) 30–35.

[119] M. Su, Z. Wang, H. Guo, X. Li, S. Huang, W. Xiao, L. Gan, Enhancement of the cyclability of a Si/graphite@ graphene composite as anode for lithium-ion batteries, Electrochim. Acta 116 (2014) 230–236.

[120] A. Wang, F. Liu, Z. Wang, X. Liu, Self-assembly of silicon/carbon hybrids and natural graphite as anode materials for lithium-ion batteries, RSC Adv. 6 (2016) 104995–105002.
[121] R. Zhou, H. Guo, Y. Yang, Z. Wang, X. Li, Y. Zhou, N-doped carbon layer derived from polydopamine to improve the electrochemical performance of spray-dried Si/graphite composite anode material for lithium ion batteries, J. Alloys Compd. 689 (2016) 130–137.
[122] Z. Wang, Z. Mao, L. Lai, M. Okubo, Y. Song, Y. Zhou, X. Liu, W. Huang, Sub-micron silicon/pyrolyzed carbon@ natural graphite self-assembly composite anode material for lithium-ion batteries, Chem. Eng. J. 313 (2017) 187–196.
[123] H. Chen, X. Hou, F. Chen, S. Wang, B. Wu, Q. Ru, H. Qin, Y. Xia, Milled flake graphite/plasma nano-silicon@ carbon composite with void sandwich structure for high performance as lithium ion battery anode at high temperature, Carbon 130 (2018) 433–440.
[124] H. Wolf, Z. Pajkic, T. Gerdes, M. Willert-Porada, Carbon–fiber–silicon-nanocomposites for lithium-ion battery anodes by microwave plasma chemical vapor deposition, J. Power Sources 190 (2009) 157–161.
[125] X. Zhu, H. Chen, Y. Wang, L. Xia, Q. Tan, H. Li, Z. Zhong, F. Su, X. Zhao, Growth of silicon/carbon microrods on graphite microspheres as improved anodes for lithium-ion batteries, J. Mater. Chem. A 1 (2013) 4483–4489.
[126] M. Ko, S. Chae, J. Ma, N. Kim, H.-W. Lee, Y. Cui, J. Cho, Scalable synthesis of silicon-nanolayer-embedded graphite for high-energy lithium-ion batteries, Nat. Energy 1 (2016) 1–8.
[127] N. Kim, S. Chae, J. Ma, M. Ko, J. Cho, Fast-charging high-energy lithium-ion batteries via implantation of amorphous silicon nanolayer in edge-plane activated graphite anodes, Nat. Commun. 8 (2017) 1–10.
[128] J. Yu, H. Zhan, Y. Wang, Z. Zhang, H. Chen, H. Li, Z. Zhong, F. Su, Graphite microspheres decorated with Si particles derived from waste solid of organosilane industry as high capacity anodes for Li-ion batteries, J. Power Sources 228 (2013) 112–119.
[129] Y. Ren, M. Li, Si-SiOx-cristobalite/graphite composite as anode for li-ion batteries, Electrochim. Acta 142 (2014) 11–17.
[130] C. Ma, C. Ma, J. Wang, H. Wang, J. Shi, Y. Song, Q. Guo, L. Liu, Exfoliated graphite as a flexible and conductive support for Si-based Li-ion battery anodes, Carbon 72 (2014) 38–46.
[131] C. Gao, H. Zhao, P. Lv, T. Zhang, Q. Xia, J. Wang, Engineered Si sandwich electrode: Si nanoparticles/graphite sheet hybrid on Ni foam for next-generation high-performance lithium-ion batteries, ACS Appl. Mater. Interfaces 7 (2015) 1693–1698.
[132] S.Y. Kim, J. Lee, B.-H. Kim, Y.-J. Kim, K.S. Yang, M.-S. Park, Facile synthesis of carbon-coated silicon/graphite spherical composites for high-performance lithium-ion batteries, ACS Appl. Mater. Interfaces 8 (2016) 12109–12117.
[133] S. Jeong, X. Li, J. Zheng, P. Yan, R. Cao, H.J. Jung, C. Wang, J. Liu, J.-G. Zhang, Hard carbon coated nano-Si/graphite composite as a high performance anode for Li-ion batteries, J. Power Sources 329 (2016) 323–329.
[134] H. Yu, X. Liu, Y. Chen, H. Liu, Carbon-coated Si/graphite composites with combined electrochemical properties for high-energy-density lithium-ion batteries, Ionics 22 (2016) 1847–1853.
[135] J. Wu, W. Tu, Y. Zhang, B. Guo, S. Li, Y. Zhang, Y. Wang, M. Pan, Poly-dopamine coated graphite oxide/silicon composite as anode of lithium ion batteries, Powder Technol. 311 (2017) 200–205.
[136] F.-S. Li, Y.-S. Wu, J. Chou, N.-L. Wu, A dimensionally stable and fast-discharging graphite–silicon composite Li-ion battery anode enabled by electrostatically self-assembled multifunctional polymer-blend coating, Chem. Commun. 51 (2015) 8429–8431.

[137] N. Lin, T. Xu, T. Li, Y. Han, Y. Qian, Controllable self-assembly of micro-nanostructured Si-embedded graphite/graphene composite anode for high-performance Li-ion batteries, ACS Appl. Mater. Interfaces 9 (2017) 39318–39325.

[138] X. Li, P. Yan, X. Xiao, J.H. Woo, C. Wang, J. Liu, J.-G. Zhang, Design of porous Si/C–graphite electrodes with long cycle stability and controlled swelling, Energ. Environ. Sci. 10 (2017) 1427–1434.

[139] X. Li, D. Yang, X. Hou, J. Shi, Y. Peng, H. Yang, Scalable preparation of mesoporous silicon@ C/graphite hybrid as stable anodes for lithium-ion batteries, J. Alloys Compd. 728 (2017) 1–9.

[140] Q. Hao, D. Zhao, H. Duan, Q. Zhou, C. Xu, Si/Ag composite with bimodal micro-nano porous structure as a high-performance anode for Li-ion batteries, Nanoscale 7 (2015) 5320–5327.

[141] Y. Yang, D. Chen, B. Liu, J. Zhao, Binder-free Si nanoparticle electrode with 3D porous structure prepared by electrophoretic deposition for lithium-ion batteries, ACS Appl. Mater. Interfaces 7 (2015) 7497–7504.

[142] W. Ren, Y. Wang, Z. Zhang, Q. Tan, Z. Zhong, F. Su, Carbon-coated porous silicon composites as high performance Li-ion battery anode materials: can the production process be cheaper and greener? J. Mater. Chem. A 4 (2016) 552–560.

[143] X. Wang, L. Sun, X. Hu, R.A. Susantyoko, Q. Zhang, Ni–Si nanosheet network as high performance anode for Li ion batteries, J. Power Sources 280 (2015) 393–396.

[144] K. Stokes, H. Geaney, M. Sheehan, D. Borsa, K.M. Ryan, Copper silicide nanowires as hosts for amorphous Si deposition as a route to produce high capacity lithium-ion battery anodes, Nano Lett. 19 (2019) 8829–8835.

[145] I. Kang, J. Jang, K.-W. Yi, Y.W. Cho, Porous nanocomposite anodes of silicon/iron silicide/3D carbon network for lithium-ion batteries, J. Alloys Compd. 770 (2019) 369–376.

[146] B. Han, C. Liao, F. Dogan, S.E. Trask, S.H. Lapidus, J.T. Vaughey, B. Key, Using mixed salt electrolytes to stabilize silicon anodes for lithium-ion batteries via in situ formation of Li–M–Si ternaries (M = Mg, Zn, Al, Ca), ACS Appl. Mater. Interfaces 11 (2019) 29780–29790.

[147] B. Ding, Z. Cai, Z. Ahsan, Y. Ma, S. Zhang, G. Song, C. Yuan, W. Yang, C. Wen, A review of metal silicides for lithium-ion battery anode application, Acta. Metall. Sin. (2020) 1–18.

[148] Y. Kwon, H. Kim, S.-G. Doo, J. Cho, $Sn_{0.9}Si_{0.1}$/carbon core–shell nanoparticles for high-density lithium storage materials, Chem. Mater. 19 (2007) 982–986.

[149] X. Xiao, J.S. Wang, P. Liu, A.K. Sachdev, M.W. Verbrugge, D. Haddad, M.P. Balogh, Phase-separated silicon–tin nanocomposites for high capacity negative electrodes in lithium ion batteries, J. Power Sources 214 (2012) 258–265.

[150] A. Kohandehghan, K. Cui, M. Kupsta, E. Memarzadeh, P. Kalisvaart, D. Mitlin, Nanometer-scale Sn coatings improve the performance of silicon nanowire LIB anodes, J. Mater. Chem. A 2 (2014) 11261–11279.

[151] J. Wu, Z. Zhu, H. Zhang, H. Fu, H. Li, A. Wang, H. Zhang, A novel Si/Sn composite with entangled ribbon structure as anode materials for lithium ion battery, Sci. Rep. 6 (2016) 29356.

[152] K. Wang, Y. Huang, D. Wang, Y. Zhao, M. Wang, X. Chen, H. Wu, Controlled synthesis of hollow Si–Ni–Sn nanoarchitectured electrode for advanced lithium-ion batteries, RSC Adv. 6 (2016) 23260–23264.

[153] D. Yang, J. Shi, J. Shi, H. Yang, Simple synthesis of Si/Sn@ CG anodes with enhanced electrochemical properties for Li-ion batteries, Electrochim. Acta 259 (2018) 1081–1088.

[154] S. Chae, M. Ko, S. Park, N. Kim, J. Ma, J. Cho, Micron-sized Fe–Cu–Si ternary composite anodes for high energy Li-ion batteries, Energ. Environ. Sci. 9 (2016) 1251–1257.

[155] Y.M. Kim, J. Ahn, S.-H. Yu, D.Y. Chung, K.J. Lee, J.-K. Lee, Y.-E. Sung, Titanium silicide coated porous silicon nanospheres as anode materials for lithium ion batteries, Electrochim. Acta 151 (2015) 256–262.

[156] H. Raj, S. Singh, A. Sil, TiO_2 shielded Si nano-composite anode for high energy Li-ion batteries: the morphological and structural study of electrodes after charge-discharge process, Electrochim. Acta 326 (2019) 134981.

[157] A. Belgibayeva, I. Taniguchi, Synthesis and characterization of SiO_2/C composite nanofibers as free-standing anode materials for Li-ion batteries, Electrochim. Acta 328 (2019) 135101.

[158] C. Wang, J. Ren, H. Chen, Y. Zhang, K.K. Ostrikov, W. Zhang, Y. Li, Synthesis of high-quality mesoporous silicon particles for enhanced lithium storage performance, Mater. Chem. Phys. 173 (2016) 89–94.

[159] K. Wang, B. Xue, Y. Tan, J. Sun, Q. Li, S. Shi, P. Li, Recycling of micron-sized Si powder waste from diamond wire cutting and its application in Li-ion battery anodes, J. Clean. Prod. 239 (2019) 117997.

[160] R. Fang, W. Xiao, C. Miao, P. Mei, Y. Zhang, X. Yan, Y. Jiang, Fabrication of Si–SiO_2@Fe/NC composite from industrial waste AlSiFe powders as high stability anodes for lithium ion batteries, Electrochim. Acta 324 (2019) 134860.

[161] J. Wang, P. Nie, B. Ding, S. Dong, X. Hao, H. Dou, X. Zhang, Biomass derived carbon for energy storage devices, J. Mater. Chem. A 5 (2017) 2411–2428.

[162] R.R. Gaddam, D. Yang, R. Narayan, K. Raju, N.A. Kumar, X. Zhao, Biomass derived carbon nanoparticle as anodes for high performance sodium and lithium ion batteries, Nano Energy 26 (2016) 346–352.

[163] W. Long, B. Fang, A. Ignaszak, Z. Wu, Y.-J. Wang, D. Wilkinson, Biomass-derived nanostructured carbons and their composites as anode materials for lithium ion batteries, Chem. Soc. Rev. 46 (2017) 7176–7190.

[164] W. Tang, Y. Zhang, Y. Zhong, T. Shen, X. Wang, X. Xia, J. Tu, Natural biomass-derived carbons for electrochemical energy storage, Mater. Res. Bull. 88 (2017) 234–241.

[165] N. Liu, K. Huo, M.T. McDowell, J. Zhao, Y. Cui, Rice husks as a sustainable source of nanostructured silicon for high performance Li-ion battery anodes, Sci. Rep. 3 (2013) 1–7.

[166] L. Wang, B. Gao, C. Peng, X. Peng, J. Fu, P.K. Chu, K. Huo, Bamboo leaf derived ultrafine Si nanoparticles and Si/C nanocomposites for high-performance Li-ion battery anodes, Nanoscale 7 (2015) 13840–13847.

[167] I.J. Fernandes, D. Calheiro, A.G. Kieling, C.A. Moraes, T.L. Rocha, F.A. Brehm, R.C. Modolo, Characterization of rice husk ash produced using different biomass combustion techniques for energy, Fuel 165 (2016) 351–359.

[168] Y.-C. Zhang, Y. You, S. Xin, Y.-X. Yin, J. Zhang, P. Wang, X.-S. Zheng, F.-F. Cao, Y.-G. Guo, Rice husk-derived hierarchical silicon/nitrogen-doped carbon/carbon nanotube spheres as low-cost and high-capacity anodes for lithium-ion batteries, Nano Energy 25 (2016) 120–127.

[169] J. Ryu, D. Hong, S. Choi, S. Park, Synthesis of ultrathin Si nanosheets from natural clays for lithium-ion battery anodes, ACS Nano 10 (2016) 2843–2851.

[170] S. Gao, D. Yang, Y. Pan, L. Geng, S. Li, X. Li, P.-F. Cao, H. Yang, From natural material to high-performance silicon-based anode: towards cost-efficient silicon-based electrodes in high-performance Li-ion batteries, Electrochim. Acta 327 (December 2019) 135058.

[171] X. Liu, Y. Gao, R. Jin, H. Luo, P. Peng, Y. Liu, Scalable synthesis of Si nanostructures by low-temperature magnesiothermic reduction of silica for application in lithium ion batteries, Nano Energy 4 (2014) 31–38.

[172] H. Zhong, H. Zhan, Y.-H. Zhou, Synthesis of nanosized mesoporous silicon by magnesium-thermal method used as anode material for lithium ion battery, J. Power Sources 262 (2014) 10–14.

[173] T. Wada, J. Yamada, H. Kato, Preparation of three-dimensional nanoporous Si using dealloying by metallic melt and application as a lithium-ion rechargeable battery negative electrode, J. Power Sources 306 (2016) 8–16.

[174] J. Liang, X. Li, Q. Cheng, Z. Hou, L. Fan, Y. Zhu, Y. Qian, High yield fabrication of hollow vesica-like silicon based on the Kirkendall effect and its application to energy storage, Nanoscale 7 (2015) 3440–3444.

[175] W. Wang, Z. Favors, R. Ionescu, R. Ye, H.H. Bay, M. Ozkan, C.S. Ozkan, Monodisperse porous silicon spheres as anode materials for lithium ion batteries, Sci. Rep. 5 (2015) 8781.

[176] N. Liu, Z. Lu, J. Zhao, M.T. McDowell, H.-W. Lee, W. Zhao, Y. Cui, A pomegranate-inspired nanoscale design for large-volume-change lithium battery anodes, Nat. Nanotechnol. 9 (2014) 187–192.

[177] B. Campbell, R. Ionescu, M. Tolchin, K. Ahmed, Z. Favors, K.N. Bozhilov, C.S. Ozkan, M. Ozkan, Carbon-coated, diatomite-derived nanosilicon as a high rate capable Li-ion battery anode, Sci. Rep. 6 (2016) 33050.

[178] Y. Zheng, H. Fang, F. Wang, H. Huang, J. Yang, G. Zuo, Fabrication and characterization of mesoporous Si/SiC derived from diatomite via magnesiothermic reduction, J. Solid State Chem. 277 (2019) 654–657.

[179] S. Kim, C. Hwang, S.Y. Park, S.-J. Ko, H. Park, W.C. Choi, J.B. Kim, D.S. Kim, S. Park, J.Y. Kim, High-yield synthesis of single-crystal silicon nanoparticles as anode materials of lithium ion batteries via photosensitizer-assisted laser pyrolysis, J. Mater. Chem. A 2 (2014) 18070–18075.

[180] S.K. Kumar, S. Ghosh, S.K. Malladi, J. Nanda, S.K. Martha, Nanostructured silicon–carbon 3D electrode architectures for high-performance lithium-ion batteries, ACS Omega 3 (2018) 9598–9606.

Part II

Mechanical properties

Chapter 3

Computational study on the effects of mechanical constraint on the performance of silicon nanosheets as anode materials for lithium-ion batteries

Qifang Yin[a,b] and Haimin Yao[a]

[a]*Department of Mechanical Engineering, The Hong Kong Polytechnic University, Kowloon, Hong Kong SAR, PR China,* [b]*Department of Mechanical Engineering, School of Civil Engineering, Wuhan University, Wuhan, PR China*

3.1 Introduction

Recent years have been witnessing the birth, growth, and prosperity of lithium-ion batteries (LIBs). Research on LIBs began in the 1960s, and the commercialization of LIBs was realized in the 1990s. Currently, LIBs are widely used to power a wide range of consumer electronics, such as smartphones and electrical vehicles. With the rising population and development of industries, traditional LIBs, which use carbon as the anode material and metal oxide as the cathode material, cannot meet the ever-increasing performance requirements any longer. The LIBs with high energy density, high charging rate, and high safety are in urgent need. Among various candidates of anode materials, silicon (Si) stands out and is deemed as one of the most promising anode materials for the next-generation LIBs due to its extraordinarily theoretical high capacity ($\sim$4200 mA h g^{-1}), whereas the huge volume change during lithiation and delithiation processes and the consequent rapid degradation of electrochemical performance impede its wider application. To make the best use of the high capacity of Si and meanwhile suppress the side effects of large volume change, a lot of efforts have been devoted to the revelation of the atomistic details of the lithiation and delithiation processes of Si [1–8], the examination of the mechanical properties of lithiated Si [9–13], and the optimization of the performance of the Si-based LIBs [14–17]. The Si-based materials are not freestanding in the LIB. Instead, they are either encompassed or attached to other peripheral materials, which impose mechanical constraints in response to the lithiation-induced volume change. This chapter is specifically devoted to the investigation of the effects of mechanical constraint on the electrochemical performances of Si nanosheet anode using molecular dynamics (MD) simulation approach. The rest of this chapter is structured as follows. Section 3.2 introduces the MD techniques to be applied to simulate the dynamic lithiation and delithiation of Si anode. Section 3.3 describes the application of such techniques to the study on the effects of mechanical constraint on the electrochemical performances of Si nanosheet anode. The whole chapter is finally concluded by providing the guidelines for optimizing the electrochemical performances of Si by properly controlling the mechanical constraint.

3.2 Realization of lithiation and delithiation of silicon (Si) in molecular dynamics (MD) simulation

3.2.1 Introduction to MD simulation and ReaxFF potential

Molecular dynamics simulation is a computational tool to predict the positions and velocities of atoms in a substance system in the framework of Newton's laws of motion. The parameters characterizing the physical states of the system, such as pressure and temperature, can be deduced from the calculated positions and velocities of the atoms according to the laws of thermodynamics. MD simulation is a powerful approach allowing us to visualize what is happening at the atomistic length scale that cannot be observed directly from experiments at present. The application of MD simulation also enables us to reveal the essential mechanisms underlying the observed phenomena and predict the properties of materials. This section is specifically devoted to the techniques for simulating Si anode in LIBs by MD simulations. For more fundamentals of MD simulations, readers are referred to the related Refs. [18, 19].

The Li—Si interactions are crucial for the performances of Si anode in LIBs. To simulate the Li—Si interactions with MD simulation, the following assumptions and approximations have to be adopted. First, the electrochemical interactions between Li and Si are simplified as the collisions between rigid spheres based on Newton's laws of motion. Therefore the quantum effects during the interaction, if available, are unable to be captured. Second, as MD simulation does not include electrons in the simulating system, the transferring of charges in the system has to be realized alternatively by adjusting the partial charges on each atom in such a way that the total electric potential is minimized. A potential drawback of this treatment is that it may lead to a noninteger number of charges on an atom. Third, due to the limitation of the calculation capability, a typical simulation system has dimensions of a few nanometers. To simulate a system with a larger scale, the technique of representative volume element (RVE) in combination with the periodic boundary conditions (PBCs) is applied.

In MD simulation, the interatomic interactions in a simulated system are defined by potential functions. To include the effects of charges on the Li—Si interactions, the potential function of Reactive Force Field (ReaxFF) [1] is applied to describe the interactions between Li and Si atoms. For details of the ReaxFF, one can refer to reference [20]. Unless otherwise stated, all simulations reported below are performed with Large-scale Atomic/Molecular Massively Parallel Simulator (LAMMPS) [18], with with pre- and post-processing arrangement of atoms realized by the Open Visualization Tool (OVITO) [21].

3.2.2 MD simulation of the lithiation process of Si

Currently, there are two prevalent methods in MD to simulate the dynamic lithiation process of Si. The first method was initially proposed by Chevrier and Dahn [22] and therefore is hereinafter called "Chevrier-Dahn Lithiation." Briefly speaking, this method achieves the lithiation by repeating the following sequence until a desired Li—Si ratio is reached: (1) insert one Li atom into the largest spherical void in Si or lithiated Si, and (2) relax the system for equilibrium. The process is schematically illustrated in Fig. 3.1. Although it is computationally economical, this method fails to include the diffusion of Li atoms in the Si atoms.

The second method resembles the in situ lithiation experiment in TEM (see Fig. 3.2A and B) [8], in which lithiation is achieved by bringing Li_2O/Li into contact with the Si directly. It should be pointed out that the lithiation process simulated here is different from what happens in the real batteries as the roles of liquid electrolyte, binder and conductive additive are too complicated to be considered in MD simulation. As our focus is on the interaction between the Li and Si atoms, this method is still acceptable. In comparison to the former "Chevrier-Dahn Lithiation," this method enables us to visualize the atomistic details of the diffusion of Li atoms in Si, although it is computationally more cost intensive.

However, this method for simulating lithiation of Si only worked initially at a temperature as high as 600–1200 K [8, 15, 17, 23], which greatly limits the applicability of the related simulation results. To overcome this problem, we made the following modification to the parameters in LAMMPS. First, the implementation of the charge equilibration (QEq) method [24] is switched

FIG. 3.1 Schematics showing the "Chevrier-Dahn Lithiation" scheme.

98 PART | II Mechanical properties

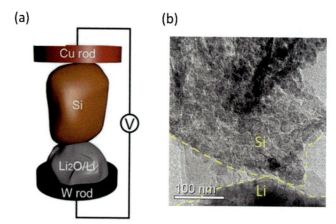

FIG. 3.2 The experimental setup for in situ lithiation. (A) A schematic illustration of the setup. (B) TEM image of in situ lithiation of Si. *((Part figure (A) is adapted from C.F. Shen, M.Y. Ge, L.L. Luo, X. Fang, Y.H. Liu, A.Y. Zhang, J.P. Rong, C.M. Wang, C.W. Zhou, In situ and ex situ TEM study of lithiation behaviors of porous silicon nanostructures. Scient. Rep. 6 (2016) 31334; (B) is an unpublished image; for details, readers are referred to J.J. Tang, Q.F. Yin, Q. Wang, Q.Q. Li, H.T. Wang, Z.L. Xu, H. Yao, J. Yang, X.Y. Zhou, J.K. Kim, L.M. Zhou, Two-dimensional porous silicon nanosheets as anode materials for high performance lithium-ion batteries, Nanoscale 11 (2019) 10984–10991.))*

on to calculate the effects of charges by including either the "fix qeq/shielded" [25] or the "fix qeq/reax" [26] command in the input file. The utilization of either command would result in an immediate minimization of the electrostatic energy of the system by involving a reassignment of charges to atoms periodically based on the electronegativity of elements and the immediate atomic configuration during the simulation. Second, the upper cut-off of the Taper correction (for "fix qeq/reax" command, see reference [26]) or the global cut-off for charge-charge interactions (for "fix qeq/shielded" command, see Ref. [25]) is set as 9.0 Å. These modifications allow us to simulate the lithiation of Si at a more realistic temperature of 300 K.

Based on the method and simulation setup introduced above, the dependence of lithiation on the crystalline direction of Si was studied. Lithiation of the amorphous and crystalline Si was carried out with the models shown in Fig. 3.3. For the crystalline Si, lattices with different orientations were considered to study the directional dependence. In each model, the leftmost boundary was fixed to eliminate rigid displacement and the rightmost boundary is set free. PBCs were applied to the remaining boundaries. The temperature was constantly controlled at 300 K with the Berendsen thermostat [27], which rescaled atom velocities at each time step. The time step was set as 2 fs. Before lithiation, all models were relaxed to the equilibrium state for at least 250 ps (in MD time, the same hereinafter).

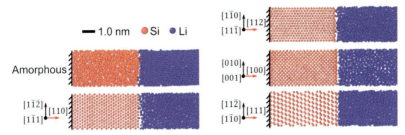

FIG. 3.3 Atomic models for simulating lithiation of amorphous and crystalline Si.

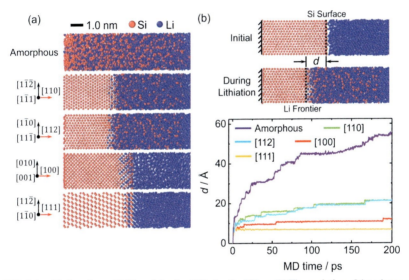

FIG. 3.4 (A) Snapshots of MD models after lithiation for 200 ps. (B) The evolution of the advancing distance of the Li frontier, d, as a function of the time.

The configurational snapshots of the systems after 200 ps are displayed in Fig. 3.4A. It can be seen that in all simulated cases a sharp phase boundary exists between the lithiated and the unlithiated Si, which is consistent with other simulations or experimental observation [8, 28, 29]. The lithiation process in the crystalline Si starts with the migration of Li atoms to the interstitial positions of the crystalline Si, followed by the cleavage of the Si—Si bonds and the formation of amorphous Li_xSi phase in the Li-rich region [6, 7]. Fig. 3.4B plots the traveled distance of the advancing frontier of Li atoms as a function of the MD time. The stair-like curves manifest the intermittent advancing motion of the frontier during lithiation. Among all the simulated cases, the average advancing velocities of the frontier along the directions of [110] and [112] are comparable and higher than those along the directions [100] and [111]. This is consistent with the experiments [28] and can be attributed to the fact that the densities of Si atoms in {111} or {100} planes are higher than those in {110} and {112} planes. Moreover, the lithiation rate in the crystalline Si, irrespective of the lithiation directions, is much smaller than that in the amorphous Si. This is also in agreement with the experimental observation reported in the literature [30].

3.2.3 MD simulation of the delithiation process of Si

There are three methods to simulate the delithiation process of Li$_x$Si. The first one was also proposed by Chevrier and Dahn [22], hereafter it is called "Chevrier-Dahn Delithiation." As a reverse process of the Chevrier-Dahn Lithiation, it randomly removes Li atoms from the lithiated Si and then has the system relaxed (see Fig. 3.5). Certainly, this method fails to capture the diffusion process of Li.

The second method for simulating the delithiation process is to harness an auxiliary Al$_2$O$_3$ phase to extract the Li atoms out of Li$_x$Si [16] based on the expected reaction

$$Li_xSi + Al_2O_3 \rightarrow Li_xAl_2O_3 + Si$$

However, whether this chemical reaction can happen in reality remains questionable. Since Al$_2$O$_3$ has been used as the coating material for Si anode [31–33], this reaction may not happen easily in reality [32]. Although it was used to simulate the dynamic diffusion of Li atoms in the Si phase, it should be used with caution.

The third method is akin to the "Chevrier-Dahn Delithiation," but the removal of Li atoms takes place near the surface of Li$_x$Si rather than at a random spot inside as in the "Chevrier-Dahn Delithiation." This seems more realistic compared to the other methods and therefore is adopted in our simulation. Fig. 3.6A illustrates how this algorithm applies to an Li$_x$Si nanosheet with a thickness of L. The left boundary at $z=0$ is fixed, and no atom can pass through it. The shaded area near the right boundary $z=L$, called drain region, indicates the region where Li atoms are removed. The thickness of the drain region should be much smaller than L. The delithiation rate is defined as the number of Li atoms removed from the system per unit MD time, which corresponds to the current intensity per unit mass of Si in the anode $i = Ne_0/m_{Si}t_0$, where

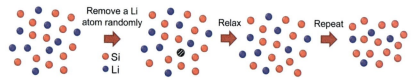

FIG. 3.5 Schematics showing the "Chevrier-Dahn Delithiation" scheme.

Computational study on the effects of mechanical constraint **Chapter | 3** **101**

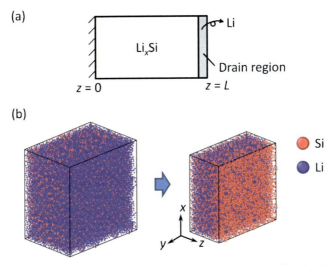

FIG. 3.6 Simulation of the delithiation process of Li$_x$Si. (A) Illustration of the algorithm to simulate delithiation. (B) Snapshots of an Li$_{15}$Si$_4$ nanosheet consisting of 11,347 Si atoms and 42,552 Li atoms before and after delithiation at a constant rate of 2 fs^{-1}. The thickness of the nanosheet is around 8 nm and the thickness of the drain region for removal of Li atoms is 0.3 nm.

N stands for the number of Li atoms removed within MD time t_0, e_0 the charge of an electron and m_{Si} the mass of Si in the anode.

Based on the above algorithm, the delithiation of an amorphous Li$_{15}$Si$_4$ nanosheet is simulated with the model shown in Fig. 3.6B to reveal the spatial distribution of the Li atoms in the anode after a delithiation period at a constant delithiation rate. It should be pointed out that the constant delithiation rate can be maintained only for a limited period as the Li concentration near the surface will decrease as the delithiation proceeds. There is a critical moment after which the Li concentration is not enough to maintain the given delithiation rate. If this scenario happens, the delithiation process is halted. PBCs were used along x and y directions to represent an infinitely large sheet. Delithiation would happen along the positive z-direction, that is, the surface normal to $+z$ direction would be the active surface from which Li atoms are removed from the system at a given rate. Two delithiation rates, 2 fs^{-1} and 20 fs^{-1}, were considered, corresponding to current intensities of 1.5×10^{13} A g^{-1} and 1.5×10^{14} A g^{-1}, respectively. The use of such an extremely high delithiation rate was to complete the simulations in an acceptable time [16]. The simulation was performed at a temperature of 300 K and zero external pressure was applied along x and y directions. The right panel in Fig. 3.6B depicts the atomic configuration when the delithiation with the lower rate stopped, in which Li atoms are almost depleted near the surface of the anode. Fig. 3.7 compares the Li atomic concentrations in the anode as a function of relative position (the position normalized

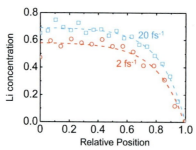

FIG. 3.7 The Li concentration remaining in the anodes after delithiation at constant delithiation rates. Dashed lines are exponential fitting curves of the calculated data (scatters).

with the total thickness) after delithation at two constant rates [16]. It can be seen that the number of Li atoms left in the anode with the high delithiation rate is larger than that with the low delithiation rate, indicating that a lower delithiation rate could induce a higher delithiation capacity, which is qualitatively consistent with experiments.

3.3 Predicting properties of lithiated Si with MD simulation

With the above-mentioned techniques for achieving lithiation/delithiation processes at room temperature with MD simulations, typical properties of lithiated Si were predicted in the following.

3.3.1 Lithiation-induced volume change of Si

Lithiation of Si features drastic volume expansion. Here the volume of amorphous Si lithiated to different extents was computed with ReaxFF potential. RVE models containing a given number (475) of Si atoms and varying numbers of Li atoms were constructed to simulate the relaxed volume of the lithiated Si Li_xSi with x ranging from 0 to 4.4 (see insets in Fig. 3.8). PBCs were applied along all the three directions to simulate an infinitely large bulk sample. During the simulation, the temperature was first increased to 1200 K to eliminate any pores inside and then slowly decreased to 300 K, and the sample reached an equilibrium state finally. In this quenching-like process, the temperature was controlled by the Berendsen thermostat, and zero external pressure was maintained by Berendsen barostat [27]. The final equilibrium volumes of the samples are shown in Fig. 3.8 with a black solid line, where the relative volume is defined as the ratio of the volume of Li_xSi to that of pristine amorphous Si. It can be seen that the variation of the relative volume with the lithiation index x approximately follows a linear manner, as shown by the dashed line in red. The volume of $Li_{4.4}Si$ is more than four times that of the pristine Si. This result is consistent with the results obtained from experiments [34, 35] and first-principle calculation [7].

Computational study on the effects of mechanical constraint **Chapter | 3** **103**

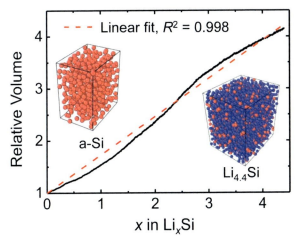

FIG. 3.8 The relative volume of lithiated Si as a function of lithiation extent represented by the value of x in Li$_x$Si.

3.3.2 Young's modulus of the pristine and lithiated Si

Young's modulus is an important quantity characterizing the stiffness of a material. To estimate Young's modulus of the lithiated Si, we conducted virtual tensile tests with MD simulation on three amorphous lithiated Si samples (Li$_{4.4}$Si, Li$_2$Si, and LiSi), amorphous Si (a-Si), and crystalline Si (c-Si) at a temperature of 300 K. RVE models containing 3741 Si atoms and varying numbers of Li atoms were established to undergo the uniaxial tensile test. PBCs were exerted along all three directions. Before the tensile tests, all models were relaxed to equilibrium. Other simulation setups follow those in Section 3.3.1. The calculated stress-strain curves are shown in Fig. 3.9, from which Young's moduli can be deduced as listed in Table 3.1 in comparison to the first-principle simulation [9] and experimental results [36], the MD simulation based on the ReaxFF potential underestimates the stiffness of the lithiated Si.

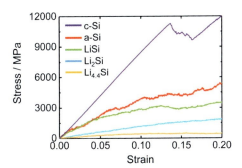

FIG. 3.9 Simulated stress-strain curves of amorphous lithiated Si with different atomic fractions of Li.

TABLE 3.1 Young's moduli of pristine Si and lithiated Si obtained from different methods.

Materials	ReaxFF (GPa)	First principle calculation[a] (GPa) [9]	Nanoindentation[b] (GPa) [36]
Crystalline Si	85.6	152	∼92
Amorphous Si	41.5	90.1	–
LiSi	40.8	54.5	∼70
Li_2Si	11.3	42.6	∼68
$Li_{3.75}Si$	–	33.9	∼12
$Li_{4.4}Si$	5.0	32.1	–

[a] Calculated through $E_{a-Li_xSi} = (Ax+B)/(1+x)$, where $A = 18.90$ and $B = 90.13$. This formula is obtained by curve fitting of the calculated results with $R^2 = 0.93$.
[b] Directly read from the plot in the literature.

3.3.3 Other properties of lithiated Si

3.3.3.1 The charge distribution

Besides, ReaxFF potential was applied to simulate the static charge distribution of lithiated Si. A simulation system consisting of 475 Si atoms and 2090 Li atoms was first constructed with PBCs applied along the three directions. Then, the temperature of the system was increased to 1200 K to create amorphous $Li_{4.4}Si$. Finally, energy minimization was conducted until the relative change of the energy between the successive iterations was less than 0.01%. The calculated charge distribution is shown in Fig. 3.10. The total net positive charges are 504.9 e_0, so each Li atom donates 0.24 e_0 to Si atoms on average, which is about one-third of the value as calculated by the first principle study [37]. Therefore, further optimization of the ReaxFF potential is needed for more accurate charge prediction.

3.3.3.2 The electromotive force between Si and Li electrodes

Besides the physical and mechanical properties of lithiated Si, MD simulations can give electrochemical properties such as the electromotive force (EMF) between Li_xSi and Li electrodes. The EMF of a battery reflects its capability to drive electric devices. To obtain the EMF, we need to first calculate the difference of the free energy between the lithiated Si and the pristine one, namely [22]

$$\Delta E_f(x) = E_{Li_xSi} - (xE_{Li} + E_{Si}) \quad (3.1)$$

Computational study on the effects of mechanical constraint **Chapter | 3 105**

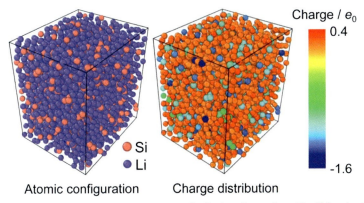

FIG. 3.10 Atomic configuration and the charge distribution of amorphous Li$_{4.4}$Si (e_0: the charge of an electron).

where x is the ratio of the number of Li atoms over the number of Si atoms in lithiated Si, E_{Li_xSi} is the free energy of Li$_x$Si divided by the number of Si atoms, and E_{Li} and E_{Si} are the free energy per single Li and Si atom in its initial crystalline form, respectively. Then, the EMF could be calculated from

$$\mathcal{E} = -\frac{d\Delta E_f(x)}{dx} \tag{3.2}$$

We adopted the same simulation models applied in Section 3.3.1 to calculate the free energy and EMF as a function of x in Li$_x$Si. The results are plotted in Fig. 3.11. The black circles are the simulated ΔE_f with its exponential fitting (the dashed line) defined by

$$\Delta E_f = Ae^{-\frac{x}{x_0}} + E_0 \tag{3.3}$$

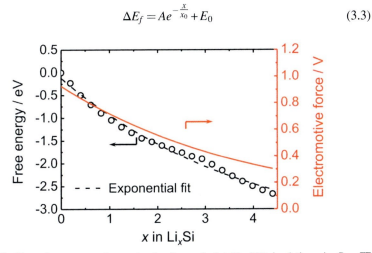

FIG. 3.11 Formation energy and open-circuit voltage calculated by MD simulation using ReaxFF potential.

where $A = 3.645\,V$, $x_0 = 3.938$ and $E_0 = -3.767\,V$. The red curve shows the calculated EMF based on Eqs. (3.2) and (3.3). It can be seen that EMF decreases from 0.9 V to 0.3 V as x increases. These results exhibit deviation from those given by the first-principle calculations (free energy: $-0.8 \sim 0.2\,eV$ [22], EMF: $0.05 \sim 0.5\,V$ [14]). Therefore, further optimization of the ReaxFF potential is needed for a more accurate prediction of the electromotive force.

In summary, the properties predicted by MD simulations such as volume expansion and elastic modulus of lithiated Si are reasonable and qualitatively consistent with those obtained by first-principle calculations and experiments. However, it should be admitted that the charge distribution [37] and the EMF given by the MD simulations deviate from those by first-principle calculations [14]. Nevertheless, we believe that MD simulation will be useful for a comprehensive understanding of the Li—Si reaction at the atomistic scale. For example, MD simulations with ReaxFF can estimate the relationship between the mechanical constraint on Si anode and the lithiation processes, which is presented in the next section.

3.4 Effects of mechanical constraint on the performance of Si nanosheets as the anode materials for LIB

In LIBs, the active materials such as Si particles are normally embedded in a matrix of accessory materials including binder and conductive additive. The previous simulation works often assume that the Si particle is free-standing and neglect the mechanical constraint by the accessory materials in the electrode. In fact, the matrix may play a nontrivial role in determining the performance of the electrode, because mechanical constraints from the matrix may suppress the volume expansion of Si and consequently reduce the electrochemical performance of the batteries. It was reported that that two-dimensional Si nanosheets exhibit distinct capacities if different binder materials are applied ($\sim 1500\,mAh\,g^{-1}$ and $\sim 865\,mAh\,g^{-1}$) [38, 39], implying the important role played by the accessory materials. Nevertheless, little light has been shed on the effects of mechanical constraint on the electrochemical performance of electrodes such as capacity and charging rate. We carried out molecular dynamics simulations to investigate the effects of the mechanical constraint on the performances such as capacity and lithiation rate of Si nanosheets as the active material for LIBs. Our study will be mainly focused on Si nanosheets because the large active surface area and small thickness (often smaller than 10 nm) of nanosheets would result in a high lithiation/delithiation rate [40, 41] and stable SEI layer [38]. More importantly, a scalable fabrication approach of Si nanosheets was recently developed, implying a great promise of Si nanosheets as an anode material for LIBs [38, 42].

3.4.1 Modeling and computational details

In experiments, when the electrode of LIBs is prepared, active materials, say, Si nanosheets normally should be blended with accessory materials including binder (e.g., sodium alginate), conductive additive (e.g., carbon black) and solvent (e.g., deionized water). The obtained slurry is then cast on a current collector and dried for use. Clearly, in such a composite electrode, the Si nanosheets are not free-standing. Instead, they are embedded in a composite matrix consisting of binder and carbon black with structure schematically shown in Fig. 3.12. During the lithiation and delithiation processes, the volume change of the Si nanosheets will be mechanically constrained by the surrounding matrix. As a consequence, the performance of the Si in the electrode might be affected.

To reveal the effect of mechanical constraint on the lithiation of Si nanosheet, an MD model was constructed to simulate the lithiation process of Si nanosheet. Fig. 3.13A shows the unit cell of the model which includes an Si nanosheet sandwiched by two thick Li plates. Originally, the Si sheet was crystalline and the lithiation would take place along the [112] direction, which was normal to the plane of the sheet. PBCs were applied along the in-plane directions to simulate an infinitely large nanosheet. Three different sheet thicknesses were considered including 3.6 nm, 4.3 nm, and 5.0 nm. The dimensions along $[11\bar{1}]$ direction and $[1\bar{1}0]$ direction were fixed as 7.3 nm and 6.2 nm, respectively. Due to the limitation of MD simulation, here the accessory materials including binder and carbon black were not included in our model, but their mechanical constraint to the Si nanosheet was considered by introducing two virtual walls situated at the surfaces of Si sheets before lithiation, as shown

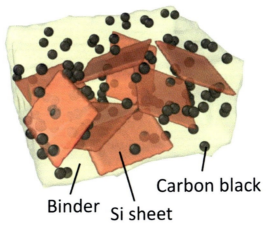

FIG. 3.12 Schematics of the structure of composite anode consisting of Si nanosheet as the active material, sodium alginate as the binder, and carbon black as the conductive additive.

108 PART | II Mechanical properties

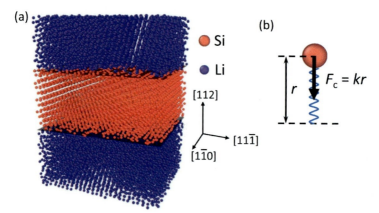

FIG. 3.13 (A) Illustration of the MD simulation model. The black planes indicate the original positions of the surface Si atoms. (B) The virtual tether force is applied to the surface Si atoms to simulate the mechanical constraint by the surrounding accessory materials.

in Fig. 3.13A. During lithiation, Si atoms on the surface that cross the virtual walls would undergo a tethering force given by $F_{tether} = kr$, where k is the spring constant of the virtual tether and r stands for the distance from the corresponding virtual wall, as illustrated in Fig. 3.13B. The original thicknesses of the Si nanosheet and the surrounding plates are denoted as h_0 and λh_0, respectively. The ratio λ can be roughly taken as half of the volumetric ratio between the accessory material and Si.

Let the Si nanosheet have a virtual uniform expansion along the thickness direction, which results in a displacement of the surface of the nanosheet into the accessory material, denoted as Δr. If the surfaces of the surrounding plates are fixed (as shown in Fig. 3.14A), the force experienced by a single Si atom on the surface is $A_0 E \Delta r / \lambda h_0$, where E is the effective elastic modulus of the composite matrix, and A_0 is the sectional area of a single Si atom.

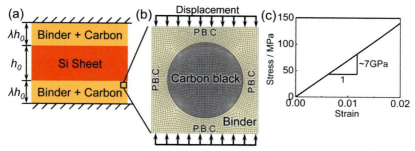

FIG. 3.14 (A) A simplified model of anode consisting of Si nanosheets embedded in a composite matrix. (B) Representative volume element (RVE) model of the composite accessory materials consisting of binder and carbon black for FEA simulation. (C) The stress-strain curve.

Therefore, the derivative of the force with respect to the displacement, which represents the spring constant k, is given by

$$k = \frac{EA_0}{\lambda h_0} \qquad (3.4)$$

In Eq. (3.4), the upper bound and lower bound of the effective elastic modulus E of the composite matrix can be estimated through the rule-of-mixture. $E_{upper} = E_b V_b + E_c V_c$, $E_{lower} = \frac{E_b E_c}{E_b V_c + E_c V_b}$ where E_b and E_c are the elastic moduli of binder and carbon black particle, respectively, and V_b and V_c are the volume fraction of the two components, respectively. Taking the mass ratio between the binder and carbon black as 1:1 and the densities of binder (e.g., sodium alginate) and carbon black as $1.6\,g\,cm^{-3}$ and $2.3\,g\,cm^{-3}$, respectively, the volumetric ratio between them is estimated to be 0.59:0.41. The Young's moduli of the binder (sodium alginate) and carbon black are taken as 2.7 GPa and 100 GPa [43], respectively. The upper bound and lower bound of the effective elastic modulus are 42.6 GPa and 4.5 GPa, respectively.

To make a more accurate estimation of the effective elastic modulus of the composite matrix, a finite element-based RVE model was constructed (see Fig. 3.14B), which consists of a circle included in a square, simulating a carbon black particle embedded in a binder matrix. The side length of the square and the diameter of the included circle in the RVE model were taken as 140 nm and 100 nm, respectively, giving rise to a volumetric ratio of 0.59:0.41 between the binder and carbon black. Perfect bonding was assumed on the interface between the binder and carbon black. PBCs were applied on all four sides. Uniaxial compressive loading on the model was simulated by applying a uniform strain along the vertical direction. The simulation was performed using 4-node bilinear plane strain quadrilateral solid elements (CPE4R) with commercial FEA software ABAQUS (Dassault Systèmes). The Young's moduli of the binder (sodium alginate) and carbon black were taken as 2.7 GPa and 100 GPa, respectively, while the Poisson's ratios for both materials were taken as 0.3. The calculated stress-strain curve of the composite matrix is shown in Fig. 3.14C, from which the effective elastic modulus is estimated to be around 7 GPa, and it will be used as a typical value in the analysis hereinafter.

In a real electrode, if the mass ratio among Si/carbon black/sodium alginate is 3:1:1, by taking the densities of Si, sodium alginate and carbon black as $2.3\,g\,cm^{-3}$, $1.6\,g\,cm^{-3}$ and $2.3\,g\,cm^{-3}$, respectively, λ can be estimated to be 0.4. Substituting $E = 7$ GPa and $\lambda = 0.4$ into Eq. (3.4) and taking $A_0 = 7$ Å^2 and $h_0 = 4$ nm, k is estimated to be $0.3\,N\,m^{-1}$. In our MD simulations, different k ranging from 10^{-3} to $10^3\,N\,m^{-1}$ were considered to study the effect of mechanical constraint on the lithiation of Si nanosheets.

All MD simulations reported in this section were conducted at the temperature of 300 K with Berendsen thermostat [27] and the time step was set as 2 fs. Before the onset of the lithiation process, the system was relaxed for at least 250 ps until the maximum relative change in the system potential energy of the last 1000 steps did not exceed 10^{-4}. Lithiation was deemed as completed if one of the following criteria is satisfied: (1) the increment of the number of inserted Li atoms is less than 1% in the last 10,000 MD steps, and (2) the ratio of atom number x in Li$_x$Si reaches 4.4. The mass capacity c_{m0} of the anodes is computed through $c_{m0} = \frac{N_{Li} e_0}{m_{Si}}$, where N_{Li} represents the number of Li atoms that have been inserted into the Si sheet at the end of lithiation [16], e_0 the charge of an electron, and m_{Si} the mass of the Si.

3.4.2 Results and discussion

Fig. 3.15A depicts the snapshots of a free-standing Si nanosheet ($k=0$) at different lithiation extents. Here, the lithiation extent is characterized by the ratio of the number of inserted Li$^+$ to that at full lithiation; that is, 100% corresponds to the fully lithiated Si (Li$_{4.4}$Si) and a mass capacity of 4200 mAh g^{-1} and 60% to Li$_{2.6}$Si and 2500 mAh g^{-1}, etc. During the simulated lithiation process, a sharp boundary between the lithiated and unlithiated regions is observed, which is consistent with the previous results [8, 28, 29, 44]. Fig. 3.15B shows the normalized mass capacities as a function of k for different sheet thicknesses. Under weak constraint with $k \leq 10^{-2}$ N m^{-1}, full lithiation can be achieved; while under strong constraint with $k \geq 10$ N m^{-1}, partial lithiation (~20%) is achieved, giving rise to capacity around ~800 mAh g^{-1}. The dependence of mass

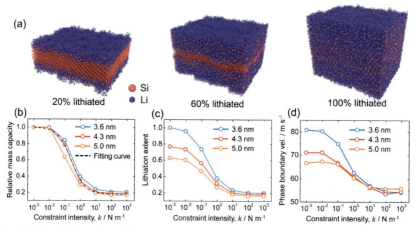

FIG. 3.15 (A) Snapshots of a free-standing ($k=0$) Si nanosheet at different lithiation extents. (B) Calculated dependence of the mass capacity (normalized by the theoretical value of 4200 mAh g^{-1}) on the constraint parameter k. The black dashed line is the fitting curve for the case of 4.3 nm (the red symbols). (C) Lithiation extents of Si nanosheets of different thicknesses under different levels of constraint after a given MD time (52.6 ps) of lithiation. (D) The average velocity of the phase boundary within the initial 20 ps of lithiation in nanosheets of different thicknesses under different levels of constraint.

capacity on the mechanical constraint shown in Fig. 3.15B can be described by an empirical equation

$$\alpha = \alpha_0 + \frac{1-\alpha_0}{1+k/k_0} \quad (3.5)$$

where α is the relative mass capacity, and α_0 and k_0 are two constants. Here, α_0 represents the relative mass capacity under rigid constraint ($k \to \infty$). For the nanosheet with a thickness of 4.3 nm, curve fitting ($R^2 > 0.99$) of the data points in Fig. 3.15B determines that $\alpha_0 = 0.184$ and $k_0 = 0.265\,\mathrm{N\,m^{-1}}$.

For a physical sample with the mass ratio between Si/carbon black/binder being 3:1:1, k has been estimated to be around $0.3\,\mathrm{N\,m^{-1}}$ which, according to Fig. 3.15B, would lead to partial lithiation (~50%) and capacity around $2000\,\mathrm{mAh\,g^{-1}}$. For an electrode with the mass ratio between Si/carbon black/binder is 8:1:1, k is estimated to be around $1.0\,\mathrm{N\,m^{-1}}$ which, according to Fig. 3.15B, leads to capacity as low as $1400\,\mathrm{mAh\,g^{-1}}$. This may explain the phenomenon that high mass loading of Si usually affects the performance of the electrode [45]. Compared to sodium alginate, other binders such as polyvinylidene fluoride (PVDF) and carboxymethyl cellulose (CMC) have similar density but lower Young's moduli (200–400 MPa for PVDF [46, 47] and ~1.0 GPa for CMC [48]), resulting in less mechanical constraint and therefore higher mass capacity. Additionally, Fig. 3.15B shows the effect of Si sheet thickness on the capacity. Within the range of the thickness considered, only little influence on the capacity is observed. This is consistent with the earlier findings that significant thickness-dependent capacity occurs only when the thickness is relatively large (>50 nm) [49–51].

Fig. 3.15C shows the lithiation extents reached within a prescribed time (52.6 ps in MD time, which is the time for a 3.6 nm thick, unconstrained Si nanosheet to reach full lithiation). It can be seen that Si nanosheet under stronger constraint reaches lower-leveled lithiation, implying that mechanical constraint would reduce the diffusivity of Li^+ in Si and therefore reduce the lithiation rate. This is further demonstrated by the average migration velocity of the phase boundary [23] within the initial 20 ps of lithiation as plotted in Fig. 3.15D. It can be seen that the lithiation rate, unlike the capacity, exhibits a strong dependence on the sheet thickness. For example, under constraint with $k = 0.1\,\mathrm{N\,m^{-1}}$ which is close to the real cases, a sheet of 5.0 nm thickness achieves ~47% lithiation in 52.6 ps (MD time), while a 3.6-nm thick sheet reaches ~74% lithiation in the same time. These results indicate that thickness can also effectively affect the diffusivity of Li in Si of the anode material: with the same constraint intensity, larger thickness leads to smaller Li diffusivity. This can be understood because what affects Li diffusivity is the external stress [23]. In our model, take h_0 as the thickness of the sheet before lithiation, and λh_0 as the thickness at any moment during lithiation, where λ is a parameter, we can obtain the lithiation-induced variation of the thickness $\Delta h = (\lambda - 1)h_0$. Because the force exerted on the Si sheet is proportional to Δh, thicker sheet would experience larger force from the accessory material in the out-of-plane directions. This will result in larger hydrostatic compressive stress in the sheet, which leads to smaller Li diffusivity. Based on the above simulation results, two remedial strategies can be proposed to improve the performance of the Si nanosheet anode: one is to choose a softer binder and the other one to reduce the thickness of Si nanosheets.

It is worth noting in Fig. 3.15B that the relative mass capacities reach a steady value when the constraint intensity is relatively large ($>10^2$ N m^{-1}). This trend implies that lithiation can always happen regardless of the magnitude of the constraint intensity, and there exists a minimum capacity even for very large constraint intensity. This phenomenon may be attributed to the facile lithiation to the surface sites of Si [52], which, in turn, manifests the promise of the Si nanosheet for its high surface area. The effects of the mechanical constraint on the lithiation time and Li diffusivity may help in understanding the relationship between the constraint intensity and the impedance of the electrode material [53, 54].

Considering that the mechanical constraint takes effect when the volume of Si varies during lithiation, the effect of constraint on the performance can be greatly reduced if the lithiation-induced volume change of Si can be absorbed. This might be achieved by introducing porosity into the Si sheets. To examine the effect of porosity on easing the side effects of mechanical constraint, the above MD simulations were repeated on a series of porous Si sheets containing a through-hole at the center, as shown in Fig. 3.16A. Holes with diameters ranging from 1 to 5 nm were considered, and the corresponding porosity varied from 1.7% to 45%. The thickness of the sheets was fixed at 3.6 nm.

The calculated mass capacity is plotted in Fig. 3.16C as a function of porosity for $k = 0.1$–100 N m^{-1}. As expected, higher porosity gives rise to higher mass capacity, irrespective of the external mechanical constraint. Therefore, the side effect of mechanical constraint can be alleviated by introducing porosity. For example, for a Si sheet under constraint with $k = 0.1$ N m^{-1}, the mass capacity can reach that at the unconstrained level if a porosity of 28% is introduced. Although larger porosity is preferable for larger mass capacity, it may not be necessarily preferred if volumetric capacity is a matter of concern. Fig. 3.16D shows the variation of the volumetric capacity on the porosity. Different from the mass capacity, the volumetric capacity does not monotonically increase with the porosity. If $k = 0.1$ N m^{-1}, for example, there exists an optimal porosity equal to $\sim$10%, at which the maximum volumetric capacity is achieved. In contrast, if $k > 10$ N m^{-1}, the volumetric capacity monotonically increases with the increasing porosity. The existence of the maximum point for $k = 0.1$ N m^{-1} and the relationships between the mass and volumetric capacities and different constraint intensities will be addressed shortly. For the discussed polymer binder materials with k less than 1 N m^{-1}, higher porosity may not necessarily lead to higher volumetric capacity, and therefore porous structure should be applied with caution if volumetric capacity is a matter of concern.

Computational study on the effects of mechanical constraint **Chapter | 3** **113**

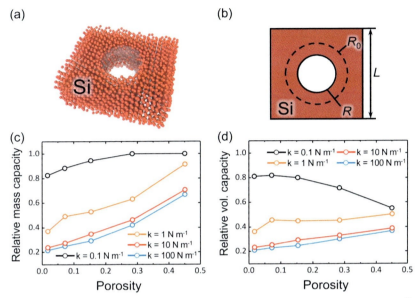

FIG. 3.16 (A) A simulation model for porous Si. (B) A theoretical model for porous Si. (C) Relative mass capacity and (D) relative volumetric capacity of the constrained porous Si nanosheet with different porosities.

To address the existence of the maximum point of the volumetric capacity when $k = 0.1\,N\,m^{-1}$, a unit cell of the Si porous sheet can be modeled as a square with a through-hole in the center as illustrated by Fig. 3.16B. The side length of the square is L and the radius of the hole is R. During the mechanically constrained lithiation process, it can be assumed that the Si atoms inside the dashed line circle (R_0 is the radius) can be fully lithiated (i.e., $Li_{4.4}Si$ phase forms after lithiation stops), while those outside the circle can be partially lithiated only. Thus, when lithiation stops, the effective mass capacity, c_m, of this Si porous sheet can be expressed as

$$c_m = \frac{(\pi R_0^2 - \pi R^2)\rho h c_{m0} + (L^2 - \pi R_0^2)\rho h c_{m0}\alpha}{(L^2 - \pi R^2)\rho h} \quad (3.6)$$

where ρ is the density of the Si material, h the thickness of the sheet before lithiation, c_{m0} the mass capacity of unconstrained Si, and α the relative mass capacity of the constrained solid (nonporous) sheet (which is a function of k, see Fig. 3.15B and Eq. 3.5). After lithiation, the hole should be filled by the expansion of the Si material inside R_0, which gives

$$(\pi R_0^2 - \pi R^2)h\beta = \pi R_0^2 H \quad (3.7)$$

where H is the thickness of the sheet after lithiation and $\beta \sim 400\%$ the volumetric expansion of unconstrained, fully lithiated Si electrode. As porosity is defined as $\Phi = \pi R^2/L^2$, the relative mass capacity of the porous Si sheet, $\bar{c}_m$, can be derived as

$$\bar{c}_m = \frac{c_m}{c_{m0}} = \alpha + (1-\alpha)\left(\frac{\Phi}{1-\Phi}\right)\left(\frac{H/h}{\beta - H/h}\right) \quad (3.8)$$

Assume that the relationship between H/h and α is linear (this assumption is reasonable according to Section 3.3.1), and for $\alpha = 1$, $H/h = \beta$; $\alpha = \alpha_0$, $H/h = 1$. Then, we have

$$\frac{H}{h} = \frac{\beta - 1}{1 - \alpha_0}\alpha + \frac{1 - \alpha_0\beta}{1 - \alpha_0} \quad (3.9)$$

Combining Eqs. (3.8) and (3.9), we can express for the porous Si sheet the relative mass capacity and the relative volumetric capacity, $\bar{c}_v$, as

$$\bar{c}_m = \alpha + \left(\alpha + \frac{1 - \alpha_0\beta}{\beta - 1}\right)\frac{\Phi}{1 - \Phi} \quad (3.10)$$

$$\bar{c}_v = \bar{c}_m(1 - \Phi) = \alpha + \frac{1 - \alpha_0\beta}{\beta - 1}\Phi \quad (3.11)$$

It should be pointed out that Eqs. (3.10) and (3.11) apply when $\bar{c}_m \le 1$ with porosity

$$\Phi \le \Phi_0 = \left(\frac{\beta - 1}{\beta}\right)\left(\frac{1-\alpha}{1-\alpha_0}\right) \quad (3.12)$$

when $\Phi > \Phi_0$, $\bar{c}_m$ maintains 1 and $\bar{c}_v$ decreases as Φ increases. For $k = 0.1\,\text{N}\,\text{m}^{-1}$, it can be obtained from Eq. (3.5) that $\alpha = 0.776$. Substituting it and $\beta = 400\%$, $\alpha_0 = 0.184$ into Eq. (3.12), we have $\Phi_0 = 20.6\%$ which is qualitatively close to the optimal point shown in Fig. 3.16D. Deviations may be due to the aforementioned assumptions and simplification. In contrast, for $k = 1\,\text{N}\,\text{m}^{-1}$, Φ_0 is estimated to be 59.3%, which is beyond the range of porosity we considered. Even larger constraint intensities will lead to Φ_0 as high as 75%. Therefore, in Fig. 3.16D the optimal point only appears on the curve of $k = 0.1\,\text{N}\,\text{m}^{-1}$. To further visualize the effects of the mechanical constraint and porosity on the capacity of the Si porous sheet, we plotted the relative mass and volumetric capacities given by Eqs. (3.10) and (3.11) on the k-Φ plane, as shown in Fig. 3.17.

3.5 Conclusions

In this chapter, we showed that MD simulation is a powerful tool to obtain insights into the details of Li—Si interactions during the lithiation and delithiation processes. The ReaxFF potential of Li—Si interaction is demonstrated reliable to give reasonable predictions to the properties of lithiated Si and the phenomena such as the anisotropic lithiation and rate-dependent delithiation.

Computational study on the effects of mechanical constraint **Chapter | 3** 115

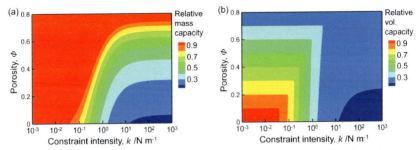

FIG. 3.17 The dependence of the (A) relative mass capacity $\bar{c}_m$ and (B) relative volumetric capacity $\bar{c}_v$ on the constraint intensity (k) and porosity (Φ).

With the developed ReaxFF potential, we demonstrated that the mechanical constraint from the accessory materials will affect the electrochemical performances of Si nanosheets. In particular, stiffer matrix materials can lead to lower capacity and lithiation rate. Three methods for enhancing the performances of the anode thus are proposed here: (1) choosing a softer binder material, (2) reducing the thickness of the Si nanosheet, and (3) introducing porous structure. Besides, the porous structure should be introduced with caution so that the volumetric capacity would not be decreased. Although the results are derived based on Si, to some extent they can be generalized to the other electrode materials that experience similar large volume expansion during alloying with Li, such as Ge, Sn, SnO_2, or even cathode materials. We believe that our results can serve as a guideline for the fabrication of the Si-based anode material and inspire the design for the next-generation LIBs.

In the past decades, considerable efforts have been dedicated to alleviating the degradation and prolonging the cycle life of the Si-based anode. Apart from the delicate nano-structured materials [55–58], innovative strategies such as the use of conductive binder [59], implementation of gradient Si concentration [60], and generation of all-solid-state batteries [61, 62] are also potential routes to the high-performance LIBs. Although great progress has been achieved, it is suggested that further efforts remain in need for the realization of low-cost, high-performance, and environmental-friendly LIBs for the next generation.

References

[1] A. Ostadhossein, E.D. Cubuk, G.A. Tritsaris, E. Kaxiras, S. Zhang, A.C. van Duin, Stress effects on the initial lithiation of crystalline silicon nanowires: reactive molecular dynamics simulations using ReaxFF, Phys. Chem. Chem. Phys. 17 (2015) 3832–3840.

[2] K.J. Zhao, W.L. Wang, J. Gregoire, M. Pharr, Z.G. Suo, J.J. Vlassak, E. Kaxiras, Lithium-assisted plastic deformation of silicon electrodes in lithium-ion batteries: A first-principles theoretical study, Nano Lett. 11 (2011) 2962–2967.

[3] H. Jung, M. Lee, B.C. Yeo, K.R. Lee, S.S. Han, Atomistic observation of the lithiation and delithiation behaviors of silicon nanowires using reactive molecular dynamics simulations, J. Phys. Chem. C 119 (2015) 3447–3455.

[4] H.S. Lee, B.J. Lee, Structural changes during lithiation and delithiation of Si anodes in Li-ion batteries: A large scale molecular dynamics study, Met. Mater. Int. 20 (2014) 1003–1009.
[5] M.E. Stournara, X.C. Xiao, Y. Qi, P. Johari, P. Lu, B.W. Sheldon, H.J. Gao, V.B. Shenoy, Li segregation induces structure and strength changes at the amorphous Si/Cu interface, Nano Lett. 13 (2013) 4759–4768.
[6] W.H. Wan, Q.F. Zhang, Y. Cui, E.G. Wang, First principles study of lithium insertion in bulk silicon, J. Phys. Condens. Matter 22 (2010) 415501.
[7] P. Johari, Y. Qi, V.B. Shenoy, The mixing mechanism during lithiation of Si negative electrode in Li-ion batteries: an ab initio molecular dynamics study, Nano Lett. 11 (2011) 5494–5500.
[8] S.P. Kim, D. Datta, V.B. Shenoy, Atomistic mechanisms of phase boundary evolution during initial lithiation of crystalline silicon, J. Phys. Chem. C 118 (2014) 17247–17253.
[9] V.B. Shenoy, P. Johari, Y. Qi, Elastic softening of amorphous and crystalline Li—Si Phases with increasing Li concentration: A first-principles study, J. Power Sources 195 (2010) 6825–6830.
[10] H. Sitinamaluwa, J. Nerkar, M.C. Wang, S.Q. Zhang, C. Yan, Deformation and failure mechanisms of electrochemically lithiated silicon thin films, RSC Adv. 7 (2017) 13487–13497.
[11] B. Ding, X.Y. Li, X. Zhang, H. Wu, Z.P. Xu, H.J. Gao, Brittle versus ductile fracture mechanism transition in amorphous lithiated silicon: From intrinsic nanoscale cavitation to shear banding, Nano Energy 18 (2015) 89–96.
[12] X.J. Wang, F.F. Fan, J.W. Wang, H.R. Wang, S.Y. Tao, A. Yang, Y. Liu, H.B. Chew, S.X. Mao, T. Zhu, S.M. Xia, High damage tolerance of electrochemically lithiated silicon, Nat. Commun. 6 (2015) 8417.
[13] S.M. Khosrownejad, W.A. Curtin, Crack growth and fracture toughness of amorphous Li—Si anodes: Mechanisms and role of charging/discharging studied by atomistic simulations, J. Mech. Phys. Solids 107 (2017) 542–559.
[14] M.E. Stournara, Y. Qi, V.B. Shenoy, From ab initio calculations to multiscale design of Si/C core-shell particles for li-Ion anodes, Nano Lett. 14 (2014) 2140–2149.
[15] K.J. Kim, Y. Qi, Vacancies in Si can improve the concentration-dependent lithiation rate: Molecular dynamics studies of lithiation dynamics of Si electrodes, J. Phys. Chem. C 119 (2015) 24265–24275.
[16] K.J. Kim, J. Wortman, S.Y. Kim, Y. Qi, Atomistic simulation derived insight on the irreversible structural changes of Si electrode during fast and slow delithiation, Nano Lett. 17 (2017) 4330–4338.
[17] S.Y. Kim, A. Ostadhossein, A.C. van Duin, X. Xiao, H. Gao, Y. Qi, Self-generated concentration and modulus gradient coating design to protect Si nano-wire electrodes during lithiation, Phys. Chem. Chem. Phys. 18 (2016) 3706–3715.
[18] S. Plimpton, Fast parallel algorithms for short-range molecular-dynamics, J. Comput. Phys. 117 (1995) 1–19.
[19] J.M. Haile, Molecular Dynamics Simulation: Elementary Methods, Wiley, 1997, p. 512.
[20] K. Chenoweth, A.C.T. van Duin, W.A. Goddard, ReaxFF reactive force field for molecular dynamics simulations of hydrocarbon oxidation, J. Phys. Chem. A 112 (2008) 1040–1053.
[21] A. Stukowski, Visualization and analysis of atomistic simulation data with OVITO-the open visualization tool, Model. Simulat. Mater. Sci. Eng. 18 (2010), 015012.
[22] V.L. Chevrier, J.R. Dahn, First principles model of amorphous silicon lithiation, J. Electrochem. Soc. 156 (2009) A454–A458.
[23] B. Ding, H. Wu, Z.P. Xu, X.Y. Li, H.J. Gao, Stress effects on lithiation in silicon, Nano Energy 38 (2017) 486–493.

[24] H.M. Aktulga, J.C. Fogarty, S.A. Pandit, A.Y. Grama, Parallel reactive molecular dynamics: Numerical methods and algorithmic techniques, Parallel Comput. 38 (2012) 245–259.
[25] LAMMPS, n.d., Fix qeq/shielded command. Available from: https://lammps.sandia.gov/doc/fix_qeq.html.
[26] LAMMPS, n.d., Fix qeq/reax command. Available from: https://lammps.sandia.gov/doc/fix_qeq_reax.html.
[27] H.J.C. Berendsen, J.P.M. Postma, W.F. Vangunsteren, A. Dinola, J.R. Haak, Molecular-dynamics with coupling to an external bath, J. Chem. Phys. 81 (1984) 3684–3690.
[28] X.H. Liu, J.W. Wang, S. Huang, F.F. Fan, X. Huang, Y. Liu, S. Krylyuk, J. Yoo, S.A. Dayeh, A.V. Davydov, S.X. Mao, S.T. Picraux, S.L. Zhang, J. Li, T. Zhu, J.Y. Huang, In situ atomic-scale imaging of electrochemical lithiation in silicon, Nat. Nanotechnol. 7 (2012) 749–756.
[29] M.T. McDowell, S.W. Lee, J.T. Harris, B.A. Korgel, C.M. Wang, W.D. Nix, Y. Cui, In situ TEM of two-phase lithiation of amorphous silicon nanospheres, Nano Lett. 13 (2013) 758–764.
[30] C. Kim, M. Ko, S. Yoo, S. Chae, S. Choi, E.H. Lee, S. Ko, S.Y. Lee, J. Cho, S. Park, Novel design of ultra-fast Si anodes for Li-ion batteries: crystalline Si@amorphous Si encapsulating hard carbon, Nanoscale 6 (2014) 10604–10610.
[31] L.L. Luo, H. Yang, P.F. Yan, J.J. Travis, Y. Lee, N. Liu, D.M. Piper, S.H. Lee, P. Zhao, S.M. George, J.G. Zhang, Y. Cui, S.L. Zhang, C.M. Ban, C.M. Wang, Surface-coating regulated lithiation kinetics and degradation in silicon nanowires for lithium ion battery, ACS Nano 9 (2015) 5559–5566.
[32] H.T. Nguyen, M.R. Zamfir, L.D. Duong, Y.H. Lee, P. Bondavalli, D. Pribat, Alumina-coated silicon-based nanowire arrays for high quality Li-ion battery anodes, J. Mater. Chem. 22 (2012) 24618–24626.
[33] X.C. Xiao, P. Lu, D. Ahn, Ultrathin multifunctional oxide coatings for lithium ion batteries, Adv. Mater. 23 (2011) 3911–3915.
[34] L.Y. Beaulieu, K.W. Eberman, R.L. Turner, L.J. Krause, J.R. Dahn, Colossal reversible volume changes in lithium alloys, Electrochem. Solid State Lett. 4 (2001) A137–A140.
[35] L.Y. Beaulieu, T.D. Hatchard, A. Bonakdarpour, M.D. Fleischauer, J.R. Dahn, Reaction of Li with alloy thin films studied by in situ AFM, J. Electrochem. Soc. 150 (2003) A1457–A1464.
[36] B. Hertzberg, J. Benson, G. Yushin, Ex-situ depth-sensing indentation measurements of electrochemically produced Si—Li alloy films, Electrochem. Commun. 13 (2011) 818–821.
[37] V.L. Chevrier, J.R. Dahn, First principles studies of disordered lithiated silicon, J. Electrochem. Soc. 157 (2010) A392–A398.
[38] X.H. Zhang, X.Y. Qiu, D.B. Kong, L. Zhou, Z.H. Li, X.L. Li, L.J. Zhi, Silicene flowers: A dual stabilized silicon building block for high-performance lithium battery anodes, ACS Nano 11 (2017) 7476–7484.
[39] J. Ryu, D. Hong, S. Choi, S. Park, Synthesis of ultrathin Si nanosheets from natural clays for lithium-ion battery anodes, ACS Nano 10 (2016) 2843–2851.
[40] J.H. Liu, X.W. Liu, Two-dimensional nanoarchitectures for lithium storage, Adv. Mater. 24 (2012) 4097–4111.
[41] V.V. Kulish, O.I. Malyi, M.F. Ng, P. Wu, Z. Chen, Enhanced Li adsorption and diffusion in silicon nanosheets based on first principles calculations, RSC Adv. 3 (2013) 4231–4236.
[42] J.L. Lang, B. Ding, S. Zhang, H.X. Su, B.H. Ge, L.H. Qi, H.J. Gao, X.Y. Li, Q.Y. Li, H. Wu, Scalable synthesis of 2D Si nanosheets, Adv. Mater. 29 (2017) 1701777.
[43] J. Robertson, Diamond-like amorphous carbon, Mater. Sci. Eng. R-Rep. 37 (2002) 129–281.
[44] M.T. McDowell, I. Ryu, S.W. Lee, C.M. Wang, W.D. Nix, Y. Cui, Studying the kinetics of crystalline silicon nanoparticle lithiation with in situ transmission electron microscopy, Adv. Mater. 24 (2012) 6034–6041.

[45] Z. Karkar, D. Mazouzi, C.R. Hernandez, D. Guyomard, L. Roue, B. Lestriez, Threshold-like dependence of silicon-based electrode performance on active mass loading and nature of carbon conductive additive, Electrochim. Acta 215 (2016) 276–288.
[46] H. Mendoza, S.A. Roberts, V.E. Brunini, A.M. Grillet, Mechanical and electrochemical response of a LiCoO2 cathode using reconstructed microstructures, Electrochim. Acta 190 (2016) 1–15.
[47] E.K. Rahani, V.B. Shenoy, Role of plastic deformation of binder on stress evolution during charging and discharging in lithium-ion battery negative electrodes, J. Electrochem. Soc. 160 (2013) A1153–A1162.
[48] J. Li, R.B. Lewis, J.R. Dahn, Sodium carboxymethyl cellulose—A potential binder for Si negative electrodes for Li-ion batteries, Electrochem. Solid State Lett. 10 (2007) A17–A20.
[49] B. Liang, Y.P. Liu, Y.H. Xu, Silicon-based materials as high capacity anodes for next generation lithium ion batteries, J. Power Sources 267 (2014) 469–490.
[50] T. Takamura, S. Ohara, M. Uehara, J. Suzuki, K. Sekine, A vacuum deposited Si film having a Li extraction capacity over 2000 mAh/g with a long cycle life, J. Power Sources 129 (2004) 96–100.
[51] J.P. Maranchi, A.F. Hepp, P.N. Kumta, High capacity, reversible silicon thin-film anodes for lithium-ion batteries, Electrochem. Solid State Lett. 6 (2003) A198–A201.
[52] O. Malyi, V.V. Kulish, T.L. Tan, S. Manzhos, A computational study of the insertion of Li, Na, and Mg atoms into Si(111) nanosheets, Nano Energy 2 (2013) 1149–1157.
[53] R. Ruffo, S.S. Hong, C.K. Chan, R.A. Huggins, Y. Cui, Impedance analysis of silicon nanowire lithium ion battery anodes, J. Phys. Chem. C 113 (2009) 11390–11398.
[54] C. Ho, I.D. Raistrick, R.A. Huggins, Application of A-C techniques to the study of lithium diffusion in tungsten trioxide thin films, J. Electrochem. Soc. 127 (1980) 343–350.
[55] Y.Z. Li, K. Yan, H.W. Lee, Z.D. Lu, N. Liu, Y. Cui, Growth of conformal graphene cages on micrometer-sized silicon particles as stable battery anodes, Nat. Energy 1 (2016) 15029.
[56] N. Liu, H. Wu, M.T. McDowell, Y. Yao, C.M. Wang, Y. Cui, A yolk-shell design for stabilized and scalable Li-ion battery alloy anodes, Nano Lett. 12 (2012) 3315–3321.
[57] J.J. Tang, Q.F. Yin, Q. Wang, Q.Q. Li, H.T. Wang, Z.L. Xu, H.M. Yao, J. Yang, X.Y. Zhou, J.K. Kim, L.M. Zhou, Two-dimensional porous silicon nanosheets as anode materials for high performance lithium-ion batteries, Nanoscale 11 (2019) 10984–10991.
[58] C.K. Chan, H.L. Peng, G. Liu, K. McIlwrath, X.F. Zhang, R.A. Huggins, Y. Cui, High-performance lithium battery anodes using silicon nanowires, Nat. Nanotechnol. 3 (2008) 31–35.
[59] S. Lawes, Q. Sun, A. Lushington, B.W. Xiao, Y.L. Liu, X.L. Sun, Inkjet-printed silicon as high performance anodes for Li-ion batteries, Nano Energy 36 (2017) 313–321.
[60] Z.B. Guo, L.M. Zhou, H.M. Yao, Improving the electrochemical performance of Si-based anode via gradient Si concentration, Mater. Des. 177 (2019) 107851.
[61] J.F.M. Oudenhoven, L. Baggetto, P.H.L. Notten, All-solid-state lithium-ion microbatteries: A review of various three-dimensional concepts, Adv. Energy Mater. 1 (2011) 10–33.
[62] F. Le Cras, B. Pecquenard, V. Dubois, V.P. Phan, D. Guy-Bouyssou, All-solid-state lithium-ion microbatteries using silicon nanofilm anodes: High performance and memory effect, Adv. Energy Mater. 5 (2015) 1501061.

Chapter 4

Mechanical properties of silicon-based electrodes

D.Y.W. Yu, P.-K. Lee, S. Wang, and T. Tan

School of Energy and Environment, City University of Hong Kong, Kowloon, Hong Kong

4.1 Introduction

Lithium-ion batteries (LIBs), first manufactured in 1991 by Sony, have improved much over the past years. The usable energy density has more than doubled since the first inception, a large part of it is due to advances in cell components—better binder, thinner current collector, and stronger can material lead to reduction of inactive mass in the battery. Better control of water content in electrolyte also allows batteries to be charged to 4.3 or 4.35 V, thereby increases the utilization of the cathode, etc. [1]. Nowadays, energy density of LIB reaches as high as 248 Wh kg^{-1} and 677 Wh L^{-1} for Panasonic NCR18650BF [2]. Battery capacity and energy density are now limited by the charge storage capacity of the active materials used for their cathode and anode, typically a lithium transition metal oxide such as $LiCoO_2$ and graphite, respectively. Further increase in energy density for high-energy applications will have to come about by changing to other active materials with larger capacity and Li storage capability.

Graphite is an excellent anode material for accommodation of Li ions, as it has a layered structure for Li to intercalate and de-intercalate. One-sixth of Li can be accommodated into the graphite layer per carbon atom (LiC_6), with a volume expansion of 12% [3]. The reversible capacity of graphite is moderate (∼350 mAh g^{-1}), and the volume change is small. With the help of a stable solid electrolyte interphase (SEI) formed on the surface, graphite can be stably charged and discharged for many cycles [4].

To increase the capacity of the anode, one will need to find other Li-ion active materials that can accommodate more Li at low operational voltage. Intercalation processes are no longer sufficient for this purpose because there is limited material structure that can hold larger number of lithium. Researchers have therefore turned to alloying process with the following reaction:

$$M + xLi^+ + xe^- \rightarrow Li_xM \qquad (4.1)$$

where M is an element that can alloy with Li, and x is the maximum number of Li that can be alloyed with M. Many elements such as Si, Ge, Sn, Sb, and Pb are known to alloy with higher amount of lithium, which are therefore good candidates for next-generation anode materials [3].

Based on Eq. (4.1), theoretical capacities of the different elements can be calculated if the alloying products are known. Take Si for example, the Li-Si phase diagram (Fig. 4.1) identifies a stable phase of $Li_{22}Si_5$ with the highest amount of Li. With an uptake of 4.4 Li per Si, the theoretical gravimetric capacity is calculated to be 4200 mAh g^{-1}. Though, electrochemical tests indicate that the maximum composition that is observed during 100% depth of discharging is $Li_{15}Si_4$, corresponding to a capacity of 3570 mAh g^{-1} [6].

When Li is alloying with Si to form different Li-Si alloys, the crystal lattice has to expand in order to accommodate the Li atoms. The amount of volume expansion can be estimated from structural information of different phases of lithiated silicon. Fig. 4.2 shows the atomic arrangement of crystalline $Li_{12}Si_7$, Li_7Si_3, $Li_{13}Si_4$, $Li_{15}Si_4$, and $Li_{21}Si_5$ [3]. Since their structural parameters, unit cell volumes, densities, and number of atoms per unit cell are readily available (see Table 4.1) [8], one can use these information to calculate the molar volume $V_{molar,x}$ for each phase

$$V_{molar,x} = MW_x/\rho \qquad (4.2)$$

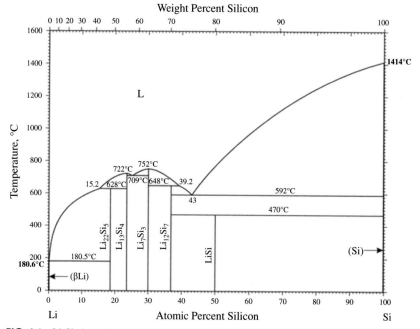

FIG. 4.1 Li-Si phase diagram [5]. *(Reproduced with permission from Springer Nature.)*

Mechanical properties of silicon-based electrodes **Chapter | 4 121**

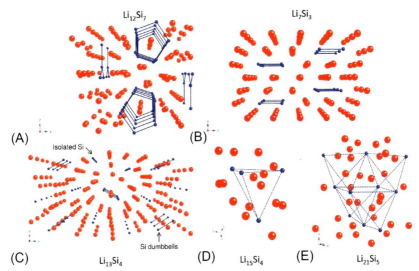

FIG. 4.2 Structure of various lithiated silicon [7]. *(Reproduced with permission from American Chemical Society.)*

where MW_x is the molecular weight and ρ is the density of Li_xSi. Then, we can use Eq. (4.3) to estimate the amount of volume expansion (ΔV) upon lithiation, with $V_{molar,Si}$ corresponding to the molar volume of pure Si without Li:

$$\Delta V = \frac{V_{molar,x} - V_{molar,Si}}{V_{molar,Si}} \qquad (4.3)$$

During electrochemical charge and discharge, the obtained capacity reflects the amount of Li in Si. So, the corresponding capacities for the different phases of Li_xSi can also be calculated (see Table 4.2) [8,9]. A plot of ΔV vs. capacity shows that the volume expansion is linear with respect to capacity (Fig. 4.3). While lithium is inserted into Si, the volume increases according to Li amount. This is an intrinsic volume expansion associated with accommodation of Li into the Si lattice that is unavoidable. Alloying to $Li_{22}Si_5$ will theoretically result in a volume increase of the lattice of 312%. If we assume a spherical particle with isotropic expansion, full lithiation will lead to about 60% increase in the radius. For a thin film, since the expansion in the lateral direction is restricted, the volume change of the lattice will translate to a thickness change of the film.

Note that the theoretical volume change calculated earlier assumes that the reaction product is crystalline. In reality, during electrochemical lithiation, crystalline Si will become amorphous until reaching a phase of $Li_{15}Si_4$. So, the amount of volume change is expected to be slightly higher than the crystalline phases. Nonetheless, the Si lattice is still expected to expand linearly with capacity as the density of amorphous Li_xSi is similar to that of crystalline Li_xSi [10].

TABLE 4.1 Structural parameters for Si and lithium silicide compounds.

Materials	Space group	Structural parameter (Å, ?)	Unit cell volume (Å³)	Density (g/mL)	Z (atom)	Bond length of Si-Si; Li-Si; Li-Li (Å)
Si (Li₀Si₁)	Fd-3m (227)	a = 5.42979	160.08	2.331	8	2.351
		α = 90				
Li₁Si₁ (Li₁Si₁)	I4₁/a (88)	a = 9.35300	502.39	1.852	16	2.417
		α = 90				2.609
						2.725
Li₁₂Si₇ (Li₁.₇₁Si₁)	Pnma (62)	a = 8.566	2413.08	1.541	8	2.359
		b = 19.701				2.589
		c = 14.299				2.546
		α = 90				
Li₂Si (Li₂Si₁)	C2/m (12)	a = 7.700	204.44	1.364	4	2.372
		b = 4.410				2.590
		c = 6.560				2.920
		α = 90				
		β = 113.4				
		γ = 90				

Li$_7$Si$_2$ (Li$_{3.5}$Si$_1$)	Pbam (55)	a = 7.990	538.37	1.292	4	2.383
		b = 15.210				2.308
		c = 4.430				2.492
		α = 90				
Li$_{15}$Si$_4$ (Li$_{3.75}$Si$_1$)	I-43 d (220)	a = 10.6852	1219.97	1.179	3	4.548
		α = 90				2.652
						2.706
Li$_{22}$Si$_5$ (Li$_{4.40}$Si$_1$)	F-43 m (216)	a = 18.75	6591.8	1.181	16	4.556
		α = 90				2.573
						2.696
Li (Li$_1$Si$_0$)	Im-3 m (229)	a = 3.43879	40.66	0.567	2	–
		α = 90				–
						2.978

Reproduced from C.H. Doh, M.W. Oh, B.C. Han, Lithium alloying potentials of silicon as anode of lithium secondary batteries, Asian J. Chem. 25 (10) (2013) 5739–5743. This work is licensed under a Creative Commons Attribution 4.0 International License.

TABLE 4.2 Physical parameters of Si and Si-Li alloys.

	Density (g cm^{-3})	MW (g mol^{-1})	Molar volume (cm^3 mol^{-1})	Volume expansion (%)	Capacity (mAh g^{-1})
Si	2.331	28.0855	12.05	–	–
LiSi	1.852	35.0265	18.91	57	954
Li$_{12}$Si$_7$ (Li$_{1.71}$Si)	1.541	39.984	25.95	115	1636
Li$_2$Si	1.364	41.9675	30.77	155	1909
Li$_7$Si$_2$ (Li$_{3.5}$Si)	1.292	52.379	40.54	236	3340
Li$_{15}$Si$_4$ (Li$_{3.75}$Si)	1.179	54.11425	45.90	281	3579
Li$_{22}$Si$_5$ (Li$_{4.4}$Si)	1.181	58.6259	49.64	312	4200

Adapted from C.H. Doh, M.W. Oh, B.C. Han, Lithium alloying potentials of silicon as anode of lithium secondary batteries, Asian J. Chem. 25 (10) (2013) 5739–5743; M.K. Datta, P.N. Kumta, In situ electrochemical synthesis of lithiated silicon–carbon based composites anode materials for lithium ion batteries, J. Power Sources 194 (2) (2009) 1043–1052.

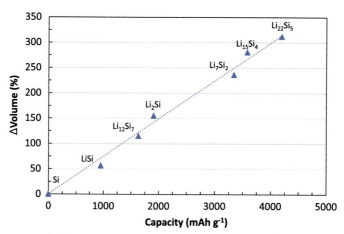

FIG. 4.3 Theoretical volume expansion of Si with respect to capacity.

The actual volume change of a Si-based electrode, though, depends on many factors. First of all, the volume expansion process of crystalline Si is not isotropic, which gives preference to expansion in certain directions [11, 12]. In addition, the lithiation process undergoes a core-shell mechanism, where the surface of a particle is first lithiated, leaving an unreacted core [13]. This will create interfacial stress in the particles that lead to pulverization. Last but not least, when a composite electrode is made up of many Si particles with conductive carbon black and binder is lithiated (discharged), the overall amount of volume expansion will depend on the porosity of the electrode and also the interaction between the binder and the particles. These will be discussed in the next sections.

4.2 Process of volume expansion and crack formation

While the amount of volume expansion discussed in the previous section is determined without consideration of the orientation of the Si lattice, various researchers have observed that the volume expansion of crystalline Si (c-Si) is anisotropic [11, 12]. For example, Lee et al. used scanning electron microscopy (SEM) to monitor the morphological changes of c-Si nanopillars upon lithiation (Fig. 4.4). They first made cylindrical pillars with diameter of about 300–400 nm with $\langle 100 \rangle$, $\langle 110 \rangle$, and $\langle 111 \rangle$ axial orientations out of Si wafers by etching. The nanopillars are then subjected to potential sweep until 120 or 10 mV followed by a constant potential hold for at least 20 h to ensure equilibrium. They observed that there is a difference in shape change for the three different pillars. Specifically, the pillar with $\langle 100 \rangle$ axial direction became crossed shape, whereas the $\langle 110 \rangle$ and $\langle 111 \rangle$ pillars became elliptical and hexagonal, respectively (Fig. 4.4A–L). Based on the initial orientation of the sidewalls of the pillars (Fig. 4.4 m), they concluded that the lithiation of silicon is primary along the $\langle 110 \rangle$ direction perpendicular to the pillar axis. Similar results are also reported by Goldmen et al. who studied the volume expansion of regular arrays of micron-sized bars with different orientations [12].

Lee et al. attributed the observed anisotropic expansion of c-Si to the difference in diffusivity of Li along the different orientations. Though, Zhao et al. argued with a theoretical model that the anisotropic morphology changes are due to a difference in reaction rates, i.e. rate of rearrangement of atoms during reaction, of the various surfaces of c-Si, rather than diffusivity of Li [14].

For a c-Si particle, the anisotropy in expansion will give rise to a strain gradient and induce a stress within the Si particle. If the stress is too large, cracks will be formed in the lithiated Si region. Further Li insertion favors the propagation of cracks to reduce the tensile stress at the crack tips. This was demonstrated by Liang et al. who monitored the morphological change of crystalline SiNP (Si nanoparticles) with a size of 160 nm during lithiation with transmission electron microscopy (TEM). Initially, the outside of the Si particle is lithiated, leaving an inner core that is unreacted (Fig. 4.5) [13]. Because of the anisotropy in volume expansion, intensified stress is built up at the corners of the c-Si particles near neighboring domains, leading to fracturing.

In comparison, amorphous Si (a-Si) particles are more resistant to crack formation where the expansion is isotropic. This was demonstrated by McDowell et al. who used in-situ TEM to visualize the dimensional change of a-Si particles [15]. During lithiation, isotropic expansion of the particle is observed. With the

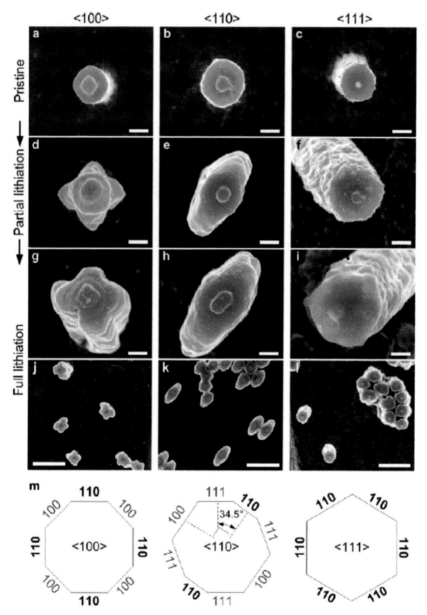

FIG. 4.4 (A–L) Morphology changes of c-Si nanopillars with different orientations during lithiation. (M) Crystallographic orientations of the sides of the Si pillars [11]. Scale bar: 200 nm (A–I); 2 μm (J, L). *(Reproduced with permission from American Chemical Society.)*

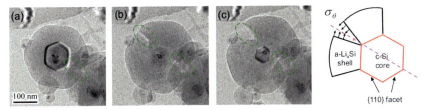

FIG. 4.5 (A) Partially lithiated Si particle showing the inner c-Si core and the amorphous Li$_x$Si region outside. (B, C) Formation and propagation of crack with further lithiation. (D) Model showing that the cracks are formed along one of the diameter planes [13]. *(Reproduced with permission from American Chemical Society.)*

absence of preferred lattice planes for lithiation, fracture of the material is absent. Even with a particle size of up to 870 nm, a-Si particles do not fracture when lithium is inserted into the lattice (Fig. 4.6). Regarding the amount of volume expansion, they have sampled 26 nanoparticles and observed the overall volume expansion after lithiation to range from 101% to 332% with an average of 204%. Even though volume expansion can be monitored through TEM, the values are highly influenced by experimental conditions such as the thickness of SEI layer and the quality of electrical contacts, which affect the overall degree of lithiation.

While large particles will undergo pulverization due to the large strain energy from the electrochemical reaction that drives crack propagation, there exists a critical size below which the strain can be accommodated without fracture. Liu et al. studied the volume expansion of spherical crystalline Si particles with different sizes ranging from ∼20 nm to ∼1 μm in diameter using in-situ TEM [16]. They observed that the first surface crack usually occurs when the lithiated silicon shell has a thickness of about 100–200 nm. After compiling all these experimental data, they concluded that there is a critical particle size (D_c) of about 150 nm below which cracks are not observed (see Fig. 4.7). They then developed a model that suggests that the crack will not propagate, if the driving force of strain energy release rate is less than the resistance of surface energy. A similar conclusion was also reached by Zhao et al. who developed an analytical solution based on concurrent reaction and plasticity to explain the critical size [14]. Ma et al. further modeled the fracture failure of Si particle

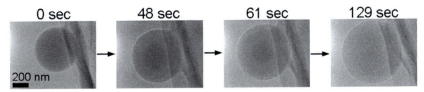

FIG. 4.6 Lithiation process of an amorphous Si particle, showing the absence of cracks even up to full lithiation [15]. *(Reproduced with permission from American Chemical Society.)*

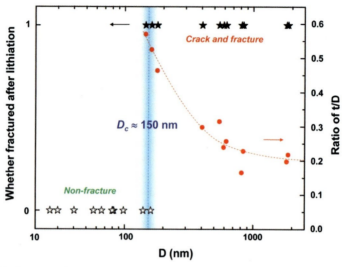

FIG. 4.7 Summary of TEM results on occurrence of fracture and the ratio of Li_xSi shell thickness to diameter (t/D) when first crack appears [16]. *(Reproduced with permission from American Chemical Society.)*

based on surface effects [17]. They then calculated the critical silicon particle sizes for nanoparticles, nanowires, and nanofilms, which are of the same order of magnitude as experimental values.

The cracking phenomena of a-Si thin film were also studied by Li et al. [18]. Amorphous thin films of Si ranging from 100 to 1000 nm were deposited by magnetron sputtering onto stainless steel disks, and the electrodes were subjected to cycling tests in coin cells with 1M $LiPF_6$ in ethylene carbonate/dimethyl carbonate (EC/DMC) = 1:1. After 5 or 10 cycles, the cells were disassembled, and the a-Si electrodes were observed under scanning electron microscopy (SEM) to see the morphology changes during cycling. While the as-deposited film was flat, interconnected cracks were observed in the films after cycling. Formation of cracks were attributed to the large tensile stress as high as several GPa built up in the film during delithiation. They observed that the crack patterns vary with the thickness of the Si thin film—smaller crack area is observed for thinner films. This is attributed to smaller force on the film with smaller thickness. They also conclude that below a critical thickness of about 100–200 nm, the film will not crack, which is consistent with critical diameter of Si particles. The work demonstrates the seriousness of the volume expansion and the generated stress during lithiation and delithiation of Si electrodes.

To further understand how crack propagates in a Si thin film during lithiation and delithiation, Shi et al. performed cross-sectional SEM on Si thin films [19]. Phosphorus-doped Si(100) wafers with a thickness of 500 μm were used

Mechanical properties of silicon-based electrodes Chapter | 4 129

for the test. After undergoing electrochemical tests at different states of charge (delithiation), the wafer was washed with dimethyl carbonate (DMC) and then monitored with an SEM. Cross-sections of the Si wafer were done with a focused ion beam with Ga ion to observe the electrode fracture. They also used finite element method (FEM) to simulate the stress/strain contour and the progression of the crack during delithiation. In their case, because of the use of a Si wafer, the amount of active Si and the region of lithiated material increase with the number of cycles. Cross-sectional observations are shown in Fig. 4.8. At the beginning, while the surface of the Si wafer is lithiated and delithiated, a crack

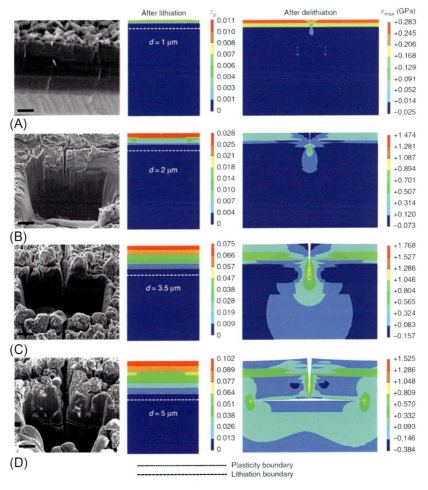

FIG. 4.8 Experimental observation and theoretical simulation of crack generated in a Si wafer after charge and discharge for (A) 3, (B) 8, (C) 30, and (D) 50 cycles [19]. Scale bar: 0.5 μm (A); 3 μm (B–D).

perpendicular to the surface is initiated. With progressive lithiation and delithiation, as the reaction front moves into the bulk of the wafer, the crack moves further into the bulk until a depth of around 5–8 µm, where lateral cracks start to be observed. The lateral cracks were attributed to the large tensile residual stresses built up in the film with delithiation, and eventually propagate along the weaker a-Si/c-Si interface in the (100) direction as opposed to into the film. The experimental results were also verified by FEM simulation. These lateral cracks will cause delamination of the film and eventually capacity degradation.

4.3 Measuring volume change of Si-based battery electrodes

For battery applications, the active material will have to be made into electrodes. Since Si will undergo volume expansion when alloying with Li with possibility of cracking, neighboring particles or fragments will push into each other leading to irregular local expansion. We therefore ask the following question: How much of the material volume change will translate to that of the electrode? To answer this, one would need an effective way to measure the volume change of an electrode in a whole.

Traditionally, the volume change of an electrode is measured by taking cross-sectional SEM images of an electrode before and after lithiation. There are several disadvantages to this approach. First, it is only possible to study the electrode at a fixed state of charge because the measurement is destructive which requires disassembling the battery. If multiple states of charge are desired, multiple cells have to be made, which is time-consuming. Second, the cross-sectioning process is complicated and requires an air-tight transfer vessel to prevent reaction of the lithiated electrode with air and moisture. If the electrode is delaminating from the current collector, the film will peel off during the cross-sectioning process. Therefore in-situ methods to monitor the state of the electrode during charge and discharge are desirable.

In-situ SEM observations of Si-based materials have previously been reported [20–22]. For example, Hovington et al. used a pouch cell on a custom-made sample holder with a solid polymer electrolyte (PEO-LiTSI) to observe the cross-sectional change of electrodes with Si and SiO_x particles [20]. Chen et al. converted a coin cell into an SEM holder by cutting away the edge of the cell to expose the electrodes using Li[TFSA]/[C2mim][FSA] (1:5 M ratio) ionic liquid as the electrolyte [22].

These methods can give real-time observation in change of volume or thickness of the electrode during charge and discharge. However, the drawbacks of in-situ SEM tests are: first, they require specialized battery setup, which is cumbersome to fabricate. Second, the fabrication process is specific to the type of electrode, making it difficult to compare performance of different types of electrodes. Third, because SEM imaging is conducted typically in vacuum, common carbonate-based liquid electrolytes cannot be used, so the cells are not tested in conditions that correspond to a real battery.

An alternative way to monitor the volume change of an electrode during charge and discharge is to measure its thickness change. Since most battery electrodes are planar, in the form of a thin film with only the active materials, or composite electrodes with Si particles, conductive agent, and binder, a volume change will therefore translate to a change in the thickness of the electrode (out-of-plane) as the in-plane direction is restricted by the electrode configuration.

The thickness change of an electrode can be measured with an in-situ dilatometer. Our group has adopted one from EL-cell (model ECD-2). A schematic of the cell is shown in Fig. 4.9. Specifically, the cell consists of a counter electrode (Li metal in our case), a T-shaped glass frit as a solid porous spacer, a working electrode, a stainless steel membrane, and also linear voltage displacement transducer (LVDT) height sensor on the top. The working principle is as follows: during charge and discharge, current will be conducted to the working electrode through the stainless steel membrane. The volume expansion and contraction of the working electrode will push on the membrane, and then the electrode thickness change will be monitored by the height sensor. The T-shaped glass frit is an essential part of the setup because it is stationary, and it separates the volume change of the lithium metal (counter electrode) from the measurement. The resolution of the LVDT is 50 nm. Therefore even for a thin film of a thickness of 1 µm, a volume expansion of the order of 5% can be observed.

The dilatometer is a useful equipment to help understand the mechanical changes in a battery electrode during charge and discharge, so that we can develop technologies and mitigation strategies to increase electrode stability to achieve the targeted cycle life. Two parameters are particularly critical for battery applications: first is how big the actual electrode thickness change is during lithiation and delithiation, and second is how reversible the thickness change is upon delithiation. Consider a fixed-volume prismatic cell with stacked electrode where the anode is Si. If the total volume expansion during lithiation is large, this can lead to large stress within the cell and cell swelling. So, ideally, one would like to reduce the amount of electrode thickness change without sacrificing the capacity. In addition, if the thickness change is not reversible, i.e. thickness does not return to original level, one would expect

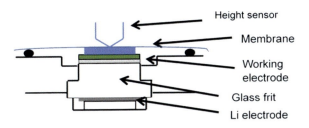 

FIG. 4.9 Schematic and photo of an in-situ dilatometer [23].

the porosity of the electrode to change, which affects the electrical connectivity within the electrode, leading to capacity losses.

One should note that the dilatometer is only able to measure the thickness change of the thickest part of the electrode. It cannot monitor local volume expansions. Nonetheless, with proper electrode fabrication and control experiments, meaningful results can still be obtained.

4.4 Improving stability of Si electrodes

4.4.1 Binder

A composite electrode is typically made up of active materials, conductive carbon black, and binder. If the active material undergoes volume expansion, one would expect that the fracture strength of the binder to be an important factor affecting how reversible the lithiation and delithiation process is, because it is the only medium in the electrode that can effectively hold the particles together even during particle shrinkage. In particular, a binder with low fracture strength will have problem with Si as it will break down under the large strain during lithiation [24].

To illustrate the importance of binder, our group took commercially available Si particles (Sigma Aldrich) with typical particle sizes ranging from 1 to 10 μm and made into electrodes with acetylene black (AB) and carboxymethyl cellulose (CMC) binder with 10, 20, or 30 wt% binder (a composition of Si:AB:CMC = 8:1:1, 6:2:2, and 4:3:3) and tested the electrodes with 1M LiPF$_6$ in fluoroethylene carbonate/diethyl carbonate (FEC/DEC) = 1:1 with Li metal as the counter electrode [23]. For simplicity, we kept the ratio of the AB and CMC constant. The electrodes were then discharged to 0.01 V and then charged back to 1 V with a current rate of 250 mA g^{-1}. The 1st cycle discharge and charge curves of the 8:1:1 and 4:3:3 electrodes are shown in Fig. 4.10. One

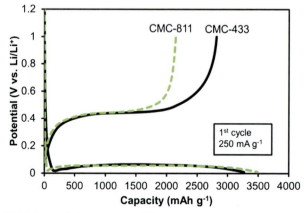

FIG. 4.10 Initial charge-discharge curves of Si electrode with 10 and 30 wt% CMC [23].

can see that both electrodes can be discharged to a capacity of about 3500 mAh $(g\ Si)^{-1}$, but the charge capacity is significantly different, with 2200 and 2800 mAh $(g\ Si)^{-1}$ for the 8:1:1 and 4:3:3 electrodes, respectively. The discharge capacity is close to that of the theoretical capacity corresponding to the formation of $Li_{3.75}Si$, so both electrodes are fully lithiated at the end of 1st discharge. However, the electrode with 10 wt% binder has a large amount of lithium trapped in the electrode after 1st charge, which is associated with the breakdown of the electrode.

The reversibility of the electrodes can be further investigated by a "depth of discharge" test, where we took the electrodes and limited the first discharge capacity to 500, 1000, 1500, 2000, 2500, 3000, and 3500 mAh $(g\ Si)^{-1}$ and measured the corresponding charge capacity until 1 V. The representative charge-discharge curves for the 4:3:3 electrodes are shown in Fig. 4.11. The results are summarized in Fig. 4.12, where the measured charge capacity is plotted against the applied discharge capacity. For the 4:3:3 and 6:2:2 electrodes, both electrodes show a linear relationship between the measured charge capacity and the applied discharge capacity with a slope of 90%. The 10% irreversibility is attributed to the Li trapped inside the material due to the amorphorization of Si during first charge, and also some in the carbon black and binder. Interestingly, the behavior of the 8:1:1 electrode is different. In particular, if the 8:1:1 electrode is discharged up to about 2000 mAh $(g\ Si)^{-1}$, the electrode can be charged similar to 4:3:3 and 6:2:2 with a charge capacity of about 1800 mAh $(g\ Si)^{-1}$. However, if the electrode is overdischarged beyond 2000 mAh $(g\ Si)^{-1}$, reversibility of the electrode becomes poor.

The observable irreversibility of the 8:1:1 electrode is due to the breakdown of the binder, as indicated by in-situ dilatometer measurements. Fig. 4.13 shows the thickness change of the 8:1:1 electrode during 1st cycle with two different

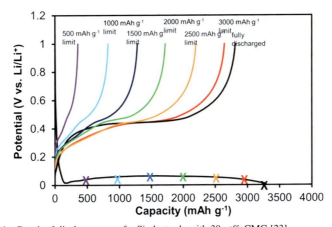

FIG. 4.11 Depth of discharge test of a Si electrode with 30 wt% CMC [23].

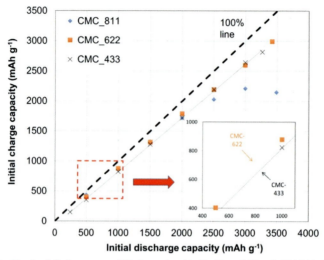

FIG. 4.12 Depth of discharge test of Si electrode with 10, 20, and 30 wt% CMC binder [23].

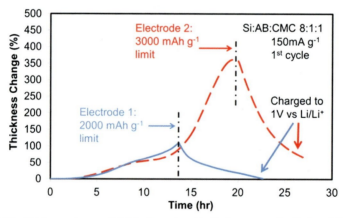

FIG. 4.13 Thickness change of 8:1:1 electrode with discharge capacity limits of 2000 and 3000 mAh g^{-1} [23].

discharge capacity limitations: 2000 and 3000 mAh g^{-1}. When the electrode is discharged to a capacity limit of 2000 mAh (g Si)$^{-1}$, the electrode experiences a thickness change of about 100%. This thickness change is reversible upon delithiation. In comparison, when the same type of electrode is discharged to a capacity of 3000 mAh (g Si)$^{-1}$ (discharge time of 20 h), the thickness change is 360%, but does not return to zero when the potential reaches 1 V. This irreversible thickness change indicates that the binder is not strong enough to pull back the particles after expansion. This result also suggests that to improve the mechanical and electrochemical stability of a Si electrode, one should increase

Mechanical properties of silicon-based electrodes **Chapter | 4** **135**

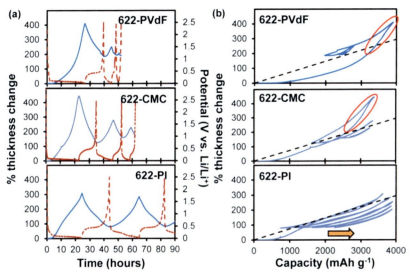

FIG. 4.14 (A) In-situ dilatometer results from Si electrodes with CMC, PVdF, and PI. (B) Thickness change vs. cumulative capacity of Si electrodes with CMC, PVdF, and PI [25]. *(Reproduced with permission from John Wiley and Sons.)*

the amount of binder, optimize the binder in terms of strength and adhesion, and limit the discharge capacity.

To compare the effect of different binders on the stability of the electrodes, our group performed in-situ dilatometer measurements on Si electrodes with 20 wt% CMC, polyvinylidene fluoride (PVdF), and polyimide (PI) as the binder (electrode configuration of Si:AB:binder = 6:2:2) [25]. Fig. 4.14A shows the thickness change and voltage vs. time for the three different electrodes under a constant current of 150 mA g^{-1} for 3 cycles. All 3 electrodes have the same electrode density of about 1.2 g cm^{-3}. One can see that the electrode with PVdF shows the shortest measurement time, followed by CMC and polyimide, with better capacity reversibility for PI. A detailed study of the thickness change curves indicates that even though all electrodes give similar initial discharge capacity of about 3500 mAh g^{-1}, their corresponding thickness changes are different. In particular, CMC and PVdF show thickness change of more than 400% after lithiation, whereas that for PI is only about 300%. In fact, at the end of lithiation, both CMC and PVdF electrodes experience a drastic increase in thickness (after about 2500 mAh g^{-1}), which is attributed to the breakdown of the binders (see the thickness change vs. cumulative capacity plot in Fig. 4.14B). On the other hand, PI does not show the drastic increase in thickness at the end of lithiation. In addition, the reversible thickness change for the PI electrode is higher upon charging. This is because PI is a high elastic modulus polymer so it can exert a compressive force within the electrode to pull the particles back during delithiation.

Recently there have been a few research papers on binder development for Si anode, indicating that apart from binder strength, adhesion with the active materials is also an important factor affecting the electrode stability [26]. For example, Bridel et al. compared Si/C/CMC composite electrodes with Na carboxyl methyl cellulose made in an acidic vs. neutral solution. The electrode made under acidic condition shows much better cycle performance than that made in neutral condition. They used solid state nuclear magnetic resonance spectroscopy to show that the CMC chains will bind to the surface of Si via a covalent-hydrogen bond in an acid medium (Fig. 4.15), leading to the better mechanical stability of the electrode [27].

Magasinski et al. studied poly(acrylic acid) (PAA) binder for Si anode, where PAA consists of a carboxylic acid group in the repeating unit shown in Fig. 4.16A that can form strong hydrogen bond in-between the polymer chain, and between the polymer and Si particle. They attributed the better stability of the Si electrode with PAA binder to these strong interactions [28]. Kovalenko et al. also showed better mechanical stability of Si electrode with sodium alginate (Fig. 4.16B), which can form stable hydrogen bonding and ion-dipole interactions between the binder and the Si particles [29].

Since the interactions between polymer chains, and also between Si and binder are important, one strategy to improve electrode stability is to optimize the functional groups of the binder. For example, Ryou et al. added catechol group (C) to PAA and sodium alginate (Alg) and used them as binders for Si electrodes (Fig. 4.16C) [30]. They measured the bond strength between binder and Si with atomic force microscopy and found that the bond strength increases from ~73 to ~750 pN with the addition of catechol group. They attributed the improvement in capacity and cycle performance to the stronger interaction of PAA-C and Alg-C. Liu et al. on the other hand uses guar gum (Fig. 4.16D) with many hydroxyl groups to form hydrogen bonds with the surface of Si nanoparticles. They showed that electrodes with guar gum are more stable than that with sodium alginate [31].

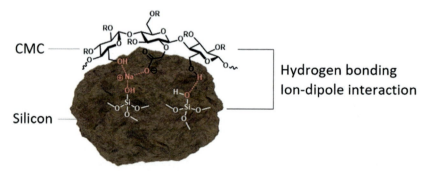

FIG. 4.15 Schematic diagram of structural characteristics and interactions of CMC with silicon with hydrophilic surface [26]. *(Reproduced with permission from Royal Society of Chemistry.)*

Mechanical properties of silicon-based electrodes **Chapter | 4** **137**

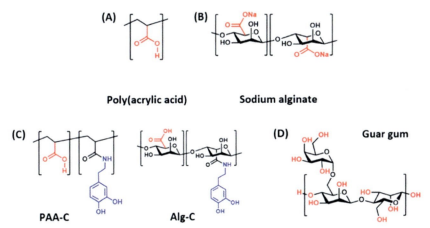

FIG. 4.16 Chemical structures of polymeric binders: (A) poly(acrylic acid), (B) sodium alginate, (C) catechol-functionalized PAA and alginate, and (D) guar gum [26]. *(Reproduced with permission from Royal Society of Chemistry.)*

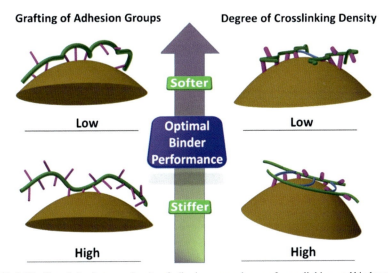

FIG. 4.17 Correlation between density of adhesion group, degree of cross-linking, and binder performance [32]. *(Reproduced with permission from American Chemical Society.)*

Cao et al. recently published a work on Si electrode with catechol-functionalized chitosan cross-linked by glutaraldehyde (CS-CG+GA) as the binder, where the catechol group bonds with surface of the Si particles and the cross-link creates a 3D network (Fig. 4.17) [32]. Their results indicate that there is an optimal adhesion strength and degree of cross-linking. Low density of adhesion group with low degree of cross-linking will lead to a soft binder, whereas high density of adhesion group with high degree of cross-linking will make a stiff binder. Electrochemical performance is poor in both of these cases.

Another direction of research is to use self-healing polymers, in particular one with low glass transition temperature, for Si application. The idea is to create a "viscous" medium so that electrode cracks induced by the Si expansion can be recovered. For example, Wang et al. made a self-healing polymer using diacid and triacid with diethylenetriamine and urea as shown in Fig. 4.18 [33]. The self-healing ability of the polymer originates from the hydrogen bonds between the urea molecules. The SEM images verified that the crack diminishes in size, allowing stable operation of Si microparticles.

All in all, binder is an important part of Si electrodes for battery applications. With the advance of organic chemistry and polymer science, new binders with different functional groups that can both maintain good adhesion with the active materials and improve mechanical stability of the electrode will be developed in the future. This will allow reversible lithiation and delithiation of Si-based electrodes.

4.4.2 Particle size

Apart from binder engineering, many researchers use nanomaterials with low electrode porosity to increase stability of Si electrodes, as the void between the particles can accommodate the large volume change during charge and discharge [34–36]. However, small particles have large surface area which will increase the amount of side reaction between the electrolyte and Li-Si alloy on the surface of the particles, lower Coulombic efficiency during cycling, and reduce available capacity. In addition, the low electrode density reduces the volumetric energy density of the electrode. Take graphite electrode for example, the practical capacity is about 350 mAh g^{-1} with a packing density of about 1.6 g cm^{-3}. So the volumetric energy density is 560 mAh cm^{-3}. In order for a Si electrode to achieve higher volumetric density than graphite, packing density of the electrode will have to be more than 0.2 g cm^{-3}. One compromise to reduce surface area and increase density would therefore be to agglomerate the primary Si nanoparticles into secondary clusters to take advantage of the smaller particle size.

This has shown to be a viable strategy to improve the cycle performance of Si electrodes. Our group has previously used high-energy ballmilling method to convert large micron-sized Si (m-Si) particles into Si secondary particle cluster (SiSPC) [37]. The SEM images of the particles are shown in Fig. 4.19A–D. The initial Si particles are 5–10 μm in size. After ballmilling at 500 rpm for 6 h, the particles are broken down into secondary particles of about 3–10 μm in size with primary particles of the order of 100–200 nm. Because of clustering effect, the tap density of the SiSPC is close to that of the m-Si, whereas its BET surface area is smaller than commercially available nanoSi (n-Si) (see Fig. 4.19E for the physical appearance of the material with 0.3 g Si and Table 4.3 for its physical parameters).

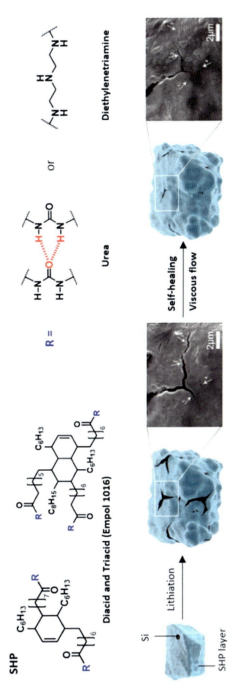

FIG. 4.18 (Top) Chemical structure of self-healing polymer (SHP) and (bottom) schematic diagram of the healing mechanism and SEM images of the SHP-Si anode [26]. *(Reproduced with permission from Royal Society of Chemistry.)*

FIG. 4.19 SEM images of (A) m-Si, (B–D) SiSPC (with different magnifications); (E) appearance with 0.3 g of each material; (F) N$_2$ adsorption and desorption isothermals of different Si materials [37]. *(Reproduced with permission from American Chemical Society.)*

TABLE 4.3 BET surface area and tap density of Si particles [37].

	m-Si	SiSPC	n-Si
BET surface area (m^2 g^{-1})	1.4	12.6	26.5
Tap density (g cm^{-3})	0.94	0.81	0.17

Reproduced with permission from American Chemical Society.

Polyimide used as the binder (20 wt% unless specified otherwise) for the test and the cycle performance of various Si electrodes in 1M LiPF$_6$ in FEC/DEC = 1:1 as the electrolyte is shown in Fig. 4.20. Initial charge capacity of the SiSPC is as high as 3154 mAh g^{-1}, which is the same as the capacity of micron-sized Si. In contrast, initial charge capacity of nanoSi (n-Si) particles is only 1998 mAh g^{-1}. The result suggests that there is less surface oxide and SEI formation on the SiSPC as compared with n-Si due to smaller surface area. One can see that SiSPC can improve the cycle performance and Coulombic efficiency. The capacity retention of 78.3% is obtained with 20 wt% PI as binder after 70 cycles with an average Coulombic efficiency of 99.25%. This is higher than the capacity retention and Coulombic efficiency of other reference materials. Cycle stability can be further improved by increasing the PI binder content to 30 wt%, with 89.8% capacity retention after 70 cycles.

Fig. 4.21 shows the in-situ dilatometer profile of the SiSPC electrode with 30 wt% PI. One can see that, despite the large change in electrode thickness of 320% upon lithiation when discharged to about 3700 mAh g^{-1}, 90% of the change in thickness is reversible after delithiation. The nanosized primary particles in the secondary cluster reduce particle cracking. In addition, the polyimide binder is able to exert a compressive force on the particles during charging to maintain the integrity of the electrode. The SiSPC is effective even at a higher current rate of 3500 mA g^{-1} (1C rate), where 95% of the initial capacity is retained after 500 cycles [37].

4.4.3 Active matrix with oxygen

To improve the stability of Si-based material and reduce volume expansion, one of the strategies is to incorporate oxygen into the lattice, as in silicon monoxide (SiO). SiO is a metastable phase that can undergo charge and discharge with a theoretical capacity of 2615 mAh g^{-1} when 17.2 Li reacts with 4 SiO to form 3 Li$_{3.75}$Si and lithium silicate [38, 39]. Because the volume expansion due to the formation of Li$_4$SiO$_4$ and Li$_2$O is only about 100%, the addition of oxygen into the Si lattice will decrease the overall particle expansion to about 160% [40, 41], as opposed to 312% for Si. Electrode mechanical failure induced by SiO

142 PART | II Mechanical properties

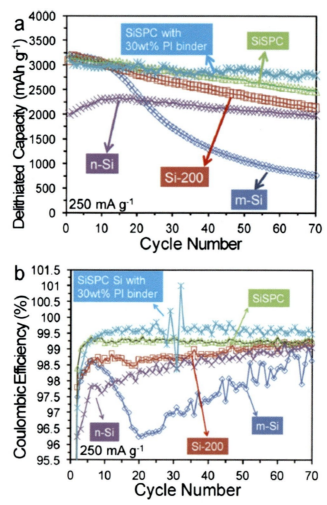

FIG. 4.20 (A) Cycle performance and (b) coulombic efficiency of various Si materials at 250 mA g^{-1} [37]. *(Reproduced with permission from American Chemical Society.)*

particles is therefore less severe than that by pure Si particles. In addition, the coverage of Li$_4$SiO$_4$ and Li$_2$O after lithiation prevents direct contact between lithium silicide and electrolyte, thus avoids excessive growth of solid electrolyte interphase [42].

The effect of the oxygen matrix on the electrochemical performance of Si is investigated, as shown in the following text. To eliminate the effect of particle size, Si and SiO with average particle size of 1 μm are used for the test. Both Si and SiO materials are mixed with AB and PAA binders in NMP in a ratio of 6:1:2, and the slurry coated onto Cu foil to form electrodes in the same

Mechanical properties of silicon-based electrodes **Chapter | 4** **143**

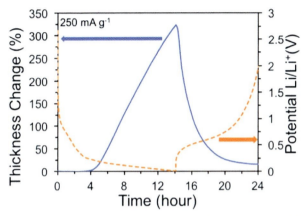

FIG. 4.21 In-situ dilatometer and voltage profiles of SiSPC electrode with 30 wt% PI [37]. *(Reproduced with permission from American Chemical Society.)*

way. The electrodes were cut into 16 mm discs and subjected to roll press. Typical thickness and packing density of the Si and SiO electrodes are around 33–34 μm and 1.1 g cm^{-3}, respectively. The electrodes were then made into 2032 coin cells with Li as the counter electrode and 1M LiPF$_6$ in FEC/DEC = 1:1 as the electrolyte, and tested at a current rate of 150 mA g^{-1} between 0 and 2 V.

Fig. 4.22A and B shows a comparison of the charge-discharge curves of the Si and SiO electrodes. Similar to previously reported results, Si electrode can

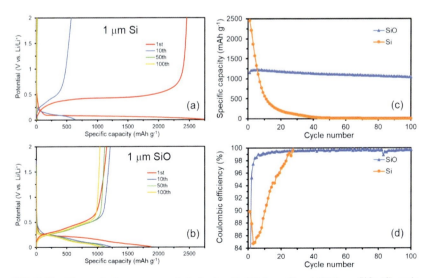

FIG. 4.22 Charge-discharge curves of electrode with (A) 1 μm Si and (B) 1 μm SiO; (C) cycle performance and (D) coulombic efficiency of the electrodes.

TABLE 4.4 Charge-discharge performance of electrodes with 1 µm Si and SiO.

Electrode	Si	SiO
1st discharge capacity (mAh g^{-1})	2735.1	1879.9
1st charge capacity (mAh g^{-1})	2457.3	1154.4
1st CE (%)	89.8	61.4
Capacity retention rate (100th/1st, %)	0.3	90.2

give an initial discharge and charge capacity of 2735.1 and 2457.3 mAh g^{-1}, respectively, with a first-cycle Coulombic efficiency of 89.8% (see Table 4.4). Though, upon cycling, capacity drastically decreases. Even after 10 cycles, available capacity is dropped to about 600 mAh g^{-1} (Fig. 4.22C). Coulombic efficiency is only between 84 and 90% during the first 10 cycles, indicating that there are many irreversible losses during cycling. In comparison, SiO gives an initial discharge and charge capacity of 1879.9 and 1154.4 mAh g^{-1}, respectively, with a first-cycle Coulombic efficiency of 61.4%. The lower available capacity and first-cycle Coulombic efficiency are due to the formation of inactive Li-Si-O compounds during initial discharge. Unlike Si, cycle capacity of SiO is rather stable, with capacity retention of about 90.2% after 100 cycles. Charge-discharge curves of SiO do not change much during cycling (Fig. 4.22B). In addition, Coulombic efficiency during cycling remains more than 99% for most of the time, indicating SiO has better stability and also forms a more stable SEI layer than Si.

To understand the mechanical change of the Si and SiO electrodes during charge and discharge, the electrodes were subjected to dilatometer tests. Fig. 4.23 shows a comparison of the thickness change of the Si and SiO electrodes during initial cycle with respect to the amount of Li inserted into the material. For Si, lithiation leads to a thickness increase of 230%, while 44% of the thickness increase remains after delithiaton. In comparison, lithiation of SiO leads to a thickness increase of only 127% (with similar amount of Li inserted), while 18% of the thickness increase remains after delithiaton. One can see that with the same amount of Li insertion, the thickness change is smaller for SiO with the presence of O in the lattice, i.e. oxygen reduces the amount of expansion of the lattice. In addition, more of the thickness change of SiO is reversible, contributing to the better cycle performance of SiO compared to Si.

Even though 1 µm SiO shows good stability during cycling, mechanical stress build-up can still lead to particle cracking and capacity fading, in particular for larger particles. Our group has observed this for SiO particles with 5 µm particle size [43]. When 5 µm SiO with a surface layer of carbon (SiO@C) was

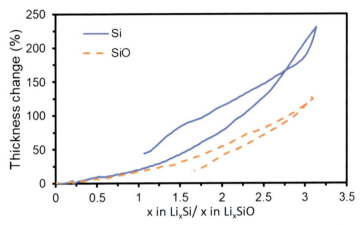

FIG. 4.23 Thickness change of Si and SiO electrodes during initial cycle by in-situ dilatometry.

made into electrodes with PAA as binder, drastic capacity fading is observed with cycling within the initial 30 cycles (see Fig. 4.24). Coulombic efficiency during the initial 30 cycles is between 94 and 99%, suggesting there are much losses probably due to SEI formation during particle cracking. The performance is significantly worse than electrodes with 1 μm SiO particles tested under the same electrode configuration and electrolyte. The SiO@C electrode was then subjected to in-situ dilatometer test (see Fig. 4.25) to see the thickness change during charge-discharge. During initial lithiation, a capacity of about 1900 mAh g^{-1} is observed, with a 158% increase in electrode thickness, which is attributable to particle cracking and electrode degradation. This is verified by postmortem SEM observation, which shows cracks on the electrode surface after 100 cycles.

To help contain the particle during charge-discharge, one strategy is to perform surface coating of the particle. The surface coating material will have to be able to sustain the large mechanical strain, thus our group used a high-modulus polyimide (PI) material for the purpose. We have previously shown that PI coating is able to maintain the cycle stability of Sb material for LIB [44] by first mixing polyimide precursor (9.4 wt%) with 1–2 μm Sb powder and heat treating at 400°C in Ar atmosphere to polymerize the surface coating. When the same PI surface coating technique is applied to 5 μm SiO@C particles, an improvement in cycle stability is recorded (see Fig. 4.24), however, capacity still decreases during the first 30 cycles. Coulombic efficiency is improved to between 98.5 and 99.5% within the initial 30 cycles, yet there are still some losses. We suspect this is due to poor adhesion between the surface of SiO and PI, so the surface coating is not well covered.

We then developed a novel method to improve the adhesion between the SiO@C particle surface and PI by employing a self-assembled monolayer (SAM) in between. Specifically, the SiO@C is first treated by UV ozone plasma

146 PART | II Mechanical properties

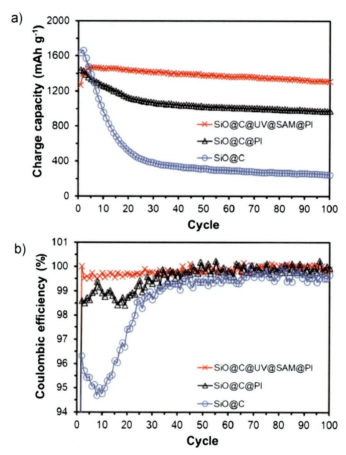

FIG. 4.24 (A) Cycle performance and (b) coulombic efficiency of SiO@C electrode with or without particle coating (binder = PAA) [43].

treatment to increase the amount of hydroxyl functional group on the surface. Then a SAM monolayer of (3-aminopropyl)triethoxysilane (APTES) is deposited on the surface of the particle with a condensation reaction between the -OH group of the particle and the silane group of APTES. Then, a PI layer is covalently bonded on the surface with the amine functional group of the APTES, forming a surface-coated SiO (SiO@C@UV@SAM@PI). With the SAM layer, the materials give much more stable capacity retention during cycling (Fig. 4.24). Coulombic efficiency is also improved to above 99.5% during all the cycles due to the protection of the surface by the PI layer.

In-situ dilatometry tests indicate that the thickness change of electrode with SiO@C@UV@SAM@PI is significantly smaller than that with just SiO@C even with similar charge-discharge capacity (Fig. 4.25). The intrinsic volume expansion of the SiO is expected to be the same, so the SAM@PI layer is able

Mechanical properties of silicon-based electrodes **Chapter | 4** **147**

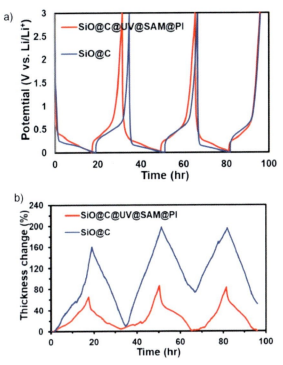

FIG. 4.25 (A) Voltage-time profiles and (B) thickness-time profiles of SiO@C with and without surface coating during the initial 3 cycles at 100 mA g^{-1} [43].

to keep the particles together and prevent the fragments from interacting with each other. A postmortem SEM image of the SIO@C@UV@SAM@PI after 100 cycles shows that the electrode is still flat, similar to that of a pristine electrode, indicating that the surface coating strategy is able to suppress mechanical change and improve reversibility (Fig. 4.26). The electrode can be cycled to

FIG. 4.26 Postmortem SEM images of electrode with (A) SiO@C and (B) SiO@C with SAM-PI coating after 100 cycles [43].

more than 300 cycles with a stable capacity. In addition, good cycle performance with LiFePO$_4$ cathode was also demonstrated [43].

We see that good cycle stability of SiO can be sustained with the smaller particle size and also surface coating. Though, the main disadvantage of SiO as anode material is its low initial Coulombic efficiency (65.1%–82.1%) [45]. Li-ion is trapped after the first lithiation, and the majority of the irreversibility originates from the formation of Li$_2$O and lithium silicate [46]. This remains a problem to be solved before the practical use of SiO for battery applications.

4.4.4 Silicon thin films

Thin-film batteries are important for providing power to on-board chips and sensors for various applications. A recent review on Si thin-film battery was published by Mukanova et al. [47]. Because the electrodes typically do not contain binder in them, there needs to be other ways to accommodate the volume change and suppress mechanical degradation during cycling.

Maranchi et al. were one of the first researchers who published the results on Si thin films [48]. They deposited 250 nm and 1 μm amorphous Si thin film on Cu substrate by magnetron sputtering and observed excellent cycle performance for 30 cycles for the 250 nm film, whereas the capacity started to decay after about 12 cycles for the 1 μm film. The good cycle performance of the 250 nm film is attributed to good adhesion between Si and Cu substrate. Also, they attributed the poor cycle performance of the 1 μm film to residual stress.

To further study the volume expansion of Si thin films, our group has also deposited 1 μm Si film on Cu current collector by magnetron sputtering [49]. The Si thin film is amorphous in structure, as verified by X-ray diffraction. When the electrodes are cycled in a coin cell with 1M LiPF$_6$ in FEC/DEC = 1:1, fast capacity fading is observed, which is verified to be due to electrode cracking and delamination.

To reduce the mechanical issue, we deposited a 500 nm polyimide (PI)-capping layer on the 1 μm Si thin film. Polyimide is a well-known high-modulus polymer that was shown earlier as effective binder for Si. Even as a capping layer, it is effective to maintain stability of the Si thin film. Even though the capacity of the thin film is reduced to about 2610 mAh g^{-1} with the inactive PI, about 97.5% of the capacity is retained after 50 cycles. Coulombic efficiency is also improved to close to 100% (see Fig. 4.27).

A comparison of the thickness change of the Si thin film with and without PI capping layer is shown in Fig. 4.28A. Initially, both Si thin films show linear increase in thickness with lithiation. Beyond about Li$_{1.4}$Si, the Si thin film demonstrates a larger increase in thickness compared with the one with PI capping layer. At the end of lithiation (to about Li$_{3.75}$Si), the thickness change of the Si thin film with and without PI capping is 280% and 345%, showing that the capping layer is able to reduce thickness change. In addition, the thickness change

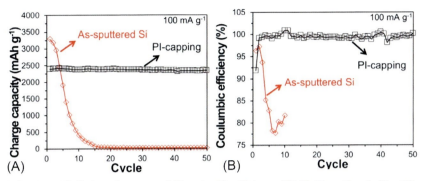

FIG. 4.27 (A) Cycle performance and (B) coulombic efficiency of Si thin film with and without PI capping layer [49].

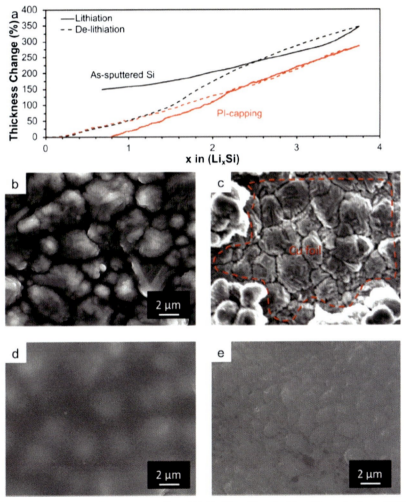

FIG. 4.28 (A) In-situ dilatometry profiles of Si thin film with and without PI capping layer; SEM images of Si thin (B) before and (C) after cycling; SEM images of Si thin film with PI capping layer (D) before and (E) after cycling [49].

is reversible for the one with PI capping layer, whereas the one without capping layer shows irreversible changes after delithiation. Postmortem SEM observations verify significant electrode delamination after cycling for film without PI layer, while no cracks are observed with PI layer (see Fig. 4.28B-E). Thus the PI capping layer is able to improve mechanical integrity of the electrode.

Apart from PI, LiPON coating was also previously shown to improve electrochemical stability of Si thin film by providing mechanical support [50–52].

4.4.5 Inactive matrix with titanium

During lithiation, the lattice of Si will have to expand in order to accommodate the Li atoms. This intrinsic expansion is not avoidable if we have pure Si. In order to reduce the volume change, one possibility is to incorporate inactive foreign atoms that can act as buffer. Our group has explored the effect of titanium (Ti) addition into the lattice on the stability of Si [53]. In order to clearly see the effect of Ti without the influence of binder and carbon additives, Si-Ti thin films are deposited on Cu current collector by magnetron sputtering. The deposited films are amorphous in nature, as verified by X-ray diffraction.

The charge-discharge curves of different Si-Ti thin films are shown in Fig. 4.29A. The addition of Ti leads to a decrease in capacity, which is expected because Ti does not alloy with Li. The first cycle efficiency of Si-Ti is as high as 90%, similar to Si, indicating that Ti is not trapping Li in the lattice. Though, if we calculate the amount of alloyed Li per Si, we see a decreasing trend with increasing Ti content (decreasing Si content) as shown in Fig. 4.29B, suggesting that Ti partially blocks the lithiation of Si.

Despite the decrease in capacity, Ti addition improves cycle performance and Coulombic efficiency of Si. Specifically, Si thin film with 20% Ti addition delivers a stable capacity with a capacity retention of 94.8% after 50 cycles, with close to 100% CE (Fig. 4.30a, B). The 80%Si-20%Ti film also shows

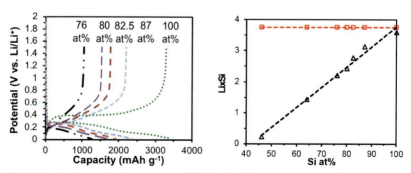

FIG. 4.29 (A) Charge-discharge curves of Si-Ti with different percentage of Si. (B) Degree of lithiation of Si-Ti with different amounts of Si (*red*—assuming no influence from Ti; *black*—experimentally measured results) [53]. *(Reproduced with permission from John Wiley and Sons.)*

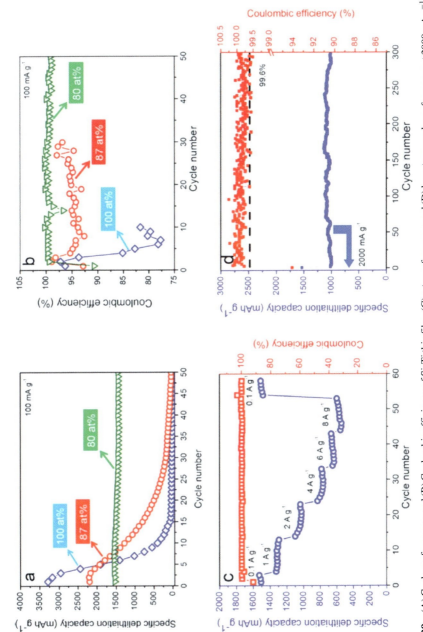

FIG. 4.30 (A) Cycle performance and (B) Coulombic efficiency of Si-Ti thin film; (C) rate performance and (D) long-term cycle performance at 2000 mA g^{-1} of the 80%Si-20%Ti film [53]. *(Reproduced with permission from John Wiley and Sons.)*

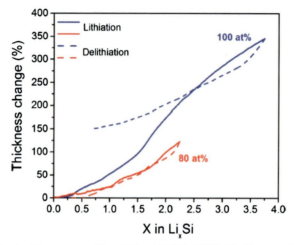

FIG. 4.31 In-situ dilatometry profiles of Si and 80%Si-20%Ti thin films at 100mAg^{-1} [53]. *(Reproduced with permission from John Wiley and Sons.)*

excellent rate performance of up to 8 A g^{-1} and good long-term cycling capability (Fig. 4.30C, D).

Characterization by Raman spectroscopy indicates that the interaction between Si and Ti is recoverable after cycling. Moreover, the addition of Ti reduces the overall thickness change of the electrode during lithiation and allows for a highly reversible thickness change during delithiation (see Fig. 4.31). These results indicate that Ti is not a spectator atom inside the Si matrix, but actively plays the role as an atomic glue to hold the Si atoms together, ultimately leading to better cycle performance for both half cells and full cells. We expect the advantage of Ti to be also observable in composite electrodes with Si-Ti particles, which can be a direction for future development.

4.5 Concluding remarks

In this chapter, we have shown that the volume change of Si is an important factor affecting the mechanical stability of the electrode for battery applications. Much work has been done to quantify the changes and to develop mitigation technologies to make the Si-based electrodes more reversible. Many strategies including binder optimization, particle size optimization, incorporation of active and inactive matrixes, and capping layer are viable to reduce volume changes and provide mechanical support to the electrode. These provide a platform that can facilitate commercialization of Si-based anode for high energy density LIBs in the future. Though, cyclability of Si-based materials is not only due to mechanical effect, but also other factors such as chemical reaction with electrolyte and unstable SEI formation have to be investigated and optimized concurrently.

References

[1] U. Heider, R. Oesten, M. Jungnitz, Challenge in manufacturing electrolyte solutions for lithium and lithium ion batteries quality control and minimizing contamination level, J. Power Sources 81 (1999) 119–122.
[2] Panasonic NCR18650BF Batteries Datasheet. https://na.industrial.panasonic.com/products/batteries/rechargeable-batteries/lineup/lithium-ion. (Accessed 22 May 2020).
[3] W.J. Zhang, A review of the electrochemical performance of alloy anodes for lithium-ion batteries, J. Power Sources 192 (6) (2011) 13–24.
[4] V.A. Agubra, J.W. Fergus, The formation and stability of the solid electrolyte interface on the graphite anode, J. Power Sources 268 (2014) 153–162.
[5] H. Okamoto, Li-Si (lithium-silicon), J. Phase Equilib. Diffus. 30 (1) (2009) 118–119.
[6] C.J. Wen, R.A. Huggins, Chemical diffusion in intermediate phases in the lithium-silicon system, J. Solid State Chem. 37 (3) (1981) 271–278.
[7] B. Key, R. Bhattacharyya, M. Morcrette, V. Seznec, J.M. Tarascon, C.P. Grey, Real-time NMR investigations of structural changes in silicon electrodes for lithium-ion batteries, J. Am. Chem. Soc. 131 (26) (2009) 9239–9249.
[8] C.H. Doh, M.W. Oh, B.C. Han, Lithium alloying potentials of silicon as anode of lithium secondary batteries, Asian J. Chem. 25 (10) (2013) 5739–5743.
[9] M.K. Datta, P.N. Kumta, In situ electrochemical synthesis of lithiated silicon–carbon based composites anode materials for lithium ion batteries, J. Power Sources 194 (2) (2009) 1043–1052.
[10] H. Kim, C.Y. Chou, J.G. Ekerdt, G.S. Hwang, Structure and properties of Li–Si alloys: a first-principles study, J. Phys. Chem. C 115 (5) (2011) 2514–2521.
[11] S.W. Lee, M.T. McDowell, J.W. Choi, Y. Cui, Anomalous shape changes of silicon nanopillars by electrochemical lithiation, Nano Lett. 11 (7) (2011) 3034–3039.
[12] J.L. Goldman, B.R. Long, A.A. Gewirth, R.G. Nuzzo, Strain anisotropies and self-limiting capacities in single-crystalline 3D silicon microstructures: models for high energy density lithium-ion battery anodes, Adv. Funct. Mater. 21 (13) (2011) 2412–2422.
[13] W. Liang, H. Yang, F. Fan, Y. Liu, X.H. Liu, J.Y. Huang, T. Zhu, S. Zhang, Tough germanium nanoparticles under electrochemical cycling, ACS Nano 7 (4) (2013) 3427–3433.
[14] K. Zhao, M. Pharr, Q. Wan, W.L. Wang, E. Kaxiras, J.J. Vlassak, Z. Suo, Concurrent reaction and plasticity during initial lithiation of crystalline silicon in lithium-ion batteries, J. Electrochem. Soc. 159 (3) (2012) A238–A243.
[15] M.T. McDowell, S.W. Lee, J.T. Harris, B.A. Korgel, C. Wang, W.D. Nix, Y. Cui, In situ TEM of two-phase lithiation of amorphous silicon nanospheres, Nano Lett. 13 (2) (2013) 758–764.
[16] X.H. Liu, L. Zhong, S. Huang, S.X. Mao, T. Zhu, J.Y. Huang, Size-dependent fracture of silicon nanoparticles during lithiation, ACS Nano 6 (2) (2012) 1522–1531.
[17] Z. Ma, T. Li, Y.L. Huang, J. Liu, Y. Zhou, D. Xue, Critical silicon-anode size for averting lithiation-induced mechanical failure of lithium-ion batteries, RSC Adv. 3 (20) (2013) 7398–7402.
[18] J. Li, A.K. Dozier, Y. Li, F. Yang, Y.T. Cheng, Crack pattern formation in thin film lithium-ion battery electrodes, J. Electrochem. Soc. 158 (6) (2011) A689–A694.
[19] F. Shi, Z. Song, P.N. Ross, G.A. Somorjai, R.O. Ritchie, K. Komvopoulos, Failure mechanisms of single-crystal silicon electrodes in lithium-ion batteries, Nat. Commun. 7 (1) (2016) 1–8.
[20] P. Hovington, M. Dontigny, A. Guerfi, J. Trottier, M. Lagacé, A. Mauger, C.M. Julien, K. Zaghib, In situ scanning electron microscope study and microstructural evolution of nano silicon anode for high energy Li-ion batteries, J. Power Sources 248 (2014) 457–464.

[21] C.Y. Chen, T. Sano, T. Tsuda, K. Ui, Y. Oshima, M. Yamagata, M. Ishikawa, M. Haruta, T. Doi, M. Inaba, S. Kuwabata, In situ scanning electron microscopy of silicon anode reactions in lithium-ion batteries during charge/discharge processes, Sci. Rep. 6 (2016) 36153.

[22] C.Y. Chen, A. Sawamura, T. Tsuda, S. Uchida, M. Ishikawa, S. Kuwabata, Visualization of Si anode reactions in coin-type cells via operando scanning electron microscopy, ACS Appl. Mater. Interfaces 9 (41) (2017) 35511–35515.

[23] P.K. Lee, Y. Li, D.Y.W. Yu, Insights from studying the origins of reversible and irreversible capacities on silicon electrodes, J. Electrochem. Soc. 164 (1) (2017) A6206–A6212.

[24] H. Wu, Y. Cui, Designing nanostructured Si anodes for high energy lithium ion batteries, Nano Today 7 (5) (2012) 414–429.

[25] D.Y.W. Yu, M. Zhao, H.E. Hoster, Suppressing vertical displacement of lithiated silicon particles in high volumetric capacity battery electrodes, ChemElectroChem 2 (8) (2015) 1090–1095.

[26] T.W. Kwon, J.W. Choi, A. Coskun, The emerging era of supramolecular polymeric binders in silicon anodes, Chem. Soc. Rev. 47 (6) (2018) 2145–2164.

[27] J.S. Bridel, T. Azais, M. Morcrette, J.M. Tarascon, D. Larcher, Key parameters governing the reversibility of Si/carbon/CMC electrodes for Li-ion batteries, Chem. Mater. 22 (3) (2010) 1229–1241.

[28] A. Magasinski, B. Zdyrko, I. Kovalenko, B. Hertzberg, R. Burtovyy, C.F. Huebner, T.F. Fuller, I. Luzinov, G. Yushin, Toward efficient binders for Li-ion battery Si-based anodes: polyacrylic acid, ACS Appl. Mater. Interfaces 2 (11) (2010) 3004–3010.

[29] I. Kovalenko, B. Zdyrko, A. Magasinski, B. Hertzberg, Z. Milicev, R. Burtovyy, I. Luzinov, G. Yushin, A major constituent of brown algae for use in high-capacity Li-ion batteries, Science 334 (6052) (2011) 75–79.

[30] M.H. Ryou, J. Kim, I. Lee, S. Kim, Y.K. Jeong, S. Hong, J.H. Ryu, T.S. Kim, J.K. Park, H. Lee, J.W. Choi, Mussel-inspired adhesive binders for high-performance silicon nanoparticle anodes in lithium-ion batteries, Adv. Mater. 25 (11) (2013) 1571–1576.

[31] J. Liu, Q. Zhang, T. Zhang, J.T. Li, L. Huang, S.G. Sun, A robust ion-conductive biopolymer as a binder for Si anodes of lithium-ion batteries, Adv. Funct. Mater. 25 (23) (2015) 3599–3605.

[32] P.F. Cao, G. Yang, B. Li, Y. Zhang, S. Zhao, S. Zhang, A. Erwin, Z. Zhang, A.P. Sokolov, J. Nanda, T. Saito, Rational design of a multifunctional binder for high-capacity silicon-based anodes, ACS Energy Lett. 4 (5) (2019) 1171–1180.

[33] C. Wang, H. Wu, Z. Chen, M.T. McDowell, Y. Cui, Z. Bao, Self-healing chemistry enables the stable operation of silicon microparticle anodes for high-energy lithium-ion batteries, Nat. Chem. 5 (12) (2013) 1042.

[34] F. Jeschull, F. Scott, S. Trabesinger, Interactions of silicon nanoparticles with carboxymethyl cellulose and carboxylic acids in negative electrodes of lithium-ion batteries, J. Power Sources 431 (2019) 63–74.

[35] J.R. Szczech, S. Jin, Nanostructured silicon for high capacity lithium battery anodes, Energ. Environ. Sci. 4 (1) (2011) 56–72.

[36] H. Li, X. Huang, L. Chen, Z. Wu, Y. Liang, A high capacity nano Si composite anode material for lithium rechargeable batteries, Electrochem. Solid St. 2 (11) (1999) 547–549.

[37] P.K. Lee, T. Tan, S. Wang, W. Kang, C.S. Lee, D.Y.W. Yu, Robust micron-sized silicon secondary particles anchored by polyimide as high-capacity, high-stability Li-ion battery anode, ACS Appl. Mater. Interfaces 10 (40) (2018) 34132–34139.

[38] T. Chen, J. Wu, Q. Zhang, X. Su, Recent advancement of SiOx based anodes for lithium-ion batteries, J. Power Sources 363 (2017) 126.

[39] H. Yamamura, K. Nobuhara, S. Nakanishi, H. Iba, S. Okada, Investigation of the irreversible reaction mechanism and the reactive trigger on SiO anode material for lithium-ion battery, J. Cerma. Soc. Jpn. 119 (2011) 855.
[40] T. Kim, S. Park, S.M. Oh, Solid-state NMR and electrochemical dilatometry study on Li$^+$ uptake/extraction mechanism in SiO electrode, J. Electrochem. Soc. 154 (12) (2007) A1112–A1117.
[41] S.C. Jung, H.J. Kim, J.H. Kim, Y.K. Han, Atomic-level understanding toward a high-capacity and high-power silicon oxide (SiO) material, J. Phys. Chem. C 120 (2) (2016) 886–892.
[42] L. Zhang, J. Deng, L. Liu, W. Si, S. Oswald, L. Xi, M. Kundu, G. Ma, T. Gemming, S. Baunack, F. Ding, C.L. Yan, O.G. Schimidt, Hierarchically designed SiOx/SiOy bilayer nanomembranes as stable anodes for lithium ion batteries, Adv. Mater. 26 (26) (2014) 4527–4532.
[43] T. Tan, P.K. Lee, N. Zettsu, K. Teshima, D.Y.W. Yu, Highly stable lithium-ion battery anode with polyimide coating anchored onto micron-size silicon monoxide via self-assembled monolayer, J. Power Sources 453 (2020) 227874.
[44] S. Wang, P.K. Lee, X. Yang, A. Rogach, A. Amstrong, D.Y.W. Yu, Polyimide-cellulose interaction in Sb anode enables fast charging lithium-ion battery application, Mater. Today Energy 9 (2018) 295–302.
[45] Z. Wang, Y. Fu, Z. Zhang, S. Yuan, K. Amine, V. Battaglia, G. Liu, Application of stabilized lithium metal powder (SLMP®) in graphite anode—a high efficient prelithiation method for lithium-ion batteries, J. Power Sources 260 (2014) 57–61.
[46] T. Tan, P.K. Lee, D.Y.W. Yu, Probing the reversibility of silicon monoxide electrodes for lithium-ion batteries, J. Electrochem. Soc. 166 (3) (2019) A5210–A5214.
[47] A. Mukanova, A. Jetybayeva, S.T. Myung, S.S. Kim, Z. Bakenov, A mini-review on the development of Si-based thin film anodes for Li-ion batteries, Mater. Today Energy 9 (2018) 49–66.
[48] J.P. Maranchi, A.F. Hepp, P.N. Kumta, High capacity, reversible silicon thin-film anodes for lithium-ion batteries, Electrochem. Solid St. 6 (9) (2003) A198–A201.
[49] P.K. Lee, M.H. Tahmasebi, T. Tan, S. Ran, S.T. Boles, D.Y.W. Yu, Polyimide capping layer on improving electrochemical stability of silicon thin-film for Li-ion batteries, Mater. Today Energy 12 (2019) 297–302.
[50] Y.H. Jouybari, F. Berkemeier, Enhancing silicon performance via LiPON coating: a prospective anode for lithium ion batteries, Electrochim. Acta 217 (2016) 171–180.
[51] A.R. Jiménez, R. Nölle, R. Wagner, J. Hüsker, M. Kolek, R. Schmuch, M. Winter, T. Placke, A step towards understanding the beneficial influence of a LIPON-based artificial SEI on silicon thin film anodes in lithium-ion batteries, Nanoscale 10 (4) (2018) 2128–2137.
[52] A. Al-Obeidi, D. Kramer, S.T. Boles, R. Mönig, C.V. Thompson, Mechanical measurements on lithium phosphorous oxynitride coated silicon thin film electrodes for lithium-ion batteries during lithiation and delithiation, Appl. Phys. Lett. 109 (7) (2016), 071902.
[53] P.K. Lee, M.H. Tahmasebi, S. Ran, S.T. Boles, D.Y.W. Yu, Leveraging titanium to enable silicon anodes in lithium-ion batteries, Small 14 (41) (2018) 1802051.

Chapter 5

Effect of insertion of an elastic buffer layer on stability of patterned amorphous silicon thin film Li-ion anode

Sameer S. Damle[a], Siladitya Pal[b], Prashant N. Kumta[a,b,c,d], and Spandan Maiti[a,b,c]

[a]Department of Chemical and Petroleum Engineering, University of Pittsburgh, Pittsburgh, PA, United States, [b]Department of Bioengineering, University of Pittsburgh, Pittsburgh, PA, United States, [c]Mechanical Engineering and Materials Science, University of Pittsburgh, Pittsburgh, PA, United States, [d]Center for Complex Engineered Multifunctional Materials, University of Pittsburgh, Pittsburgh, PA, United States

5.1 Introduction

Lithium ion batteries with improved capacity and cycling performance are being envisaged as the most promising energy storage devices for applications ranging from cell phones, camcorders, and laptop computers to high energy density batteries for deployment in electric vehicles including next generation aircrafts, helicopters, and drones. There has been tremendous research efforts over the years since the commercialization of the first Li-ion battery in 1994 targeting the development of high energy density battery systems. Any improvements in energy density required careful research into identifying new anode materials rather than attempting to improve the existing carbon based anodes. As a result, major efforts since 1998 and early 2000 have been directed towards finding better alternatives to traditional graphite-based anodes in lithium ion batteries (LIBs) that exhibit a limited theoretical capacity of ~370 mAh/g. Silicon holds the promise of being the integral alternative anode material for next-generation high energy density LIBs due to its ten-fold higher theoretical capacity (~4200 mAh/g) than carbon. However, Si undergoes colossal alloying-induced volume change(s), known to compromise the mechanical integrity of the anode structure after just a few electrochemical cycles [1–6].

Thus in order to enable large-scale deployment of Si-based anodes in LIBs, development of strategies to improve their cycling performance is of paramount importance. Towards this end, different silicon- based anode configurations, such as nanotubes, nanorods, thin film-based micro-patterns, and particulate composites, have been explored to date, to achieve high specific capacity with stable performance over many cycles [2, 4, 7–13]. We, in this chapter, particularly focus on studying the cycling stability of thin film micro-patterned amorphous silicon (*a*-Si) with a soft elastic buffer layer sandwiched between the silicon thin film and the metallic current collector.

As mentioned earlier, a number of strategies have been pursued to delay and/or suppress altogether the delamination of the active thin film from the passive current collector with the primary goal to improve the cycling performance of the anode system. These approaches can be broadly classified into three categories. In the first category, researchers controlled the size of the patterns to a flaw-tolerant dimension. As thin films are prone to vertical cracking, a typical approach is to create patterned anodes with a width less than the average vertical crack spacing referred to as the critical size for a given film thickness [14–17]. Another type of approach seeks stability through surface modification of the thin film including the interface between silicon and the current collector. He et al. showed that stability of a patterned anode can be improved by coating the pattern with atomic layer deposition coating of alumina [18]. Recently, Cho et al. [19] have suggested an improvement of electrochemical cycling performance of patterned electrodes by fabricating them on a rough substrate rather than on a smooth surface.

A third promising approach that has not gained much attention is surface engineering of the current collector. Analytical study by Song et al. concluded that for cylindrical LIBs, current collector with low thickness and low elastic modulus minimizes stresses in the active material layer [20]. Numerical studies by Zhang et al. [21] concluded that a soft and flexible current collector material would also be ideal for the electrode. Yu et al. [22] showed that the capacity retention of a patterned thin film is greatly enhanced by incorporating an elastomeric layer as a part of the current collector. Work by Zhang et al. demonstrated that including a graphene buffer layer between the Cu current collector and *a*-Si thin film benefits the capacity retention of the electrode [23]. Datta et al. earlier reported that the presence of a 50-nm thick carbon layer deposited by sputtering in between the active silicon thin film and the copper current collector significantly improves the cycling performance of the thin film anode [24]. Tong et al. also studied multilayer Si/Cu thin films and concluded that the C layer improves the anode structural stability by helping to accommodate the volume expansion of Si [25]. Choi et al. [26] used a current collector made by coating porous polyolefin polymer membrane with a thin layer of copper. Presence of the polymer membrane provides flexibility to the current collector, which helps the electrode performance by allowing it to freely expand and contract during electrode cycling. A common aspect in all these three [22, 24, 25] reports is the introduction of an elastic layer along with the usual silicon thin film and metallic current collector. Indeed, in a recent article exploring the role of mechanical properties of the current collector on the mechanical stability of thin film-based anodes, we also observed that an elastic substrate with low modulus can retard the interfacial delamination significantly [27]. However, the mechanisms operative in the presence of a soft elastic layer that gives rise to improved cycling performance are not well understood. Therefore, the aim of the present chapter is to provide a mechanistic understanding of the role of insertion of an elastic layer on the delamination of a patterned a-Si thin film anode.

Adhesion of the thin film pattern to the current collector at the thin film—collector interface applies mechanical constraints on the Si pattern. Presence of these mechanical constraints thus prevents stress-free expansion of the active material and, hence, results in generation of stresses in the anode configuration. Larger mechanical constraints correspondingly lead to higher energy absorbed by the material. Here we show that insertion of an elastic soft buffer layer can reduce these mechanical constraints and effectively increase the anode integrity. We take an energy-based approach towards this end and introduce a quantity called total mechanical energy of system "Ψ_{mech}" as a measure of the mechanical performance of the anode system. It is defined as the amount of mechanical energy absorbed by the anode at a given time during electrochemical cycling. We investigate in detail the effect of elastic modulus (E_b) of the intermediate elastic layer on Ψ_{mech}. We employ a recently developed thermodynamics-based numerical framework [28] to calculate the energetic contributions of various mechanical events occurring at the anode during lithiation or alloying with Li. Our study is restricted for clarity to a single square-shaped pattern made of silicon thin film and a thin elastic layer residing on top of a copper current collector for numerical tractability, see Fig. 5.1. However, our main conclusion is that the modulus mismatch between the silicon thin film and the sandwiched elastic layer is essentially the key design parameter important for achieving improvement in the mechanical stability. This conclusion is generally enough to be applicable to a broad spectrum of micro-patterned layered anode architectures.

The chapter is organized as follows. In the next section, we briefly present the modeling framework used in this study. General energy balance of the Li ion half-cell is also explained. A more detailed discussion of the theoretical foundation of our model can be found elsewhere [28]. Finite element-based numerical implementation of this theoretical development is addressed in Section 5.3. Details of the computational model of the layered pattern utilized for subsequent simulation studies are discussed in Section 5.4. Results of a detailed parametric study of the elastic buffer layer mechanical properties are presented in Section 5.5. The mechanistic understanding of the results is discussed in Section 5.6, followed by the conclusions and prescriptions for future work.

160 PART | II Mechanical properties

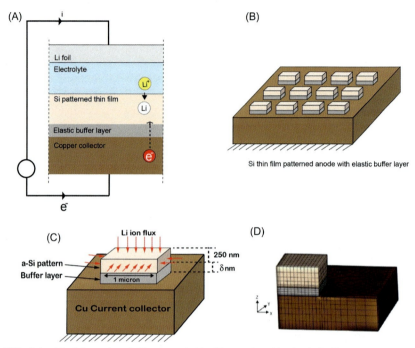

FIG. 5.1 (A) Schematics of Si-patterned thin film anode with elastic buffer layer half-cell. (B) Schematic of patterned anode configuration with *a*-Si pattern, intermediate buffer layer, and current collector. (C) Mathematical domain representing a single pattern with intermediate buffer layer attached to a current collector. Li flux is applied from all the surfaces exposed to electrolyte. (D) Finite element mesh of 1/4th of the mathematical domain.

5.2 Model description

General schematic of an anode half-cell containing Li foil as the counter electrode is shown in Fig. 5.1A. We have considered a patterned anode configuration in this chapter (Fig. 5.1A), which has three distinct layers of materials: silicon thin film (top), elastic buffer layer (middle), and the copper current collector (bottom). We consider that only the Si film is electrochemically active while the buffer layer and the Cu substrate together act as current collectors. Accordingly, the transport of lithium ions and the attendant large volume expansion have been modeled only in the silicon layer. Electrochemical reaction has been simulated at all surfaces exposed to the electrolyte. Faradaic reaction kinetics at these solid/electrolyte surfaces has been modeled through the Butler-Volmer equation (Eq. 5.1, Table 5.1). Silicon undergoes colossal volumetric expansion (∼300%) during electrochemical cycling [1] and attendant alloying-induced elastic softening [29]. In addition, experimental observations suggest that the Li-Si alloy undergoes severe elasto-plastic deformation when the effective stress exceeds the yield strength [31]. Both of these phenomena

TABLE 5.1 Governing equations.

Surface reaction kinetics at the electrolyte/electrode interface

$J_{Li} = \frac{i_0}{F}\left[\exp\left(\frac{\alpha_a F}{RT}\eta_s\right) - \exp\left(\frac{\alpha_c F}{RT}\eta_s\right)\right]$ (Butler-Volmer equation) (5.1)

$J_{Li} = Cm/FAn$ (5.2)

$i_0 = Fk(c_e)^{\alpha_a}(C_{max} - C_s)^{\alpha_a}(C_s)^{\alpha_c}$ (5.3)

Electrode deformation

$F = F_e F_p F_\theta$ (multiplicative decomposition of deformation gradient) (5.4)

$F_\theta = (1 + \eta c)^{1/3} I$ (5.5)

$\nabla_X \cdot P = 0$ (linear momentum balance) (5.6)

Coupled transport inside anode bulk

$\partial_t c = D\nabla\left[\nabla c + \frac{c}{RT}\nabla\left(\frac{\partial \Psi_1}{\partial c}\right) - \frac{\eta c}{RT}\nabla p\right]$ (5.7)

$p = 1/3 \; tr[J_\theta^{2/3} M_e F_\theta^{-1}]$ (5.8)

Elastic softening of silicon due to lithium alloying [29]

$E_0^{Si}(Li_{fraction}) = E_0^{Si} - \omega \times Li_{fraction}$ (5.9)

Traction separation law at the Si pattern/buffer and buffer/current collector interface [30]

$t_c = \frac{t_e}{\delta_e}\hat{t}, \; \hat{t} = [\beta^2 \delta + (1-\beta^2)(\delta \cdot n)n]$ (traction at the interface) (5.10)

have been taken into account through a multiplicative decomposition of the total deformation gradient F into an elastic part (F_e), plastic part (F_p), and a concentration-dependent component (F_θ), as described by Eq. (5.4) in Table 5.1. The transport equation for the lithium is accordingly described by Eq. (5.7). Note that the effect of mechanical stress has been incorporated through the last term in the bracket. The expression for a "pressure-like" term p is given by Eq. (5.8) in Table 5.1. Elastic softening of Si has been modeled by a linear reduction of elastic modulus dependent on the lithium volume fraction (Eq. 5.9, Table 5.1).

We assume that the buffer layer deforms elastically while the Cu substrate deforms in an elasto-plastic manner. In the present study, we have considered J_2 plasticity theory with linear strain hardening and associated flow rule. We presume that both the Si film/buffer interface and the buffer/Cu substrate interface are susceptible to delamination. Therefore a cohesive zone modeling technique has been used at both these interfaces to allow for possible nucleation and subsequent propagation of cracks. We consider that the electrochemical reaction occurs at all the surfaces of the a-Si thin film that are exposed to the electrolyte (see Fig. 5.1C) and is described by the Butler-Volmer equation. However, for mathematical simplicity, the newly exposed surface of Si created by the interfacial delamination and propagation is not considered for the additional electrochemical reaction. In the present study, we have considered galvanostatic cycling of thin film. Accordingly, the voltage is calculated by estimating the overpotential through the Butler-Volmer equation (Eq. 5.1, Table 5.1). The governing equations of the model are listed in brief in Table 5.1. For the definition of each variable, see the list of symbols, given in the Appendix.

5.2.1 Energy balance of the Li ion half-cell discharge process

Schematic of the Li ion half-cell discharge voltage curve under galvanostatic conditions is shown in Fig. 5.2. The anode is cycled between the voltage limits of V_{max} and V_{min} at a discharge current of i with a total discharge time of t. During the galvanostatic half-cell discharge, the maximum energy (Ψ_{outmax}) can be derived from the cell, if the discharge process is carried out at an infinitesimally small current. The maximum energy, Ψ_{outmax}, is represented by the product of area under the open circuit potential (U_{ocp})—time curve and the discharge current of i.

$$\Psi_{outmax} = i \int_0^t U_{ocp} \cdot dt \qquad (5.11)$$

However, due to the energy loss associated with the overpotential (η_s), the energy derived from the half-cell (Ψ_{out}) is less than Ψ_{outmax} and is represented by the product of discharge current i and the area under the output voltage ($V(t)$)—time curve. Thus

$$\Psi_{outmax} = \Psi_{out} + \Psi_{\eta_s} \qquad (5.12)$$

and,

$$\Psi_{out} = i \int_0^t V(t) \cdot dt \qquad (5.13)$$

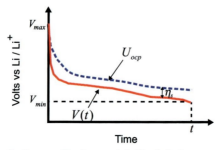

FIG. 5.2 Schematics of voltage profile of an anode half-cell discharge process. Open circuit voltage (U_{ocp}), output voltage ($V(t)$), and the overpotential (η_s) are indicated.

Part of Ψ_{η_s} is lost in terms of heat (Ψ_q) and consists of the reaction overpotential, concentration overpotential, and the resistance overpotential. The rest of this lost energy is contributed by the overpotential related to the transport of Li inside the electrode material ($\Psi_{Li transport}$) and the energy required for causing the mechanical changes (Ψ_{Mech}) in the electrode system. Hence,

$$\Psi_{\eta_s} = \Psi_q + \Psi_{Li transport} + \Psi_{mech} \tag{5.14}$$

The mechanical contribution Ψ_{Mech} can be further partitioned into three different categories: total elastic strain energy (Ψ_E) stored in the system, total plastic dissipation (Ψ_P) associated with the plastic flow of the material, and total dissipated interfacial energy (Ψ_I), which results from the delamination.

The rate of elastic strain energy accumulated at each layer is evaluated as

$$\dot{\Psi}_E^{(i)} = \int_{\Omega^i} S^i : \dot{E}^i d\Omega \quad \text{with } i \in [\text{Si, buffer, Cu}] \tag{5.15}$$

Therefore the total elastic energy of the anode at a given time, t, can be estimated as:

$$\Psi_E = \int_0^t \dot{\Psi}_E^{Si} dt + \int_0^t \dot{\Psi}_E^{buffer} dt + \int_0^t \dot{\Psi}_E^{Cu} dt \tag{5.16}$$

The rate of plastic dissipation in Si thin film and Cu substrate is calculated as

$$\dot{\Psi}_P^{(i)} = \int_{\Omega_i} M_e^i : L_P^i d\Omega \quad i \in [\text{Si, Cu}] \tag{5.17}$$

Therefore total plastic energy at a given time, t, is calculated as:

$$\Psi_P = \int_0^t \dot{\Psi}_P^{Si} dt + \int_0^t \dot{\Psi}_P^{Cu} dt \tag{5.18}$$

We note that buffer layer is elastic and does not exhibit any plasticity. The energy expended in propagation of delamination in the Si/buffer and buffer/Cu interface is evaluated as:

$$\dot{\Psi}_I^{(i)} = \int_{\Gamma_i} t^{(i)} \cdot \dot{\delta}^{(i)} d\Gamma \quad i \in [\text{Si/buffer, buffer/Cu}] \tag{5.19}$$

Therefore the total energy dissipation due to delamination propagation at any given time, t, is given as

$$\Psi_I = \int_0^t \dot{\Psi}_I^{Si/buffer} dt + \int_0^t \dot{\Psi}_I^{buffer/Cu} dt \tag{5.20}$$

and,

$$\Psi_{Mech} = \Psi_E + \Psi_P + \Psi_I \tag{5.21}$$

When an anode system is lithiated, it begins to deform mechanically owing to the transport of the lithium atoms into it, and commences storing the mechanical part of the free energy by elastic deformation and plastic dissipation. As Ψ_{Mech} increases further, a portion of the stored mechanical energy is released through the nucleation and propagation of delamination at the interface.

We will use our numerical framework to precisely evaluate these quantities and study the effect of the different parameters on Ψ_{Mech} evolution in the specific anode configuration detailed in this chapter.

5.3 Computational method

A custom finite element-based computational code has been developed to solve the coupled lithium transport and mechanical equilibrium equations as described in Table 5.1. Using this code, we determined the concentration of lithium $c(X,t)$ in silicon and the resulting displacement $u(X,t)$ of the anode for a given capacity, C at a given C-rate. Variational forms of the coupled equilibrium and transport equations were also obtained to perform finite element semi-discretization in space. A Newton-Raphson scheme-based linearization procedure was used to solve the resulting algebraic equations in an iterative manner. A backward Euler implicit time stepping algorithm was employed to solve the resulting temporal ordinary differential equations numerically. Finite element discretization of the solid domain was performed with 8-noded brick elements. The interface between the pattern/buffer and buffer/current collector was accordingly modeled by the 8-noded cohesive elements of zero thickness. Various energy components that contribute towards the total mechanical energy were postprocessed from the simulation data at the end of each time step.

5.4 Problem description

To study the effect of a thin buffer layer between the electrochemically active patterned thin film and the current collector on the mechanical stability of the electrochemically active patterned thin film-based anode, we have considered a domain as shown schematically in Fig. 5.1B. The square-shaped a-Si patterns are each 250 nm thick and 1 μm wide. In our simulations, the gap between the adjacent patterns is assumed to be sufficient such that the patterns do not interfere with each other in the fully alloyed lithiated state. Accordingly, we isolate a model domain with only one pattern, as shown in Fig. 5.1C, for this study. We further consider the advantage of the pattern configuration symmetry and consider only 1/4th of the domain for the simulation purpose (Fig. 5.1D). The intermediate buffer also has the same width as the pattern. The thickness of the buffer layer (h_b) is kept constant at 100 nm. The thickness of the current collector is assumed to be three times that of the pattern to avoid any influence of the boundary on the simulation results. The domain has two interfaces namely, the *a*-Si/buffer interface and the buffer/current collector interface. Both these interfaces can delaminate during the lithiation process of alloying under electrochemical cycling. We assume that the anode is cycled under galvanostatic conditions and, hence, consider a constant flux through the surfaces of the active material that are exposed to the electrolyte. It should be noted that some possible buffer layer materials, such as carbon, show considerable commensurate Li intercalation capacity. Consideration of contribution of the buffer layer to the anode Li capacity would effectively reduce the contribution of the *a*-Si film pattern to the overall anode Li capacity and, hence, the accompanying associated volumetric expansion. Since the aim of this chapter is to understand the effect of mechanical properties of the elastic buffer layer on the patterned *a*-Si anode mechanical stability, in order to consider maximum volumetric expansion associated with the *a*-Si film, we further assume that the buffer layer and the current collector have no lithium diffusion and, hence, no associated diffusion-induced volumetric expansion, i.e., $F_\theta = I$.

In the galvanostatic lithiation of the anode by alloying, the *a*-Si anode capacity is assumed to be 3000 mAh/g. All the studies performed in this chapter are also conducted considering a C/2.5 charge rate. The anode discharge simulation is also performed in the voltage window of 1.2–0.02 V, similar to those used in the published experiments. The rate constants for lithiation and delithiation process due to alloying and de-alloying of Li are listed in Table 5.2. The diffusivity of Li inside the bulk *a*-Si is considered to be 10^{-16} m^2/s [33] while the expansion coefficient is assumed to be 4.5×10^{-6} m^3/mol. The compressible Neo-Hookean constitutive model has been used for all the materials to simulate large deformation-induced mechanical response. The buffer layer is taken to be elastic for all the simulations as indicated earlier, while an elasto-plastic behavior of the copper current collector is assumed throughout the process. Lithiation-induced softening of *a*-Si is taken into account in the model using the $E_0^{Si}(Li_{fraction}) = E_0^{Si} - m \times Li_{fraction}$ relationship. The materials properties and the interface properties considered for the simulations are accordingly compiled in Table 5.2.

5.5 Results and discussion

Using the computational framework described in Section 5.2, we have investigated the effect of buffer layer on the stability of *a*-Si pattern anode configuration. As it is known, complete delamination of the *a*-Si pattern from the rest of the anode electrode components will lead to complete loss of electrical connectivity to the *a*-Si pattern and will render the anode completely nonfunctional for electrochemical cycling. Also, upon partial delamination, the delaminated area will get exposed to the electrolyte leading to growth of the solid electrolyte interphase layer causing additional capacity loss. Hence, to characterize the mechanical integrity of the patterned thin film electrode, we track the

TABLE 5.2 Materials properties.

Young's modulus of a-Si (E_0^{Si})	90 GPa
Rate of change of elastic modulus of a-Si with Li fraction (m)	71.25 GPa/Li$_{fraction}$ [29]
Poisson's ratio of a-Si (v_0^{Si})	0.28
Yield strength of a-Si (σ_Y^{Si})	1 GPa [28]
Young's modulus of buffer layer (E_b)	0.1 to 200 GPa
Poisson's ratio of the buffer (v_0^{Buffer})	0.28
Young's modulus of Cu (E_0^{Cu})	100 GPa
Poisson's ratio of Cu (v_0^{Cu})	0.34
Yield strength of Cu (σ_Y^{Cu})	300 MPa [27]
Hardening modulus of Si (H^{Si})	1 GPa
Hardening modulus of Cu (H^{Cu})	5 GPa
Interface fracture toughness (G_c)	15 J/m^2 [32]
Interface fracture strength (σ)	2 GPa
k (lithiation)	1.55×10^{-13} (ms^{-1})(mol m^{-3})$^{-0.5}$ [27]
k (delithiation)	60×10^{-13} (ms^{-1})(mol m^{-3})$^{-0.5}$ [27]

delamination of the a-Si pattern from the rest of the anode configuration. A systematic parametric study of the effect of the elastic buffer layer stiffness is thus, also undertaken. To obtain a mechanistic understanding of the effect of the buffer modulus, we find contributions of different components of the total mechanical energy of the anode configuration Ψ_{mech} (Eq. 5.21) and analyze their variation with change in the buffer elastic modulus. The results of this study are presented here. The interfacial properties considered are 2 GPa for the fracture strength and 15 J/m^2 for the fracture toughness unless otherwise specified.

5.5.1 Effect of buffer layer stiffness on the a-Si pattern anode stability

First, we demonstrate the effect of insertion of an elastic buffer layer on the stability of the a-Si pattern anode. We compare the mechanical integrity of the a-Si thin film pattern directly deposited on a Cu current collector against the mechanical integrity of the anode configuration, which has an elastic buffer inserted between the *a*-Si thin film pattern and the Cu current collector (Fig. 5.1B).

We consider a buffer layer of 100 nm thickness while varying the elastic modulus of the buffer layer (E_b) from 0.1 to 200 GPa. Fig. 5.3 shows the evolution of delamination of the *a*-Si thin film pattern during the half-cell discharge

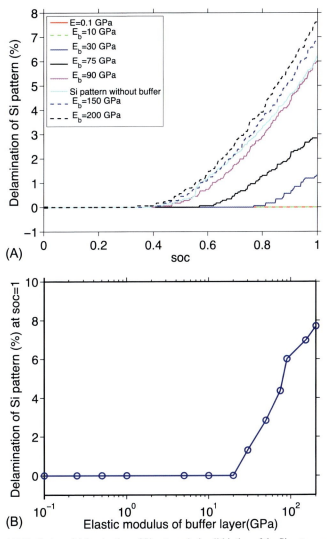

FIG. 5.3 (A) Evolution of delamination of Si pattern during lithiation of the Si pattern with elastic buffer layer of different stiffness of 100 nm thickness and without the presence of buffer layer ($\sigma = 2$ GPa, $G_c = 15$ J/m^2). (B) Amount of delamination at the end of lithiation for anode configuration with different elastic moduli.

process. The *soc* (state of charge) of 0 corresponds to the start of lithiation or alloying and $soc=1$ corresponds to end of lithiation or the alloying process. It should be noted that *soc* of the anode is linearly related to the time in the half-cell galvanostatic discharge process. The percentage of the *a*-Si thin film delamination is calculated with respect to the referential interfacial area. Thus, a percentage of zero signifies no delamination and a percentage value of 100 represents complete delamination. The results presented here show that in the absence of the elastic buffer layer, 6.2% delamination of Si pattern from the Cu current collector is observed. While alternatively, when the elastic buffer layer is present, the delamination of the *a*-Si pattern varies depending on the elastic modulus of the buffer layer. Also, it is to be noted that earlier the time when delamination is initiated, higher is the amount of delamination observed at the end of the lithiation or alloying process. For the case when the buffer elastic modulus is less than 20 GPa, the *a*-Si thin pattern shows no delamination at all. The amount of delamination increases as the elastic modulus of the buffer is increased beyond 20 GPa and reaches about 7.6% for buffer stiffness of 200 GPa as summarized in the inset plot in Fig. 5.3A.

In Fig. 5.4A–C, we have shown the deformation contours for the three different cases at the end of lithiation or alloying. The cases shown are: *a*-Si pattern on Cu current collector, *a*-Si pattern with an elastic buffer layer of 0.1 GPa stiffness, and *a*-Si pattern with an elastic buffer layer of 200 GPa stiffness. The corresponding delamination state of the *a*-Si pattern/buffer or *a*-Si pattern/ Cu

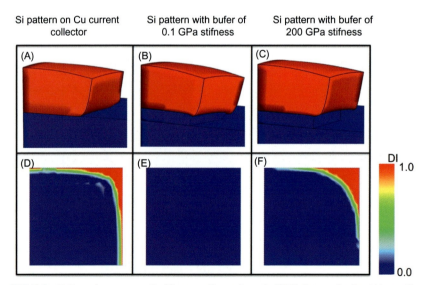

FIG. 5.4 Deformation contour of *a*-Si pattern film at the end of lithiation or alloying (A) on a Cu current collector ($E_{Cu}=100$ GPa), (B) with elastic buffer layer of 0.1 GPa and (C) 200 GPa stiffness. Delamination of *a*-Si pattern corresponding to these cases is represented by Delamination Index (DI) in (D), (E) and (F), respectively ($\sigma=2$ GPa, $G_c=15$ J/m^2).

interface is shown in Fig. 5.4E and Dd, respectively, where the red color indicates delamination and blue color indicates intact interface. Fig. 5.4A and D shows that at the end of lithiation or alloying, the a-Si pattern partially delaminates from the current collector at the corner of the Si/Cu interface. The observed delamination of the a-Si pattern on copper is about 6.2% of the interfacial area. Delamination of 7.6% is observed when an elastic buffer layer of 200 GPa stiffness is present below the a-Si pattern (Fig. 5.4C and F). In both the cases, the a-Si pattern appears to have delaminated at the corner of the a-Si/Cu and a-Si/buffer interface, respectively. However, when the stiffness of the elastic buffer is 0.1 GPa, the elastic buffer layer appears to have deformed significantly (Fig. 5.4B) as compared to the buffer layer with 200 GPa stiffness, and the a-Si pattern/buffer interface remains intact (Fig. 5.4E). Thus the results in Figs. 5.3 and 5.4 clearly indicate that the presence of an elastic buffer layer can alter the mechanical stability of the thin film pattern during electrochemical cycling depending on the mechanical properties of the buffer layer.

In order to further investigate the improved stability of the thin film pattern upon insertion of the elastic buffer layer, we studied the total mechanical energy absorbed by the configuration Ψ_{mech} (Eq. 5.21). In Fig. 5.5A we plot the evolution of the total mechanical energy in the system. It can be seen that the total mechanical energy in the configuration increases as the corresponding volumetric expansion increases with progression of lithiation or alloying of the anode. The inset plot in Fig. 5.5B shows the Ψ_{mech} of the anode configuration at $soc=1$ for different stiffness values of the buffer. From the inset plot, it can be seen that as the stiffness of the elastic buffer layer is reduced from 200 to 30 GPa, there is no significant change in Ψ_{mech}. However, upon reducing the buffer stiffness from 30 to 0.1 GPa, the Ψ_{mech} reduces sharply. Thus changing the elastic buffer stiffness from 0.1 to 200 GPa leads to approximately 20 fold higher energy absorption by the anode configuration. We further explore the contributions by the different energy components to Ψ_{mech} for anode configurations with different stiffness values, E_b, for the buffer.

In Fig. 5.6A, we have plotted the variation of the total mechanical energy Ψ_{mech} and the three different energies contributing to it (Ψ_E, Ψ_P, and Ψ_I, see Eqs. 5.16, 5.18, and 5.20 for details). Fig. 5.6B shows the variation of elastic strain energy in the different components of the anode configuration (i.e., a-Si film, elastic buffer layer, and the current collector) along with the elastic strain energy in the configuration. The behavior of the curves in Fig. 5.6A and B can be explained by dividing the plots into three different regimes based on the buffer elastic modulus E_b namely: ≤ 0.5 GPa, 0.5–20 GPa, and >20 GPa.

In the first regime ($E_b \leq 0.5$ GPa), the energy due to plastic dissipation is zero, which indicates that the a-Si film does not undergo any plastic deformation. Since the pattern has not delaminated from the buffer layer, $\Psi_{interface}$ is also zero and $\Psi_{mech} = \Psi_e$. The elastic strain energy is found to increase as the buffer stiffness value is increased from 0.1 to 0.5 GPa. In fact, inspection of regime 1 in Fig. 5.6B indicates that the buffer layer stores almost all of

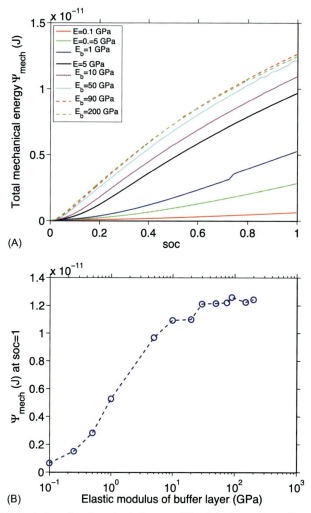

FIG. 5.5 (A) Evolution of total mechanical energy (Ψ_{Mech}) in the anode configuration during lithiation of the Si pattern with underlying buffer of 100 nm thickness and different stiffness ($\sigma = 2$ GPa, $G_c = 15$ J/m^2). (B) Total mechanical energy at the end of lithiation or alloying in the anode configuration with different buffer stiffness values.

the anode elastic strain energy. Also, the elastic strain energy in the current collector is negligible. Thus the buffer layer accommodates most of the strain resulting from the volumetric expansion of the a-Si film keeping the total mechanical energy low and at the same time retaining the integrity of the anode.

The second regime ranges for 0.5 GPa $< E_b \leq 20$ GPa. Fig. 5.6A reveals that Ψ_p increases sharply as the buffer is made stiffer. This indicates that as the buffer layer stiffness is increased, the mechanical constraints on the film are

Effect of insertion of an elastic buffer layer **Chapter | 5** **171**

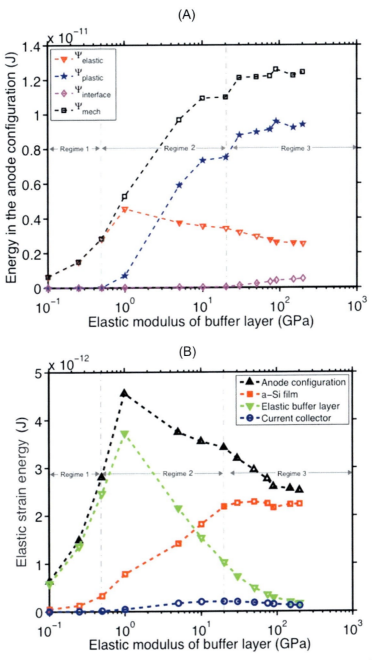

FIG. 5.6 (A) Variation of the total mechanical energy, elastic strain energy, dissipated plastic energy, and the energy dissipated due to fracture at interfaces in the anode configuration at $soc=1$. (B) Variation of the elastic strain energy in the anode configuration and its components (a-Si film, elastic buffer layer, and current collector) at $soc=1$. Values are reported for anode configurations with different elastic buffer layer stiffness ($\sigma=2\,\text{GPa}$, $G_c=15\,\text{J/m}^2$).

more stringent and consequently result in the *a*-Si thin film pattern undergoing relatively higher plastic deformation. The observed $\Psi_{interface}$ is almost zero and it supports the observation of no delamination in Fig. 5.3. Plot of Ψ_e shows that the elastic strain energy in the anode configuration reduces as the buffer modulus is changed from 1 to 20 GPa. Information in Fig. 5.6B reveals that even though the elastic strain energy in the *a*-Si film increases, the strain energy in the buffer layer reduces. Also, the current collector gains some strain energy, which reveals that the buffer layer is not able to completely accommodate the lithiation or alloying-induced volumetric strain. However, the energy dissipation due to plastic deformation of the *a*-Si film prevents the initiation of delamination at Si pattern/buffer interface up to a certain extent.

Regime 3 in Fig. 5.6 is displayed for $E_b > 20$ GPa. Fig. 5.6A shows that Ψ_p is almost constant throughout this regime. Observation of increase in $\Psi_{interface}$ is consistent with the observation of increase in interface delamination as the buffer modulus increases to 200 GPa in Fig. 5.3. Even though Ψ_e is found to reduce, the rate of reduction of elastic strain energy in the configuration ($\partial \Psi_e / \partial E_b$) is about 12 times less indicating that the increase in buffer stiffness beyond 20 GPa does not affect Ψ_e significantly. Careful observation of regime 3 in Fig. 5.6B reveals that the strain energy in the *a*-Si film is almost constant. Also, the strain energy in the current collector reduces, as the buffer is made stiffer. This can be explained as follows by relating it to the increase in the amount of delamination upon increasing E_b, the buffer stiffness. During lithiation or alloying, as the *a*-Si pattern delaminates partially from the buffer layer, it reduces the mechanical constraints on the pattern. Less mechanical constraints lead to less stresses in the buffer layer. As a result, after delamination initiation, less stress is transferred to the current collector from the elastic buffer. As the delamination increases, relatively less stress is transferred to the current collector and, hence, less strain energy is stored in the current collector, as the buffer is made stiffer.

Therefore it can be concluded that the insertion of a softer buffer layer between *a*-Si pattern and the Cu current collector facilitates accommodation of the strains caused by the diffusion-induced volumetric expansion of the *a*-Si pattern by reducing the mechanical constraints on the *a*-Si pattern. Reduction in the total mechanical energy of the anode configuration prevents the dissipation of excess mechanical energy through the interface delamination and thus improves the anode configuration integrity.

5.5.2 Effect of interface properties on the *a*-Si pattern anode stability

In order to study the effect of interface properties on the stability of the *a*-Si patterned anode, we reduced the interface fracture strength from 2 to 1 GPa, while keeping the fracture toughness constant at 15 J/m². Fig. 5.7 shows the amount of delamination of Si pattern from the buffer layer at $soc = 1$ or full

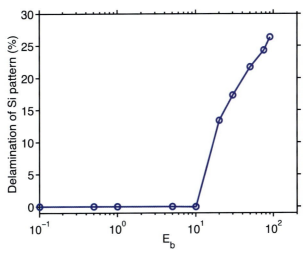

FIG. 5.7 Amount of delamination at the end of lithiation for anode configuration with different buffer elastic moduli ($\sigma = 1$ GPa, $G_c = 15$ J/m^2).

lithiation or alloying. Comparing it with the inset plot in Fig. 5.3 shows that for the same value of buffer stiffness, E_b, the delamination is higher. When the fracture strength was 2 GPa, the a-Si pattern showed no delamination when $E_b < 20$ GPa. However, upon reducing the fracture strength to 1 GPa, 13% interface area of a-Si pattern is delaminated for $E_b = 20$. Below the buffer stiffness of 10 GPa, the interface is found to be intact at $soc = 1$.

Fig. 5.8A shows the variation of Ψ_{mech} and its different components, while Fig. 5.8B shows the variation of Ψ_e and its components with the stiffness, E_b, of the buffer. Similar to Fig. 5.6A and B, Fig. 5.8A and B is divided into 3 regimes. Trends explained in Section 5.5.1 for these 3 regimes are also applicable here. Similar to Fig. 5.6, regime 1 exists for $E_b \leq 0.5$ GPa. However, the area covered by regimes 2 and 3 is different as the fracture strength is reduced to 1 GPa.

Reduction in fracture strength allows the fracture initiation to occur at a relatively lower soc or alloying. As the delamination is initiated at a lower soc for fracture strength of 1 GPa, the total mechanical energy absorbed by the anode configuration for the same buffer elastic modulus at initiation of delamination is also relatively lower. Hence in Fig. 5.8A, Ψ_{mech} is lower as compared to that in Fig. 5.6A. Thus the results indicate that having a higher fracture strength delays the delamination initiation by letting the anode configuration absorb more Ψ_{mech}. Hence, the results displayed herein clearly indicate that having good adhesion between the Si pattern and the elastic buffer layer is also important for improving the anode mechanical integrity.

174 PART | II Mechanical properties

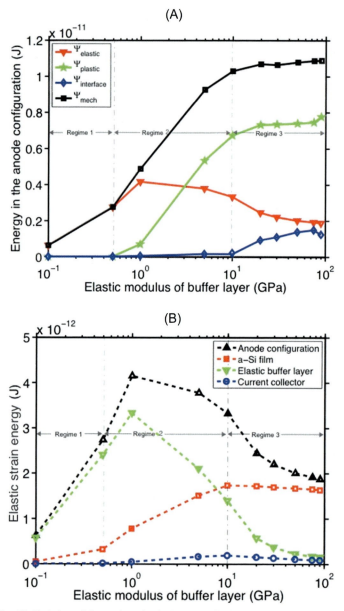

FIG. 5.8 (A) Variation of the total mechanical energy, elastic strain energy, dissipated plastic energy, and the energy dissipated due to fracture at interfaces in the anode configuration at $soc=1$. (B) Variation of the elastic strain energy in the anode configuration and its components (a-Si film, elastic buffer layer, and current collector) at $soc=1$. Values are reported for anode configurations with different elastic buffer layer stiffness ($\sigma=1\,\text{GPa}$, $G_c=15\,\text{J/m}^2$).

5.6 Discussion

The simulations performed in this chapter to study the effect of presence of an elastic buffer layer on the mechanical stability of the a-Si thin film pattern closely mimic the electro-chemico-mechanical boundary conditions prevalent in an actual Li ion half-cell. The mechanisms operational in the delamination of the a-Si thin film pattern (island) on a substrate have been explained in detail in our previous publication [27]. In this work, we introduce a thin elastic buffer layer between the metallic current collector and the a-Si patterned thin film to study its effect on the stability of the patterned a-Si anode configuration.

Electrochemical cycling of the active material requires the presence of other components in the anode configuration for different reasons. For example, a Cu current collector is needed in the a-Si thin film anode for electron conduction, while the PVDF binder and carbon black additive are customary and required in an anode containing Si nanoparticles as active material to keep the Si nanoparticles intact and also provide adequate electronic conductivity [7]. If the active material can be cycled without the presence of these components, the only stress that the active material would exhibit, depends on the parameters such as the mechanical properties and geometry of the active material, Li diffusivity in the electrochemically active material, charge rate, etc. as studied in Ref. [34–39]. However, due to the presence of other components that make up the anode configuration, additional mechanical constraints are imposed on the electrochemically active material during electrochemical cycling of the active material. Presence of these mechanical constraints (e.g. adhesion of a-Si pattern to the current collector) develops additional stress in the anode configuration resulting in higher mechanical energy absorption by the anode. Results from the inset in Fig. 5.5 indicate that increasing the buffer elastic modulus increases the energy absorbed by the anode Ψ_{mech}. Also, as Ψ_{mech} increases, energy dissipated through plastic deformation also increases (see Fig. 5.6).

Another energy dissipative mechanism available in the anode configuration is fracture. Fracture at the a-Si pattern/buffer interface releases energy and partially reduces the mechanical constraints on the a-Si pattern. Thus the presence of a softer elastic buffer layer ($E_b < 20\,\text{GPa}$) limits the mechanical energy absorbed by the configuration by reducing the mechanical constraints and prevents energy dissipation through fracture.

Outcomes of simulations presented in Section 5.5.1 can be directly related and correlated to the excellent cycling performance shown by Datta et al. [24]. The report showed that the presence of a thin buffer layer of carbon between the Cu current collector and the a-Si thin film electrode improves the capacity retention and prevents cracking of the thin film into islands as otherwise observed in bare a-Si thin film identically deposited on the Cu current collector. The concept of presence of a buffer layer to alleviate the volumetric expansion-induced stresses in an electrochemically active material on an electrode is applicable not only in thin film electrodes but also in electrodes containing particles. Zhang et al. [23] have shown that presence of a layered material such as graphene in a composite with Si nanoparticles buffers the stresses resulting from alloying of Li-Si and thus preserves the integrity of the composite electrode. In another study, Zhu et al. [40] have shown that the cellulose fiber can act as a natural buffer in a Sn-coated cellulose fiber anode for Na ion battery.

Another interesting aspect of reduction in Ψ_{mech} can be elucidated by Eqs. (5.12) and (5.14). Reduction in Ψ_{mech} results in reduction of over-potential energy, Ψ_η. Thus reduction in the overpotential energy loss increases the Ψ_{out} or in other words, a higher output voltage will be observed in the voltage-capacity plot of anode electrochemical cycling. Similar phenomenon has been experimentally observed by [22, 41].

The results presented in Section 5.5.2 thus indicate that having a better adhesion of the active material layer with the buffer layer is highly important to improve the anode mechanical integrity. Having a higher fracture strength also allows relatively higher Ψ_{mech} to be stored in the anode configuration before the delamination is initiated as the barrier to release energy through fracture is increased by increasing the fracture strength.

5.7 Conclusions

A multiphysics modeling framework has been utilized to understand the mechanical stability of the patterned a-Si thin film anode in the presence of a buffer layer during lithiation of the anode. Effect of the buffer layer thickness and the mechanical properties has been studied by taking delamination of a-Si thin film pattern as the criteria for ascertaining the anode stability. We concluded that having a soft buffer (e.g., carbon, conductive polymer) between the patterned film and the current collector reduces the mechanical constraints on the electrochemically active material and is thus beneficial for contributing to the stability of the patterned a-Si anode configuration. Apart from the properties of the thin buffer layer, the adhesion between the different anode components is also important in prolonging the anode life. Thus, the present study sheds light on the important factors that govern the stability of the patterned a-Si anode configuration having an elastic buffer layer by providing an understanding of mechanisms that lead to the failure of the anode. It is expected that these findings would lead to the development of much improved next-generation Si thin film anode configurations.

Appendix

A	Surface area of Si in contact with electrolyte (m^2)	
c	Concentration of Li in the reference configuration ($mol\,m^{-3}$).	
c_{max}	Maximum Li intercalation concentration in anode material ($mol\,m^{-3}$).	
c_e	Li ion concentration in electrolyte ($mol\,m^{-3}$)	

c_s	Li ion concentration on the surface of the active material (mol m^{-3})
C	Capacity of the electrode material (mAh g^{-1})
C-rate	Charge rate
D	Diffusivity of Li atom in the anode material (m^2/s)
E_0^{Si}	Elastic modulus of silicon without softening (Pa)
E_b	Elastic modulus of the intermediate layer
F	Faraday's constant (A h mol^{-1})
F	Deformation gradient
F$_e$	Elastic component of the deformation gradient
F$_p$	Plastic component of the deformation gradient
F$_\theta$	Expansion component of the deformation gradient
i_0	Exchange current density (A m^{-2})
I	Identity tensor
J_{Li}	Li atom flux (mol m^{-2} s^{-1})
J_θ	Jacobian of expansion component of the deformation gradient
k	Kinetic rate constant (m s^{-1})(mol m^3)$^{-0.5}$
Li$_{fraction}$	Fraction of lithium concentration in Li$_x$Si alloy
m	Mass of the active material (Si) in the electrode configuration (kg)
M$_e$	Mandel stress tensor
n	Duration of the lithium alloying/de-alloying half cycle (h)
p	Hydrostatic pressure (Pa)
P	First Piola-Kirchoff stress tensor
R	Universal gas constant (J mol^{-1} K^{-1})
T	Temperature (K)
t_e	Effective traction
t_c	Traction at the cohesive surface
α_a, α_c	Anodic and cathodic transfer coefficient
δ	Displacement jump vector
η	Partial volume of Li in Si (m^3/mol)
ω	Rate of change of elastic modulus of a-Si with Li fraction in the Li$_x$Si alloy (Pa Li$_{fraction}^{-1}$)

Acknowledgments

The authors would like to acknowledge the financial support from the US Department of Energy's Office of Vehicle Technologies BATT program (Contract DE-AC02-05CHI1231), subcontract 6151369, and the National Science Foundation (CBET-0933141). PNK would also like to acknowledge the Edward R. Weidlein Endowed Chair Professorship Funds for partial support of this work. In addition, PNK and SM would like to thank the Center for Complex Engineered Multifunctional Materials (CCEMM) for providing a graduate fellowship to SSD as well as the instruments authors needed to perform the simulations reported in this work.

References

[1] L.Y. Beaulieu, et al., Colossal reversible volume changes in lithium alloys, Electrochem. Solid St. 4 (9) (2001) A137.
[2] M.-H. Park, et al., Silicon nanotube battery anodes, Nano Lett. 9 (11) (2009) 3844–3847.
[3] L.-F. Cui, et al., Crystalline-amorphous core-shell silicon nanowires for high capacity and high current battery electrodes-Cui, Nano Lett. 9 (1) (2009) 491–495.
[4] R. Teki, et al., Nanostructured silicon anodes for lithium ion rechargeable batteries, Small 5 (20) (2009) 2236–2242.
[5] A. Magasinski, et al., High-performance lithium-ion anodes using a hierarchical bottom-up approach, Nat. Mater. 9 (4) (2010) 353–358.
[6] A.S. Aricò, et al., Nanostructured materials for advanced energy conversion and storage devices, Nat. Mater. 4 (5) (2005) 366–377.
[7] U. Kasavajjula, C. Wang, A.J. Appleby, Nano- and bulk-silicon-based insertion anodes for lithium-ion secondary cells, J. Power Sources 163 (2) (2007) 1003–1039.
[8] J.P. Maranchi, A.F. Hepp, P.N. Kumta, High capacity, reversible silicon thin-film anodes for Lithium-ion batteries, Electrochem. Solid St. 6 (9) (2003) A198–A201.
[9] C.K. Chan, et al., High-performance lithium battery anodes using silicon nanowires, Nat. Nanotechnol. 3 (1) (2008) 31–35.
[10] Y. Yao, et al., Interconnected silicon hollow nanospheres for lithium-ion battery anodes with long cycle life, Nano Lett. 11 (7) (2011) 2949–2954.
[11] Y. Zhang, et al., Composite anode material of silicon/graphite/carbon nanotubes for Li-ion batteries, Electrochim. Acta 51 (23) (2006) 4994–5000.
[12] M.K. Datta, P.N. Kumta, In situ electrochemical synthesis of lithiated silicon–carbon based composites anode materials for lithium ion batteries, J. Power Sources 194 (2) (2009) 1043–1052.
[13] H. Guan, et al., Coaxial Cu-Si@C array electrodes for high-performance lithium ion batteries, Chem. Commun. (Camb.) 47 (44) (2011) 12098–12100.
[14] H. Haftbaradaran, et al., Modified Stoney equation for patterned thin film electrodes on substrates in the presence of interfacial sliding, J. Appl. Mech. 79 (3) (2012), 031018.
[15] H. Haftbaradaran, et al., Method to deduce the critical size for interfacial delamination of patterned electrode structures and application to lithiation of thin-film silicon islands, J. Power Sources 206 (2012) 357–366.
[16] S.K. Soni, et al., Stress mitigation during the lithiation of patterned amorphous Si Islands, J. Electrochem. Soc. 159 (1) (2012) A38–A43.
[17] X. Xiao, et al., Improved cycling stability of silicon thin film electrodes through patterning for high energy density lithium batteries, J. Power Sources 196 (3) (2011) 1409–1416.
[18] Y. He, et al., Alumina-coated patterned amorphous silicon as the anode for a lithium-ion battery with high coulombic efficiency, Adv. Mater. 23 (42) (2011) 4938–4941.
[19] G. Cho, et al., Improved electrochemical properties of patterned Si film electrodes, Microelectron. Eng. 89 (2012) 104–108.

[20] Y. Song, et al., Diffusion induced stresses in cylindrical lithium-ion batteries: analytical solutions and design insights, J. Electrochem. Soc. 159 (12) (2012) A2060–A2068.
[21] J. Zhang, et al., Diffusion induced stress in layered Li-ion battery electrode plates, J. Power Sources 209 (2012) 220–227.
[22] C. Yu, et al., Silicon thin films as anodes for high-performance lithium-ion batteries with effective stress relaxation, Adv. Energy Mater. 2 (1) (2012) 68–73.
[23] Y.Q. Zhang, X.H. Xia, X.L. Wang, Y.J. Mai, S.J. Shi, Y.Y. Tang, L. Li, J.P. Tu, Silicon/graphene-sheet hybrid film as anode for lithium ion batteries, Electrochem. Commun. 23 (2012) 17–20.
[24] M.K. Datta, et al., Amorphous silicon–carbon based nano-scale thin film anode materials for lithium ion batteries, Electrochim. Acta 56 (13) (2011) 4717–4723.
[25] Z.X. Yongfeng Tong, C. Liu, G.'a. Zhang, J. Wang, Z.G. Wu, Magnetic sputtered amorphous Si/C multilayer thin films as anode materials for lithium ion batteries, J. Power Sources 247 (2013) 78–83.
[26] J.-Y. Choi, et al., Silicon nanofibrils on a flexible current collector for bendable lithium-ion battery anodes, Adv. Funct. Mater. 23 (17) (2013) 2108–2114.
[27] S.S.D. Siladitya Pal, S.H. Patel, M.K. Datta, P.N. Kumta, S. Maiti, Modeling the delamination of amorphous-silicon thin film anode for lithium-ion battery, J. Power Sources (2014) 149–159.
[28] S. Pal, et al., Modeling of Lithium segregation induced delamination of a-Si thin film anode in Li-ion batteries, Comput. Mater. Sci. 79 (2013) 877–887.
[29] V.B. Shenoy, P. Johari, Y. Qi, Elastic softening of amorphous and crystalline Li–Si phases with increasing li concentration: a first-principles study, J. Power Sources 195 (19) (2010) 6825–6830.
[30] M. Ortiz, A. Pandolfi, Finite deformation irreversible cohesive elements for three-dimensional crack-propagation analysis, Int. J. Numer. Methods Eng. 44 (1999) 1267–1282.
[31] V.A. Sethuraman, M.J. Chon, M. Shimshak, V. Srinivasan, P.R. Guduru, *In situ* measurements of stress evolution in silicon thin films during electrochemical lithiation and delithiation, J. Power Sources 195 (2010) 5062–5066.
[32] J.P. Maranchi, et al., Interfacial properties of the a-Si/Cu:active–inactive thin-film anode system for lithium-ion batteries, J. Electrochem. Soc. 153 (6) (2006) A1246.
[33] N. Ding, et al., Determination of the diffusion coefficient of lithium ions in nano-Si, Solid State Ion. 180 (2–3) (2009) 222–225.
[34] R. Deshpande, Y.-T. Cheng, M.W. Verbrugge, Modeling diffusion-induced stress in nanowire electrode structures, J. Power Sources 195 (15) (2010) 5081–5088.
[35] R. Deshpande, et al., Diffusion induced stresses and strain energy in a phase-transforming spherical electrode particle, J. Electrochem. Soc. 158 (6) (2011) A718–A724.
[36] R. Deshpande, Y. Qi, Y.-T. Cheng, Effects of concentration-dependent elastic modulus on diffusion-induced stresses for battery applications, J. Electrochem. Soc. 157 (8) (2010) A967–A971.
[37] S.J. Harris, et al., Mesopores inside electrode particles can change the Li-ion transport mechanism and diffusion-induced stress, J. Mater. Res. 25 (08) (2011) 1433–1440.
[38] Y.-T. Cheng, M.W. Verbrugge, Evolution of stress within a spherical insertion electrode particle under potentiostatic and galvanostatic operation, J. Power Sources 190 (2) (2009) 453–460.
[39] Y.-T. Cheng, M.W. Verbrugge, Diffusion-induced stress, interfacial charge transfer, and criteria for avoiding crack initiation of electrode particles, J. Electrochem. Soc. 157 (4) (2010) A508–A516.
[40] H. Zhu, et al., Tin anode for sodium-ion batteries using natural wood fiber as a mechanical buffer and electrolyte reservoir, Nano Lett. 13 (7) (2013) 3093–3100.
[41] V.A. Sethuraman, et al., In situ measurements of stress-potential coupling in lithiated silicon, J. Electrochem. Soc. 157 (11) (2010) A1253–A1261.

Part III

Electrolytes and surface electroyle interphase (SEI) issues

Chapter 6

SEI layer and impact on Si-anodes for Li-ion batteries

Partha Saha[a,b], Tandra Rani Mohanta[a], and Abhishek Kumar[a]
[a]*Department of Ceramic Engineering, National Institute of Technology, Rourkela, Odisha, India*
[b]*Centre for Nanomaterials, National Institute of Technology, Rourkela, Odisha, India*

6.1 Introduction

Li-ion batteries (LIBs) have made astounding progress since Sony Inc. commercialized the technology in 1991 [1, 2]. Over the last two decades, a conscientious effort and systematic experimental/theoretical study have made it possible to identify the plethora of positive and negative electrodes wherein Li-ions can (de)intercalate and/or (de)alloy in tandem within the stipulated electrochemical window of nonaqueous electrolytes constituting lithium salt and aprotic solvents [3, 4]. However, to date, the maximum energy density of $\sim$200 Wh kg^{-1} could be achieved for a LIB using graphite anode and transition metal layered oxide cathode (e.g., 111 Li-nickel manganese cobalt (NMC) oxide) in the carbonate-based electrolyte, which is likely insufficient for stringent energy requirements in vehicle electrification of targeted energy density $\sim$500 Wh kg^{-1} [5].

The most significant impediment for a giant leap in LIB technology is the limited electrochemical stability of nonaqueous electrolytes, mainly carbonate-based solvents (i.e., ethylene carbonate (EC), propylene carbonate (PC), ethyl methyl carbonate (EMC), etc.), limiting exploration of high energy density cathodes (>5.0 V versus Li/Li$^+$), and formation of organic/inorganic by-products at electrode/electrolyte interface known as solid electrolyte interphase (SEI) during Li-ions transport [6–9]. SEI formation not only increases the overall resistance (impedance) of the cell but also consumes solvated lithium-ions and salt/solvent radical species as decomposition products leading to capacity fade, limiting cycle life and calendar life [6, 10, 11]. Sometimes, the chemical/electrochemical reactions leading to SEI formation are exothermic, and the presence of flammable liquids leads to localized heat generation and thermal runway [12]. However, one advantage of SEI formation is that the surface passivation of electrodes (primarily anodes) by radical species restrict further movement of electrons (known as electron tunneling effect) beyond ~1 nm SEI thickness limiting recurring decomposition of electrolytes, however, allowing Li-ions transport unharmed keeping the electrochemical cell alive for many cycles [13, 14].

An ideal SEI should be mechanically flexible and must possess a high cation transference number allowing ions but limiting electron movement [15]. Once a stable Li-ion conducting SEI layer is formed on the electrode surface, the inorganic, organic, and insoluble Li salts, which comprise the SEI composition, prevent further solvent decomposition, allowing Li-ions movement in and out from the electrodes. Therefore, a stable Li-ion conducting but electronically insulator SEI layer formation is imperative and considered an all-important phenomenon for LIBs [16].

Over the course of four decades, a large body of work and progressive buildup of knowledge has made it possible to understand the chemical compositions, formation, and growth of SEI on the negative electrodes (see Fig. 6.1). However, the intricacies of SEI on different surface and subsurface structures attribute the complexity and challenges for the researcher to predict the behavior in different testing conditions. In the pioneering work, A.N. Dey [17] first observed the surface passivation of lithium metal soaked in organic electrolytes. Two years later, Peled [18] proposed the formation of a solid surface layer covering the alkali/alkaline earth metal and nonaqueous electrolyte interphase and coined the term "solid electrolyte interphase (SEI)." Nazri and Muller [19] demonstrated the presence of Li_2CO_3 as a by-product present in SEI on lithium metal surface. At the same time, Aurbach et al. [20] claimed the presence of lithium alkyl dicarbonate due to carbonate solvent decomposition as the main SEI components. Kanamura et al. [21] showed that SEI compact layer on the lithium metal surface consists of LiF and Li_2O, whereas the top porous layer contains organic compounds that probably laid the foundation of bi-layer SEI structure in the early nineties and later confirmed by others [22–25]. At the same time, Besenhard et al. [26] showed that Li^+/solvents co-intercalation into graphite layers builds up a protective surface layer that prevents further solvent co-intercalation. The entire gamut of work was nicely summarized by Peled [27], who proposed a mosaic structure of SEI present on lithium/graphite surface consisting of multiple inorganic and organic products from electrolyte decomposition (Li_2O, LiF, Li_2CO_3, polyolefins, semicarbonates, etc.). Aurbach et al. [28] also proposed that the SEI passivation layer consisting Li_2CO_3, $ROCO_2Li$ on the anode (graphite) and oxide cathodes (Li_xCoO_2, $LiNiO_2$, Li_2MnO_4) impair their thermodynamic stability and the main reason for capacity fade in alkyl carbonate-based electrolytes, which are usually a mixture of cyclic carbonate viz. ethylene carbonate (EC) and a linear carbonate, viz., dimethyl carbonate (DMC), diethyl carbonate (DEC), or ethyl methyl carbonate (EMC). The SEI formation on the electrode's surface mainly stems from

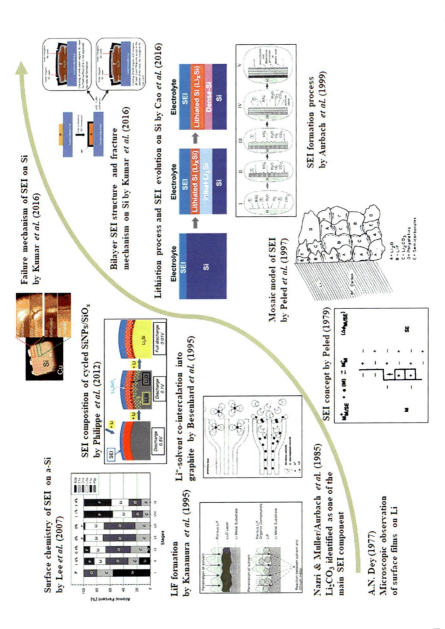

FIG. 6.1

(Continued)

the decomposition of a lithium salt (LiPF$_6$) or carbonate solvents. The lithiation voltages of graphite ($\sim$0.16 V versus Li$^+$/Li) [9] and Si ($\sim$0.4 V versus Li$^+$/Li) [29, 30] are below the threshold stability voltage of carbonate-solvents and electrolyte additives (see Table 6.1) [31]. Therefore, solvents and salt decompositions become inevitable and cannot be avoided in liquid electrolyte [8]. Aurbach et al. [32, 33] showed the anion of the lithium salt increases the impedance

---

FIG. 6.1,CONT'D A chronicles of SEI formation on the negative electrodes for Li-ion batteries. In 1977, Dey first observed the Li metal passivation in organic electrolytes by microscopy. Nazri and Muller and Aurbach et al. identified Li$_2$CO$_3$ as one of the main components in SEI in 1985 and 1987, respectively. Kanamura et al. show the formation of Li$_2$O, LiF due to the decomposition of lithium salt/solvents, and Besenhard et al. proposed a model where solvent molecules co-intercalate in the graphite layers to further decompose and form an SEI in 1995. Peled et al. proposed the mosaic structure of SEI formation in 1997. Aurbach et al. illustrated the SEI formation process starting from electrolyte reduction on the graphite electrode surface in 1999. Lee et al. show the composition of the surface SEI layer on the Si thin electrode during lithiation/delithiation phenomena in 2007. Philippe et al. show that formation of Li$_2$O, Li$_4$SiO$_4$, SiO$_x$F$_y$ as an inorganic hard layer on Si/SiO$_x$ anode surface; Cao et al. show three stages of SEI formation on Si surface, and Kumar et al. explain the bi-layer SEI structure and fracture mechanisms on silicon between 2012 and 2016. Kumar et al. also show by in operando AFM analysis the buckling, delamination, and associated fracture mechanism of SEI layer on silicon in 2016. *(Reprinted with permission from A.N. Dey, Lithium anode film and organic and inorganic electrolyte batteries, Thin Solid Films 43 (1977) 131–171. Copyright 1977 Elsevier B.V. Reprinted with permission from G. Nazri, Composition of surface layers on Li electrodes in PC, LiClO$_4$ of very low water content, J. Electrochem. Soc. 132 (1985) 2050; D. Aurbach, Identification of surface films formed on lithium in propylene carbonate solutions, J. Electrochem. Soc. 134 (1987) 1611. Copyright 1985, 1987 Institute of Physics Publishing. Reprinted with permission from K. Kanamura, XPS analysis of lithium surfaces following immersion in various solvents containing LiBF$_4$, J. Electrochem. Soc. 142 (1995) 340; J.O. Besenhard, M. Winter, J. Yang, W. Biberacher, Filming mechanism of lithium-carbon anodes in organic and inorganic electrolytes, J. Power Sources 54 (1995) 228–231. Copyright 1995 Institute of Physics Publishing and Elsevier. B.V. Reprinted with permission from E. Peled, Advanced model for solid electrolyte interphase electrodes in liquid and polymer electrolytes, J. Electrochem. Soc. 144 (1997) L208. Copyright 1997 Institute of Physics Publishing. Reprinted with permission from D. Aurbach, B. Markovsky, M.D. Levi, E. Levi, A. Schechter, M. Moshkovich, Y. Cohen, New insights into the interactions between electrode materials and electrolyte solutions for advanced nonaqueous batteries, J. Power Sources 81–82 (1999) 95–111. Copyright 1999 Elsevier B.V. Reprinted with permission from Y.M. Lee, J.Y. Lee, H.-T. Shim, J.K. Lee, J.-K. Park, SEI layer formation on amorphous Si thin electrode during precycling, J. Electrochem. Soc. 154 (2007) A515. Copyright 2007 Institute of Physics Publishing. Reprinted with permission from B. Philippe, R. Dedryvère, J. Allouche, F. Lindgren, M. Gorgoi, H. Rensmo, D. Gonbeau, K. Edström, Nanosilicon electrodes for lithium-ion batteries: interfacial mechanisms studied by hard and soft X-ray photoelectron spectroscopy, Chem. Mater. 24 (2012) 1107–1115; C. Cao, H.-G. Steinrück, B. Shyam, K.H. Stone, M.F. Toney, In situ study of silicon electrode lithiation with X-ray reflectivity, Nano Lett. 16 (2016) 7394–7401; R. Kumar, A. Tokranov, B.W. Sheldon, X. Xiao, Z. Huang, C. Li, T. Mueller, In situ and operando investigations of failure mechanisms of the solid electrolyte interphase on silicon electrodes, ACS Energy Lett. 1 (2016) 689–697. Copyright 2012, 2016 American Chemical Society. Reprinted with permission from R. Kumar, A. Tokranov, B.W. Sheldon, X. Xiao, Z. Huang, C. Li, T. Mueller, In situ and operando investigations of failure mechanisms of the solid electrolyte interphase on silicon electrodes, ACS Energy Lett. 1 (2016) 689–697. Copyright 2016 American Chemical Society.)*

TABLE 6.1 Comparison of the calculated and experimental reduction potential of solvents.

Solvents	Calculated (V vs Li/Li$^+$)	Experimental (V vs Li/Li$^+$)
PC	1.24	1.00–1.60
EC	1.46	1.36
DMC	0.86	1.32
DEC	1.33	1.32
VC	0.25	1.30–1.40
FEC	1.40	1.20–1.30

Information adapted from X. Zhang, R. Kostecki, T.J. Richardson, J.K. Pugh, P.N. Ross, electrochemical and infrared studies of the reduction of organic carbonates, J. Electrochem. Soc. 148 (2001) A1341; U.S. Vogl, S.F. Lux, E.J. Crumlin, Z. Liu, L. Terborg, M. Winter, R. Kostecki, The mechanism of SEI formation on a single crystal Si(100) electrode, J. Electrochem. Soc. 162 (2015) A603–A607; G.M. Veith, M. Doucet, R.L. Sacci, B. Vacaliuc, J.K. Baldwin, J.F. Browning, Determination of the solid electrolyte interphase structure grown on a silicon electrode using a fluoroethylene carbonate additive, Sci. Rep. 7 (2017) 6326; Y. Jin, N.-J.H. Kneusels, L.E. Marbella, E. Castillo-Martínez, P.C.M.M. Magusin, R.S. Weatherup, E. Jónsson, T. Liu, S. Paul, C.P. Grey, Understanding fluoroethylene carbonate and vinylene carbonate based electrolytes for Si anodes in lithium ion batteries with NMR spectroscopy, J. Am. Chem. Soc. 140 (2018) 9854–9867; C. Xu, F. Lindgren, B. Philippe, M. Gorgoi, F. Björefors, K. Edström, T. Gustafsson, Improved performance of the silicon anode for Li-ion batteries: understanding the surface modification mechanism of fluoroethylene carbonate as an effective electrolyte additive, Chem. Mater. 27 (2015) 2591–2599.

of the SEI layer on lithium metal with the following order $PF_6^- > BF_4^- \approx SO_3CF_3^- > AsF_6^- > ClO_4^-$ suggesting anions strongly influences the formation of SEI. The solvents used for electrolyte formulation also played a key role for SEI species formation whereas, lithium alkoxide (ROLi) and lithium carboxylates (RCOOLi) were predominant in ethereal solvents (THF, DME, methyl formate) but Li_2CO_3, Li dicarbonates [(ROCO$_2$Li)$_2$], and semicarbonates (ROCO$_2$Li) were evident in carbonate-based electrolyte [20, 32–34].

The vast amount of knowledge gained from SEI formation on graphite anode and the microscopic and spectroscopic analytical tools used to determine the SEI composition was immensely helpful during SEI study on silicon (Si) anode [15, 35]. Although the mosaic model of SEI on graphite is well-established and accepted, SEI formation on Si follows a different trajectory due to repeated formation and destruction of SEI layer during (de)lithiation andpresence of native oxide (SiO$_2$) layer terminated hydroxyl (—OH) group makes things complexed [36, 37]. Preliminary work by Choi et al. [38] suggested the formation of EC-derivatives such as metastable linear alkyl carbonates [—Si—OCH$_2$CH$_2$O-CO$_2$Li, Si—CH$_2$CH$_2$OCO$_2$Li, R(OCO$_2$Li)$_2$] as SEI layer on Si anode. Later,

Lee et al. [38] proposed the continuous evolution of SEI structure on thin-film Si anode in which reduction of anions and organic solvents contributed to the SEI layer formation. The major components found in the SEI layer were Li_2CO_3, lithium alkyl carbonate, LiF, Li_xPF_y [39]. However, Chan et al. [40] argued that SEI formation was a voltage-dependent phenomenon and had severe implications during the first cycle charging/discharging. Afterward, the SEI layer compositions remain the same, with Li_2CO_3 and LiF as the primary species. It was also found that due to continuous volumetric expansion/contraction of lithiated/delithiated Si phase(s), the SEI composition is dynamic in nature and continuously evolves with time [41–45]. Chen et al. [40] observed a variety of solvents/anions reduction products, including PEO type oligomers, Li_2CO_3, lithium alkyl carbonate, LiF, Li_xPF_y, $Li_xPF_yO_z$, LiOH, etc. in SEI layer at different state of charge/depth of discharge profiles. Yen et al. [46] showed the absence of silicon and carbon fluorides but the presence of siloxane (—Si—O—Si—) species on carbon-coated Si anode. Initial exploratory work suggested that SEI layer on pristine silicon, oxide terminated silicon (Si/SiO_x, $1 \leq x \leq 2$) and silicon covered with a thin layer of carbon are structurally same, however, differ in composition owing to different surface/subsurface reactions between electrode/salt and electrode/solvents in variety of electrolytes.

Recently Philippe et al. [47] performed depth profile analyses of Si/C composite electrode covered with a ~50 nm thin layer of SiO_2 using XPS and soft/hard X-rays studies and observed the formation of Li_2O and Li_4SiO_4 as SEI phase and partially lithiated Li_xSi_y phase(s) and un-reacted Si during first cycle charging. The amount of Li_2O increases continuously until the end of the lithiation, and the chemical composition between SEI and Li_xSi_y layer contains mixed-phase(s) of Li_2O, Li_xSiO_y, and SiO_2. Cao et al. [48, 49] and Kumar et al. [50–52] also made a similar observation and proposed the formation of bi-layer SEI structure composed of porous and soft organic layer (alkyl carbonates, ROLi, alkyl bicarbonates, etc.), and stable and hard inorganic layer (LiF, Li_2CO_3, Li_2O, Li_4SiO_4, SiO_xF_y) formed onto the Si surface. It was also found that the mechanical and electrical properties of the bi-layer SEI structure vary along with the thickness of Si anodes. The "soft" organic outer layer possesses higher electrical resistance, while the "hard and dense" inorganic inner layer permits quick Li-ions transport. *In operando* AFM and in situ microscopic analysis demonstrated that SEI layers undergo crack formation during initial lithiation, and buckling or delamination of the initial SEI layer strongly influence the morphology of the SEI that forms during subsequent cycling [23, 50, 51, 53, 54]. However, the composition, thickness, and morphology of SEI on Si anodes are not fixed and affected by the solvents and salts that constitute the electrolyte [55]. Effective modification of SEI by using electrolyte additives, surface-functionalization, and protective layers may improve the SEI stability and electrochemical performance of Si anodes [8]. Therefore, it appears that

SEI formation on anodes, especially on Si-based anodes surface is a complex phenomenon, and the interplay between salts and solvents concentration, composition, testing condition, temperature, even the polymer binders being used will dictate the reduction mechanisms of various species and eventually SEI formation.

Keeping in mind that SEI formation on crystalline and amorphous Si, Si/SiO$_x$ surface is a complicated and complex process, the book chapter deals with a detailed understanding of SEI evolution and formation on Si anodes in liquid electrolytes and explains the stability and decomposition products of lithium salts, carbonate-based solvents, along with the formation of radical species including inorganic and organic products during electrochemical cycling of Li-ions with Si/SiO$_x$ anodes. In addition, we address the role of two main electrolyte additives (VC, FEC) and how they impact the SEI formation and stability of Si anodes. We also discuss the SEI properties, formation and growth mechanism, and their mechanical stability. The computational work of SEI formation using various electrolytes and their stability in Si anode will also be discussed. Finally, we cover the role of various novel architecture in particular "core-shell," "yolk-shell" morphology, numerous carbon coating strategies to improve the Li-ion cyclability, and the effect of supramolecular binders towards stability of SEI compared to traditionally used binders.

6.2 Energetics of SEI

The difference between the Fermi level of the electrodes and the highest occupied molecular orbital (HOMO) or the lowest unoccupied molecular orbital (LUMO) levels of the electrolyte governs the thermodynamic stability of the electrolyte [8]. The SEI formation happens due to thermodynamic instability of aprotic electrolytes during battery operation with the uptake of lithium and electrolyte components and construction of an ionically conducting and an electronically insulating passivation layer, leading to capacity fade. The SEI formation in an electrochemical cell is nicely explained by Goodenough and Kim (Fig. 6.2) from electronic band structure perspectives, which occurs when the redox potential of the electrodes falls outside the electrochemical window (E_g) of the electrolyte [2]. If the LUMO of the electrolyte is higher than the electrochemical potential (μ_A) of the anode, the electrolyte will remain stable in the battery; otherwise, the electrolyte will be reduced unless a surface passivation layer creates a barrier to electron transfer from the anode to the LUMO. Similarly, the electrolyte will be stable if the HOMO is lower than the electrochemical potential (μ_C) of the cathode; otherwise, if μ_C is below the HOMO, it will oxidize the electrolyte unless a passivation layer blocks electron transfer from the electrolyte HOMO to the cathode.

190 PART | III Electrolytes and SEI issues

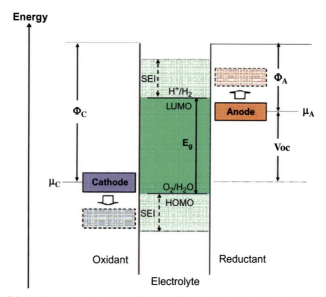

FIG. 6.2 Schematic open-circuit energy diagram of an electrolyte. Φ_A and Φ_C are the anode and cathode work functions. E_g is the electrolyte's electrochemical stability window. μ_A and μ_C are the redox potential of the anode and cathode, respectively. *(Reprinted with permission from J.B. Goodenough, Y. Kim, Challenges for rechargeable Li batteries, Chem. Mater. 22 (2010) 587–603. Copyright 2010 American Chemical Society.)*

However, Peljo and Girault [56] argued that the potential of electrolyte reduction at negative potentials and potential of solvent oxidation at positive potentials are more appropriate to use as electrochemical stability window (E_g) of electrolytes since HOMO and LUMO gap defines the intrinsic electronic band structure of individual solvent molecules and often the redox potential gets affected in the presence of auxiliary electrolyte components. Nevertheless, LUMO of most electrolyte solvents used in commercial batteries are higher than the lithiated graphite (∼0.1 eV) and lithiated silicon (∼0.1–0.2 eV) voltage, which leads to electrolyte decomposition and SEI layer formation [43, 57]. In comparison to the SEI on the cathode, the SEI on the anode is more unstable since the lithiation voltages are beyond the electrochemical window of the carbonate-based solvents used in LIBs [2]. Due to its importance to battery performance and durability, extensive investigations have been conducted on anode SEI films. While many review articles exist in the literature on SEI formation on graphite anode [8, 11, 28], however, collated information of SEI formation on Si anodes is scanty [55]. Therefore, the focus of the present exercise is to provide a detailed understanding of SEI formation, growth, and failure mechanisms of silicon anode during lithiation/delithiation phenomena.

6.3 SEI compositions via electrolyte and salt decomposition

Commercial LIBs possess nonaqueous liquid electrolytes composed primarily of a lithium salt ($LiPF_6$, $LIPO_4$, LiTFSI, etc.) dissolved in a solvent mixture of organic carbonates (EC, DEC, DMC). During the charge-discharge cycle, electron transfer from anode may trigger the decomposition of solvents and salt near the electrode/electrolyte interface resulting in the formation and growth of SEI film as the cycle progressed. However, the chemical composition of the SEI is not only affected by the electrolyte but also depends on the composition and morphology of Si surface. Silicon surface always remains covered with a native oxide layer (SiO_x) featuring oxygen terminating dangling bonds, contributing to the significant difference of the SEI formation between Si and graphite [55]. The few nanometers (~1–100 nm) thick SEI layer on the Si surface is complex and heterogeneous in nature reported by many groups [23, 37, 53, 58–62] with an inner layer composed of inorganic compounds such as Li_2O, Li_2CO_3, LiF and the outer shell at the electrolyte interface contains alkyl carbonates and ROLi where R (–alkyl group) depends on the solvents being used. Various ex-situ analyses, including scanning force spectroscopy, atomic force microscopy, time of flight-secondary-ions mass spectroscopy (TOF-SIMS), and X-rays photoelectron spectroscopy (XPS) results validated the two-layer structure of SEI, an organic layer near the electrolyte and an inorganic layer near the electrode [45, 63]. Furthermore, the SEI growth followed a "bottom-up" mechanism where the organic components (e.g., polyethylene glycol and organic lithium salt) formed first on the electrode; then, the inorganic components (e.g., LiF, Li_2CO_3, and Li_2O) formed underneath the organic layer and pushed the as-formed layer up as the SEI grows (see Fig. 6.3A and B). It was shown that most solvent molecules would be reduced via a one-electron reduction mechanism, which favors organic species formation. The electrode surface will be covered by the porous organic film; further, as the organic layer grows thicker, anions from salt decomposition and solvents (such as EC) will transport through the porous layer to the electrode surface. Thus, two-electron reduction reactions will take place, favoring the formation of inorganic species, such as LiF, Li_2O, and Li_2CO_3. Balbuena and co-workers reported that the formation of SEI primarily contains inorganic compounds such as LiF, Li_2O, Li_2CO_3, and organic species such as lithium ethylene dicarbonate (Li_2EDC) [64, 65]. At higher cell potential, the EC or PC molecules undergo a "2 electrons" reduction pathway, where it either reacts with a Li-ion or with another Li-coordinated solvent molecule forming either lithium carbonate or the alkyl carbonate, respectively (see Scheme 6.1) [43, 66, 67].

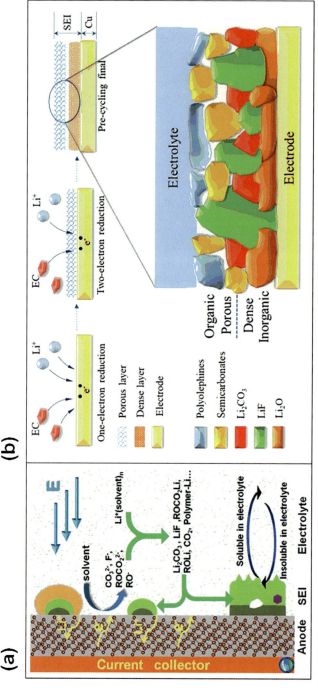

FIG. 6.3 (A) Dynamic formation scheme of the SEI of the anode. The formed SEI can deposit on the surface of the electrode, dissolve and drift in the electrolyte. (B) Schematic of complete SEI formation mechanism combining the two-layer structure chemistry and "bottom-up" growth dynamics. *(Panel (A): Reprinted with permission from J. Zheng, H. Zheng, R. Wang, L. Ben, W. Lu, L. Chen, L. Chen, H. Li, 3D visualization of inhomogeneous multi-layered structure and Young's modulus of the solid electrolyte interphase (SEI) on silicon anodes for lithium ion batteries, Phys. Chem. Chem. Phys. 16 (2014) 13229–13238. Copyright 2014 Royal Society of Chemistry. Panel (B): Reprinted with permission from Z. Liu, P. Lu, Q. Zhang, X. Xiao, Y. Qi, L.-Q. Chen, A bottom-up formation mechanism of solid electrolyte interphase revealed by isotope-assisted time-of-flight secondary ion mass spectrometry. J. Phys. Chem. Lett. 9 (2018) 5508–5514. Copyright 2018 American Chemical Society.)*

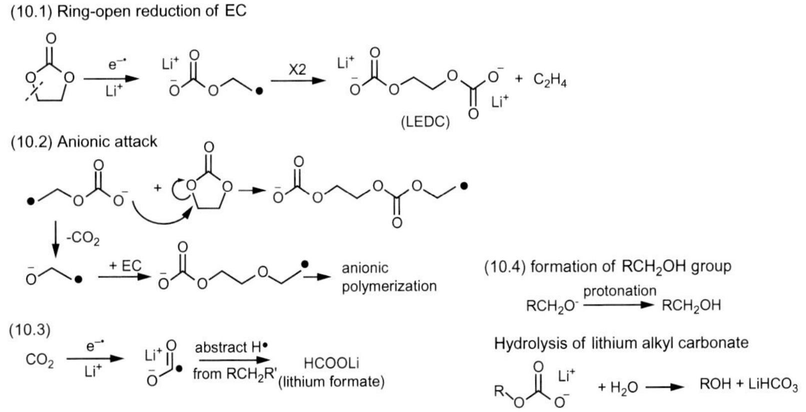

SCHEME 6.1 Schematic illustration of electrochemical reduction and decomposition of ethylene carbonate (EC) and formation of organic species. *(Reprinted with permission from K.W. Schroder, A.G. Dylla, S.J. Harris, L.J. Webb, K.J. Stevenson, Role of surface oxides in the formation of solid-electrolyte interphases at silicon electrodes for lithium-ion batteries, ACS Appl. Mater. Interfaces 6 (2014) 21510–21524; Y. Jin, N.-J.H. Kneusels, P.C.M.M. Magusin, G. Kim, E. Castillo-Martínez, L. E. Marbella, R.N. Kerber, D.J. Howe, S. Paul, T. Liu, C.P. Grey, Identifying the structural basis for the increased stability of the solid electrolyte interphase formed on silicon with the additive fluoroethylene carbonate, J. Am. Chem. Soc. 139 (2017) 14992–15004. Copyright 2014, 2017 American Chemical Society.)*

Similarly, DEC undergoes electro-reduction to form alkoxy and carboxyl radicals (see Scheme 6.2) [67]. At low cell potential, the reduction of EC or PC molecule proceeds via "one electron" pathway, where the reduced solvent molecule reacts with another radical anion forming alkyl carbonates [ROCO$_2$Li, (CH$_2$OCO$_2$Li)$_2$] [43]. An effective SEI film must allow Li-ions transfer back and forth from the electrode surface. However, with each subsequent cycle, Li-ions get trapped inside the SEI layer due to side reactions with electrolyte decomposition products instead of intercalating or alloying with the electrode. The above uncharacteristic phenomena lead to irreversible capacity loss from the first cycle, low coulombic efficiency at the subsequent cycles and compromise the battery performance. Depending on the reactivity of the solvent, and salt a great variety of radical and neutral species germinates after the first cycle from the electrode surface to the surface adsorbed molecule via electron transfer. For example, LiPF$_6$ salt decomposed on Si surfaces produces F$^-$ radicals, which recombine with Li-ions and forms LiF as the SEI layer on the Si surface [13, 14, 64, 65, 68–72]. LiPF$_6$ has also been dissociated into lithium fluorophosphates (Li$_x$POF$_y$ and Li$_x$PF$_y$) [66]. In addition, radical O species are also formed either from the decomposition of ppm level moisture present or from the salt leads to the formation of Li$_2$O, Li$_4$SiO$_4$, and Li$_2$CO$_3$ [14, 71]. EC molecules decomposed and formed lithium ethyl carbonate/CH$_3$CH$_2$OCO$_2$Li (LEC) via a single electron transfer or combined with a

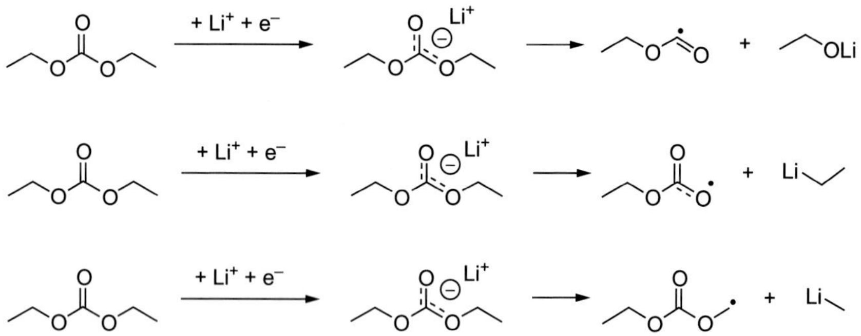

SCHEME 6.2 Electro-reduction of diethyl carbonate (DEC) to form lithium ethoxy and a radical ethyl carboxylate. *(Reprinted with permission from K.W. Schroder, H. Celio, L.J. Webb, K.J. Stevenson, Examining solid electrolyte interphase formation on crystalline silicon electrodes: influence of electrochemical preparation and ambient exposure conditions, J. Phys. Chem. C 116 (2012) 19737–19747. Copyright 2012 American Chemical Society.)*

second radical forming lithium butylene dicarbonate/CH$_2$CH$_2$OCO$_2$Li)$_2$ (LBDC) [59]. In addition, lithium ethylene dicarbonate (CH$_2$OCO$_2$Li)$_2$ (LEDC), Li$_2$CO$_3$, and PEO (—OCH$_2$CH$_2$O—) type oligomers are also being detected by spectroscopic analyses [60]. Dimethyl carbonate (DMC) is primarily decomposed into Li$_2$CO$_3$ and lithium methyl carbonate/CH$_3$OCO$_2$Li (LMC) along with minor alkyl carbonate (RCO$_2$Li) via a single electron transfer reduction mechanism [59]. It has been emphasized that the stability of electrolyte decomposition products is much more important than the stability of the electrolyte components for stability and controlled SEI growth on the outer surface of the Si anode surface [14]. Based on the theoretical calculation, it has been shown that electron transfer through LiF/Li$_2$O based SEI films decay rapidly to zero beyond ∼0.3 nm thickness on a pristine Si surface. On the contrary, electrons can transfer at a very slow rate up to ∼1 nm SEI thickness for a lithiated Si anode surface [69]. It is also found that at the same SEI thicknesses, LiF films offer a higher impedance to electron transfer than Li$_2$O films [69]. Generally, low polar solvents like DEC slow down the decomposition of LiPF$_6$ salt and therefore prevent Li$^+$ solvating in the SEI. However, LiPF$_6$ eventually dissociates into PF$_6^-$, F$^-$ radicals that further react with Li-ions or other fragmented species of the solvent/additive and form SEI film on silicon surface [71]. Vinylene carbonate (VC)—a known electrolyte additive decomposes into CO$_2$ and other radicals that react further with the PF$^-$ anions. Fluoroethylene carbonate (FEC) is also decomposed along with LiPF$_6$ salt and produces O radicals as a product, which acts as precursors for the eventual formation and growth of Li$_2$O on the lithiated Si surface [71]. Recently it has been shown that solvent is a key factor for the SEI film formation and stability of crystalline silicon (c-Si)-based electrodes [42]. It was observed that the capacity of c-Si drops to ∼110 mAh g^{-1} by the 100th cycle from ∼3060 mAh g^{-1} in the first cycle in 1 M solution of lithium bis (trifluoromethanesulfonyl) amide (LiTFSA) in propylene carbonate (PC) based electrolyte [73]. On the contrary, 1-[(2-methoxyethoxy)methyl]-1-methylpiperidinium bis-(trifluoromethanesulfonyl) amide (PP1MEM-TFSA) ionic liquid-based electrolyte exhibited markedly improved performance with a capacity of ~1050 mAh g^{-1} was achieved even after 100th cycle [74]. Raman mapping images of the c-Si electrodes after 10th cycle in delithiated state showed that localized lithiation/delithiation in PC based electrolyte yielded localized stress generation and severe fragmentation of electrode from Cu current collector leading to capacity fade. On the contrary, uniform Li-ion insertion/extraction over the entire c-Si surface in PRIMEA-TFSA based electrolyte resulted in moderate volume expansion and contraction and an improved capacity [74]. Authors argued that PRIMEA-TFSA-derived SEI films are thin, therefore, allowing better Li-ion conductivity over the entire c-Si surface and enhances the utilization ratio of active material contributing significant improvement in the cycling performance and suppressing electrode fragmentation at the same time. On the other hand, formation of thick SEI layer was responsible for inhibiting the Li-ion movement in c-Si electrode using PC-based electrolyte which impeded Li-ion diffusion to the entire electrode film concurrent with a decrease in the utilization of the active material [73]. The above results reveal that salt and solvent play vital roles for the SEI film formation and stability and tell us that selection of proper solvent(s) will be a key determinant for the long-term stability and successful demonstration of Si-based anodes in LIBs. In order to understand how, various inorganic lithium salts and nonaqueous solvents influence the SEI film formation and what exactly constitutes the decomposition products, collated information deemed appropriate and informative and presented here for the readers. Table 6.2 tabulates the decomposition products of pristine nonaqueous solvents and inorganic lithium salts being used in LIBs along with lithium/or graphite electrodes. Similarly, Table 6.3 tabulates the principal inorganic and organic SEI products that formed on graphite, Si, Si/SiO$_x$, and Si/C based anodes for LIBs using various salt and carbonate-based solvent mixtures.

6.4 Electrolyte additives and their roles during SEI formation

The purpose of an additive is to form a stable SEI layer on Si surface during initial cycle such that the SEI can prevent further electrolyte component degradation [103]. There are numerous electrolyte additives that have been proposed to improve the electrochemical stability of Si-based anodes owing to the formation of a stable SEI layer in LiPF$_6$/carbonate-based electrolytes. Two of the most sought after electrolyte additives, i.e., vinylene carbonate (VC) and fluoroethylene carbonate (FEC), have been widely studied and reported preferentially reduced on anode surface producing stable SEI leading to improved cycle life in LIBs; however, the exact reasons for stable SEI formation are still debatable [104, 105]. Therefore, we will confine our study to VC and FEC based

TABLE 6.2 Decomposition product of primary solvents and salts used in LIBs with lithium and graphite electrodes.

Solvent type	Specific solvent	Uncontaminated solution (decomposition product)
Alkyl carbonate	PC	$CH_3CH(OCO_2Li)CH_2OCO_2Li$ and propylene
	EC	$(CH_2OCO_2Li)_2$ and ethylene
	DMC	$ROCO_2Li$ $(CH_3OCO_2Li) + CH_3OLi$
	DEC	$CH_3CH_2OCO_2Li + CH_3CH_2OLi$
	EMC	CH_3OLi, CH_3CO_2Li
Esters	MF	$HCO_2Li, ROLi (CH_3OLi)$
	γ-BL	$CH_3(CH_2)_2COOLi$
	THF	$ROLi (CH_3(CH_2)_3OLi)$
	DME	$ROLi(CH_3OLi)$
Ethers	1,3-DN	$HCO_2Li, CH_3CH_2OCH_2OLi$
	DEE	$CH_3CH_2OLi, (CH_2OLi)_2$
	Diglyme (DG)	$CH_3OLi, CH_3OCH_2CH_2OLi, (CH_2OLi)_2$
Alkyl carbonate mixture	EC-PC	EC reduction products dominate
	EC-DMC	
	EC-DEC	
Salts		
$LiAsF_6$		$LiF, Li_xAsF_{3x}^-$
$LiClO_4$		$Li_2O, LiCl, LiClO_3, LiClO_2$
$LiBF_4$		$LiF, Li_xBF_y, Li_xBF_yO_z$
$LiPF_6$		$LiF, Li_xPF_y, Li_xPF_yO_z$
$LiSO_3CF_3$		$Li_xS_yO_z, LiF, ROF_yLi_z$
$LiN(SO_2CF_3)_2$		$LiF, Li_xS_yO_z, Li_3N, RCF_yLi_z, Li_2NSO_2CF_3$

Propylene carbonates (PC), Ethylene carbonates (EC), Dimethyl carbonates (DMC), Methylformate (MF), γ-butyrolactone (γ-BL), Tetrahydrofuran (THF), 2-Methyltetrahydrofuran (2Me-THF), Dimethoxyethene (DME), 1–3 Dioxolane (1,3-DN), diethoxyethane (DEE), $(CH_3OCH_2CH_2)_2O$ (Diglyme).
Information adapted from W. Schalkwijk, B. Scosati, Advances in Lithium-Ion Batteries, 2002.

TABLE 6.3 Principal inorganic and organic SEI products formed on graphite, silicon, silicon/silica, and silicon/carbon-based anodes for LIBs.

SEI products	Formation/origin/reactions	References
LiF	Decomposition of salts such as $LiPF_6$, $LiAsF_6$ or $LiBF_4$. It is a major salt reduction product. HF contaminant also reacts with semi carbonates to give LiF byproduct. The amount of LiF increases during storage. Reduction of FEC additive $LiPF_6 \rightarrow LiF + PF_5$ $LiBF_4 \rightarrow LiF + BF_3$ $PF_5 + 2xLi^+ + 2e^- \rightarrow Li_xPF_{5-x} + xLiF$ $LiPF_6 + H_2O \rightarrow LiF + 2HF + POF_3$ $Li_2CO_3 + HF \rightarrow LiF$ $Li_2O + 2HF \rightarrow 2LiF + 2H_2O$ $PF_5 + H_2O \rightarrow PF_3O + 2HF$ $PF_5 + H_2O \rightarrow PF_4OH + HF$ $PF_4OH \rightarrow POF_3 + HF$ $ROCO_2Li, Li_2CO_3 + HF \rightarrow LiF + ROCO_2H + H_2CO_3$ $ROLi + HF \rightarrow LiF + ROH$ $LiPF_6 + Li_2CO_3 \rightarrow 3LiF + POF_3 + CO_2$ $Li_2CO_3 + PF_5 \rightarrow 2LiF + POF_3 + CO_2$ $Li_2CO_3 + HF \rightarrow 2LiF + CO_2 + H_2O$ $Li_2CO_3 + 2HF \rightarrow 2LiF + H_2CO_3$ $3Li^+ + PF_6PF_5 + 2e^- \rightarrow 3LiF + PF_3$ $Li^+ + PF_6^- + H_2O \rightarrow LiF + PF_3O + 2H^+ + 2F^-$ $FEC + Li^+ + e^- \rightarrow LiF + CO_2 + CH_2CHO$	[7, 9, 10, 18, 27, 28, 32, 46, 75–82]
Li_2CO_3	Two electrons reduction of EC, PC, DMC, DEC, EMC. The reaction of $ROCO_2Li$ with H_2O or HF. It may also appear as a reaction product of semi-carbonates with HF or water or CO_2 $(CH_2O)_2CO + 2e^- + 2Li^+ \rightarrow Li_2CO_3 + CH_2=CH_2$ $(CH_3O)_2CO + 2e^- + 2Li^+ \rightarrow Li_2CO_3 + C_2H_6$ and $(C_2H_5O)_2CO + 2e^- + 2Li^+ \rightarrow Li_2CO_3 + C_4H_{10}$ $2ROCO_2Li + H_2O \rightarrow Li_2CO_3 + ROH + CO_2$ $(CH_2OCO_2Li)_2 + H_2O \rightarrow Li_2CO_3 + (CH_2OH)_2 + CO_2$ $(CH_2OCO_2Li)_2 \rightarrow Li_2CO_3 + CH_2=CH_2 + CO_2 + 0.5O_2$ $(CH_2O)_2CO + 2e^- + 2Li^+ \rightarrow CO_3^{2-} + CH_2=CH_2$ $2Li^+ + CO_3^{2-} \rightarrow Li_2CO_3$	[7, 10, 18, 22, 27, 28, 32, 33, 42, 46, 79, 82–84]

Continued

TABLE 6.3 Principal inorganic and organic SEI products formed on graphite, silicon, silicon/silica, and silicon/carbon-based anodes for LIBs—cont'd

SEI products	Formation/origin/reactions	References
Li$_2$O	$CH_3CHCH_2CO_3 + 2Li^+ + 2e^- \rightarrow CH_3CH=CH_2 + Li_2CO_3$ $R'OCO_2R'' + 2e^- + 2Li^+ \rightarrow Li_2CO_3 + R'R''$ (R'R'' = alkene for cyclic molecules e.g., EC, PC, or alkane for opened chain molecules DMC, DEC, EMC, etc.) $ROCO_2Li + e^- + Li^+ \rightarrow Li_2CO_3 + R^•$ Reduction of Li$_2$CO$_3$, degradation of SEI products during Ar$^+$ sputtering in the XPS experiment. The reaction of lithium with the silica layer remains on the silicon surface $Li_2CO_3 \rightarrow Li_2O + CO_2$ $SiO_2 + 4Li \rightarrow Si + 2Li_2O$	[7, 18, 22, 27, 78]
LiOH	Produced by the reaction of other products with water contamination. Degradation of organic SEI products. It may also result from the reaction of Li$_2$O and Li$_2$CO$_3$ with traces of moisture or with aging. $Li_2CO_3 + H_2O \rightarrow LiHCO_3 + LiOH$ $2LiOH + H_2O \rightarrow 2LiOH \cdot H_2O$ $2LiOH \cdot H_2O + CO_2 \rightarrow Li_2CO_3 + 3H_2O$ $LiF + H_2O \rightarrow HF + LiOH$ $Li_2O + H_2O \rightarrow 2LiOH$ $RCHOLi + O_2 \rightarrow CO_2 + LiOH + R$	[14, 20, 21, 33, 68, 85]
Li oxalate (Li$_2$C$_2$O$_4$)	Reduction of semi-carbonates. Reduction of CO$_2$	[86, 87]
Lithium alkyl carbonates (ROCO$_2$Li)$_2$, Lithium ethylene dicarbonate (CH$_2$OCO$_2$Li)$_2$	Two electron reduction product of EC, found mostly in the SEI of the EC based electrolytes $2(CH_2O)_2CO + 2e^- + 2Li^+ \rightarrow (CH_2OCO_2Li)_2 + CH_2=CH_2$ $CH_3C_2H_3O_2CO + 2e^- + 2Li^+ \rightarrow CH_3CH(OCO_2Li)CH_2OCO_2Li + Li_2CO_3 + C_3H_6$	[10, 33, 37, 42, 46, 61, 79, 82, 85, 87, 88]

Lithium salts of semi-carbonates (ROCO$_2$Li)	One electron reduction of EC, PC, DMC, DEC, DMC. Present in the outer layer of the SEI and are absent near Li. They occur in most PC containing electrolytes when the concentration of PC in the electrolyte is high (CH$_3$O)$_2$CO+e$^-$+Li$^+$→CH$_3$OCO$_2$Li+CH$_3$OLi CH$_3$C$_2$H$_3$O$_2$CO+e$^-$+Li$^+$→CH$_3$CH(OCO$_2$Li)CH$_2$OCO$_2$Li+CH$_3$CH= CH$_2$ ROLi+CO$_2$→ROCO$_2$Li CH$_3$CHCH$_2$CO$_3$+e$^-$→CH$_3$CHCH$_2$OCO$_2^-$ CH$_3$CHCH$_2$OCO$_2$+CH$_3$CHCH$_2$CO$_3$+2Li$^+$ + e$^-$ → CH$_3$CHCH$_2$+CH$_3$—CH(OCO$_2$Li)—CH$_2$—OCO$_2$Li	[32–34, 37, 65, 90, 93–95]
Alkoxides (ROLi), RCOLi	Reduction of ethers or EC, PC, DMC, EMC. It is soluble and may thus undergo further reactions ROCO$_2$R′+e$^-$+Li$^+$ → ROLi + R′OCO (CH$_3$O)$_2$CO+e$^-$+Li$^+$ →CH$_3$OLi+CH$_3$OCO· and CH$_3$OCO$_2$Li+CH$_3$· CH$_3$CH$_2$O(C=O)OCH$_3$+e$^-$ → CH$_3$CH$_2$O(C·−O$^-$)OCH$_3$,CH$_3$CH$_2$O (C·−O$^-$) OCH$_3$+e$^-$+2Li$^+$→LiO(C=O)CH$_3$+CH$_3$CH$_2$OLi (CH$_2$O)$_2$CO+2e$^-$+2Li$^+$→LiOCH$_2$CH$_2$OLi+CO (CH$_2$O)$_2$CO+2e$^-$+2Li$^+$→LiCH$_2$OCO$_2$Li	[10, 28, 37, 42, 46, 60, 61, 86–89]
Oligomers/polymers/polycarbonates	Polymerization of cyclic carbonates. Present in the outermost layer of the SEI, close to the electrolyte phase. This part imparts flexibility to the SEI	[19, 27, 58, 70, 90–92]
Lithium carboxylates (RCOOLi), Lithium formate (HCOOLi), Lithium succinate (LiO$_2$CCH$_2$CH$_2$CO$_2$Li), HCOLi	Products of degradation of ether-based electrolyte. Product of degradation of methylformate. Observed on SEI on graphite in carbonate-based electrolytes. Degradation of the carboxylic group of binders 2RCOOH→RCOOOCR + H$_2$O RCOONa + HF→RCOOH + NaF RCOOH +Li$^+$ + e$^-$→RCOOLi +0.5H$_2$	[7, 9, 86, 93–96]
Orthocarbonates, orthoesters, acetals. Fluorine-based alkoxy compounds	Nucleophilic attack on the carbonyl carbon by alkoxy, radicals, carbanion or fluorine-based species	[97]

Continued

TABLE 6.3 Principal inorganic and organic SEI products formed on graphite, silicon, silicon/silica, and silicon/carbon-based anodes for LIBs—cont'd

SEI products	Formation/origin/reactions	References
Li_xPF_y, $Li_xPF_yO_z$, Li_xBF_y	Decomposition of $LiPF_6$, $LiBF_4$ in presence of traces of moisture $LiPF_6 + 4e^- + 4 Li^+ \rightarrow 4 LiF + Li_xPF_y$ $Li_xPF_y + [O] \rightarrow Li_xPO_yF_z$ $LiBF_4 + 3e^- + 3 Li^+ \rightarrow 3 LiF + Li_xBF_y$ $LiPF_6 + 2H_2O \rightarrow LiPO_2F_2 + 4HF$ $LiPO_2F_2 + LiF + H_2O \rightarrow Li_2PO_3F + 2HF$ $Li_2PO_3F + LiF + H_2O \rightarrow Li_3PO_4 + 2HF$	[39, 98, 99]
SiO_xF_y, Li_4SiO_4, Li_xSiO_y	Reaction of Li with Si/SiO$_2$ form Li$_2$O and Li$_4$SiO$_4$. HF reacts with SiO$_2$ and forms SiO$_x$F$_y$ $8Li + SiO_2 + 3O_2 \rightarrow 2Li_2O + Li_4SiO_4$ $2SiO_2 + 4Li \rightarrow Si + Li_4SiO_4$ $LiPF_6 + H_2O \rightarrow LiF + HF + POF_3$ $SiO_y + XLi^+ + Xe^- \rightarrow Li_xSiO_y$ $HF \rightarrow H^+ + F^-$ $SiO_2 + F^- \rightarrow SiO_xF_y$ $SiO_2 + HF \rightarrow SiO_xF_y + H_2O$	[47, 67, 78, 100]
$Li_xPO_yF_z$	$LiPF_6$ decomposition and reaction with traces of moisture. Reduction of FEC additive	[47, 78–80, 101]
PEO	$LiPF_6$ decomposition in the presence of traces of moisture $LiPF_6 \rightarrow LiF + PF_5$ $PF_5 + H_2O \rightarrow PF_3O + 2HF$ $POF_3 + (CH_2O)_2CO \rightarrow CH_2FCH_2OCOOPF_2O$	[10, 81, 102]
$CH_2FCH_2OCOOPF_2O$, $CH_2FCH_2OPF_2O$	$CH_2FCH_2OCOOPF_2O \rightarrow CH_2FCH_2OPF_2O + CO_2$ $POF_3 + (CH_3O)_2CO + PF_6^- \rightarrow CH_2FCH_2OCOOPF_3O^- + PF_5$ $CH_2FCH_2OCOOPF_3O^- + PF_5 \rightarrow CH_2FCH_2OCOOPF_2O + PF_6^-$	

ROCOF, RCF, RC·	$ROCO_2Li + 2HF \rightarrow LiF + ROCOF + H_2O$ $RCOR' + 2HF \rightarrow RCF + FR' + H_2O$ $R' = C, Li, or H$ $RCO + 2Li^+ + 2e^- \rightarrow RC· + Li_2O$	[14, 67, 68]
Solvent/additive radicals	$EC + 2e^- \rightarrow O(C_2H_4)O^{2-} + CO_{(ads)}$ $EC + 2e^- \rightarrow O(C_2H_4)OCO^{2-}$ $EC + 1e^- \rightarrow (C_2H_4)OCO_2^-$ $EC(PC) + 2e^- \rightarrow CO_3^{2-} + alkylene (gas)$ $CO_3^{2-} + PC \rightarrow CH_3CH(OCO_2^-)CH_2OCO_2^-$ $CO_3^{2-} + EC \rightarrow (CH_2OCO_2^-)_2$ $(C_2H_4)OCO^- + 1e^- \rightarrow C_2H_4 + CO^{2-}$ $FEC + 2e^- \rightarrow OCO(C_2H_3)FO^{2-}_{(ads)}$ $OCO(C_2H_3)FO_2^-{}_{(ads)} \rightarrow F^-{}_{(ads)} + OCO(C_2H_3)O^-{}_{(ads)}$ $OCO(C_2H_3)O^-{}_{(ads)} + 2e^- \rightarrow CO_2^{-2}{}_{(ads)} + C_2H_3O^-$ $VC_{(ads)} + 2e^- \rightarrow ·OC_2H_2OCO_2^-{}_{(ads)}$ $·OC_2H_2OCO_2^-{}_{(ads)} + 2e^- \rightarrow ·OC_2H_2O_2^- + CO_2{}_{(ads)}$ $VC + CO_3^{2-} \rightarrow ·OC_2H_2OCO_2^{-2} + CO_2$ $VC + EC^{2-} \rightarrow ·OC_2H_4OCO_2C_2H_2O^{2-} + CO$ $VC + EC^{2-} \rightarrow ·OC_2H_4OC_2H_2OCO_2 + CO$ $FEC + 2e^- \rightarrow ·OCOCH_2CHFO_2^-{}_{(ads)}$ • $OCOCH_2CHFO_2^- {}_{(ads)} \rightarrow ·OCOC_2H_3O-{}_{(ads)} + F^-{}_{(ads)}$ • $OCOC_2H_3O^-{}_{(ads)} + F^-{}_{(ads)} + 2e^- \rightarrow CO^{-2}{}_{(ads)} + ·OC_2H_3O^- + F^-{}_{(ads)}$ • $OCOC_2H_3O^-{}_{(ads)} + F^-{}_{(ads)} + 2e^- \rightarrow CO_2^{-2}{}_{(ads)} + ·C_2H_3O^- + F^-{}_{(ads)}$ • $OCOC_2H_3O^-{}_{(ads)} + F^-{}_{(ads)} + e^- \rightarrow ·$ $OCOC_2H_2O_2^-{}_{(ads)} + H{}_{(ads)} + F^-{}_{(ads)}$	[28, 70–72]
Fluorinated radicals on SiO_2 surface	$SiO_2 + 6F^- + 4H^+ \rightarrow SiF_6^{-2} + 2H_2O$ $SiO_2 + 2PF6^- \rightarrow SiF_6 + 2PF_3O$	[46]

Information adapted from P. Verma, P. Maire, P. Novák, A review of the features and analyses of the solid electrolyte interphase in Li-ion batteries, Electrochim. Acta 55 (2010) 6332–6341; M. Gauthier, T.J. Carney, A. Grimaud, L. Giordano, N. Pour, H.-H. Chang, D.P. Fenning, S.F. Lux, O. Paschos, C. Bauer, F. Maglia, S. Lupart, P. Lamp, Y. Shao-Horn, Electrode-electrolyte interface in Li-ion batteries: current understanding and new insights, J. Phys. Chem. Lett. 6 (2015) 4653–4672.

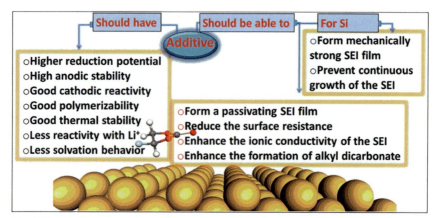

SCHEME 6.3 A summary of the desired properties for an electrolyte additive for Si-based anodes. *(Reprinted with permission from A.M. Haregewoin, A.S. Wotango, B.-J. Hwang, Electrolyte additives for lithium ion battery electrodes: progress and perspectives, Energy Environ. Sci. 9 (2016) 1955–1988. Copyright 2016 Royal Society of Chemistry.)*

additives here. It should be borne in mind that due to the volume expansion issue with Si anode, the additives must prevent the continued growth of SEI during cycling on the new Si surface if arises any. Scheme 6.3 summarizes the essential qualities an additive must possess to form a stable SEI on Si anode.

6.4.1 Role of vinylene carbonate

Various research groups experimentally verified that VC exhibits slightly inferior performance than FEC as an additive with Si for the formation of a stable SEI and suppressing the decomposition of carbonate-based electrolytes [76, 106]. The difference between VC and FEC lies in the fact that SEI originates from FEC decomposition has a higher percent of LiF compared to VC decomposition and is hinted as the reason for better SEI stability and passivation on Si surface [14, 64, 70, 107]. However, it is a matter of the fact that VC possesses slightly higher reduction potential (see Table 6.1) than the solvents such EC, DEC, DMC, etc., versus Li$^+$/Li due to the presence of a double bond [108]. It has been claimed that VC and FEC both offer outstanding capacity retention in Si nanowires (SiNWs) anode after the 250th cycle [109]. The formation of a flexible SEI layer comprising poly(VC) on SiNWs surface exhibited outstanding mechanical integrity mitigating irreversible decomposition reactions and facilitating the transformation of nanowires (NWs) to porous sponge-like network structure [109]. VC additive in LiPF$_6$/EC-DMC and LiPF$_6$/EC-EMC electrolytes has been reported inhibits the reduction of EC and suppresses the formation of LiF, LEDC, and ethylene gas [110]. VC preferentially reduced to poly(VC) and Li$_2$CO$_3$ with a thinner SEI layer than additive-free electrolyte

[111, 112]. However, the understanding role of poly (VC) versus Li$_2$CO$_3$ is vital to develop a better electrolyte additive and design a stable SEI on the anode. Early Study by Aurbach et al. [113] predicted that VC polymerizes on the lithiated graphite surfaces, thus forming poly alkyl Li-carbonate (—OCO$_2$Li groups, part of which may also have C=C double bonds) species that suppress both solvent and salt anion reduction and reducing the irreversible capacity loss. Chen et al. [114] showed that the presence of VC in 1 M LiPF$_6$ (EC: DMC 1:1 v/v ratio) electrolyte decreased the LiF content in the SEI layer on the Si anode. The significant components of the SEI layer were lithium salt (e.g., ROCO$_2$Li, Li$_2$CO$_3$, LiF), polycarbonate, and SiO$_2$. The cycle performance and coulombic efficiency of Si film anode were enhanced significantly with the presence of VC in the electrolyte. It has also been found that VC and FEC, both electrolyte additives play crucial roles in improving the thermal stability of the SEI at elevated temperatures for Li$_x$Si$_y$ phase(s) [115]. The improvement in thermal stability of additive-bearing Li$_x$Si$_y$ phase(s) attributed to the presence of a polycarbonate species and formation of a secondary SEI layer which repairs cracks in the primary pores and prevents spallation and loss of Si [115]. Soto et al. [14] observed that additive like VC is known to slowdown side reactions and form much stable, compact films and concluded polymerization reactions initiated by open VC radical anions are thermodynamically more favorable [112, 116]. They also observed that VC could generate stable polymer-based and inorganic (LiF, Li$_2$CO$_3$) films responsible for steady SEI layer growth. The advantage of VC as an additive lies with its fast polymerization with VC-derived surface films consist of poly-(vinylene carbonate), poly-VC, oligomeric VC, and a ring-opened polymeric form of VC and less prone to get reduced by a radical attack [112, 116]. Ota et al. [117] argued that VC as an additive has excellent cycling performance at elevated temperature (increase thermal stability) due to formation of a stable surface films comprising poly (VC). However, the performance deteriorated at 0°C and subzero temperature due to formation of thick SEI layer with poor ionic and electronic conductivity [117, 118]. Theoretical calculations using DFT suggest that VC reduced to an opened VC^{2-} ion through 2e$^-$ mechanism on lithiated silicon anodes (Li$_x$Si$_y$) and further to CO^{2-} and ˙OC$_2$H$_2$O$^-$ radical anions as the degree of lithiation progressed further [72, 77]. Overall, VC as an additive has its merit to form a stable SEI on Si anode and providing good cyclability in LIBs [55, 119].

6.4.2 Role of fluoroethylene carbonate

The formation of a stable SEI layer on Si surface using electrolyte additive like FEC in the carbonate-based electrolyte (1 M LiPF$_6$/EC-DMC, 1,1 v/v known as LP30) is a widely accepted strategy for improving the electrochemical performance of LIBs [38]. It is generally accepted that FEC containing electrolyte produces organic components as SEI products compared to conventional battery electrolyte [107]. However, there are still contrasting reports that FEC containing electrolyte produces inorganic SEI components [120]. The exact composition of SEI from FEC containing electrolyte is the subject of debate and required in-depth analysis. However, SEI is generally considered to consist of an inner inorganic layer (i.e., Li_2CO_3, LiF, etc.) close to the electrode surface and an outer organic layer [$(CH_2OCO_2Li)_2$, CH_3OCO_2Li, and CH_3OLi] close to the electrolyte [53, 54, 79, 121]. FEC generally reduces at a higher potential (~1.2–1.3 V) than other electrolyte components such as ethylene carbonate (0.7–0.8 V) versus Li/Li^+ [78, 122]. Reduction at higher potential give rise to a more protective SEI layer enabling much longer cycling in a conventional electrolyte [123].

It is generally accepted that crack-free conformal SEI layer covering Si surface forms uniformly during cycling preserve the volumetric expansion and yields better electrochemical stability [78, 99, 107, 122]. Various groups of researchers suggested that FEC-containing electrolytes generally produce LiF as the main product in the SEI layer besides other organic species [38, 70, 78, 99, 107, 120, 122, 123]. The origin of the LiF is either the decomposition of $LiPF_6$ salt or FEC during galvanostatic cycling [37, 120]. FEC decomposition by de-fluorination and a ring-opening reaction to LiF, —CHF—OCO_2—, HF and Li_2CO_3 reported by various groups (see Fig. 6.4A) [122, 123]. Li et al. [99] characterized the composition and structure of the SEI on Si/C composite anode with and without the presence of FEC additive, using solid-state NMR, XPS, and X-ray photoemission electron microscopy. They observed that the hydrolysis of a large amount of $LiPF_6$ generated $Li_xPO_yF_z$ species in the first cycle and Li_3PO_4 in the ensuing cycles. At the same time, there is a significant increase in Li salts after cycling with EC/DMC electrolyte. On the contrary, a porous SEI was formed, which captured small molecules such as $LiPF_6$, P—O, and Li—O species on the inner side when FEC was added. The Si particles suffered cracking post cycling with the EC-DMC electrolyte; however, they maintained a compact structure when FEC is added. It appeared FEC reduction on Li_xSi_y surfaces was independent of the degree of lithiation and occurs through three mechanisms via multielectron transfer mechanisms resulting formation of LIF moieties and either CO_3^{2-}, F^-, and CH_2CHO^- or CO^{2-}, F^- and OCH_2CHO^- anionic species oligomerize to SEI layers [70, 77]. One electron and two-electrons induced bond-breaking routes and FEC breakdown predict the formation of F^- and $C_3H_3O_3^-$, CO_2, $CHOCH_2$ species that further act as nucleophiles for reduction of solvent (EC, DMC) molecules to neutral organic species and formation of SEI layers bearing LiF [64].

It has been argued that the formation of LiF for FEC bearing EC-DMC electrolyte prevented cracking and pulverization of the Si particles during long-term cycling. Jaumann et al.[78] proposed that LiF has neither play a beneficial nor a disadvantageous role on the reversibility of Si anode in a Li-ion cell. The enhanced reversibility is attributed to the formation of a thin (~4nm) vinyl

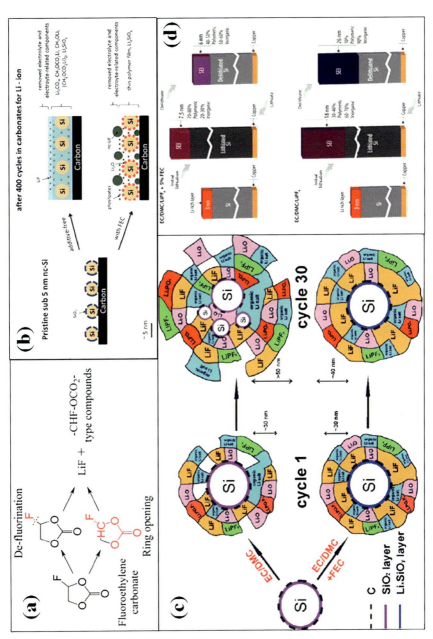

FIG. 6.4 (Continued)

polymer layer during cycling, and the intrinsic SiO$_2$ layer prevented Li$_2$O dissolution by HF attack on the silicon/carbon electrode (see Fig. 6.4B). FTIR and XPS spectra of a cycled SiNW electrodes in FEC-containing LiPF$_6$/EC-DMC electrolyte solution indicate the formation of poly(VC) (see Scheme 6.4A) as the major surface film component and remarkably stable electrochemical performance [79]. The presence of LiF and polyene-compound (see Scheme 6.4B) as SEI layer originated from free fluoride radical and acetylene unit also confirmed by XPS and TOF-SIMS analysis [123]. A thin SEI film formation in FEC bearing electrolyte predominantly containing insoluble polymeric species, LiF, and Li$_x$SiO$_y$ are also evidenced by others [37]. On the contrary, FEC-free carbonate-based electrolytes primarily involving EC showed thick SEI formation composed of LiF, LEDC, Li$_x$SiO$_y$, and Li$_2$CO$_3$ and poor electrochemical properties with Si anodes. Another theory proposed that FEC was first reduced to vinyl fluoride and further polymerized on the anode surface with poly (vinyl fluoride) formation [77, 123]. Bordes et al. [124] showed excellent capacity retention of Si-graphene anode and NCA cathode-based full cell when tested in LiPF$_6$/EC-DEC electrolyte with 5 wt% FEC additive. The improved electrochemical performance is attributed to the formation of a thin (∼30–50 nm) and stable SEI containing less CH$_2$OCO$_2$Li, Li$_2$CO$_3$, and LiF compounds. However, increasing FEC content beyond 5 wt% increased the formation of LiF along with consumption of lithium ions leading to high irreversible capacity loss. Recent reports suggest the formation of poly(-vinyl carbonates) as SEI layer in FEC based LP30 electrolyte [120]. The reduction of LiPF$_6$ first forms Li$_x$PO$_y$F$_z$ and further Li$_3$PO$_4$ during the subsequent cycles if traces of moisture present (see Fig. 6.4C) [99]. Veith et al. [58], using neutron reflectivity and XPS analyses, showed that at ∼0.9 V, FEC simply reduced to C—O containing

---

FIG. 6.4, CONT'D (A) Possible FEC decomposition reactions and products. (B) Simplified schematic illustration of the silicon-carbon nanostructure after long-term cycling depending on FEC addition. (C) Schematic of SEI on Si electrodes with and without FEC as electrolyte additive. (D) Graphical summary of SEI layer chemistry grown on silicon with and without FEC. *(Panel (A): Reprinted with permission from C. Xu, F. Lindgren, B. Philippe, M. Gorgoi, F. Björefors, K. Edström, T. Gustafsson, Improved performance of the silicon anode for Li-ion batteries: understanding the surface modification mechanism of fluoroethylene carbonate as an effective electrolyte additive, Chem. Mater. 27 (2015) 2591–2599. Copyright 2015 American Chemical Society. Panel (B): Reprinted with permission from T. Jaumann, J. Balach, M. Klose, S. Oswald, U. Langklotz, A. Michaelis, J. Eckert, L. Giebeler, SEI-component formation on sub 5 nm sized silicon nanoparticles in Li-ion batteries: the role of electrode preparation, FEC addition and binders, Phys. Chem. Chem. Phys. 17 (2015) 24956–24967. Copyright 2015 Royal Society of Chemistry. Panel (C): Reprinted with permission from Q. Li, X. Liu, X. Han, Y. Xiang, G. Zhong, J. Wang, B. Zheng, J. Zhou, Y. Yang, Identification of the solid electrolyte Interface on the Si/C composite anode with FEC as the additive, ACS Appl. Mater. Interfaces 11 (2019) 14066–14075. Copyright 2019 American Chemical Society. Panel (D): Reprinted with permission from G.M. Veith, M. Doucet, R.L. Sacci, B. Vacaliuc, J.K. Baldwin, J.F. Browning, Determination of the solid electrolyte interphase structure grown on a silicon electrode using a fluoroethylene carbonate additive, Sci. Rep. 7 (2017) 6326. Copyright 2017 Springer Nature Publishing Group.)*

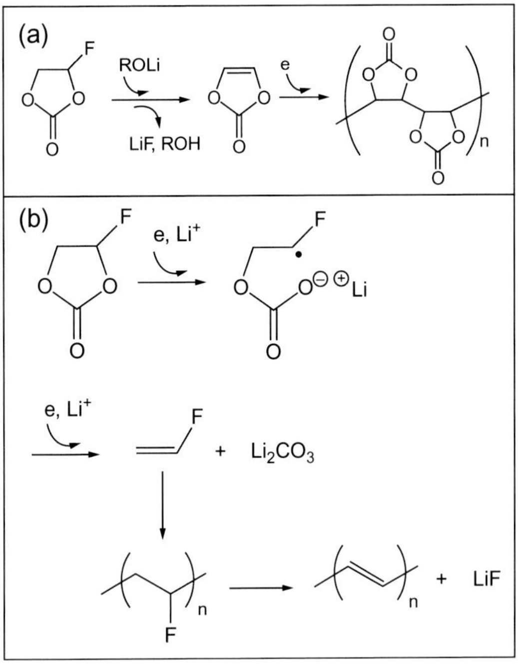

SCHEME 6.4 Possible decomposition products by FEC on (A) SiNWs electrode using 1 M LiPF$_6$ in EC: DMC (1:1 w/w) + 10 wt% FEC electrolyte. (B) 2 mm thick Si thin film electrode using 1 M LiPF$_6$ in FEC: DEC (1:1 v/v) electrolyte. *(Panel (A): Reprinted with permission from V. Etacheri, O. Haik, Y. Goffer, G.A. Roberts, I.C. Stefan, R. Fasching, D. Aurbach, Effect of Fluoroethylene carbonate (FEC) on the performance and surface chemistry of Si-nanowire Li-ion battery anodes, Langmuir 28 (2012) 965–976. Copyright 2011 American Chemical Society. Panel (B): Reprinted with permission from H. Nakai, T. Kubota, A. Kita, A. Kawashima, Investigation of the solid electrolyte interphase formed by fluoroethylene carbonate on Si electrodes, J. Electrochem. Soc. 158 (2011) A798. Copyright 2011 Institute of Physics Publishing.)*

polymeric species since FEC is thermodynamically unstable below 1.2 V versus Li/Li$^+$. During lithiation in a Si electrode, a thin layer SEI with a thickness of ~7 nm is formed for FEC based electrolyte compared to ~18 nm for non-FEC based electrolyte attributes that FEC helps to create thinner SEI with a higher percent of flexible polymeric components. The thin polymeric layer provided flexibility and ensures the formation of a stable SEI that would effectively passivate the surface against further reactions (see Fig. 6.4D). During delithiation SEI layer shrinks further by ~1.5 nm with the formation of LiF. The expansion and contraction of SEI layer with concomitant increase and decrease of thickness attributes dynamic nature for FEC bearing electrolytes [79]. However, non-FEC based electrolytes showed a thick SEI layer (~18–26 nm) during lithiation with a 40% increase in the amount of inorganic LiF content during delithiation.

6.5 SEI layer properties

The SEI layer properties depend on the SEI product(s) formed on Si surface due to the reduction of salt anions and solvent molecules. Typically, salt anions decompose into inorganic compounds (LiF, LiCl, fluorophosphate, and Li$_2$O), while solvent reduction products are lithium carbonates, lithium alkyl carbonates oligomers [27]. The salt and solvent reduction mechanism entirely depends on the cell potential and varies as the potential decreases from open circuit potential to fully lithiated phase (Li$_{22}$Si$_5$) at ~0.044 V versus Li$^+$/Li during lithiation of silicon.

However, SEI composition is not fixed and continues to undergo reduction as the cycle progresses and develop multiphasic compositions comprising various inorganic and organic species over time [59]. The electron transport characteristics through different SEI layers were investigated using density functional theory and Green's function (DFT-GF) approach [57]. It was identified that with increasing SEI layer thickness HOMO-LUMO gap of the electrolyte decreases, and electrical current decreases exponentially due to the formation of Li$_2$Si$_2$O$_5$, Li$_2$CO$_3$, LiF, and Li$_2$O [57]. The theoretical study also finds that Li$_2$CO$_3$ is electrically more insulating than LiF and Li$_2$O [57]. Wang and Chew [125] using first-principles calculation shows that, SEI layer after sufficient Li-ions segregation at LiF-Li$_x$Si$_y$ ($x > 1$) interface, deforms plastically and imparts ductility with the formation of delocalized voids. In contrast, the Li$_2$O-Li$_x$Si$_y$ SEI interface remains tightly bonded and provide significant rigidity across the entire lithiated silicon range ($0 \leq x \leq 3.75$). The above findings from theoretical calculation matched with the experimental results that LiF is beneficial over Li$_2$O as the SEI inorganic product. Li$_2$O being less strain-tolerant, will promote brittle-like failure of SEI later as the lithiation/delithiation progress [125].

6.6 SEI formation on Si/SiO$_x$ electrodes

To understand the SEI formation and irreversible capacity loss, galvanostatic cycling was performed using a Si/C electrode prepared without any binder and tested in a coin cell at C/100 current rate between 2 to 0.001 V versus Li/Li$^+$ (see Fig. 6.5A) [60]. A plateau was observed at ~0.8 V during the discharge cycle ascribed to SEI formation on the Si and/or C surface. A long plateau at ~0.15 V was characteristic of the lithiation of crystalline Si and the formation of amorphous Li$_x$Si$_y$ phases. The voltage versus capacity curve during the charge cycle was characteristic of the delithiation phenomena of Li$_{15}$Si$_4$ phase and formation of amorphous Si[60].

The irreversible capacity loss evidenced by the difference in lithiation and delithiation capacity was mainly due to SEI formation on the C and Si particle surfaces [60]. The classic example of irreversible capacity loss during discharge/charge cycle and SEI formation on silicon was further supported by

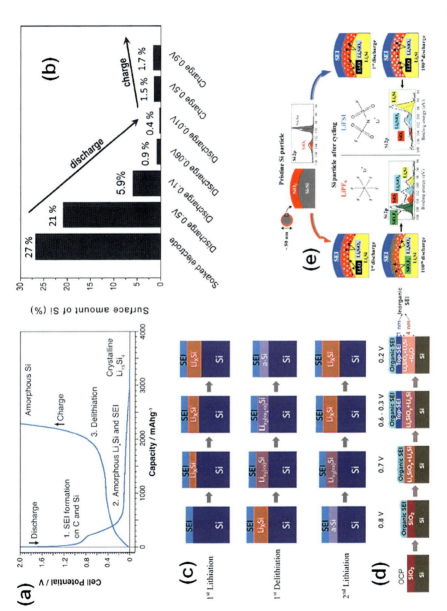

FIG. 6.5 (Continued)

XPS analysis of Si/C composite electrode. The amount of silicon (atomic percent) was quantified from the Si 2p XPS spectra during first cycle discharge and charge at a sample depth $\approx$5–10nm. Fig. 6.5B illustrates the increase (or decrease) of the SEI thickness verified by the amount of silicon present. The dramatic drops in Si 2p signal upon discharge were observed due to SEI formation and reversed upon charge slightly; however, they remain at a very low value with respect to the pristine silicon electrode [47]. The XPS analysis supported the hypothesis that SEI formation occurs during the first cycle. Cao et al. [49, 126, 127] performed in-situ X-ray reflectivity (XRR) study of 500 μm silicon wafer and identified the fundamental differences between (de) lithiation phenomena of crystalline and amorphous silicon originate during galvanostatic cycling and provide insights on the SEI nucleation, growth, and evolution phenomena. The first lithiation of crystalline silicon (c-Si) is a layer-by-layer, reaction limited two-phase process dependent on the crystallographic orientation of silicon, whereas delithiation of Li_xSi_y resulting in the formation of amorphous silicon (a-Si) and subsequent lithiation of a-Si are reaction-limited single-phase processes. The thickness as well as electron density of the inorganic SEI layer varies during lithiation and delithiation process. Fig. 6.5C shows the mechanism of the lithiation/delithiation phenomena in c-Si. During the first lithiation, Li_xSi

---

FIG. 6.5, CONT'D (A) First electrochemical cycle of a C/Si electrode galvanostatically cycled against Li metal. (B) Evolution of silicon content at the surface of the electrodes determined from XPS spectra of the Si/C/CMC composite electrodes upon the first discharge/charge cycle (in-house XPS, 1487 eV). (C) Illustration of lithiation and delithiation process in silicon, (top panel) first lithiation process in which the thickness of Li_xSi increases but the electron density remains constant (two-phase reaction); (middle panel) first delithiation process where the thickness of Li_xSi decreases and the electron density increases and the final delithiation state is a-Si (single-phase reaction); (bottom panel) the second lithiation process. (D) Schematic illustration of the proposed potential-dependent SEI growth mechanism on native oxide-terminated Si anodes. (E) Schematic comparison of the mechanisms occurring at the surface of a silicon nanoparticle upon cycling of a Li//Si cell using either $LiPF_6$ or LiFSI salts. *(Panel (A): Reprinted with permission from A.L. Michan, M. Leskes, C.P. Grey, Voltage dependent solid electrolyte interphase formation in silicon electrodes: monitoring the formation of organic decomposition products, Chem. Mater. 28 (2016) 385–398. Copyright 2016 American Chemical Society. Panle (B): Reprinted with permission from B. Philippe, R. Dedryvère, J. Allouche, F. Lindgren, M. Gorgoi, H. Rensmo, D. Gonbeau, K. Edström, Nanosilicon electrodes for lithium-ion batteries: interfacial mechanisms studied by hard and soft X-ray photoelectron spectroscopy, Chem. Mater. 24 (2012) 1107–1115. Copyright 2012 American Chemical Society. Panel (C): Reprinted with permission from C. Cao, H.-G. Steinrück, B. Shyam, M.F. Toney, The atomic scale electrochemical lithiation and delithiation process of silicon, Adv. Mater. Interfaces 4 (2017) 1700771. Copyright 2017 WILEY-VCH Verlag GmbH & Co. KGaA, Weinheim. Panel (D): Reprinted with permission from C. Cao, I.I. Abate, E. Sivonxay, B. Shyam, C. Jia, B. Moritz, T.P. Devereaux, K.A. Persson, H.-G. Steinrück, M.F. Toney, Solid electrolyte interphase on native oxide-terminated silicon anodes for Li-ion batteries, Joule 3 (2019) 762–781. Copyright 2019 Elsevier B.V. Panel (E): Reprinted with permission from B. Philippe, R. Dedryvère, M. Gorgoi, H. Rensmo, D. Gonbeau, K. Edström, Improved performances of nanosilicon electrodes using the salt LiFSI: a photoelectron spectroscopy study, J. Am. Chem. Soc. 135 (2013) 9829–9842. Copyright 2013 American Chemical Society.)*

forms with increasing thickness and a constant electron density i.e. two-phase reaction (top panel Fig. 6.5C). However, during delithiation of Li_xSi, the thickness of Li_xSi decreases and the electron density increases (middle panel Fig. 6.5C) and finally converts into a-Si which is a meta stable phase. Further lithiation of a-Si has two steps, with the first step being the lithiation of a-Si (bottom panel Fig. 6.5C) and followed by the lithiation of bulk c-Si from the 500 µm thick Si substrate (bottom panel Fig. 6.5C). The XRR data also sheds light of Li-dip observed during delithiation step of first and second cycles caused by limited Li-ions diffusion in the SEI layer and increasing internal resistance, leading to large over potentials and capacity loss of silicon with increasing C-rates. However, most often the silicon used for the fabrication of LIBs comes with thin layer coverage of native oxide (SiO_2). Therefore, understanding SEI formation and evolution on the oxide terminated silicon surface (Si/SiO_2) will be key findings for future battery researchers. Using a combination of in-situ synchrotron XRR, LSV, ex-situ XPS, and first-principles calculations, Cao et al. [48] studied the SEI evolution on an oxide terminated (001)-Si wafer using 1 M $LiPF_6$/EC-DMC electrolyte sans binder/conductive carbon. The major finding of the work was that the inorganic SEI layer consists of a bottom-SEI layer formed at ∼0.7 V near the Si/SiO_2 interface, and a top SEI layer forms at ∼0.6 V in-between organic electrolyte and the bottom-SEI electrolyte layer. The bottom-SEI primarily contains Li_xSiO_y, whereas the top SEI contains LiF. On further lithiation, Li_2O formation occurs at the bottom-SEI layer around ∼0.2 V (see Fig. 6.5D). The lithiation up to ∼0.2 V yields 2.3 times increase in the thickness of SiO_2 layer. Importantly, the merit of the work was that it highlights the importance of a native oxide layer on silicon surface suppressing the continued growth of the inorganic SEI layer [67]. Although the presence of a thin SiO_2 layer over the Si surface considered a boon for long term cycling and formation of a stable SEl layer, most of the work reported till date confined using commonly used lithium salt ($LiPF_6$) in carbonate-based (EC, DEC, DMC, EMC) electrolytes. However, it is known that traces of moisture if present may leads to dissociation of $LiPF_6$ and formation of hydrofluoric acid (HF) which may lead to serious consequences during thermal runway of the battery pack. Therefore, identification of safe and alternative lithium salt compatible with silicon anode and stable for thousands of cycles will be key finding for future battery technology. Philippe et al. [47, 100] showed that bis(fluorosulfonyl)imide (LiFSI) with the chemical formula $Li[N(SO_2F)_2]$ is a better alternative for nano silicon anode than $LiPF_6$ salt in the carbonate-based electrolyte. It was found that during the first discharge cycle, both the salts react with the native SiO_2 layer present on the silicon surface and form Li_2O and Li_4SiO_4, leading to the formation of SEI layer and Li_xSi phase(s). However, during long-term cycling, HF formation owing to decomposition of $LiPF_6$ salt leads to the disappearance of Li_2O from the surface and formation of a new fluorinated species (SiO_xF_y) by the reaction of SiO_2 with HF (see Fig. 6.5E). However, LiFSI prevents the formation of HF, and therefore, fluorinated species SiO_xF_y was not formed at the silicon surface. Prevention of LIFSI salt decomposition resulted in better electrochemical performance over $LiPF_6$ salt bearing electrolyte in Li/Si cells.

6.7 SEI growth model

A high resolution in-situ atomic force microscopy (AFM) technique was used to ascertain the formation of solid electrolyte interphase (SEI) growth mechanism on silicon electrodes. During the first cycle lithiation, a mesoporous organic layer with a low elastic modulus form as SEI on silicon surface around ~0.6 V versus Li^+/Li. The initial SEI phase is referred to as "organic" SEI consist primarily of carbon-rich electrolyte decomposition products. Continuing lithiation and decomposition of the electrolyte increases the SEI layer thickness, which filled the mesopores of the organic SEI layer (see Fig. 6.6A). Therefore, the solvated ions are required to pass through a thick SEI to the silicon electrode, limiting the growth rate. At lower potential (≤ 0.3 V), a dense inorganic SEI layer formation occurs, allowing Li-ion diffusion but limiting both electrolyte diffusion and electrical conductivity (see Fig. 6.6B) [53, 54]. However, after completion of the first cycle (lithiation/delithiation) silicon films break into islands concomitant with new SEI layer formation. The silicon islands formation leads to an irreversible volume expansion of approximately 15%, leading to capacity loss from the very first cycle that cannot be recovered. All the above findings were purely experimental observations based on initial SEI formation, notwithstanding the fact that the stability of SEI for long-term cycling is also a critical factor for the success of LIBs. During the first cycle lithiation of silicon, the rapid formation of SEI can be stabilized, which depends upon the nature of the surface. However, a fast initial lithiation rate led to the formation of a thinner and smoother film, indicating the initial SEI structure is mostly dependent on the parameters used in the first cycle. The passivating behavior over a long term cycling process will plausibly depend on the initial SEI formation and its morphology. In other words, optimizing the SEI properties during the initial lithiation process could lead to improved battery performance [54].

6.8 Mechanical deformation and associated strain of SEI layer

Sheldon and co-workers [50] performed in operando atomic force microscopy (AFM) to monitor the SEI film formation and associated lateral mechanical strains during lithiation/delithiation of continuous silicon films and silicon islands of different sizes on a copper current collector. During lithiation, the mesoporous SEI layer appears at ~0.6 V and passivates the surface with lithium-rich species (e.g., Li_2CO_3), with minimal strains in both continuous and island silicon. As the potential drops below ~0.3 V, silicon expands with the formation of lithiated phases (Li_xSi_y). However, this deformation is very different for continuous and patterned silicon configurations. The continuous

SEI layer and impact on Si-anodes for Li-ion batteries **Chapter | 6** **213**

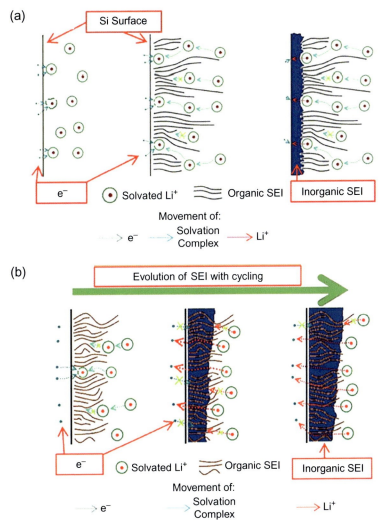

FIG. 6.6 (A) SEI growth model: SEI decomposition at higher potential, resulting in organic products. Continuing decomposition increases the SEI thickness and decreases mesoporosity, which reduces the growth rate as the solvation complex now has to diffuse to the electrode through SEI that is thicker and denser (ultimately, larger complexes are unable to reach the surface at all). At a lower voltage, dense SEI forms, which allows Li-ion diffusion and passivates by limiting electrical conductivity. (B) Model for the SEI growth with cycling, inorganic SEI formed at lower potentials continues to fill the mesoporous structure during subsequent cycles. *(Reprinted with permission from A. Tokranov, R. Kumar, C. Li, S. Minne, X. Xiao, B.W. Sheldon, Control and optimization of the electrochemical and mechanical properties of the solid electrolyte interphase on silicon electrodes in lithium ion batteries, Adv. Energy Mater. 6 (2016) 1502302; A. Tokranov, B.W. Sheldon, C. Li, S. Minne, X. Xiao, In situ atomic force microscopy study of initial solid electrolyte interphase formation on silicon electrodes for Li-ion batteries, ACS Appl. Mater. Interfaces 6 (2014) 6672–6686. Copyright 2014, 2016 American Chemical Society.)*

silicon films expand in the out-of-plane direction due to in-plane constraint from the substrate. However, the thickness and associated volumetric expansion of patterned silicon islands move parallel to the substrate giving rise to significant strains in the SEI layer (see Fig. 6.7A). During electrochemical cycling, the SEI experiences considerable mechanical strain, which further results in the degradation of these films leading to limiting life cycle of silicon-based electrodes. The crack formation due to lateral strain will enable additional SEI formation by exposing the underlying silicon to the electrolyte. However, to fulfill the above criteria, another additional mechanism needs to be fulfilled beyond crack formation, viz.; (i) crack deflection and debonding between the dense inorganic inner and soft organic outer layer of the SEI, (ii) crack penetration in the dense inorganic inner SEI layer and partial delamination, and (iii) crack penetration at the underlying electrode surface (see Fig. 6.7B) [128]. Two possible explanations for the above phenomena based on the bi-layer SEI structure could be that either the surface crack penetrates through the top organic layer and doesn't reach the SEI and silicon interface or the crack travels all the way to the SEI/silicon surface [51]. In both ways surface, re-passivation, and formation of fresh SEI layer lead to the consumption of lithium and irreversible capacity loss. Topographical evolution and mechanical properties of thin-film polycrystalline Si (c-Si) was also tracked by in-situ electrochemical AFM during initial lithiation [129]. A uniform flattening was observed for Si attributed to SEI formation followed by nonuniform expansion of the individual particles upon lithiation. The roughness of the sample surface gets reduced by the formation of the initial SEI layers followed by anisotropic lithiation for different orientations contribute towards increase roughness by ongoing SEI growth.

Based on the above hypotheses, Kumar et al. [51] later showed using in-situ AFM technique that cracks or delamination began to form during electrochemical cycling (for example, surface cracks appeared at 0.05 V) in liquid electrolyte (see Fig. 6.8A and B). The crack opening and closing orthogonal to the patterned Si surface during electrochemical cycling allowed deep insight and understanding of mechanistic failure mechanisms of SEI. Later they microscopically observed that crack delamination occurs along with the SEI interface between the outer and inner silicon (see Fig. 6.8C and D), which corroborated that the inner layer remains intact with Si surface, while the outer SEI layer delamination occurs during cycling confirmed by finite element modeling analysis [23]. With the increase in cycle numbers, delamination of the outer layer increases resulting in spallation. However, no additional growth of the inner layer was observed after spallation indicating passivation. Zhang et al. [128] proposed that in patterned silicon islands, a shear lag zone (SLZ) is developed by interfacial shear situated near the edges that permit a lateral extension parallel to the surface of the substrate. In SLZ only silicon undergoes lateral extension, whereas the center of these formed islands is quite similar to that of continuous films in which expansion occurs in the orthogonal direction to the substrate. The lateral expansion and contraction in the SLZ of silicon lead

SEI layer and impact on Si-anodes for Li-ion batteries **Chapter | 6** 215

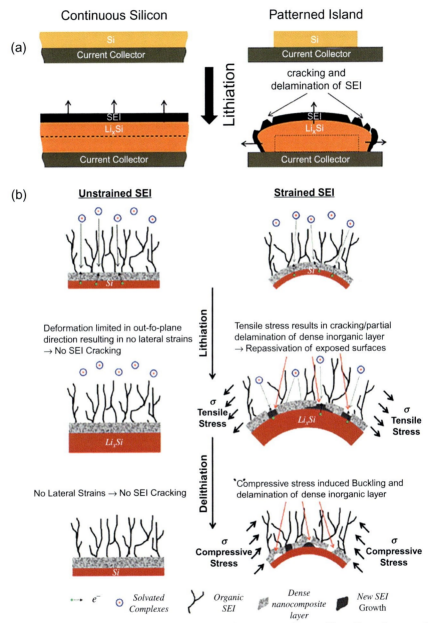

FIG. 6.7 (A) Comparison of deformation behavior of SEI on a continuous silicon film and patterned silicon island during lithiation, (B) schematic representation of the SEI evolution with and without strain. In practical electrode geometries such as silicon particles, expansion (contraction) of the underlying silicon stretches (compresses) the SEI layer and causes in-plane tensile (compressive) stress in the film. *(Reprinted with permission from R. Kumar, P. Lu, X. Xiao, Z. Huang, B.W. Sheldon, Strain-induced lithium losses in the solid electrolyte interphase on silicon electrodes, ACS Appl. Mater. Interfaces 9 (2017) 28406–28417. Copyright 2017 American Chemical Society.)*

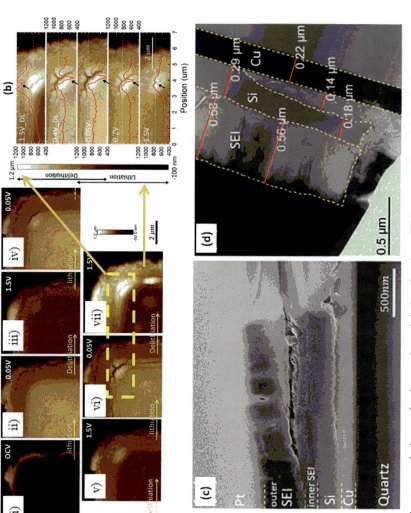

FIG. 6.8 (A) In situ AFM surface topography images showing opening and closing of cracks on SEI layer; cracks formed during (i–iii) the first cycle, (iii–v) the second cycle, (v–vii) the third cycle on Si island electrode. (B) Crack evolution at the corner of a cycled patterned silicon island showing SEI cracking and delamination behavior. (D) TEM images showing the cross-section in the shear lag zone (corner). (Panels (A, B): *Reprinted with permission from R. Kumar, A. Tokranov, B.W. Sheldon, X. Xiao, Z. Huang, C. Li, T. Mueller, In situ and operando investigations of failure mechanisms of the solid electrolyte interphase on silicon electrodes, ACS Energy Lett. 1 (2016) 689–697. Copyright 2016 American Chemical Society. Panels (C, D): Reprinted with permission from K. Guo, R. Kumar, X. Xiao, B.W. Sheldon, H. Gao, Failure progression in the solid electrolyte interphase (SEI) on silicon electrodes, Nano Energy 68 (2020) 104257. Copyright 2020 Elsevier B.V.*)

to tensile and compressive strain inside the SEI layer, respectively. When the expansion of lithiated silicon in SLZ occurs, it applies large in-plane strains to the SEI. It has been shown that temperature variation can significantly affect the stability of the SEI on the Si electrode. SEI layer either dissolves or detaches at higher temperature (~50°C) during the prolonged resting period [130]. Shi et al. [131] performed in-situ AFM on silicon single crystal electrode having different crystallographic orientations [Si (111) and Si (100)] to elucidate the interfacial SEI evolution and anisotropic dynamics. They found that SEI film growth showed different interfacial properties along with growth directions. For Si (111) system, Li—Si alloying reactions result in an expansion of step edges with spherical structures, and the swelling spreads along $\langle 121 \rangle$ direction of the horizontal axis. However, for Si (100) system, sudden swelling concurrence with conical structure growth in $\langle 100 \rangle$ direction observed perpendicular to the surface. Operando color microscopy and ex-situ atomic force microscopy (AFM) techniques were also used to study the lithiation process in a-Si and methylated a-Si thin layers. It was observed that lithiation of a-Si is uniform, whereas methylated a-Si is spatially nonuniform, starting at a confined number and expanding radically, yielding circular lithiation spots. This nonuniformity was attributed to the higher resistivity of methylated a-Si compared to that of a-Si [132].

6.9 Computational work on SEI formation

The following section outlines the computational work performed using ab initio molecular dynamics (AIMD), density functional theory (DFT) to understand the decomposition behavior of EC, FEC, and other electrolyte solvents and their role for SEI stability on silicon anode. Solid-electrolyte interphase (SEI) films forms owing to electrochemical reduction and breakdown of the organic solvent-based electrolyte and electrolyte additive molecules during battery cycling. The SEI blocks electron transfer from the anode to the electrolyte yet permits lithium-ion (Li$^+$) transport enabling the electrochemical cell remains active for long cycles [2]. SEI films on traditional graphite anodes are not static but grow thicker during cycling. However, in the case of silicon, SEI films undergo cracking, delamination and exhibit dynamic evolution due to volumetric expansion associated with fully lithiated phase(s). It has been found that Li$_2$CO$_3$ and lithium ethylene dicarbonate (LEDC) formed as SEI products are electrochemically stable but thermodynamically unstable on Li/Li$_x$Si$_y$ surfaces and finally decomposed to the Li$_2$O thin layer covering inside of the lithium metal surface underneath the SEI layer [13]. The SEI formation on Si/SiO$_x$ is complex in nature, and the electrolyte solvent molecules are the dominant contributor to the SEI final product [13, 60]. Especially, decomposition products of EC and electrolyte additives (VC, FEC) are important to understand the stability of the electrochemical cell for cycling stability [8, 37, 61, 62, 76, 77, 79].

Balbuena et al. [65] performed the ab initio molecular dynamics (AIMD) simulation to observe the reduction mechanism of ethylene carbonate (EC) solvent molecules on lithiated silicon surfaces. They observed that EC reduction occurs by two different mechanisms, which are independent of the crystallographic plane or functional group terminating silicon atoms on the surface. $Li_{13}Si_4$ surfaces are found highly reactive and trigger the reduction of EC molecules through an electron transfer mechanism with the final reduction products CO^{2-} and $O(C_2H_4)O^{2-}$. It is also documented that as the lithiation of Si anode progresses, pristine Si surface is continually gets exposed to the electrolyte, and the SEI layer continues to grow. However, after a certain SEI thickness forms (~1 nm), electron tunneling from the anode is hindered, limiting electrolyte decomposition despite Li-ion diffusion between the Si anode and the electrolyte [72].

Atomistic MD simulation of $Li_2EDC|EC:DMC$ (3,7)-$LiPF_6$ electrolyte interface reveals that EC molecules preferentially adsorbed onto the SEI surface while DMC molecules were depleted from the interfacial layer. The activation energies for the Li^+ solvation − desolvation reactions are 0.42–0.46 eV for the SEI-electrolyte interface [41, 133]. It is also documented that cyclic carbonates like EC/PC are more likely to undergo reduction than the linear counterparts (DEC, DMC, etc.) and forms $(CH_2OCO_2Li)_2$ and Li_2CO_3 in EC/DEC or EC/PC/DEC solvent systems by one-electron transfer mechanisms and the subsequent intermediate products react with other alkyl carbonate molecule present [42, 134]. Recent studies show the formation of variety of organic/inorganic species, including lithium alkoxides (RCOLi), lithium carboxylate (RCOOLi), alkyl lithium carbonate ($ROCO_2Li$), lithium carbonate (Li_2CO_3), LiF, Li_xPF_y, and $Li_xPO_yF_z$ at different proportions as SEI compositions on silicon electrode when tested using 1 M $LiPF_6$ salt in EC: DEC electrolyte with FEC as electrolyte additive [91, 135]. However, the presence of FEC leads to an SEI with increased concentrations of inorganic species, specifically Li_2CO_3, LiF, Li_2O, upon exposure to air due to fluorination and combustion processes [68, 107]. The porous SEI continues to grow as the solvent transport and desolvation of Li-ions occurs on silicon surface resulting in the formation of oxygen molecules and thick Li_2O stratum at electrode/SEI interface [67, 107]. Similar observation made by Balbuena and co-workers, who suggested a kinetically rapid formation of neutral radical carbonate and fluoride-ions via a ring-opening mechanism leading to the formation of LiF [13, 14, 64, 65, 69–72]. The FEC reduction follows a one-electron lithium-assisted reduction of the fluoromethyl group to fluoride and neutral radical carbonate to form LiF as the main reduction product [64, 70]. Scheme 6.5 illustrates the probable reactions that produce LiF and ethylene and carbonate as products (reaction 1) or alkoxy products (reaction 2). However, Chen et al. argued that FEC is reduced on silicon surface through the opening of the five-member ring leading to the formation of lithium poly(vinyl carbonate), LiF, and dimers [120]. Recently, Zhang et al. [91] showed by density functional theory calculation that initial

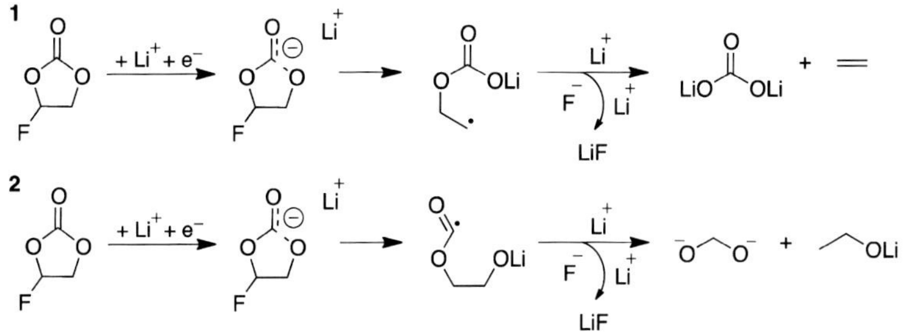

SCHEME 6.5 Reaction 1: electroreduction of fluoroethylene carbonate (FEC) to form lithium fluoride, lithium carbonate, and ethylene; Reaction 2: electroreduction of FEC to form lithium fluoride, methylenedioxyl ion (or alternately carbon dioxide), and lithium ethoxide. *(Reprinted with permission from K. Schroder, J. Alvarado, T.A. Yersak, J. Li, N. Dudney, L.J. Webb, Y.S. Meng, K.J. Stevenson, The effect of fluoroethylene carbonate as an additive on the solid electrolyte interphase on silicon lithium-ion electrodes, Chem. Mater. 27 (2015) 5531–5542. Copyright 2015 American Chemical Society.)*

bond-breaking mechanisms of FEC involve the formation of Li_2CO_3 and vinyl fluoride, and the final decomposition products are LiF, Li_2O, and FEC oligomers.

6.10 Coating strategies and core-shell/yolk-shell morphology for stable SEI formation

In an attempt to overcome the practical limitations of silicon microparticles (SiMPs) cracking, during lithiation, much attention has been devoted to the design and fabrication of silicon nanostructures, such as silicon nanowires [136, 137], nanotubes [138, 139], three-dimensional porous silicon [140–143], and silicon in composites with carbon or oxides [144–149] using effective coating strategies. The results were encouraging, in particular, Si/SiO_x composites [150, 151], and $Si/SiO_2/C$ nanostructured composites [152–154], have demonstrated excellent electrochemical performance. The superior electrochemical performances of the above architectures attributed to the formation of ultrathin carbon or SiO_x shell coating on the outside of the silicon, offering a protective surface during the formation of a stable solid electrolyte interphase (SEI) and preserving the anode from irreversible reactions with the electrolyte [155–157]. Cho et al. [140] manufactured interconnected porous SiNPs, which demonstrated capacity $\sim$2158 mAh g^{-1} at $\sim$3C current rate with $\sim$72% capacity retention at 100 cycles. Silicon nanoparticles (SiNPs) coated with polyaniline (PANi) also demonstrated excellent performance with over 90% capacity retention and a discharge capacity of $\sim$550 mAh g^{-1} at $\sim$1.5C current rate, after 5000 cycles [146].

There are numerous in the scientific literature reporting that conformal coating of carbon on silicon particles can improve electrochemical performance compared to elemental silicon due to the incorporation of carbon acting as a buffer layer to accommodate large volumetric expansion during lithiation/delithiation as well as helps to form a stable SEI layer [158–160]. Cui et al. [161] showed that by rationally designing silicon particles embedded within carbon shell with sufficient void space between the particles and shell ("yolk-shell" structure) can accommodate the large volume expansion of silicon during Li-ions cycling and provides stable capacity. The "yolk-shell" type Si/C architecture exhibited excellent capacity ∼2833 mAh g^{-1} at ∼0.1C current rate, cycle life (1000 cycles with 74% capacity retention), and Coulombic efficiency (99.84%). Guo et al. [162] extended the similar "yolk-shell" concept and designed tunable p-SiNP@HC particles using resorcinol-formaldehyde derived phenolic resin route (see Fig. 6.9A). The optimized p-SiNPs@HC showed specific capacity ∼720 mAh g^{-1} at ∼1C current rate. Recently, dual yolk-shell morphology of Si/void/SiO$_2$/void/C nanostructures has been proposed where carbon layer provided enhanced conductivity, and SiO$_2$ layer acted as robust mechanical support (see Fig. 6.9B) [163, 164]. The dual yolk-shell morphology was able to deliver a capacity of ∼956 mA hg^{-1} at 0.1C current rate after 430 cycles [163]. The core-shell architecture of SiNPs@TiO$_{2-x}$/C was found to exhibit a capacity of ∼939 mAh g^{-1} at ∼3C current rate, due to the presence of initial TiO$_{2-x}$ in the shell, and the formation of in-situ Li$_x$TiO$_2$ provided the electron/Li-ion path in the SiNPs@TiO$_{2-x}$/C architecture and thereby improved electrochemical performance [148]. Cui and coworkers further developed Si/C based hierarchical hybrid architecture inspired from pomegranate structure where individual Si@C particles were embedded by a thick carbon coating leaving enough breathing space for individual Si@C particles to breadth in and out during (de)lithiation (see Fig. 6.9C). The hybrid Si/C structure demonstrated superior volumetric capacity (1270 mAh cm^{-3}) and cyclability (97% capacity retention after 1000 cycles) due to the formation of a stable SEI layer [165].

Cui et al. [166] also demonstrated that SiNTs surrounded by an ion-permeable SiO$_x$ shell can cycle over 6000 times while retaining more than 85% of their initial capacity. The outer surface of the SiNTs is prevented from outward expansion by the SiO$_x$ shell, while the expanding inner surface does not get exposed to the electrolyte, resulting in a stable SEI. Based on this work, Cui et al. [137] also developed core-shell C/SiNWs by chemically depositing amorphous silicon onto carbon nanofibers (CNFs) using SiH$_4$ precursor. It was observed that a thick *a-Si* coating (∼50 nm) on to carbon surface and the core-shell *carbon-a-Si NWs* demonstrated stable capacity ∼800 mAh g^{-1} at ∼1C current rate. Song et al. [138] grown vertically aligned hollow SiNTs arrays using ZnO nano rods as a sacrificial template, which demonstrated capacity ∼1900 mAh g^{-1} at ∼1C current rate. The excellent capacity at high current rate was due to the structural integrity coming from the empty space within the SiNTs that was able to accommodate expansion/contraction during (de)lithaition without collapsing the SiNTs structure. Yoo et al. [139] developed carbon-coated SiNWs, which exhibited capacity ∼1000 mAh g^{-1} at 0.1C

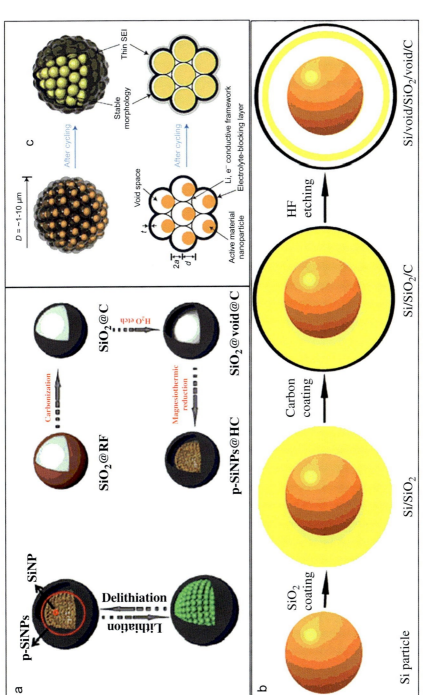

FIG. 6.9

(Continued)

current rate up to 90 cycles. So far, researchers have utilized conductive carbon coatings, generation of spheres and hollow shells of carbon onto silicon and silicon-alloy materials to generate a composite anode for use in LIBs [161, 167–170]. It was also found that porous carbon layer thickness (~10 mm) is critical for fast Li-ion transport kinetics and formation of stable SEI film on SiNPs (80 nm) [171].

6.11 Binders

Si is a promising anode for LIBs due to its superior gravimetric capacity (~4200 mAh g^{-1}). However, the widespread success of Si is only feasible if the major problems associated with it viz. delamination of binder and Si from the current collector, unstable SEI formation, and pulverization of electrode owing to massive volumetric expansion (~300%) can be circumvented during repeated cycling (see Fig. 6.10A).

6.11.1 Problem with conventional binders

The recurring volumetric expansion/contraction the Si particles undergo during cycling not only leads to fracture but also give rise to the formation of an unstable and thick solid electrolyte interphase (SEI) layer owing to extremely low lithiation/delithiation potential (~0.115–0.1 V versus Li$^+$/Li) where carbonate-based aprotic electrolytes are not stable and eventually reduced into organic/inorganic species. The SEI evolution on Si follows a dynamic pattern with fresh SEI layer formation, and destruction occurs with each discharging and charging half-cycle before the surface gets stabilized. However, the continuous evolution of fresh SEI

FIG. 6.9, CONT'D (A) Schematic of the lithiation/delithiation reaction in the p-SiNPs@HC anode for LIBs (left) and illustration of the p-SiNPs@HC design (right). (B) Schematic illustration of the fabrication process for the dual yolk-shell structure. (C) Schematic of the pomegranate-inspired design. Three dimensional view (top) and simplified two-dimensional cross-section view (below) of one pomegranate microparticle before and after electrochemical cycling (in the lithiated state). The nanoscale size of the active-material primary particles prevents fracture on (de) lithiation, whereas the micrometer size of the secondary particles increases the tap density and decreases the surface area in contact with the electrolyte. The self-supporting conductive carbon framework blocks the electrolyte and prevents SEI formation inside the secondary particle, while facilitating lithium transport throughout the whole particle. *(Panel (A): Reprinted with permission from S. Guo, X. Hu, Y. Hou, Z. Wen, Tunable synthesis of yolk–shell porous silicon@carbon for optimizing Si/C-based anode of lithium-ion batteries, ACS Appl. Mater. Interfaces 9 (2017) 42084–42092. Copyright 2017 American Chemical Society. Panel (B): Reprinted with permission from L.Y. Yang, H.Z. Li, J. Liu, Z.Q. Sun, S.S. Tang, M. Lei, Dual yolk-shell structure of carbon and silica-coated silicon for high-performance lithium-ion batteries, Sci. Rep. 5 (2015) 10908. Copyright 2015 Springer Nature Publishing Group. Panel (C): Reprinted with permission from N. Liu, Z. Lu, J. Zhao, M.T. McDowell, H.-W. Lee, W. Zhao, Y. Cui, A pomegranate-inspired nanoscale design for large-volume-change lithium battery anodes, Nat. Nanotechnol. 9 (2014) 187–192. Copyright 2014 Springer Nature Publishing Group.)*

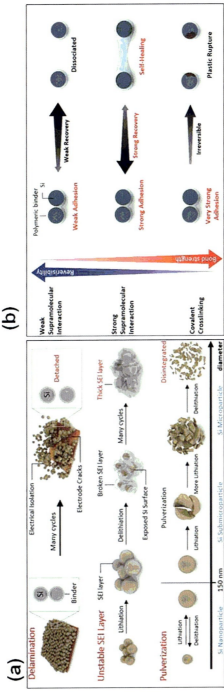

FIG. 6.10 (A) Three representative failure mechanisms of Si anodes: delamination, unstable SEI layer, and pulverization. (B) classification of chemical bonds and supramolecular interactions and their adhesion behaviors in Si anodes. *(Reprinted with permission from T.-W. Kwon, J.W. Choi, A. Coskun, The emerging era of supramolecular polymeric binders in silicon anodes, Chem. Soc. Rev. 47 (2018) 2145–2164. Copyright 2018 Royal Society of Chemistry.)*

layers may trap Li-ions and consumes electrolyte salt/solvents impairing the Coulombic efficiency and ultimately exhausting the electrolyte. The above factors posed a severe impediment to the commercialization of LIBs and call for the development of either suitable binders which will prevent detachment of the active phase from the current collector or a viable coating strategy to avoid the volumetric strain.

There are several reports which addressed that nanostructuring of Si and engineering various Si/C architectures (core-shell, yolk-shell, pomegranate structure, dual carbon coating, etc.) can prevent the severe volumetric expansion of silicon anodes [155, 165, 166]. The nanostructuring and coating strategy may salvage the rapid capacity fade of Si to some extent in a half-cell format, but the usage of conventional binder such as polyvinylidene difluoride (PVDF) cannot completely shield Si to overcome complications caused by the volumetric strain.

The mechanical strain ($\sim 10\%$) generated during cycling of graphite with Li-ion remains within the elastic limit of graphite, preventing any mechanical deformation of the composite electrode, which generally uses PVDF and super P carbon black as the binder and conductive agent, respectively [172]. However, PVDF, with its low elastic modulus (~ 1000 MPa), is not suitable for Si anode. The tremendous mechanical stress/strain generated within the composite electrode (Si+binder+conductive agent) surpasses the elastic limit of PVDF, leading to failure and irreversible capacity loss within a few cycles [172]. A variety of polymeric binders including natural polysaccharides and their derivatives, carboxymethyl cellulose (CMC) [173–178], poly(acrylic acid) (PAA) [96, 179], sodium alginate [180], and conductive polymers [181–184] have been explored and found to show good electrochemical performance in Si anodes. It has been established that features like crosslinking [185–191], self-healing [192–194], stiff backbone [179, 180, 195], and adhesive interaction [179, 180, 196] of polymeric binders with Si are few important characteristics for stable Li-ion cycling preventing Si particles fragmentation and stable SEI formation. However, all the above characteristics cannot be achieved using a single binder [189, 197]. Therefore, binder engineering is a way forward and may allow significant improvements in the stability of Si-based anodes at electrode-level and improve the cell performance significantly [189].

6.11.2 Polymeric binders with supramolecular chemistry

The basic role of a binder is to provide strong adhesion between the electrode components and the current collector during cell operation such that electrode particles remain anchored and attached with each other. Enhancement of the adhesion strength of polymeric binders with Si particles can also protect the SEI formation on the Si surface. The adhesion strength of polymers is closely associated with their chemical structure and degree of polymerization. The chemical bonds can be divided into three categories in terms of their reversibility and strength: weak supramolecular interactions, strong supramolecular interactions, including coordination bonds, and covalent bonds (see Fig. 6.10B). Weak supramolecular interactions can readily dissociate under small mechanical stress, whereas strong supramolecular interactions will recover dissociated bonds. Binders exhibiting strong supramolecular interactions would be an ideal candidate for Si anodes owing to their self-healing ability. The balance between the self-healing effect and bond strength is essential for superior cycling performance. Table 6.4 summarizes various classes of binder, their performance, and the type of bonds develops with Si.

Polymer binders that contain hydroxyl, carboxyl groups and their derivatives, viz. polyacrylic acid (PAA) [96, 179], sodium carboxymethyl cellulose (NaCMC) [173–178], guar gum [172, 200], etc. have been explored for silicon anodes and their performance compared with traditionally used PVDF binder. It has been experimentally demonstrated that the polar functional groups present in the above polymeric binders help to tether silicon particles via hydrogen and/or covalent bonding between the silanol (—SiOH) groups present on the silicon surface [96]. High elastic modulus based natural polysaccharide (alginate) and dopamine grafted PAA binder with linear polymeric chains possessing hydroxyl and carboxyl groups also shown promising results with silicon anodes [195]. However, Song et al. [206] argued the linear chain polymeric binders are prone to failure during prolonged galvanostatic cycling with silicon, and irreversible capacity loss may occur. They proposed three-dimensional polymeric binders between PVA and PAA thermally cross-linked at elevated temperature offer much-improved performance in silicon anodes. In this regard, they prepared a polymeric binder by in-situ cross-linking of water-soluble poly(acrylic acid) (PAA) and polyvinyl alcohol (PVA) precursors [206]. The deformable polymer binder effectively develops a strong adhesion onto silicon particles by hydroxy and covalent bonding leading to excellent cycling stability and Coulombic efficiency at high current rates accommodating the considerable volumetric strain upon lithiation/delithiation (see Fig. 6.11A). The PAA/PVA-Si anode exhibited a gravimetric capacity of 2283 mAh g^{-1} with 63% capacity retention after 100 cycles at a current density of 400 mA g^{-1}, higher than that of CMC-Si and PVDF-Si based anodes. Silicon microparticles (SiMPs) are always advantageous over Si nanoparticles (SiNPs) from the perspective of low cost and ease of availability [192, 194]. However, pulverization and colossal mechanical stress accompanied by SiMPs pose a tremendous challenge to work with in LiBs. Choi et al. [186] presented a new strategy that highly elastic binder surrounding SiMPs can keep the pulverized Si particles coalesced during charge/discharge cycling and allow to retain the conductive pathways to the Si particles. The idea was inspired by the working principle of moving pulleys, where the tension of the rope is greatly reduced by distributing and equalizing the localized force to sustain the mechanical stress caused by Si expansion exhibiting nonlinear stiffening behavior by the crosslinked polymers (see Fig. 6.11B). Results showed that the incorporation of 5 wt% polyrotaxane

TABLE 6.4 Summary of various binders and their electrochemical performance in Si anodes.

Polymeric binders	Formula	Bond formation/nature of the interaction	Results	References
Polyvinylidene difluoride (PVDF)	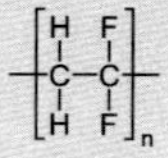 [−C(H)(F)−C(H)(F)−]$_n$	Weak van der Waals interactions, not strong enough to preserve the particle-particle cohesion	Capacity fades during initial cycling, not strong enough to preserve the particle-particle cohesion under the massive volume change of Si	[191]
K$_{100}$	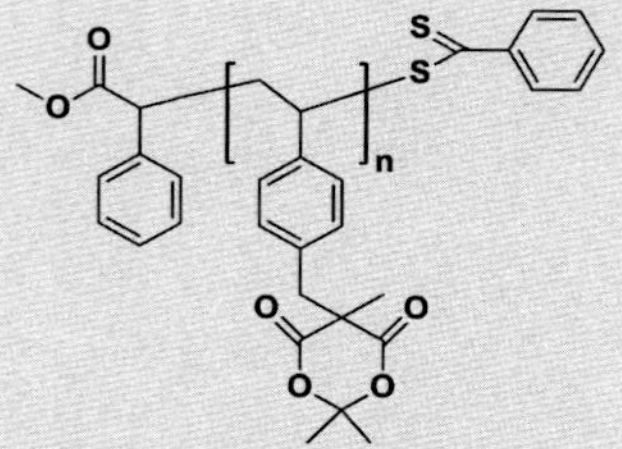 (structure shown)	5-Methyl-5-(4-vinylbenzyl) Meldrum's acid monomer based. Strong covalent crosslinking that can form a stiff three-dimensional polymer network	Better cycling performance compared to that of PVDF–Si. Performance gradually deteriorated over cycling	[191]
C$_{100}$	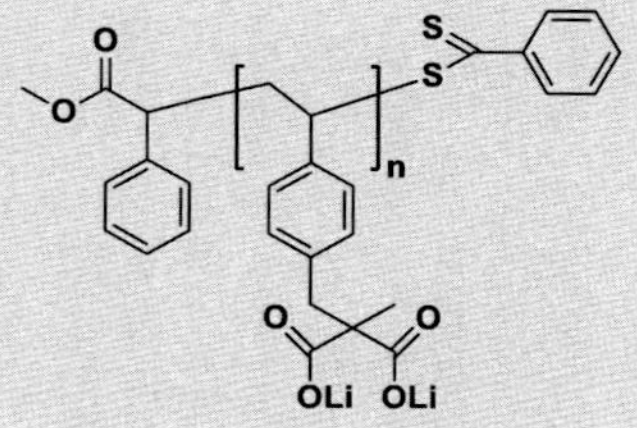 (structure shown)	Weaker bond strength than K$_{100}$ reversible ion-dipole interactions, strong supramolecular interaction	Lithium 2-methyl-2-(4-vinylbenzyl) malonate backbone. Good cycling performance due to the self-healing effect	[191]

Binder	Structure	Description	Performance	Ref.
Sodium carboxymethyl cellulose (CMC)		Linear polysaccharide, a derivative of cellulose, consisting of β-linked glucopyranoses with carboxymethyl substitution. Hydroxyl and carboxylate groups of CMC can form ion-dipole and hydrogen bonding with the silanol groups of the SiO2 layer found on Si surface. Strong supramolecular interactions can be dissociated and reformed again, achieving the desired self-healing effect	The high performance of CMC is due to the extended conformation of the stiff cellulose backbone and the electrostatic repulsion of carboxylate groups. The extended conformation is expected to increase the number of interaction points between polymers and Si, leading to stable electrode integrity	[173–178]
Poly(acrylic acid) (PAA)		Carboxylic acids in PAA can form strong hydrogen bonding interactions between binder-binder and binder-Si. Higher density of carboxylic acid moieties, linear polymeric binder	Better cycling performance compared to CMC–Si. Exhibited prolonged cycle life due to strong adhesion with SiO via hydrogen bonding, limited swelling in the electrolyte solution	[96, 179]
Sodium alginate (Alg)		Natural polysaccharide, consisting of β-D-mannuronic acid and α-L-guluronic acid. Carboxylate and hydroxyl functional groups form hydrogen bonding and ion-dipole interactions between polymer chains and with Si particles, strong adhesion	Alg-Si retained 85% (1700 mAh g^{-1}) of the original capacity after 100 cycles at current density of 4200 mA g^{-1} while CMC-Si and PVDF-Si suffered from severe capacity fading (lower than 1000 mAh g^{-1} before 30 cycles)	[180]

Continued

TABLE 6.4 Summary of various binders and their electrochemical performance in Si anodes—cont'd

Polymeric binders	Formula	Bond formation/nature of the interaction	Results	References
Alginate-carboxymethyl chitosan (Alg-C-chitosan)		Cross-linked by a condensation reaction, self-healing effect, electrostatic interaction between carboxylate (—COO$^-$) of Alg and protonated amines (—NH$_3^+$) of C-chitosan forms a self-healing porous scaffold structure	Maintain capacity of ~750 mAhg^{-1} after 100 charge-discharge cycles with Si/Gr@C composite anode, cross-linking 3D network restricts active materials and conductive agent from sliding	[198, 199]
Catechol-functionalized binders	Alg-C, PAA-C	Forms strong reversible coordination bonds with metal oxide, which is reversible. Strong conjugated polymer binder with carboxylic acid functional groups	Alg-C and PAA-C showed dual adhesion mechanism of hydrogen bonding and catecholic interaction with the Si NP surfaces and Si/graphite blended anodes, increasing their capacities	[196]

Guar gum (GG)		Presence of large number of hydroxyl moieties on the pendant groups and the backbone, interact efficiently with Si via hydrogen bonding and facilitate Li-ions transfer by engaging the lone-pair electrons of the oxygen atoms	GG–Si anode retained 2222 mAhg^{-1} (capacity retention = 66.05%) after 100 cycles by taking advantage of a high density of polar functional groups. Stable capacity with minimum fade	[172, 200]
Urea-based binder		Strong supramolecular interactions of urea group, viscoelastic flow with relaxation time ~0.1 s, maintains desired mechanical integrity, stretchable self-healing polymer	Lower T_g, self-healing ability, stable cycling performance, crack close up via branched hydrogen bonding between urea groups, high areal capacity (3–4 mAh cm^{-2}) and stable cycling for more than 140 cycles using Si particles	[185, 192–194, 197]
β-Cyclodextrin (β-CDp)		Hyperbranched network structure and multidimensional hydrogen-bonding interactions with Si particles, dynamic crosslinking and multiple noncovalent interactions at local spots of Si	Hyperbranched polymer provides a good platform to form intimate adhesion with Si surface and preserve the mechanical integrity of the Si electrode, resulting in a stable cycle life	[187, 190]

Continued

TABLE 6.4 Summary of various binders and their electrochemical performance in Si anodes—cont'd

Polymeric binders	Formula	Bond formation/nature of the interaction	Results	References
Amylopectin		Major components of starch in agricultural products, viz. corn, potato, and rice. Moderately hyperbranched polysaccharides have branches with an α(1→6) glycosidic bond	Stable capacity and good capacity retention owing to optimum branching density deliver uniform coating on Si and minimum SEI formation	[201]
Poly(acrylic acid sodium) grafted carboxylmethyl cellulose (PAA-g-CMC)		Highly crosslinked graft polymer, multidimensional contacts with Si surface via branched structure, strong hydrogen bonding and ion–dipole interactions	Ion–dipole interactions offer better cycling stability compared to the hydrogen bonding interactions primarily due to their higher bonding strength	[95, 202, 203]
Polyvinylidene difluoride-graft-poly(tert-butylacrylate) (PVDF-g-PtBA)		Graft polymer, upon heating the electrode under vacuum, PtBA of PVDF-g-PtBA pyrolyzed to PAA that form hydrogen and covalent bonds with Si	Suppress the electrode swelling ability due to enhanced ability of adhesion through multidimensional contacts of the graft polymer network	[204]

PAA/pullulun network cross-linked heteropolymer	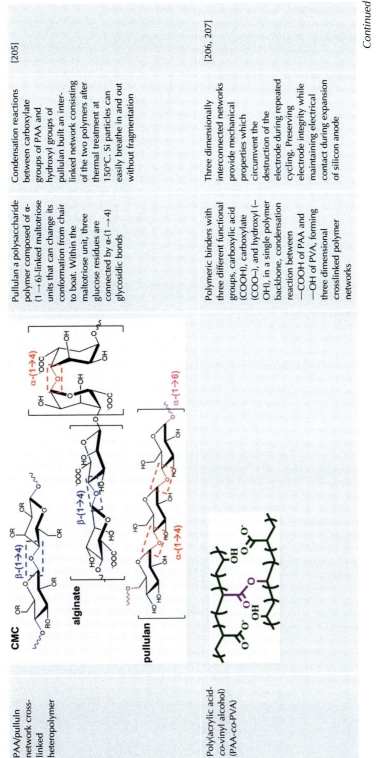	Pullulan a polysaccharide polymer composed of α-(1→6)-linked maltotriose units that can change its conformation from chair to boat. Within the maltotriose unit, three glucose residues are connected by α-(1→4) glycosidic bonds	Condensation reactions between carboxylate groups of PAA and hydroxyl groups of pullulan built an inter-linked network consisting of the two polymers after thermal treatment at 150°C. Si particles can easily breathe in and out without fragmentation	[205]
Poly(acrylic acid-co-vinyl alcohol) (PAA-co-PVA)		Polymeric binders with three different functional groups, carboxylic acid (COOH), carboxylate (COO—), and hydroxyl (—OH), in a single polymer backbone, condensation reaction between —COOH of PAA and —OH of PVA, forming three dimensional crosslinked polymer networks	Three dimensionally interconnected networks provide mechanical properties which circumvent the destruction of the electrode during repeated cycling. Preserving electrode integrity while maintaining electrical contact during expansion of silicon anode	[206, 207]

Continued

TABLE 6.4 Summary of various binders and their electrochemical performance in Si anodes—cont'd

Polymeric binders	Formula	Bond formation/nature of the interaction	Results	References
Poly(vinyl alcohol)-poly (ethylene imine) (PVA-PEI)		Water-soluble binder, in-situ thermal cross-linking by condensation, reversibly-deformable polymer network. Stable SEI film on the Si anode	Form an interconnected network that strongly bonds with Si articles, specific capacity of ~1590 mAhg^{-1} at current rate ~10 Ag^{-1}	[198, 208]
Carboxymethyl chitosan (C-chitosan)		Modified from natural polymer, self-healing effect, water soluble, prepared by carboxymethylation of chitosan	polar groups of —COOH, —NH$_2$ and —OH groups in C-chitosan adsorb on hydroxylated Si surface	[198, 209]
Chitosan-glutaraldehyde (CS-GA)		3D network, self-healing effect, cross-linked between amino group of CS and aldehyde groups of GA, strong chemical bonding with Si particles	Maintained a capacity of 1969 mAh g^{-1} at current rate of 500 mAg^{-1} over 100 cycles	[198, 210]

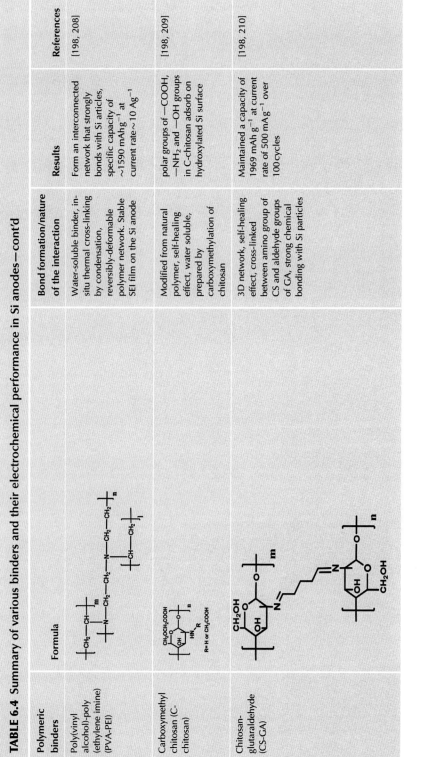

Acrylic acid-based copolymer (xCLPAH)	*(structure: Cross-linked PAA)*	Water-soluble crosslinked binder, synthesized by copolymerization of acrylic acid with the coexistence of diallyl ether as a crosslinker, strong supramolecular interactions originating from the carboxylic acid moieties	Si/graphite anode exhibits improved SEI layer stability and cyclability due to ion-dipole interactions and stretched polymer morphology from electrostatic repulsion of negatively-charged carboxylates	[211]
Xanthan gum (XG)	*(structure)*	The linear backbone of polysaccharide composed of (1,4)-b-D-glucose unit with trisaccharide side groups containing two negative carboxylates	Helical structure, the formation of the interconnected polymer network, enhances adhesion ability and electrochemical performance for Si anodes	[188, 198]
Karaya Gum (KG)	*(structure)*	Natural polymer branched structure, composed of polysaccharides and Glycoproteins, Good mechanical properties provided by the multi-branched structure	Maintain a reversible capacity of $\sim$1000 mAh g^{-1} for more than 1200 cycles at a current rate $\sim$4 A g^{-1}, superior binding strength resulting from polar function groups	[198, 212]

Continued

TABLE 6.4 Summary of various binders and their electrochemical performance in Si anodes—cont'd

Polymeric binders	Formula	Bond formation/nature of the interaction	Results	References
Gellan gum		Natural polymer with a higher degree of acetyl functional groups, reversible bonding enhance stress accommodation and self-stabilizing effect of Si/graphite electrode	Stable SEI formation by bridging between the surfaces of Si/graphite electrode via hydrogen bonding	[198, 213]
Poly(3,4-ethylene-dioxythiophene):Polystyrene-4-sulfonate) (PEDOT:PSS)		Conductive polymer, easily processable using conventional techniques due to its solubility in water	Eliminates capacity losses due to physical separation of the Si and traditional inorganic additives, excellent electrochemical stability (even when repeatedly n-doped), and mechanically robustness	[198, 214]
Gum Arabic (GA)		Natural gum composed with polysaccharides and glycoproteins, Dual-function	Enhance the tolerance to cracking from volume expansion in Si anode materials	[198, 215]

Gum arabic-polyacrylic acid (GA-PAA)		Crosslinked by a condensation reaction, Micron-sized pores structure with crack-blocking effect	Crack-blocking surface is stronger and more resilient to stress than the GA, high volumetric capacity, and long-term cycle performance	[198, 216]
Poly(acrylic acid) poly (benzimidazole) (PAA-PBI)		Reversible acid-base interaction and ionic crosslinking of polymeric binders	Ion-ion interactions between imidazolium and carboxylate moieties, resulting in robust mechanical properties and consequently enhanced battery performance	[217]
Polyfluorene (PF) based n-type conductive polymers		Functional groups, carbonyl (C=O) and methylbenzoic ester—PhCOOCH$_3$ (MB), introduced for tailoring the LUMO electronic states of PF and improve the adhesion ability	Intimate electric contact for electron conduction and mechanical integrity, resulting in high specific capacity and stable cycling performance	[181–184, 218]

Continued

TABLE 6.4 Summary of various binders and their electrochemical performance in Si anodes—cont'd

Polymeric binders	Formula	Bond formation/nature of the interaction	Results	References
Poly(acrylic acid)-poly(2-hydroxyethyl acrylate-co-dopamine methacrylate) PAA-P(HEA-co-DMA)		Water soluble rubber elastomer formed by in situ thermal condensation reaction of PAA and P(HEA-co-DMA). High content abundant carboxyl, hydrogen bonding sites and catechol groups on the side chains endow self-healing capability	Full cell areal capacity ~1.75 mAh cm^{-2} at 0.2C current rate (1C = 150 mAg$_{NCM}^{-1}$), with a Si micro particles anode and NCM cathode with 80.8% capacity retention after 120 cycles	[195]
3,6-Poly(phenanthrenequinone) (PPQ)		Conductive binder, highly conjugated and has the electron-withdrawing carbonyl group, n-doped by accepting electrons and Li$^+$ become a mixed conductor in 1st charge cycle	Lithium trapping is undesirable, leads to irreversible capacity loss, better to use a minimum quantity binder or prelithiated PPQ binder	[219]
Poly(1-pyrenemethyl methacrylate-co-triethylene oxide methyl ether methacrylate) (PPyE)		A class of methacrylate polymers based on a polycyclic aromatic hydrocarbon side moiety-pyrene, side-chain (π-π stacking) electron-conducting polymer binder	high areal capacity at ~2.5 mAh/cm^2 with Si anode, high Coulombic efficiency (~99.5%) of Si/PPyE suggests stable SEI formation with minimum electrolyte decomposition	[220]

Name	Structure	Property	Characteristics	References
Sodium poly(3,3'-(9H-fluorene-9,9-diyl) dipropionic acid (PF-COONa)	PF-COONa (with ONa, O groups on fluorene backbone)	Water soluble, conductive, consists of an n-type polyfluorene backbones and abundant carboxyl functional groups on the side chains	Strong interaction between the carboxyl groups of the binder and the hydroxyl groups on the surface of Si anodes. Capacity ~999 mAh g^{-1} over 1000 cycles of charge and discharge at 1C current rat	[218]
Polyaniline (PANi)	(PANi structure)	3D interconnected network structure	Li$^+$ ions can rapidly pass through the thin PANi layer, resulting in good electrical connection to the particles, enabling a deformable and stable SEI	[146, 221]
Polyaniline-phytic acid (PANi-PA)	(PANi-PA structure)	Hydrogen bonding between Si—O layer on the surface of Si anode materials, 3D interconnected conductive network and elastomer	Enhanced electrochemical reaction kinetics and Li storage properties, resists volume variation, suppresses fracture, and increases the electronic conductivity	[222]
Polyacrylamide (PAM)	—CH$_2$—HC(C=O)NH$_2$— $_n$	3D polymeric network, polar amide groups (—CONH$_2$), abundant in polyacrylamide chains, can interact with the hydroxyl group (—OH) on the Si surfaces (SiO$_2$) via hydrogen bonding	Maintaining its structure even under deformation, covalently cross-linked polymer network forms in the entire electrode without loss of their functionality for superior capacity retention	[223]

Continued

TABLE 6.4 Summary of various binders and their electrochemical performance in Si anodes—cont'd

Polymeric binders	Formula	Bond formation/nature of the interaction	Results	References
Poly(9,9-dioctylfluorene-co-suorenone-co-methylbenzoic acid) (PFFOMB)	PFFOMB	Tailor the lowest unoccupied molecular orbital (LUMO) electronic states	Forms continuous framework to capture Si, but also provides conduction path for electron transport	[224, 225]
Renatured DNA and alginate (reDNA/ALG)	RNA, NaAlg, RNA	Amphiphilic, interconnected fractal network structures, reinforce interparticle interactions as well as enable the homogeneous dispersion of electrode	Bind with hydrophobic carbon conductive agent in blended electrodes and binds with hydrophilic Si particles, retain 84.6% of its initial capacity after 300 cycles	[221, 226]

Alginate-carboxymethyl chitosan (Alg-C-chitosan)	[structure]	Self-healing and porous scaffold structure	Tolerate the volume change of Si and maintain an integrated electrode structure	[199, 221]
Poly(acrylic acid) ureido-pyrimidinone (PAA-UPy)	[structure]	Quadruple-hydrogen-bonded self-healing supramolecular	Repair the destruction caused by the volume variation of Si; exhibit good peeling force	[221, 227]
Poly(acrylic acid) and polyethylene glycol-co-benzimidazole (PAA-PEGPBI)	[structure]	Polymeric binder, added the crosslinked ion-conducting PEG group	Improve the processability of PBI, ionic conductive functional group yielded greater specific capacities at high rates	[221, 228]

Continued

TABLE 6.4 Summary of various binders and their electrochemical performance in Si anodes—cont'd

Polymeric binders	Formula	Bond formation/nature of the interaction	Results	References
Self healing polymer-carbon black nanoparticle (SHP-CB)		Stretchable, dynamic bonds, and embedded with microencapsulated healing agents	Similar high capacity as well as a negligible capacity loss after 20 cycles, high Coulombic efficiency (99.2% at C/2)	[194, 221]
Na alginate with Ca^{2+} (NaAlg-Ca)		Alginate hydrogel, ionic bonding between Ca^{2+} ions and —COO$^-$ stretching bands	Enhances the mechanical property of NaAlg, maintain an integrated structure without pulverization and fracture, resulting in stable cycle performance	[221, 229]
Polyamide imide (PAI)		The reductive peak of the Si-PAI electrode is evolved by the electrochemical reaction of the C=O group of an imide ring in a PAI binder with Li^+ and e^-	Disintegration of the electronic-conduction network because of the huge volume expansion of active Si materials is effectively restrained by the PAI binder	[230, 231]

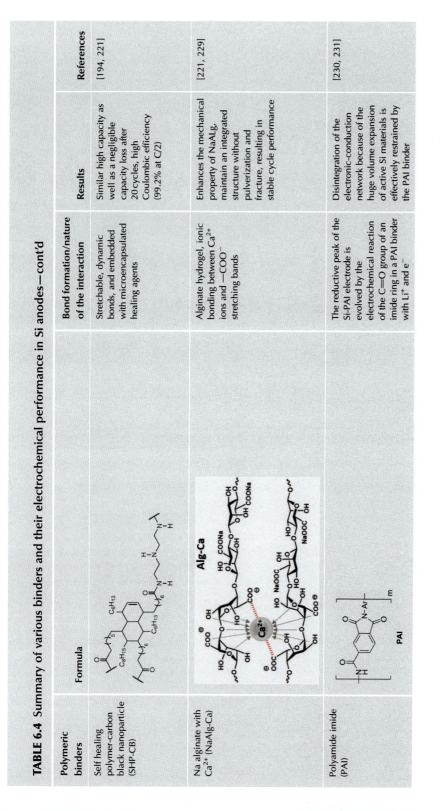

Polymer	Structure	Description	Properties	Ref.
poly(acrylic acid) and polyamide imide) (PAA-PAI)	(see figure)	Partially distributed carboxylic functionalities on the main polymeric chains in accordance with the changes in the main PAA-PAI polymeric frame provides Si with interaction sites	The individual weakness of the polymer PAA (phase separation in the slurry) and PAI (lower mechanical properties) were compensated by the hybrid PAA-PAI binder and exhibited remarkable specific capacity (1120.9 mAh g^{-1}) with excellent retention even after 300 cycles	[231, 232]
Poly(2,7-9,9-dioctylfluorene-co-2,7-9,9-(di(oxy2,5,8-trioxadecane)) fluorene-co-2,7-fluorenone-co-2,5-1-methylbenzoate ester) (PEFM)	(see figure)	Presence of the ether moieties increases the ductility of the polymers. The binder adheres with the polar silicon dioxide (SiO$_2$) surfaces of the Si particle and the copper oxide (CuO) surface of the Cu current collector due to its increased polarity	Electronic, mechanical, and electrolyte-uptake properties are optimized. PEFM polymer exhibited similar or higher levels of mechanical and swelling properties compared to nonconductive binder like PVDF	[182, 231, 233]
Poly(9,9-dioctylfluorene-co-fluorenone-comethylbenzoic ester) (PFM)	(see figure)	Methylbenzoic ester groups on PFM form chemical bonding with the hydroxide-terminated SiO2 surface via a transesterification reaction	Improved adhesion between PFM and SiO, combined with the conductive nature of PFM, considerably increases the loading of active material and improves the energy density of the lithium-ion cell	[183, 184, 231, 234]

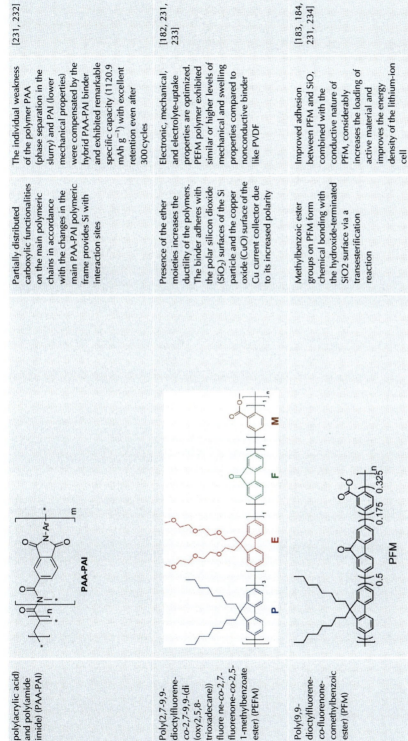

Continued

TABLE 6.4 Summary of various binders and their electrochemical performance in Si anodes—cont'd

Polymeric binders	Formula	Bond formation/nature of the interaction	Results	References
poly(1-pyrenemethyl methacrylate-co-methacrylic acid) (PPyMAA)	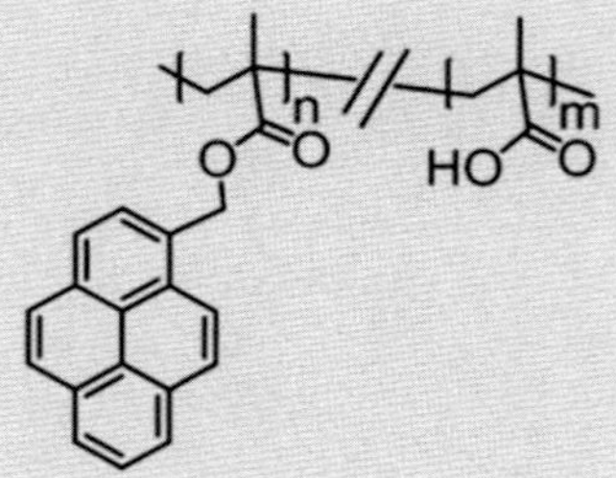 (structure shown)	Pyrene moiety and the carboxylic acid units present in the binder forms hydrogen bonding with the silanol surface groups on the Si particles	Presence of pyrene in the copolymer increases the conductivity; the binder is able to minimize surface area in the high-tap-density nano-Si	[231, 235]
Nafion	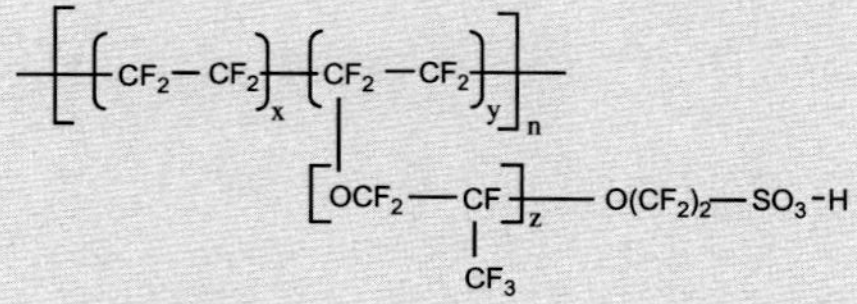 $\left[\left[CF_2-CF_2\right]_x\left[CF_2-CF_2\right]_y\right]_n\left[OCF_2-CF_z\right]_z$ $O(CF_2)_2-SO_3-H$ / CF_3	Lithium exchanged Nafion coats the electrode particles in a layer, which acts as an effective ion-conductor and provides a diffusion pathway for Li-ion	Forms an ionic conductive film between the liquid electrolyte and silicon particles and resulted in long cycling durability with a high capacity	[231, 236, 237]
Polyparaphenylene (PPP)	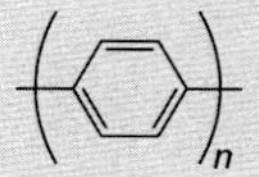 (structure shown)	Reversible n-doping/de-doping behaviors provide Li$^+$ conductivity for PPP polymer and enable Li$^+$ ions transport through PPP chains in/from the Si cores for the lithiation/de-lithiation reactions	PPP shell can effectively accommodate the volume change and prevent the aggregation of the active Si particles, and therefore increase the mechanical integrity of the electrode	[231, 238]
Poly (acrylicacid-acrylonitrile-butylacrylate)-polystyrene (PAABS)	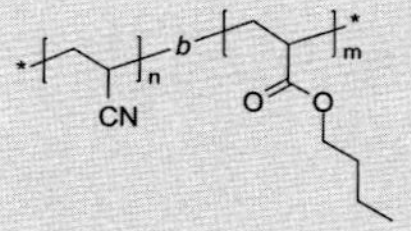 (structure shown)	Presence of carboxyl groups formed hydrogen bonds with functional groups on the Si surface	Improved adhesion strength and cyclic performance, better cyclic stability and capacity than that of the composite electrode containing the PAB or the PS-seeded PABS binder	[231, 239]

Polymer	Structure	Description	Refs
Polyacrylonitriles (PAN)		The nitrile groups have strong polarity and cause hydrogen bonds, permanent dipole-dipole, and van der Waals interactions with the Si particles and current collector (Cu foil)	PAN binders show much higher adhesion strength than the PVdF and CMC binders	[231, 240]
Galactomannan	Locust Bean Gum (LBG); α-(1,6)-galactose; β-(1,4)-mannose	Polysaccharides found in leguminous seed endosperm. H-bonding interactions occur between the less polar –OH groups (mainly from galactose) and the oxide layer of SiNPs to promote electrode stability during charge/discharge cycling	Excellent cyclability and capacity retention, superior electrolyte uptake capabilities when compared to Na-CMC	[231, 241]
Polyimide (PI)		C=O group of the imide ring in PI react with Li$^+$ and e$^-$ during Li$^+$ insertion	PI-Si electrode maintained a discharge capacity of 800 mAh g^{-1} during 196 cycles at a current density of 800 mAg^{-1}. PI binder does not essentially hinder Li-ion access to the Si particles	[231, 242–245]

Information adapted from T.-W. Kwon, J.W. Choi, A. Coskun, The emerging era of supramolecular polymeric binders in silicon anodes, Chem. Soc. Rev. 47 (2018) 2145–2164; J.-T. Li, Z.-Y. Wu, Y.-Q. Lu, Y. Zhou, Q.-S. Huang, L. Huang, S.-G. Sun, Water soluble binder, an electrochemical performance booster for electrode materials with high energy density, Adv. Energy Mater. 7 (2017) 1701185.

244 PART | III Electrolytes and SEI issues

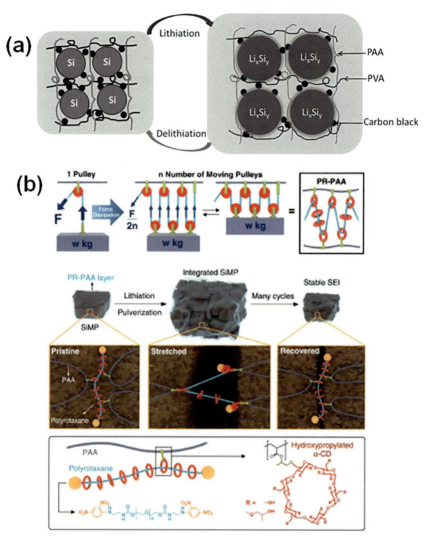

FIG. 6.11 (A) Proposed working mechanism of PAA-PVA based interpenetrated gel binder for silicon anodes. (B) Proposed stress dissipation mechanism of PR-PAA binder for SiMP anodes based on pulley principle to lower the force in lifting an object and graphical representation of the operation of PR-PAA binder to dissipate the stress during repeated volume changes of SiMPs, together with chemical structures of polyrotaxane and PAA. *(Panel (A): Reprinted with permission from J. Song, M. Zhou, R. Yi, T. Xu, M.L. Gordin, D. Tang, Z. Yu, M. Regula, D. Wang, Interpenetrated gel polymer binder for high-performance silicon anodes in lithium-ion batteries, Adv. Funct. Mater. 24 (2014) 5904–5910. Copyright 2014 WILEY-VCH Verlag GmbH & Co. KGaA, Weinheim. Panel (B): Reprinted with permission from S. Choi, T.-W. Kwon, A. Coskun, J.W. Choi, Highly elastic binders integrating polyrotaxanes for silicon microparticle anodes in lithium ion batteries, Science 357 (2017) 279. Copyright 2017 The American Association for the Advancement of Science.)*

(PR) comprising polyethylene glycol (PEG) threads and a-cyclodextrin (a-CD) rings functionalized with 2-hydroxypropyl moieties with a conventional linear binder PAA imparts extraordinary elasticity originating from the ring sliding motion of PR.

This binder combination keeps even pulverized Si particles coalesced without disintegration enabling stable cycle life in a full cell [NCA cathode: LiNi$_{0.8}$Co$_{0.15}$Al$_{0.05}$O$_2$, specific capacity = 188.7 mAh g^{-1}, n/p ratio = 1.15, current rate = 0.03C (1C = 190 mAg^{-1})] maintaining high areal capacities ($\sim$2.88 mAh cm^{-2}) [186]. A recent study suggested that Si particles fully covered by PAA form a dynamic thin SEI layer (<10 nm), which changes its chemical composition and thickness during (de)lithiation of Si [246].

6.11.3 Self-healing polymeric binders

Self-healing polymers (SHPs) is a class of polymer that exhibits excellent mechanical and electrical healing capabilities allowing delamination, cracks to recover and repeatedly heal during battery cycling [191–195, 197, 247]. A self-healing polymer (SHP) with fast flow and bonding can straightaway protect the Si surface to avoid SEI growth. Bao and coworkers [194] developed a urea-based SHP/carbon black (CB) composite using a radical polymerization technique. The three-dimensional spatial distribution of carbon black within SHP allowed a robust and stretchable polymeric binder with a high degree of stability even after a long cycle. Silicon microparticles (SiMPs) electrode with a mass loading of $\sim$0.5–0.7 mg cm^2 combined with urea-based SHP/CB exhibited excellent areal capacity ($\sim$1.5–2.1 mAh cm^{-2}). However, the thick Si/SHP/CB composite electrodes showed moderate rate capability owing to sluggish reaction kinetics. The above work was a paradigm shift that showed that low-cost metallurgical grade SiMPs, which are prone to cracking, is an excellent option for LIBs anode [185, 192, 194], dispersed in an SHP matrix with tailored porosity and mechanical strength showcasing improved interface stability and electrode kinetics distinguishable from silicon nanoparticles (SiNPs) based laboratory research [191]. An artificial SEI (aSEI) with a specific set of mechanical characteristics is henceforth designed by enclosing amorphous silicon majority anode (SIMA) within a ceramic TiO$_2$ shell with a thickness of less than 15 nm (see Fig. 6.12) [247]. In-situ TEM experiments showed that the TiO$_2$ shell exhibits greater strength than an amorphous carbon shell. Void-padded compartmentalization of Si survived the huge volume changes and electrolyte ingression, with a self-healing artificial SEI + naturally forming SEI. The half-cell capacity exceeds 990 mAh g^1 after 1500 cycles. Compressed SiMA with a three-fold increase in tap density (from 0.4 g cm^3 to 1.4 g cm^3) was further explored in a full cell against a 3 mAh cm^2 LiCoO$_2$ cathode to improve the volumetric capacity. Despite few TiO$_2$ enclosures being broken, two times volumetric ($\sim$1100 mAh cm^3) and gravimetric capacity ($\sim$762 mAh g^1) of

246 PART | III Electrolytes and SEI issues

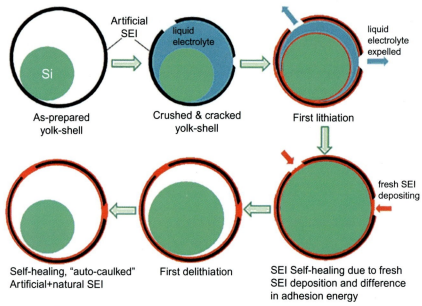

FIG. 6.12 Schematic illustrations of self-healing in a yolk-shell SiMA. *(Reprinted with permission from Y. Jin, S. Li, A. Kushima, X. Zheng, Y. Sun, J. Xie, J. Sun, W. Xue, G. Zhou, J. Wu, F. Shi, R. Zhang, Z. Zhu, K. So, Y. Cui, J. Li, Self-healing SEI enables full-cell cycling of a silicon-majority anode with a coulombic efficiency exceeding 99.9%, Energ. Environ. Sci. 10 (2017) 580–592. Copyright 2017 Royal Society of Chemistry.)*

commercial graphite anode was achieved with a stable areal capacity of $\sim 1.6\, mAh\, cm^2$ at the 100th cycle.

6.11.4 Polymer binders: Impact of various architectures

Considering linear chain polymers are prone to detach from the Si particles due to limiting anchoring points, variety of polymer structures including branched polymer, graft polymer, network polymer, *co*-polymers are explored as potential binders for Si in LIBs anode [189]. In branched polymeric binders, the mechanical stress generated during lithiation/delithiation of Si can be distributed to branched chains at the branching points. Hyperbranched β-cyclodextrin (β-CDp) was proposed as a potential branched binder that has strong adhesion towards Si due to its multidimensional contacts, thus preserving the mechanical integrity with a stable SEI formation during galvanostatic cycling [190]. A hybrid binder of β-CDp/alginate was also proposed with an increased anchoring contact point with Si owing to an extended polymer network. β-CDp binder crosslinked with Si using a dendritic crosslinker incorporating 6 adamantane units (6AD) also exhibited improved cycling performance due to size compatibility between host and guest molecules as well as self-repairable behavior while Si anode undergoes volume change [201]. Sodium salt of poly(acrylic acid) grafted carboxylmethyl cellulose (NaPAA-*g*-CMC) polymer also exhibited multidimensional contacts with Si surface via strong hydrogen bonding and ion-dipole interactions, leading to a stronger adhesion property [202]. Poly (vinylidene difluoride)-*graft*-poly(*tert*-butylacrylate) (PVDF-*g*-PtBA) was also found capable to suppress the electrode swelling upon cycling due to the enhanced adhesion through multidimensional contacts [204]. Network polymers with covalent crosslinking of polymer chains and strong supramolecular interactions were also found as an excellent scaffold limiting Si particles movement owing to self-healing effect [95, 205–207].

A 3-D crosslinked c-PAA-CMC network polymer was effective in suppressing the deformation of the electrode; hydroxyl and carboxyl groups formed strong supramolecular interactions with Si. These synergistic effects enable the c-PAA-CMC-Si anode to deliver approximately 2140 mAh g^{-1} with a good capacity retention of ~75% after 100 cycles at a current density of 300 mA g^{-1} [95]. The electrolyte uptake and the volumetric swelling of Si anode were further decreased when the chitosan (CS) binder was crosslinked by using various amounts of glutaraldehyde (GA). The highly cross-linked polymer prevented the electrolyte from infiltrating into the film due to the dense network [210]. Acrylic acid-based multifunctional monomer polymerized with various amounts of diallyl ethers as a crosslinker (xCLPAH) attained excellent capacity retention due to enhanced mechanical flexibility and increased ion-dipole interactions along with strong supramolecular interactions [211]. Xanthan gum with a helical structure showed excellent electrochemical performance for Si anodes due to side chains of the helices formed strong ion-dipole interactions with Si surface, resulting in strong adhesion properties [188]. Polyfluorene (PF) based n-type conductive binder cathodically doped with carbonyl C=O and methylbenzoic ester —PhCOOCH$_3$ functional groups endowed lowering of LUMO energy level of parent polymer and increased adhesion properties with Si surface thereby improving electrochemical properties [181]. Xu et al. [195] designed poly(acrylic acid)-poly(2-hydroxyethyl acrylate-*co*-dopamine methacrylate) [PAA-P(HEA-*co*-DMA)] based self-healing polymer having multiple interconnected network of soft rigid chains and hydrogen bonds. Excellent reversible elasticity (~400%) of synthesized binder compared to brittle PAA attributed to the presence of hydrogen bonding sites originated from ethoxy and hydroxyl groups of HEA and catechol groups of DMA molecules preventing particle disintegration, stable SEI formation, and providing excellent Li-ion cyclability with Si microparticles for hundreds of cycles.

6.12 Conclusions and future outlook

The last few years have seen tremendous advancement made in terms of developing a novel strategy to counter the volumetric expansion issues associated with commercially available microcrystalline silicon as well as nanostructuring silicon with carbonaceous phase(s). Various silicon morphology, surface feature, porosity, coating, although, able to provide stable capacity for many cycles but first cycle irreversible loss and associated SEI formation cannot be ignored. SEI formation on Si anode using conventional electrolyte is generally nonuniform, and the composition mainly composed of a combination of inorganic/organic substances originate from the decomposition of salt and solvents. A strategy to complete move over from SEI formation on Si anode in liquid electrolyte would be to explore all-solid-state batteries. However, all-solid-state batteries have its share of drawbacks; including poor interfacial contacts between electrodes/electrolyte, slow interfacial kinetics, and high resistance. Novel supramolecular binders and electrode design may provide a combinatorial strategy alleviating the mechanical stress, and provide stable SEI formation in liquid electrolyte without rupturing the electrode from the current collector, and prolong the battery life. In this regard, multiple binder addition may be another option to prevent Si particles cracking but must be verified using full cell such that cathode performance doesn't get compromised. Fluoroethylene carbonate (FEC) has already established itself as an SEI stabilizer on the anode.

Future study and the emergence of cathode SEI stabilizers may provide a boon for battery manufacturers. A complete makeover and identification of lithium salt and nonaqueous electrolyte solvents combination with wide LUMO and HOMO bandgap covering the entire spectrum of the electrochemical potential difference between electrodes may bring forth much-needed change for SEI formation chemistry on Si anode and push the limit of high voltage cathode. At present, commercial LIB started seeing Si replenish graphite by a small amount such that volumetric variance of Si can be kept at bay. Future electrification of passenger vehicles is going to witness stringent energy requirements with a US Department of Energy aggressive target of 235 $Wh\,kg^{-1}$/500 $Wh\,l^{-1}$ at US \$100 kWh^{-1} at pack level (350 $Wh\,kg^{-1}$/750 WhL^{-1} at US \$75 kWh^{-1} at cell level). The above target can only be met by a gradual increase of Si content in place of graphite anode working in tandem with next-generation layered oxide cathodes in LIBs. Another factor that must be addressed to achieve the targeted volumetric and gravimetric energy density stipulated by industry will be a systematic increase of Si mass loading at the expense of graphite, keeping the tap density around ~1.5 g/cm^3 with ~50 μm thick electrodes formulation. Also, a stable SEI must protect intrinsic irreversible capacity loss associated with each cycle and poor Coulombic efficiency such that Si/graphite anode Coulombic efficiency can balance with the cathode (equivalent to ~99.99% or better with matched areal capacity for 3 $mAhcm^{-2}$ cell) upon cycling in a full-cell for at least 1000 cycles.

References

[1] J.M. Tarascon, M. Armand, Issues and challenges facing rechargeable lithium batteries, Nature 414 (2001) 359–367.

[2] J.B. Goodenough, Y. Kim, Challenges for rechargeable Li batteries, Chem. Mater. 22 (2010) 587–603.

[3] P. Rozier, J.M. Tarascon, Review—Li-rich layered oxide cathodes for next-generation Li-ion batteries: chances and challenges, J. Electrochem. Soc. 162 (2015) A2490–A2499.
[4] S. Goriparti, E. Miele, F. De Angelis, E. Di Fabrizio, R. Proietti Zaccaria, C. Capiglia, Review on recent progress of nanostructured anode materials for Li-ion batteries, J. Power Sources 257 (2014) 421–443.
[5] J. Duan, X. Tang, H. Dai, Y. Yang, W. Wu, X. Wei, Y. Huang, Building safe lithium-ion batteries for electric vehicles: a review, Electrochem. Energy Rev. 3 (2020) 1–42.
[6] X.-B. Cheng, R. Zhang, C.-Z. Zhao, F. Wei, J.-G. Zhang, Q. Zhang, A review of solid electrolyte interphases on lithium metal anode, Adv. Sci. 3 (2016) 1500213.
[7] E. Peled, S. Menkin, Review—SEI: past, present and future, J. Electrochem. Soc. 164 (2017) A1703–A1719.
[8] K. Xu, Electrolytes and interphases in Li-ion batteries and beyond, Chem. Rev. 114 (2014) 11503–11618.
[9] D. Aurbach, Nonaqueous Electrochemistry, Marcel Dekker, New York, 1999.
[10] V.A. Agubra, J.W. Fergus, The formation and stability of the solid electrolyte interface on the graphite anode, J. Power Sources 268 (2014) 153–162.
[11] P. Verma, P. Maire, P. Novák, A review of the features and analyses of the solid electrolyte interphase in Li-ion batteries, Electrochim. Acta 55 (2010) 6332–6341.
[12] P. Biensan, B. Simon, J.P. Pérès, A. de Guibert, M. Broussely, J.M. Bodet, F. Perton, On safety of lithium-ion cells, J. Power Sources 81–82 (1999) 906–912.
[13] K. Leung, F. Soto, K. Hankins, P.B. Balbuena, K.L. Harrison, Stability of solid electrolyte interphase components on lithium metal and reactive anode material surfaces, J. Phys. Chem. C 120 (2016) 6302–6313.
[14] F.A. Soto, Y. Ma, J.M. Martinez de la Hoz, J.M. Seminario, P.B. Balbuena, Formation and growth mechanisms of solid-electrolyte interphase layers in rechargeable batteries, Chem. Mater. 27 (2015) 7990–8000.
[15] A.M. Tripathi, W.-N. Su, B.J. Hwang, In situ analytical techniques for battery interface analysis, Chem. Soc. Rev. 47 (2018) 736–851.
[16] J. Collins, G. Gourdin, M. Foster, D. Qu, Carbon surface functionalities and SEI formation during Li intercalation, Carbon 92 (2015) 193–244.
[17] A.N. Dey, Lithium anode film and organic and inorganic electrolyte batteries, Thin Solid Films 43 (1977) 131–171.
[18] E. Peled, The electrochemical behavior of alkali and alkaline earth metals in nonaqueous battery systems—the solid electrolyte interphase model, J. Electrochem. Soc. 126 (1979) 2047.
[19] G. Nazri, Composition of surface layers on Li electrodes in PC, $LiClO_4$ of very low water content, J. Electrochem. Soc. 132 (1985) 2050.
[20] D. Aurbach, Identification of surface films formed on lithium in propylene carbonate solutions, J. Electrochem. Soc. 134 (1987) 1611.
[21] K. Kanamura, XPS analysis of lithium surfaces following immersion in various solvents containing LiBF[sub 4], J. Electrochem. Soc. 142 (1995) 340.
[22] K. Edström, M. Herstedt, D.P. Abraham, A new look at the solid electrolyte interphase on graphite anodes in Li-ion batteries, J. Power Sources 153 (2006) 380–384.
[23] K. Guo, R. Kumar, X. Xiao, B.W. Sheldon, H. Gao, Failure progression in the solid electrolyte interphase (SEI) on silicon electrodes, Nano Energy 68 (2020) 104257.
[24] C.H. Lee, J.A. Dura, A. LeBar, S.C. DeCaluwe, Direct, operando observation of the bilayer solid electrolyte interphase structure: electrolyte reduction on a non-intercalating electrode, J. Power Sources 412 (2019) 725–735.

[25] Y. Liu, K. Guo, C. Wang, H. Gao, Wrinkling and ratcheting of a thin film on cyclically deforming plastic substrate: mechanical instability of the solid-electrolyte interphase in Li–ion batteries, J. Mech. Phys. Solids 123 (2019) 103–118.

[26] J.O. Besenhard, M. Winter, J. Yang, W. Biberacher, Filming mechanism of lithium-carbon anodes in organic and inorganic electrolytes, J. Power Sources 54 (1995) 228–231.

[27] E. Peled, Advanced model for solid electrolyte interphase electrodes in liquid and polymer electrolytes, J. Electrochem. Soc. 144 (1997) L208.

[28] D. Aurbach, B. Markovsky, M.D. Levi, E. Levi, A. Schechter, M. Moshkovich, Y. Cohen, New insights into the interactions between electrode materials and electrolyte solutions for advanced nonaqueous batteries, J. Power Sources 81-82 (1999) 95–111.

[29] U. Kasavajjula, C. Wang, A.J. Appleby, Nano- and bulk-silicon-based insertion anodes for lithium-ion secondary cells, J. Power Sources 163 (2007) 1003–1039.

[30] Z.-L. Xu, X. Liu, Y. Luo, L. Zhou, J.-K. Kim, Nanosilicon anodes for high performance rechargeable batteries, Prog. Mater. Sci. 90 (2017) 1–44.

[31] X. Zhang, R. Kostecki, T.J. Richardson, J.K. Pugh, P.N. Ross, Electrochemical and infrared studies of the reduction of organic carbonates, J. Electrochem. Soc. 148 (2001) A1341.

[32] D. Aurbach, I. Weissman, A. Zaban, O. Chusid, Correlation between surface chemistry, morphology, cycling efficiency and interfacial properties of Li electrodes in solutions containing different Li salts, Electrochim. Acta 39 (1994) 51–71.

[33] D. Aurbach, A comparative study of synthetic graphite and Li electrodes in electrolyte solutions based on ethylene carbonate-dimethyl carbonate mixtures, J. Electrochem. Soc. 143 (1996) 3809.

[34] D. Aurbach, Y. Gofer, M. Ben-Zion, P. Aped, The behaviour of lithium electrodes in propylene and ethylene carbonate: Te major factors that influence Li cycling efficiency, J. Electroanal. Chem. 339 (1992) 451–471.

[35] B. Moeremans, H.-W. Cheng, C. Merola, Q. Hu, M. Oezaslan, M. Safari, M.K. Van Bael, A. Hardy, M. Valtiner, F.U. Renner, In situ mechanical analysis of the nanoscopic solid electrolyte interphase on anodes of Li-ion batteries, Adv. Sci. 6 (2019) 1900190.

[36] M.J. Chon, V.A. Sethuraman, A. McCormick, V. Srinivasan, P.R. Guduru, Real-time measurement of stress and damage evolution during initial lithiation of crystalline silicon, Phys. Rev. Lett. 107 (2011) 045503.

[37] M. Nie, D.P. Abraham, Y. Chen, A. Bose, B.L. Lucht, Silicon solid electrolyte interphase (SEI) of lithium ion battery characterized by microscopy and spectroscopy, J. Phys. Chem. C 117 (2013) 13403–13412.

[38] N.-S. Choi, K.H. Yew, K.Y. Lee, M. Sung, H. Kim, S.-S. Kim, Effect of fluoroethylene carbonate additive on interfacial properties of silicon thin-film electrode, J. Power Sources 161 (2006) 1254–1259.

[39] B.S. Parimalam, B.L. Lucht, Reduction reactions of electrolyte salts for lithium ion batteries: LiPF6, LiBF4, LiDFOB, LiBOB, and LiTFSI, J. Electrochem. Soc. 165 (2018) A251–A255.

[40] C.K. Chan, R. Ruffo, S.S. Hong, Y. Cui, Surface chemistry and morphology of the solid electrolyte interphase on silicon nanowire lithium-ion battery anodes, J. Power Sources 189 (2009) 1132–1140.

[41] O. Borodin, D. Bedrov, Interfacial structure and dynamics of the lithium alkyl dicarbonate SEI components in contact with the lithium battery electrolyte, J. Phys. Chem. C 118 (2014) 18362–18371.

[42] K. Tasaki, Solvent decompositions and physical properties of decomposition compounds in Li-ion battery electrolytes studied by DFT calculations and molecular dynamics simulations, J. Phys. Chem. B 109 (2005) 2920–2933.

[43] A. Wang, S. Kadam, H. Li, S. Shi, Y. Qi, Review on modeling of the anode solid electrolyte interphase (SEI) for lithium-ion batteries, npj Comput. Mater. 4 (2018) 15.
[44] Y. Yang, X. Liu, Z. Dai, F. Yuan, Y. Bando, D. Golberg, X. Wang, In situ electrochemistry of rechargeable battery materials: status report and perspectives, Adv. Mater. 29 (2017) 1606922.
[45] J. Zheng, H. Zheng, R. Wang, L. Ben, W. Lu, L. Chen, L. Chen, H. Li, 3D visualization of inhomogeneous multi-layered structure and Young's modulus of the solid electrolyte interphase (SEI) on silicon anodes for lithium ion batteries, Phys. Chem. Chem. Phys. 16 (2014) 13229–13238.
[46] Y.-C. Yen, S.-C. Chao, H.-C. Wu, N.-L. Wu, Study on solid-electrolyte-interphase of Si and C-coated Si electrodes in lithium cells, J. Electrochem. Soc. 156 (2009) A95–A102.
[47] B. Philippe, R. Dedryvère, J. Allouche, F. Lindgren, M. Gorgoi, H. Rensmo, D. Gonbeau, K. Edström, Nanosilicon electrodes for lithium-ion batteries: interfacial mechanisms studied by hard and soft X-ray photoelectron spectroscopy, Chem. Mater. 24 (2012) 1107–1115.
[48] C. Cao, I.I. Abate, E. Sivonxay, B. Shyam, C. Jia, B. Moritz, T.P. Devereaux, K.A. Persson, H.-G. Steinrück, M.F. Toney, Solid electrolyte interphase on native oxide-terminated silicon anodes for Li-ion batteries, Joule 3 (2019) 762–781.
[49] C. Cao, H.-G. Steinrück, B. Shyam, K.H. Stone, M.F. Toney, In situ study of silicon electrode lithiation with X-ray reflectivity, Nano Lett. 16 (2016) 7394–7401.
[50] R. Kumar, P. Lu, X. Xiao, Z. Huang, B.W. Sheldon, Strain-induced lithium losses in the solid electrolyte interphase on silicon electrodes, ACS Appl. Mater. Interfaces 9 (2017) 28406–28417.
[51] R. Kumar, A. Tokranov, B.W. Sheldon, X. Xiao, Z. Huang, C. Li, T. Mueller, In situ and operando investigations of failure mechanisms of the solid electrolyte interphase on silicon electrodes, ACS Energy Lett. 1 (2016) 689–697.
[52] R. Kumar, J.H. Woo, X. Xiao, B.W. Sheldon, Internal microstructural changes and stress evolution in silicon nanoparticle based composite electrodes, J. Electrochem. Soc. 164 (2017) A3750–A3765.
[53] A. Tokranov, R. Kumar, C. Li, S. Minne, X. Xiao, B.W. Sheldon, Control and optimization of the electrochemical and mechanical properties of the solid electrolyte interphase on silicon electrodes in lithium ion batteries, Adv. Energy Mater. 6 (2016) 1502302.
[54] A. Tokranov, B.W. Sheldon, C. Li, S. Minne, X. Xiao, In situ atomic force microscopy study of initial solid electrolyte interphase formation on silicon electrodes for Li-ion batteries, ACS Appl. Mater. Interfaces 6 (2014) 6672–6686.
[55] Y. Zhang, N. Du, D. Yang, Designing superior solid electrolyte interfaces on silicon anodes for high-performance lithium-ion batteries, Nanoscale 11 (2019) 19086–19104.
[56] P. Peljo, H.H. Girault, Electrochemical potential window of battery electrolytes: the HOMO–LUMO misconception, Energ. Environ. Sci. 11 (2018) 2306–2309.
[57] L. Benitez, J.M. Seminario, Electron transport and electrolyte reduction in the solid-electrolyte interphase of rechargeable lithium ion batteries with silicon anodes, J. Phys. Chem. C 120 (2016) 17978–17988.
[58] G.M. Veith, M. Doucet, R.L. Sacci, B. Vacaliuc, J.K. Baldwin, J.F. Browning, Determination of the solid electrolyte interphase structure grown on a silicon electrode using a fluoroethylene carbonate additive, Sci. Rep. 7 (2017) 6326.
[59] B. Key, R. Bhattacharyya, M. Morcrette, V. Seznéc, J.-M. Tarascon, C.P. Grey, Real-time NMR investigations of structural changes in silicon electrodes for lithium-ion batteries, J. Am. Chem. Soc. 131 (2009) 9239–9249.

[60] A.L. Michan, M. Leskes, C.P. Grey, Voltage dependent solid electrolyte interphase formation in silicon electrodes: monitoring the formation of organic decomposition products, Chem. Mater. 28 (2016) 385–398.

[61] I. Yoon, D.P. Abraham, B.L. Lucht, A.F. Bower, P.R. Guduru, In situ measurement of solid electrolyte interphase evolution on silicon anodes using atomic force microscopy, Adv. Energy Mater. 6 (2016) 1600099.

[62] T. Yoon, C.C. Nguyen, D.M. Seo, B.L. Lucht, Capacity fading mechanisms of silicon nanoparticle negative electrodes for lithium ion batteries, J. Electrochem. Soc. 162 (2015) A2325–A2330.

[63] Z. Liu, P. Lu, Q. Zhang, X. Xiao, Y. Qi, L.-Q. Chen, A bottom-up formation mechanism of solid electrolyte interphase revealed by isotope-assisted time-of-flight secondary ion mass spectrometry, J. Phys. Chem. Lett. 9 (2018) 5508–5514.

[64] K. Leung, S.B. Rempe, M.E. Foster, Y. Ma, J.M. Martinez del la Hoz, N. Sai, P.B. Balbuena, Modeling electrochemical decomposition of fluoroethylene carbonate on silicon anode surfaces in lithium ion batteries, J. Electrochem. Soc. 161 (2013) A213–A221.

[65] J.M. Martinez de la Hoz, K. Leung, P.B. Balbuena, Reduction mechanisms of ethylene carbonate on Si anodes of lithium-ion batteries: effects of degree of lithiation and nature of exposed surface, ACS Appl. Mater. Interfaces 5 (2013) 13457–13465.

[66] G. Gourdin, J. Collins, D. Zheng, M. Foster, D. Qu, Spectroscopic compositional analysis of electrolyte during initial SEI layer formation, J. Phys. Chem. C 118 (2014) 17383–17394.

[67] K.W. Schroder, A.G. Dylla, S.J. Harris, L.J. Webb, K.J. Stevenson, Role of surface oxides in the formation of solid–electrolyte interphases at silicon electrodes for lithium-ion batteries, ACS Appl. Mater. Interfaces 6 (2014) 21510–21524.

[68] K.W. Schroder, H. Celio, L.J. Webb, K.J. Stevenson, Examining solid electrolyte interphase formation on crystalline silicon electrodes: influence of electrochemical preparation and ambient exposure conditions, J. Phys. Chem. C 116 (2012) 19737–19747.

[69] L. Benitez, D. Cristancho, J.M. Seminario, J.M. Martinez de la Hoz, P.B. Balbuena, Electron transfer through solid-electrolyte-interphase layers formed on Si anodes of Li-ion batteries, Electrochim. Acta 140 (2014) 250–257.

[70] J.M. Martínez de la Hoz, P.B. Balbuena, Reduction mechanisms of additives on Si anodes of Li-ion batteries, Phys. Chem. Chem. Phys. 16 (2014) 17091–17098.

[71] J.M. Martinez de la Hoz, F.A. Soto, P.B. Balbuena, Effect of the electrolyte composition on SEI reactions at Si anodes of Li-ion batteries, J. Phys. Chem. C 119 (2015) 7060–7068.

[72] F.A. Soto, J.M. Martinez de la Hoz, J.M. Seminario, P.B. Balbuena, Modeling solid-electrolyte interfacial phenomena in silicon anodes, Curr. Opin. Chem. Eng. 13 (2016) 179–185.

[73] K. Yamaguchi, Y. Domi, H. Usui, H. Sakaguchi, Elucidation of the reaction behavior of silicon negative electrodes in a bis(fluorosulfonyl)amide-based ionic liquid electrolyte, ChemElectroChem 4 (2017) 3257–3263.

[74] M. Shimizu, H. Usui, T. Suzumura, H. Sakaguchi, Analysis of the deterioration mechanism of Si electrode as a Li-ion battery anode using Raman microspectroscopy, J. Phys. Chem. C 119 (2015) 2975–2982.

[75] R. Dedryvère, H. Martinez, S. Leroy, D. Lemordant, F. Bonhomme, P. Biensan, D. Gonbeau, Surface film formation on electrodes in a LiCoO$_2$/graphite cell: a step by step XPS study, J. Power Sources 174 (2007) 462–468.

[76] S. Dalavi, P. Guduru, B.L. Lucht, Performance enhancing electrolyte additives for lithium ion batteries with silicon anodes, J. Electrochem. Soc. 159 (2012) A642–A646.

[77] A.L. Michan, B.S. Parimalam, M. Leskes, R.N. Kerber, T. Yoon, C.P. Grey, B.L. Lucht, Fluoroethylene carbonate and vinylene carbonate reduction: understanding lithium-ion battery electrolyte additives and solid electrolyte interphase formation, Chem. Mater. 28 (2016) 8149–8159.

[78] T. Jaumann, J. Balach, M. Klose, S. Oswald, U. Langklotz, A. Michaelis, J. Eckert, L. Giebeler, SEI-component formation on sub 5 nm sized silicon nanoparticles in Li-ion batteries: the role of electrode preparation, FEC addition and binders, Phys. Chem. Chem. Phys. 17 (2015) 24956–24967.

[79] V. Etacheri, O. Haik, Y. Goffer, G.A. Roberts, I.C. Stefan, R. Fasching, D. Aurbach, Effect of Fluoroethylene carbonate (FEC) on the performance and surface chemistry of Si-nanowire Li-ion battery anodes, Langmuir 28 (2012) 965–976.

[80] E. Markevich, G. Salitra, D. Aurbach, Fluoroethylene carbonate as an important component for the formation of an effective solid electrolyte interphase on anodes and cathodes for advanced Li-ion batteries, ACS Energy Lett. 2 (2017) 1337–1345.

[81] C.L. Campion, W. Li, W.B. Euler, B.L. Lucht, B. Ravdel, J.F. DiCarlo, R. Gitzendanner, K.M. Abraham, Suppression of toxic compounds produced in the decomposition of lithium-ion battery electrolytes, Electrochem. Solid St. 7 (2004) A194.

[82] T. Joshi, K. Eom, G. Yushin, T. Fuller, Effects of dissolved transition metals on the electrochemical performance and SEI growth in lithium-ion batteries, J. Electrochem. Soc. 161 (2014) A1915–A1921.

[83] Q. Qian, Y. Yang, H. Shao, Solid electrolyte interphase formation by propylene carbonate reduction for lithium anode, Phys. Chem. Chem. Phys. 19 (2017) 28772–28780.

[84] S. Leroy, F. Blanchard, R. Dedryvère, H. Martinez, B. Carré, D. Lemordant, D. Gonbeau, Surface film formation on a graphite electrode in Li-ion batteries: AFM and XPS study, Surf. Interface Anal. 37 (2005) 773–781.

[85] D. Aurbach, A. Zaban, Y. Gofer, Y.E. Ely, I. Weissman, O. Chusid, O. Abramson, Recent studies of the lithium-liquid electrolyte interface electrochemical, morphological and spectral studies of a few important systems, J. Power Sources 54 (1995) 76–84.

[86] A. Augustsson, M. Herstedt, J.H. Guo, K. Edström, G.V. Zhuang, J.P.N. Ross, J.E. Rubensson, J. Nordgren, Solid electrolyte interphase on graphite Li-ion battery anodes studied by soft X-ray spectroscopy, Phys. Chem. Chem. Phys. 6 (2004) 4185–4189.

[87] G.V. Zhuang, K. Xu, H. Yang, T.R. Jow, P.N. Ross, Lithium ethylene dicarbonate identified as the primary product of chemical and electrochemical reduction of EC in 1.2 M LiPF6/EC: EMC electrolyte, J. Phys. Chem. B 109 (2005) 17567–17573.

[88] D. Aurbach, M.D. Levi, E. Levi, A. Schechter, Failure and stabilization mechanisms of graphite electrodes, J. Phys. Chem. B 101 (1997) 2195–2206.

[89] J. Yang, N. Solomatin, A. Kraytsberg, Y. Ein-Eli, In-Situ spectro–electrochemical insight revealing distinctive silicon anode solid electrolyte interphase formation in a lithium–ion battery, ChemistrySelect 1 (2016) 572–576.

[90] A.M. Andersson, K. Edström, Chemical composition and morphology of the elevated temperature SEI on graphite, J. Electrochem. Soc. 148 (2001) A1100.

[91] Y. Zhang, D. Krishnamurthy, V. Viswanathan, Engineering solid electrolyte interphase composition by assessing decomposition pathways of fluorinated organic solvents in lithium metal batteries, J. Electrochem. Soc. 167 (2020) 070554.

[92] R.Z.A. Manj, F. Zhang, W.U. Rehman, W. Luo, J. Yang, Toward understanding the interaction within silicon-based anodes for stable lithium storage, Chem. Eng. J. 385 (2020) 123821.

[93] T. Yoon, M.S. Milien, B.S. Parimalam, B.L. Lucht, Thermal decomposition of the solid electrolyte interphase (SEI) on silicon electrodes for lithium ion batteries, Chem. Mater. 29 (2017) 3237–3245.

[94] Y. Ein-Eli, B. Markovsky, D. Aurbach, Y. Carmeli, H. Yamin, S. Luski, The dependence of the performance of Li-C intercalation anodes for Li-ion secondary batteries on the electrolyte solution composition, Electrochim. Acta 39 (1994) 2559–2569.

[95] B. Koo, H. Kim, Y. Cho, K.T. Lee, N.-S. Choi, J. Cho, A highly cross-linked polymeric binder for high-performance silicon negative electrodes in lithium ion batteries, Angew. Chem. Int. Ed. 51 (2012) 8762–8767.

[96] S. Komaba, K. Shimomura, N. Yabuuchi, T. Ozeki, H. Yui, K. Konno, Study on polymer binders for high-capacity SiO negative electrode of Li-ion batteries, J. Phys. Chem. C 115 (2011) 13487–13495.

[97] N. Leifer, M.C. Smart, G.K.S. Prakash, L. Gonzalez, L. Sanchez, K.A. Smith, P. Bhalla, C.P. Grey, S.G. Greenbaum, 13C solid state NMR suggests unusual breakdown products in SEI formation on lithium ion electrodes, J. Electrochem. Soc. 158 (2011) A471.

[98] K. Feng, et al., Silicon-based anodes for advanced lithium-ion batteries, in: Encyclopedia of Inorganic and Bioinorganic Chemistry, Online © 2011–2019, John Wiley & Sons, Ltd.

[99] Q. Li, X. Liu, X. Han, Y. Xiang, G. Zhong, J. Wang, B. Zheng, J. Zhou, Y. Yang, Identification of the solid electrolyte Interface on the Si/C composite anode with FEC as the additive, ACS Appl. Mater. Interfaces 11 (2019) 14066–14075.

[100] B. Philippe, R. Dedryvère, M. Gorgoi, H. Rensmo, D. Gonbeau, K. Edström, Improved performances of nanosilicon electrodes using the salt LiFSI: a photoelectron spectroscopy study, J. Am. Chem. Soc. 135 (2013) 9829–9842.

[101] R. Elazari, G. Salitra, G. Gershinsky, A. Garsuch, A. Panchenko, D. Aurbach, Li ion cells comprising Lithiated columnar silicon film anodes, TiS2 cathodes and fluoroethyene carbonate (FEC) as a critically important component, J. Electrochem. Soc. 159 (2012) A1440–A1445.

[102] F. Joho, P. Novák, SNIFTIRS investigation of the oxidative decomposition of organic-carbonate-based electrolytes for lithium-ion cells, Electrochim. Acta 45 (2000) 3589–3599.

[103] Z. Xu, J. Yang, H. Li, Y. Nuli, J. Wang, Electrolytes for advanced lithium ion batteries using silicon-based anodes, J. Mater. Chem. A 7 (2019) 9432–9446.

[104] C.-C. Chang, S.-H. Hsu, Y.-F. Jung, C.-H. Yang, Vinylene carbonate and vinylene trithiocarbonate as electrolyte additives for lithium ion battery, J. Power Sources 196 (2011) 9605–9611.

[105] S.S. Zhang, A review on electrolyte additives for lithium-ion batteries, J. Power Sources 162 (2006) 1379–1394.

[106] S. Uchida, M. Yamagata, M. Ishikawa, Effect of electrolyte additives on non-nano-Si negative electrodes prepared with polyimide binder, J. Electrochem. Soc. 162 (2014) A406–A412.

[107] K. Schroder, J. Alvarado, T.A. Yersak, J. Li, N. Dudney, L.J. Webb, Y.S. Meng, K.J. Stevenson, The effect of fluoroethylene carbonate as an additive on the solid electrolyte interphase on silicon lithium-ion electrodes, Chem. Mater. 27 (2015) 5531–5542.

[108] O. Matsuoka, A. Hiwara, T. Omi, M. Toriida, T. Hayashi, C. Tanaka, Y. Saito, T. Ishida, H. Tan, S.S. Ono, S. Yamamoto, Ultra-thin passivating film induced by vinylene carbonate on highly oriented pyrolytic graphite negative electrode in lithium-ion cell, J. Power Sources 108 (2002) 128–138.

[109] T. Kennedy, M. Brandon, F. Laffir, K.M. Ryan, Understanding the influence of electrolyte additives on the electrochemical performance and morphology evolution of silicon nanowire based lithium-ion battery anodes, J. Power Sources 359 (2017) 601–610.

[110] M. Nie, J. Demeaux, B.T. Young, D.R. Heskett, Y. Chen, A. Bose, J.C. Woicik, B.L. Lucht, Effect of vinylene carbonate and fluoroethylene carbonate on SEI formation on graphitic anodes in Li-ion batteries, J. Electrochem. Soc. 162 (2015) A7008–A7014.

[111] S.K. Heiskanen, J. Kim, B.L. Lucht, Generation and evolution of the solid electrolyte interphase of lithium-ion batteries, Joule 3 (2019) 2322–2333.

[112] C.C. Nguyen, B.L. Lucht, Comparative study of Fluoroethylene carbonate and Vinylene carbonate for silicon anodes in lithium ion batteries, J. Electrochem. Soc. 161 (2014) A1933–A1938.

[113] D. Aurbach, K. Gamolsky, B. Markovsky, Y. Gofer, M. Schmidt, U. Heider, On the use of vinylene carbonate (VC) as an additive to electrolyte solutions for Li-ion batteries, Electrochim. Acta 47 (2002) 1423–1439.

[114] L. Chen, K. Wang, X. Xie, J. Xie, Effect of vinylene carbonate (VC) as electrolyte additive on electrochemical performance of Si film anode for lithium ion batteries, J. Power Sources 174 (2007) 538–543.

[115] I.A. Profatilova, C. Stock, A. Schmitz, S. Passerini, M. Winter, Enhanced thermal stability of a lithiated nano-silicon electrode by fluoroethylene carbonate and vinylene carbonate, J. Power Sources 222 (2013) 140–149.

[116] H. Ota, Y. Sakata, Y. Otake, K. Shima, M. Ue, J.-I. Yamaki, Structural and functional analysis of surface film on Li anode in vinylene carbonate-containing electrolyte, J. Electrochem. Soc. 151 (2004) A1778.

[117] H. Ota, K. Shima, M. Ue, J.-i. Yamaki, Effect of vinylene carbonate as additive to electrolyte for lithium metal anode, Electrochim. Acta 49 (2004) 565–572.

[118] M. Haruta, T. Okubo, Y. Masuo, S. Yoshida, A. Tomita, T. Takenaka, T. Doi, M. Inaba, Temperature effects on SEI formation and cyclability of Si nanoflake powder anode in the presence of SEI-forming additives, Electrochim. Acta 224 (2017) 186–193.

[119] J. Shin, T.-H. Kim, Y. Lee, E. Cho, Key functional groups defining the formation of Si anode solid-electrolyte interphase towards high energy density Li-ion batteries, Energy Storage Mater. 25 (2020) 764–781.

[120] X. Chen, X. Li, D. Mei, J. Feng, M.Y. Hu, J. Hu, M. Engelhard, J. Zheng, W. Xu, J. Xiao, J. Liu, J.-G. Zhang, Reduction mechanism of fluoroethylene carbonate for stable solid–electrolyte interphase film on silicon anode, ChemSusChem 7 (2014) 549–554.

[121] Y. Jin, N.-J.H. Kneusels, L.E. Marbella, E. Castillo-Martínez, P.C.M.M. Magusin, R.S. Weatherup, E. Jónsson, T. Liu, S. Paul, C.P. Grey, Understanding fluoroethylene carbonate and vinylene carbonate based electrolytes for Si anodes in lithium ion batteries with NMR spectroscopy, J. Am. Chem. Soc. 140 (2018) 9854–9867.

[122] C. Xu, F. Lindgren, B. Philippe, M. Gorgoi, F. Björefors, K. Edström, T. Gustafsson, Improved performance of the silicon anode for Li-ion batteries: understanding the surface modification mechanism of fluoroethylene carbonate as an effective electrolyte additive, Chem. Mater. 27 (2015) 2591–2599.

[123] H. Nakai, T. Kubota, A. Kita, A. Kawashima, Investigation of the solid electrolyte interphase formed by fluoroethylene carbonate on Si electrodes, J. Electrochem. Soc. 158 (2011) A798.

[124] A. Bordes, K. Eom, T.F. Fuller, The effect of fluoroethylene carbonate additive content on the formation of the solid-electrolyte interphase and capacity fade of Li-ion full-cell employing nano Si–graphene composite anodes, J. Power Sources 257 (2014) 163–169.

[125] H. Wang, H.B. Chew, Nanoscale mechanics of the solid electrolyte interphase on lithiated-silicon electrodes, ACS Appl. Mater. Interfaces 9 (2017) 25662–25667.

[126] C. Cao, B. Shyam, J. Wang, M.F. Toney, H.-G. Steinrück, Shedding X-ray light on the interfacial electrochemistry of silicon anodes for Li-ion batteries, Acc. Chem. Res. 52 (2019) 2673–2683.

[127] C. Cao, H.-G. Steinrück, B. Shyam, M.F. Toney, The atomic scale electrochemical lithiation and delithiation process of silicon, Adv. Mater. Interfaces 4 (2017) 1700771.

[128] W. Zhang, T.H. Cai, B.W. Sheldon, The impact of initial SEI formation conditions on strain-induced capacity losses in silicon electrodes, Adv. Energy Mater. 9 (2019) 1803066.

[129] S. Benning, C. Chen, R.-A. Eichel, P.H.L. Notten, F. Hausen, Direct observation of SEI formation and lithiation in thin-film silicon electrodes via in situ electrochemical atomic force microscopy, ACS Appl. Energy Mater. 2 (2019) 6761–6767.

[130] C. Stetson, Y. Yin, C.-S. Jiang, S.C. DeCaluwe, M. Al-Jassim, N.R. Neale, C. Ban, A. Burrell, Temperature-dependent solubility of solid electrolyte interphase on silicon electrodes, ACS Energy Lett. 4 (2019) 2770–2775.

[131] Y. Shi, J. Wan, J.-Y. Li, X.-C. Hu, S.-Y. Lang, Z.-Z. Shen, G. Li, H.-J. Yan, K.-C. Jiang, Y.-G. Guo, R. Wen, L.-J. Wan, Elucidating the interfacial evolution and anisotropic dynamics on silicon anodes in lithium-ion batteries, Nano Energy 61 (2019) 304–310.

[132] Y. Feng, T.-D.-T. Ngo, M. Panagopoulou, A. Cheriet, B.M. Koo, C. Henry-de-Villeneuve, M. Rosso, F. Ozanam, Lithiation of pure and methylated amorphous silicon: monitoring by operando optical microscopy and ex situ atomic force microscopy, Electrochim. Acta 302 (2019) 249–258.

[133] R. Jorn, R. Kumar, D.P. Abraham, G.A. Voth, Atomistic modeling of the electrode–electrolyte interface in Li-ion energy storage systems: electrolyte structuring, J. Phys. Chem. C 117 (2013) 3747–3761.

[134] A.M. Haregewoin, E.G. Leggesse, J.-C. Jiang, F.-M. Wang, B.-J. Hwang, S.D. Lin, Comparative study on the solid electrolyte Interface formation by the reduction of alkyl carbonates in lithium ion battery, Electrochim. Acta 136 (2014) 274–285.

[135] O. Borodin, X. Ren, J. Vatamanu, A. von Wald Cresce, J. Knap, K. Xu, Modeling insight into battery electrolyte electrochemical stability and interfacial structure, Acc. Chem. Res. 50 (2017) 2886–2894.

[136] M. Ge, J. Rong, X. Fang, C. Zhou, Porous doped silicon nanowires for lithium ion battery anode with long cycle life, Nano Lett. 12 (2012) 2318–2323.

[137] L.-F. Cui, Y. Yang, C.-M. Hsu, Y. Cui, Carbon−silicon core−shell nanowires as high capacity electrode for lithium ion batteries, Nano Lett. 9 (2009) 3370–3374.

[138] T. Song, J. Xia, J.-H. Lee, D.H. Lee, M.-S. Kwon, J.-M. Choi, J. Wu, S.K. Doo, H. Chang, W.I. Park, D.S. Zang, H. Kim, Y. Huang, K.-C. Hwang, J.A. Rogers, U. Paik, Arrays of sealed silicon nanotubes as anodes for lithium ion batteries, Nano Lett. 10 (2010) 1710–1716.

[139] J.-K. Yoo, J. Kim, Y.S. Jung, K. Kang, Scalable fabrication of silicon nanotubes and their application to energy storage, Adv. Mater. 24 (2012) 5452–5456.

[140] H. Kim, B. Han, J. Choo, J. Cho, Three-dimensional porous silicon particles for use in high-performance lithium secondary batteries, Angew. Chem. Int. Ed. 47 (2008) 10151–10154.

[141] G. Kim, S. Jeong, J.-H. Shin, J. Cho, H. Lee, 3D amorphous silicon on nanopillar copper electrodes as anodes for high-rate lithium-ion batteries, ACS Nano 8 (2014) 1907–1912.

[142] M. Ge, Y. Lu, P. Ercius, J. Rong, X. Fang, M. Mecklenburg, C. Zhou, Large-scale fabrication, 3D tomography, and lithium-ion battery application of porous silicon, Nano Lett. 14 (2014) 261–268.

[143] Z. Jiang, C. Li, S. Hao, K. Zhu, P. Zhang, An easy way for preparing high performance porous silicon powder by acid etching Al–Si alloy powder for lithium ion battery, Electrochim. Acta 115 (2014) 393–398.

[144] Y. Zhu, W. Liu, X. Zhang, J. He, J. Chen, Y. Wang, T. Cao, Directing silicon–graphene self-assembly as a core/shell anode for high-performance lithium-ion batteries, Langmuir 29 (2013) 744–749.

[145] S. Chen, P. Bao, X. Huang, B. Sun, G. Wang, Hierarchical 3D mesoporous silicon@graphene nanoarchitectures for lithium ion batteries with superior performance, Nano Res. 7 (2014) 85–94.

[146] H. Wu, G. Yu, L. Pan, N. Liu, M.T. McDowell, Z. Bao, Y. Cui, Stable Li-ion battery anodes by in-situ polymerization of conducting hydrogel to conformally coat silicon nanoparticles, Nat. Commun. 4 (2013) 1–6.

[147] Y. Xu, G. Yin, Y. Ma, P. Zuo, X. Cheng, Nanosized core/shell silicon@carbon anode material for lithium ion batteries with polyvinylidene fluoride as carbon source, J. Mater. Chem. 20 (2010) 3216–3220.

[148] G. Jeong, J.-G. Kim, M.-S. Park, M. Seo, S.M. Hwang, Y.-U. Kim, Y.-J. Kim, J.H. Kim, S.X. Dou, Core–shell structured silicon nanoparticles@TiO2–x/carbon mesoporous microfiber composite as a safe and high-performance lithium-ion battery anode, ACS Nano 8 (2014) 2977–2985.

[149] X. Feng, J. Yang, Y. Bie, J. Wang, Y. Nuli, W. Lu, Nano/micro-structured Si/CNT/C composite from nano-SiO$_2$ for high power lithium ion batteries, Nanoscale 6 (2014) 12532–12539.

[150] S. Sim, P. Oh, S. Park, J. Cho, Critical thickness of SiO$_2$ coating layer on core@shell bulk@nanowire Si anode materials for Li-ion batteries, Adv. Mater. 25 (2013) 4498–4503.

[151] K.W. Lim, J.-I. Lee, J. Yang, Y.-K. Kim, H.Y. Jeong, S. Park, H.S. Shin, Catalyst-free synthesis of Si-SiOx core-shell nanowire anodes for high-rate and high-capacity lithium-ion batteries, ACS Appl. Mater. Interfaces 6 (2014) 6340–6345.

[152] L. Su, Z. Zhou, M. Ren, Core double-shell Si@SiO$_2$@C nanocomposites as anode materials for Li-ion batteries, Chem. Commun. 46 (2010) 2590–2592.

[153] Y.-S. Hu, R. Demir-Cakan, M.-M. Titirici, J.-O. Müller, R. Schlögl, M. Antonietti, J. Maier, Superior storage performance of a Si@SiOx/C nanocomposite as anode material for lithium-ion batteries, Angew. Chem. Int. Ed. 47 (2008) 1645–1649.

[154] M. Dirican, M. Yanilmaz, K. Fu, O. Yildiz, H. Kizil, Y. Hu, X. Zhang, Carbon-confined PVA-derived silicon/silica/carbon nanofiber composites as anode for lithium-ion batteries, J. Electrochem. Soc. 161 (2014) A2197–A2203.

[155] H. Wu, G. Zheng, N. Liu, T.J. Carney, Y. Yang, Y. Cui, Engineering empty space between Si nanoparticles for lithium-ion battery anodes, Nano Lett. 12 (2012) 904–909.

[156] T.H. Hwang, Y.M. Lee, B.-S. Kong, J.-S. Seo, J.W. Choi, Electrospun core–shell fibers for robust silicon nanoparticle-based lithium ion battery anodes, Nano Lett. 12 (2012) 802–807.

[157] H. Wu, Y. Cui, Designing nanostructured Si anodes for high energy lithium ion batteries, Nano Today 7 (2012) 414–429.

[158] N.S. Hochgatterer, M.R. Schweiger, S. Koller, P.R. Raimann, T. Wöhrle, C. Wurm, M. Winter, Silicon/graphite composite electrodes for high-capacity anodes: influence of binder chemistry on cycling stability, Electrochem. Solid St. 11 (2008) A76–A80.

[159] N. Dimov, S. Kugino, M. Yoshio, Mixed silicon–graphite composites as anode material for lithium ion batteries: influence of preparation conditions on the properties of the material, J. Power Sources 136 (2004) 108–114.

[160] Y. Zhang, X.G. Zhang, H.L. Zhang, Z.G. Zhao, F. Li, C. Liu, H.M. Cheng, Composite anode material of silicon/graphite/carbon nanotubes for Li-ion batteries, Electrochim. Acta 51 (2006) 4994–5000.

[161] N. Liu, H. Wu, M.T. McDowell, Y. Yao, C. Wang, Y. Cui, A yolk-Shell design for stabilized and scalable Li-ion battery alloy anodes, Nano Lett. 12 (2012) 3315–3321.

[162] S. Guo, X. Hu, Y. Hou, Z. Wen, Tunable synthesis of yolk–shell porous silicon@carbon for optimizing Si/C-based anode of lithium-ion batteries, ACS Appl. Mater. Interfaces 9 (2017) 42084–42092.

[163] L.Y. Yang, H.Z. Li, J. Liu, Z.Q. Sun, S.S. Tang, M. Lei, Dual yolk-shell structure of carbon and silica-coated silicon for high-performance lithium-ion batteries, Sci. Rep. 5 (2015) 10908.

[164] J. Xie, L. Tong, L. Su, Y. Xu, L. Wang, Y. Wang, Core-shell yolk-shell Si@C@void@C nanohybrids as advanced lithium ion battery anodes with good electronic conductivity and corrosion resistance, J. Power Sources 342 (2017) 529–536.

[165] N. Liu, Z. Lu, J. Zhao, M.T. McDowell, H.-W. Lee, W. Zhao, Y. Cui, A pomegranate-inspired nanoscale design for large-volume-change lithium battery anodes, Nat. Nanotechnol. 9 (2014) 187–192.

[166] H. Wu, G. Chan, J.W. Choi, I. Ryu, Y. Yao, M.T. McDowell, S.W. Lee, A. Jackson, Y. Yang, L. Hu, Y. Cui, Stable cycling of double-walled silicon nanotube battery anodes through solid–electrolyte interphase control, Nat. Nanotechnol. 7 (2012) 310–315.

[167] Y. Yao, N. Liu, M.T. McDowell, M. Pasta, Y. Cui, Improving the cycling stability of silicon nanowire anodes with conducting polymer coatings, Energ. Environ. Sci. 5 (2012) 7927–7930.

[168] Y. NuLi, B. Wang, J. Yang, X. Yuan, Z. Ma, Cu5Si–Si/C composites for lithium-ion battery anodes, J. Power Sources 153 (2006) 371–374.

[169] J.P. Maranchi, A.F. Hepp, A.G. Evans, N.T. Nuhfer, P.N. Kumta, Interfacial properties of the a-Si/Cu: active–inactive thin-film anode system for lithium-ion batteries, J. Electrochem. Soc. 153 (2006) A1246–A1253.

[170] C.-M. Park, J.-H. Kim, H. Kim, H.-J. Sohn, Li-alloy based anode materials for Li secondary batteries, Chem. Soc. Rev. 39 (2010) 3115–3141.

[171] W. Luo, Y. Wang, S. Chou, Y. Xu, W. Li, B. Kong, S.X. Dou, H.K. Liu, J. Yang, Critical thickness of phenolic resin-based carbon interfacial layer for improving long cycling stability of silicon nanoparticle anodes, Nano Energy 27 (2016) 255–264.

[172] R. Kuruba, M.K. Datta, K. Damodaran, P.H. Jampani, B. Gattu, P.P. Patel, P.M. Shanthi, S. Damle, P.N. Kumta, Guar gum: structural and electrochemical characterization of natural polymer based binder for silicon–carbon composite rechargeable Li-ion battery anodes, J. Power Sources 298 (2015) 331–340.

[173] W.-R. Liu, M.-H. Yang, H.-C. Wu, S.M. Chiao, N.-L. Wu, Enhanced cycle life of Si anode for Li-ion batteries by using modified elastomeric binder, Electrochem. Solid St. 8 (2005) A100.

[174] H. Buqa, M. Holzapfel, F. Krumeich, C. Veit, P. Novák, Study of styrene butadiene rubber and sodium methyl cellulose as binder for negative electrodes in lithium-ion batteries, J. Power Sources 161 (2006) 617–622.

[175] J. Li, R.B. Lewis, J.R. Dahn, Sodium carboxymethyl cellulose, Electrochem. Solid St. 10 (2007) A17.

[176] J.S. Bridel, T. Azaïs, M. Morcrette, J.M. Tarascon, D. Larcher, Key parameters governing the reversibility of Si/carbon/CMC electrodes for Li-ion batteries, Chem. Mater. 22 (2010) 1229–1241.

[177] B. Lestriez, S. Bahri, I. Sandu, L. Roué, D. Guyomard, On the binding mechanism of CMC in Si negative electrodes for Li-ion batteries, Electrochem. Commun. 9 (2007) 2801–2806.

[178] L. Chen, X. Xie, J. Xie, K. Wang, J. Yang, Binder effect on cycling performance of silicon/carbon composite anodes for lithium ion batteries, J. Appl. Electrochem. 36 (2006) 1099–1104.

[179] A. Magasinski, B. Zdyrko, I. Kovalenko, B. Hertzberg, R. Burtovyy, C.F. Huebner, T.F. Fuller, I. Luzinov, G. Yushin, Toward efficient binders for Li-ion battery Si-based anodes: polyacrylic acid, ACS Appl. Mater. Interfaces 2 (2010) 3004–3010.

[180] I. Kovalenko, B. Zdyrko, A. Magasinski, B. Hertzberg, Z. Milicev, R. Burtovyy, I. Luzinov, G. Yushin, A major constituent of brown algae for use in high-capacity Li-ion batteries, Science 334 (2011) 75.
[181] G. Liu, S. Xun, N. Vukmirovic, X. Song, P. Olalde-Velasco, H. Zheng, V.S. Battaglia, L. Wang, W. Yang, Polymers with tailored electronic structure for high capacity lithium battery electrodes, Adv. Mater. 23 (2011) 4679–4683.
[182] M. Wu, X. Xiao, N. Vukmirovic, S. Xun, P.K. Das, X. Song, P. Olalde-Velasco, D. Wang, A.Z. Weber, L.-W. Wang, V.S. Battaglia, W. Yang, G. Liu, Toward an ideal polymer binder design for high-capacity battery anodes, J. Am. Chem. Soc. 135 (2013) 12048–12056.
[183] H. Zhao, Z. Wang, P. Lu, M. Jiang, F. Shi, X. Song, Z. Zheng, X. Zhou, Y. Fu, G. Abdelbast, X. Xiao, Z. Liu, V.S. Battaglia, K. Zaghib, G. Liu, Toward practical application of functional conductive polymer binder for a high-energy lithium-ion battery design, Nano Lett. 14 (2014) 6704–6710.
[184] H. Zhao, N. Yuca, Z. Zheng, Y. Fu, V.S. Battaglia, G. Abdelbast, K. Zaghib, G. Liu, High capacity and high density functional conductive polymer and SiO anode for high-energy lithium-ion batteries, ACS Appl. Mater. Interfaces 7 (2015) 862–866.
[185] J. Lopez, Z. Chen, C. Wang, S.C. Andrews, Y. Cui, Z. Bao, The effects of cross-linking in a supramolecular binder on cycle life in silicon microparticle anodes, ACS Appl. Mater. Interfaces 8 (2016) 2318–2324.
[186] S. Choi, T.-W. Kwon, A. Coskun, J.W. Choi, Highly elastic binders integrating polyrotaxanes for silicon microparticle anodes in lithium ion batteries, Science 357 (2017) 279.
[187] Y.K. Jeong, T.-W. Kwon, I. Lee, T.-S. Kim, A. Coskun, J.W. Choi, Hyperbranched β-cyclodextrin polymer as an effective multidimensional binder for silicon anodes in lithium rechargeable batteries, Nano Lett. 14 (2014) 864–870.
[188] Y.K. Jeong, T.-W. Kwon, I. Lee, T.-S. Kim, A. Coskun, J.W. Choi, Millipede-inspired structural design principle for high performance polysaccharide binders in silicon anodes, Energ. Environ. Sci. 8 (2015) 1224–1230.
[189] T.-W. Kwon, J.W. Choi, A. Coskun, The emerging era of supramolecular polymeric binders in silicon anodes, Chem. Soc. Rev. 47 (2018) 2145–2164.
[190] T.-W. Kwon, Y.K. Jeong, E. Deniz, S.Y. AlQaradawi, J.W. Choi, A. Coskun, Dynamic crosslinking of polymeric binders based on host–guest interactions for silicon anodes in lithium ion batteries, ACS Nano 9 (2015) 11317–11324.
[191] T.-W. Kwon, Y.K. Jeong, I. Lee, T.-S. Kim, J.W. Choi, A. Coskun, Systematic molecular-level design of binders incorporating Meldrum's acid for silicon anodes in lithium rechargeable batteries, Adv. Mater. 26 (2014) 7979–7985.
[192] Z. Chen, C. Wang, J. Lopez, Z. Lu, Y. Cui, Z. Bao, High-areal-capacity silicon electrodes with low-cost silicon particles based on spatial control of self-healing binder, Adv. Energy Mater. 5 (2015) 1401826.
[193] B.C.K. Tee, C. Wang, R. Allen, Z. Bao, An electrically and mechanically self-healing composite with pressure- and flexion-sensitive properties for electronic skin applications, Nat. Nanotechnol. 7 (2012) 825–832.
[194] C. Wang, H. Wu, Z. Chen, M.T. McDowell, Y. Cui, Z. Bao, Self-healing chemistry enables the stable operation of silicon microparticle anodes for high-energy lithium-ion batteries, Nat. Chem. 5 (2013) 1042–1048.
[195] Z. Xu, J. Yang, T. Zhang, Y. Nuli, J. Wang, S.-I. Hirano, Silicon microparticle anodes with self-healing multiple network binder, Joule 2 (2018) 950–961.
[196] M.-H. Ryou, J. Kim, I. Lee, S. Kim, Y.K. Jeong, S. Hong, J.H. Ryu, T.-S. Kim, J.-K. Park, H. Lee, J.W. Choi, Mussel-inspired adhesive binders for high-performance silicon nanoparticle anodes in lithium-ion batteries, Adv. Mater. 25 (2013) 1571–1576.

[197] P. Cordier, F. Tournilhac, C. Soulié-Ziakovic, L. Leibler, Self-healing and thermoreversible rubber from supramolecular assembly, Nature 451 (2008) 977–980.
[198] J.-T. Li, Z.-Y. Wu, Y.-Q. Lu, Y. Zhou, Q.-S. Huang, L. Huang, S.-G. Sun, Water soluble binder, an electrochemical performance booster for electrode materials with high energy density, Adv. Energy Mater. 7 (2017) 1701185.
[199] Z.-H. Wu, J.-Y. Yang, B. Yu, B.-M. Shi, C.-R. Zhao, Z.-L. Yu, Self-healing alginate–carboxymethyl chitosan porous scaffold as an effective binder for silicon anodes in lithium-ion batteries, Rare Metals 38 (2019) 832–839.
[200] J. Liu, Q. Zhang, T. Zhang, J.-T. Li, L. Huang, S.-G. Sun, A robust ion-conductive biopolymer as a binder for Si anodes of lithium-ion batteries, Adv. Funct. Mater. 25 (2015) 3599–3605.
[201] M. Murase, N. Yabuuchi, Z.-J. Han, J.-Y. Son, Y.-T. Cui, H. Oji, S. Komaba, Crop-derived polysaccharides as binders for high-capacity silicon/graphite-based electrodes in lithium-ion batteries, ChemSusChem 5 (2012) 2307–2311.
[202] L. Wei, C. Chen, Z. Hou, H. Wei, Poly (acrylic acid sodium) grafted carboxymethyl cellulose as a high performance polymer binder for silicon anode in lithium ion batteries, Sci. Rep. 6 (2016) 19583.
[203] E. Attia, F. Hassan, M. Li, D. Luo, A. Elkamel, Z. Chen, Multifunctional nano-architecting of Si electrode for high-performance lithium-ion battery anode, J. Electrochem. Soc. 166 (2019) A2776–A2783.
[204] J.-I. Lee, H. Kang, K.H. Park, M. Shin, D. Hong, H.J. Cho, N.-R. Kang, J. Lee, S.M. Lee, J.-Y. Kim, C.K. Kim, H. Park, N.-S. Choi, S. Park, C. Yang, Amphiphilic graft copolymers as a versatile binder for various electrodes of high-performance lithium-ion batteries, Small 12 (2016) 3119–3127.
[205] C. Hwang, S. Joo, N.-R. Kang, U. Lee, T.-H. Kim, Y. Jeon, J. Kim, Y.-J. Kim, J.-Y. Kim, S.-K. Kwak, H.-K. Song, Breathing silicon anodes for durable high-power operations, Sci. Rep. 5 (2015) 14433.
[206] J. Song, M. Zhou, R. Yi, T. Xu, M.L. Gordin, D. Tang, Z. Yu, M. Regula, D. Wang, Interpenetrated gel polymer binder for high-performance silicon anodes in lithium-ion batteries, Adv. Funct. Mater. 24 (2014) 5904–5910.
[207] M.T. Jeena, J.-I. Lee, S.H. Kim, C. Kim, J.-Y. Kim, S. Park, J.-H. Ryu, Multifunctional molecular design as an efficient polymeric binder for silicon anodes in lithium-ion batteries, ACS Appl. Mater. Interfaces 6 (2014) 18001–18007.
[208] Z. Liu, S. Han, C. Xu, Y. Luo, N. Peng, C. Qin, M. Zhou, W. Wang, L. Chen, S. Okada, In situ crosslinked PVA–PEI polymer binder for long-cycle silicon anodes in Li-ion batteries, RSC Adv. 6 (2016) 68371–68378.
[209] L. Yue, L. Zhang, H. Zhong, Carboxymethyl chitosan: a new water soluble binder for Si anode of Li-ion batteries, J. Power Sources 247 (2014) 327–331.
[210] C. Chen, S.H. Lee, M. Cho, J. Kim, Y. Lee, Cross-linked chitosan as an efficient binder for Si anode of Li-ion batteries, ACS Appl. Mater. Interfaces 8 (2016) 2658–2665.
[211] S. Aoki, Z.-J. Han, K. Yamagiwa, N. Yabuuchi, M. Murase, K. Okamoto, T. Kiyosu, M. Satoh, S. Komaba, Acrylic acid-based copolymers as functional binder for silicon/graphite composite electrode in lithium-ion batteries, J. Electrochem. Soc. 162 (2015) A2245–A2249.
[212] Y. Bie, J. Yang, Y. Nuli, J. Wang, Natural karaya gum as an excellent binder for silicon-based anodes in high-performance lithium-ion batteries, J. Mater. Chem. A 5 (2017) 1919–1924.
[213] S. Klamor, M. Schröder, G. Brunklaus, P. Niehoff, F. Berkemeier, F.M. Schappacher, M. Winter, On the interaction of water-soluble binders and nano silicon particles: alternative

binder towards increased cycling stability at elevated temperatures, Phys. Chem. Chem. Phys. 17 (2015) 5632–5641.
[214] T.M. Higgins, S.-H. Park, P.J. King, C. Zhang, N. McEvoy, N.C. Berner, D. Daly, A. Shmeliov, U. Khan, G. Duesberg, V. Nicolosi, J.N. Coleman, A commercial conducting polymer as both binder and conductive additive for silicon nanoparticle-based lithium-ion battery negative electrodes, ACS Nano 10 (2016) 3702–3713.
[215] M. Ling, Y. Xu, H. Zhao, X. Gu, J. Qiu, S. Li, M. Wu, X. Song, C. Yan, G. Liu, S. Zhang, Dual-functional gum arabic binder for silicon anodes in lithium ion batteries, Nano Energy 12 (2015) 178–185.
[216] M. Ling, H. Zhao, X. Xiaoc, F. Shi, M. Wu, J. Qiu, S. Li, X. Song, G. Liu, S. Zhang, Low cost and environmentally benign crack-blocking structures for long life and high power Si electrodes in lithium ion batteries, J. Mater. Chem. A 3 (2015) 2036–2042.
[217] S. Lim, H. Chu, K. Lee, T. Yim, Y.-J. Kim, J. Mun, T.-H. Kim, Physically cross-linked polymer binder induced by reversible acid–base interaction for high-performance silicon composite anodes, ACS Appl. Mater. Interfaces 7 (2015) 23545–23553.
[218] D. Liu, Y. Zhao, R. Tan, L.-L. Tian, Y. Liu, H. Chen, F. Pan, Novel conductive binder for high-performance silicon anodes in lithium ion batteries, Nano Energy 36 (2017) 206–212.
[219] S.-M. Kim, M.H. Kim, S.Y. Choi, J.G. Lee, J. Jang, J.B. Lee, J.H. Ryu, S.S. Hwang, J.-H. Park, K. Shin, Y.G. Kim, S.M. Oh, Poly(phenanthrenequinone) as a conductive binder for nano-sized silicon negative electrodes, Energ. Environ. Sci. 8 (2015) 1538–1543.
[220] S.-J. Park, H. Zhao, G. Ai, C. Wang, X. Song, N. Yuca, V.S. Battaglia, W. Yang, G. Liu, Side-chain conducting and phase-separated polymeric binders for high-performance silicon anodes in lithium-ion batteries, J. Am. Chem. Soc. 137 (2015) 2565–2571.
[221] Y. Pan, S. Gao, F. Sun, H. Yang, P. Cao, Polymer Binders Constructed through Dynamic Noncovalent Bonds for High-Capacity Silicon-Based Anodes, Oak Ridge National Lab. (ORNL), Oak Ridge, TN (United States), 2019.
[222] C. Zhang, Q. Chen, X. Ai, X. Li, Q. Xie, Y. Cheng, H. Kong, W. Xu, L. Wang, M.-S. Wang, Conductive polyaniline doped with phytic acid as a binder and conductive additive for a commercial silicon anode with enhanced lithium storage properties, J. Mater. Chem. A 8 (2020) 16323–16331.
[223] H. Woo, K. Park, J. Kim, A.J. Yun, S. Nam, B. Park, 3D Meshlike polyacrylamide hydrogel as a novel binder system via in situ polymerization for high-performance Si-based electrode, Adv. Mater. Interfaces 7 (2020) 1901475.
[224] C. Li, C. Liu, K. Ahmed, Z. Mutlu, Y. Yan, I. Lee, M. Ozkan, C.S. Ozkan, Kinetics and electrochemical evolution of binary silicon–polymer systems for lithium ion batteries, RSC Adv. 7 (2017) 36541–36549.
[225] S.-L. Chou, Y. Pan, J.-Z. Wang, H.-K. Liu, S.-X. Dou, Small things make a big difference: binder effects on the performance of Li and Na batteries, Phys. Chem. Chem. Phys. 16 (2014) 20347–20359.
[226] S. Kim, Y.K. Jeong, Y. Wang, H. Lee, J.W. Choi, A "sticky" mucin-inspired DNA-polysaccharide binder for silicon and silicon–graphite blended anodes in lithium-ion batteries, Adv. Mater. 30 (2018) 1707594.
[227] G. Zhang, Y. Yang, Y. Chen, J. Huang, T. Zhang, H. Zeng, C. Wang, G. Liu, Y. Deng, A quadruple-hydrogen-bonded supramolecular binder for high-performance silicon anodes in lithium-ion batteries, Small 14 (2018) 1801189.
[228] S. Lim, K. Lee, I. Shin, A. Tron, J. Mun, T. Yim, T.-H. Kim, Physically cross-linked polymer binder based on poly (acrylic acid) and ion-conducting poly (ethylene glycol-co-benzimidazole) for silicon anodes, J. Power Sources 360 (2017) 585–592.

[229] J. Liu, Q. Zhang, Z.-Y. Wu, J.-H. Wu, J.-T. Li, L. Huang, S.-G. Sun, A high-performance alginate hydrogel binder for the Si/C anode of a Li-ion battery, Chem. Commun. 50 (2014) 6386–6389.

[230] N.-S. Choi, K.H. Yew, W.-U. Choi, S.-S. Kim, Enhanced electrochemical properties of a Si-based anode using an electrochemically active polyamide imide binder, J. Power Sources 177 (2008) 590–594.

[231] S. Huang, J. Ren, R. Liu, M. Yue, Y. Huang, G. Yuan, The progress of novel binder as a non-ignorable part to improve the performance of Si-based anodes for Li-ion batteries, Int. J. Energy Res. 42 (2018) 919–935.

[232] T. Yim, S.J. Choi, Y.N. Jo, T.-H. Kim, K.J. Kim, G. Jeong, Y.-J. Kim, Effect of binder properties on electrochemical performance for silicon-graphite anode: method and application of binder screening, Electrochim. Acta 136 (2014) 112–120.

[233] M. Wu, X. Song, X. Liu, V. Battaglia, W. Yang, G. Liu, Manipulating the polarity of conductive polymer binders for Si-based anodes in lithium-ion batteries, J. Mater. Chem. A 3 (2015) 3651–3658.

[234] H. Zhao, Y. Fu, M. Ling, Z. Jia, X. Song, Z. Chen, J. Lu, K. Amine, G. Liu, Conductive polymer binder-enabled SiO–Sn$_x$Co$_y$C$_z$ anode for high-energy lithium-ion batteries, ACS Appl. Mater. Interfaces 8 (2016) 13373–13377.

[235] H. Zhao, Y. Wei, R. Qiao, C. Zhu, Z. Zheng, M. Ling, Z. Jia, Y. Bai, Y. Fu, J. Lei, Conductive polymer binder for high-tap-density nanosilicon material for lithium-ion battery negative electrode application, Nano Lett. 15 (2015) 7927–7932.

[236] J. Xu, Q. Zhang, Y.-T. Cheng, High capacity silicon electrodes with nafion as binders for lithium-ion batteries, J. Electrochem. Soc. 163 (2015) A401.

[237] R.R. Garsuch, D.-B. Le, A. Garsuch, J. Li, S. Wang, A. Farooq, J. Dahn, Studies of lithium-exchanged nafion as an electrode binder for alloy negatives in lithium-ion batteries, J. Electrochem. Soc. 155 (2008) A721.

[238] Y. Chen, S. Zeng, J. Qian, Y. Wang, Y. Cao, H. Yang, X. Ai, Li$^+$-conductive polymer-embedded nano-Si particles as anode material for advanced Li-ion batteries, ACS Appl. Mater. Interfaces 6 (2014) 3508–3512.

[239] M.H.T. Nguyen, E.-S. Oh, Improvement of the characteristics of poly (acrylonitrile-butylacrylate) water-dispersed binder for lithium-ion batteries by the addition of acrylic acid and polystyrene seed, J. Electroanal. Chem. 739 (2015) 111–114.

[240] L. Luo, Y. Xu, H. Zhang, X. Han, H. Dong, X. Xu, C. Chen, Y. Zhang, J. Lin, Comprehensive understanding of high polar polyacrylonitrile as an effective binder for Li-ion battery nano-Si anodes, ACS Appl. Mater. Interfaces 8 (2016) 8154–8161.

[241] M.K. Dufficy, S.A. Khan, P.S. Fedkiw, Galactomannan binding agents for silicon anodes in Li-ion batteries, J. Mater. Chem. A 3 (2015) 12023–12030.

[242] Q. Yuan, F. Zhao, Y. Zhao, Z. Liang, D. Yan, Reason analysis for graphite-Si/SiO$_x$/C composite anode cycle fading and cycle improvement with PI binder, J. Solid State Electrochem. 18 (2014) 2167–2174.

[243] J.S. Kim, W. Choi, K.Y. Cho, D. Byun, J. Lim, J.K. Lee, Effect of polyimide binder on electrochemical characteristics of surface-modified silicon anode for lithium ion batteries, J. Power Sources 244 (2013) 521–526.

[244] L. Zhu, H. Hou, D. Zhao, S. Liu, W. Ye, S. Chen, M. Hanif, One step annealing treatment for performance improvement of silicon anode based on polyimide binder, Int. J. Electrochem. Sci. 10 (2015) 9547–9555.

[245] S. Uchida, M. Mihashi, M. Yamagata, M. Ishikawa, Electrochemical properties of non-nanosilicon negative electrodes prepared with a polyimide binder, J. Power Sources 273 (2015) 118–122.

[246] K.L. Browning, R.L. Sacci, M. Doucet, J.F. Browning, J.R. Kim, G.M. Veith, The study of the binder poly(acrylic acid) and its role in concomitant solid–electrolyte interphase formation on Si anodes, ACS Appl. Mater. Interfaces 12 (2020) 10018–10030.

[247] Y. Jin, S. Li, A. Kushima, X. Zheng, Y. Sun, J. Xie, J. Sun, W. Xue, G. Zhou, J. Wu, F. Shi, R. Zhang, Z. Zhu, K. So, Y. Cui, J. Li, Self-healing SEI enables full-cell cycling of a silicon-majority anode with a coulombic efficiency exceeding 99.9%, Energ. Environ. Sci. 10 (2017) 580–592.

Further reading

U.S. Vogl, S.F. Lux, E.J. Crumlin, Z. Liu, L. Terborg, M. Winter, R. Kostecki, The mechanism of SEI formation on a single crystal Si(100) electrode, J. Electrochem. Soc. 162 (2015) A603–A607.

Y.M. Lee, J.Y. Lee, H.-T. Shim, J.K. Lee, J.-K. Park, SEI layer formation on amorphous Si thin electrode during precycling, J. Electrochem. Soc. 154 (2007) A515.

Y. Jin, N.-J.H. Kneusels, P.C.M.M. Magusin, G. Kim, E. Castillo-Martínez, L.E. Marbella, R.N. Kerber, D.J. Howe, S. Paul, T. Liu, C.P. Grey, Identifying the structural basis for the increased stability of the solid electrolyte interphase formed on silicon with the additive fluoroethylene carbonate, J. Am. Chem. Soc. 139 (2017) 14992–15004.

W. Schalkwijk, B. Scosati, Advances in Lithium-Ion Batteries, Springer, Boston, MA, USA, 2002.

M. Gauthier, T.J. Carney, A. Grimaud, L. Giordano, N. Pour, H.-H. Chang, D.P. Fenning, S.F. Lux, O. Paschos, C. Bauer, F. Maglia, S. Lupart, P. Lamp, Y. Shao-Horn, Electrode–electrolyte interface in Li-ion batteries: current understanding and new insights, J. Phys. Chem. Lett. 6 (2015) 4653–4672.

A.M. Haregewoin, A.S. Wotango, B.-J. Hwang, Electrolyte additives for lithium ion battery electrodes: progress and perspectives, Energ. Environ. Sci. 9 (2016) 1955–1988.

Chapter 7

Active/inactive phases, binders, and impact of electrolyte

Chen Fang and Gao Liu
Energy Storage and Distributed Resources Division, Lawrence Berkeley National Laboratory, Berkeley, CA, United States

7.1 Active and inactive phases of Si-based materials

Unlike carbon- and graphite-based materials, silicon materials form stable silicon dioxide (SiO_2) layer on the surface of Si materials. The effects of surface oxide on Si had not come into attention to battery research until the early 2000s. One of the first research works in this field points out the SiO_2 surface layer as a major impedance factor both to lower the rate performance of Si materials and to reduce specific capacity due to particle isolation [1, 2] (Fig. 7.1). This impedance effect is significantly amplified when the Si active materials are nanoparticles, where surface effect dominates. As it is unlikely to eliminate SiO_2 layer entirely from the Si particle surface, the SiO_2 layer partially defines the Si materials and electrochemical performance of Si electrode. The impedance effects of SiO_2 layer diminish when the oxide layer thickness is below 2 nm [2]. It is also further discovered that the SiO_2 layers transformed into lithium ion-conducting silicate materials (e.g., Li_4SiO_4, $Li_2Si_2O_5$) during lithiation process [2]. Therefore the SiO_2 layer tends to irreversibly absorb lithium ion during the first and early lithiation process. When SiO_2 thickness is beyond 4 nm, it is insulating enough to prevent the Si core from lithiation reaction, when significant specific capacity reduction is observed.

The silicon oxide can also be nonstoichiometric based on the Si oxidation states, as usually expressed by SiO_x ($0 < x < 2$). The nonstoichiometric SiO_x has been made into particles as Si electrode materials (Fig. 7.2). SiO_x materials can be considered as a mixture of Si nano-domains and SiO_2 nano-domains, although the actual materials are made in different morphologies and chemical compositions [3]. The first lithiation of SiO_x particles transforms the materials into lithiated Si nano-domains and lithium silicate ion-conducting nano-domains composite particles [4]. The advantages of the SiO_x materials are the larger particle sizes, and the smaller and more uniform volume expansion

266 PART | III Electrolytes and SEI issues

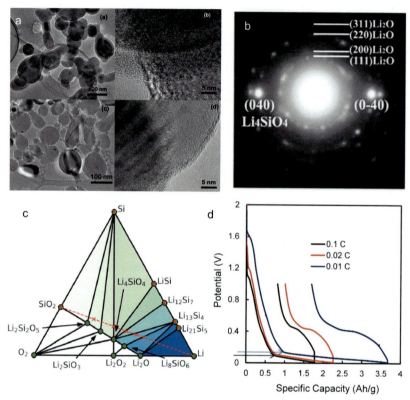

FIG. 7.1 The impact and transformation of SiO$_2$ layer on Si nanoparticles. (A) TEM images of Si nanoparticles. (a/b) A core-shell structure of SiO$_2$-covered Si particles and (c/d) after 30 min of HF etching, takes away most of the SiO$_2$ surface layer. (B) The lithiation reaction of SiO$_2$ induces a mixture of Li-Si-O glass with Li$_4$SiO$_4$ and Li$_2$O crystalline phases in the shell. (C) Si$_x$Li$_y$O phase diagram, including calculated amorphous states and crystalline phases from the Materials Project, shown as red X's and green circles, respectively. (D) The initial lithiation/delithiation curves of electrodes based on SiO$_2$-covered Si at different C rates, showing slower lithiation can overcome the insulating effect of SiO$_2$ layer. *(From (A) S. Xun, X. Song, M.E. Grass, D.K. Roseguo, Z. Liu, V.S. Battaglia, G. Liu, Improved initial performance of Si nanoparticles by surface oxide reduction for lithium-ion battery application, Electrochem. Solid State Lett. 14 (5) (2011) A61–A63. (B) Y.F. Zhang, Y.J. Li, Z.Y. Wang, K.J. Zhao, Lithiation of SiO2 in Li-ion batteries: in situ transmission electron microscopy experiments and theoretical studies. Nano Lett. 14 (12) (2014) 7161–7170. (C) E. Sivonxay, M. Aykol, K.A. Persson, The lithiation process and Li diffusion in amorphous SiO$_2$ and Si from first-principles, Electrochim. Acta 331 (2020) 9. (D) S. Xun, X. Song, L. Wang, M.E. Grass, Z. Liu, V.S. Battaglia, G. Liu, The effects of native oxide surface layer on the electrochemical performance of Si nanoparticle-based electrodes, J. Electrochem. Soc. 158 (12) (2011) A1260–A1266.)*

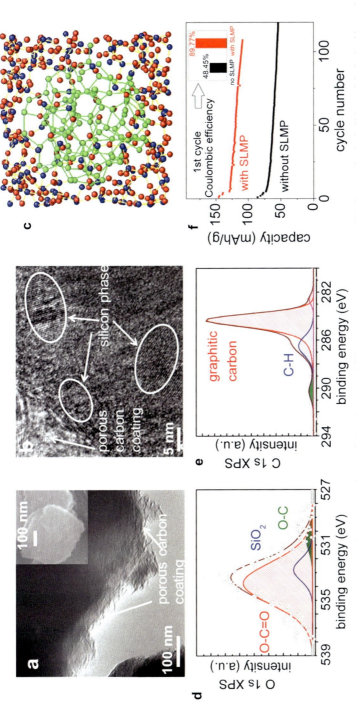

FIG. 7.2 SiO$_x$ materials with carbon coating. (A) TEM image of the carbon-coated SiO$_x$ particle. The inset shows the SEM image of a single SiO$_x$ particle, both with a scale bar of 100 nm. (B) HRTEM image of the SiO$_x$ particle. The white circles indicate the active Si nano-domains within the inactive SiO$_2$ matrix. The carbon coating is at the outer surface of the SiO$_x$ particles. (C) Atomic model of disproportionated amorphous SiO$_x$. Reconstructed heterostructure model of amorphous SiO$_x$. The inner part corresponds to an amorphous Si cluster and the outer part is amorphous SiO$_2$ matrix. The *blue*, *red*, and *green* circles denote Si and O in amorphous SiO$_2$ and Si in the Si cluster, respectively. (D/E) The observation of SiO$_2$ and carbon coating from XPS spectra was acquired simultaneously at the outer surface of the SiO$_x$ materials, with a depth penetration of 2–4 nm. Open circles are experimental data, while solid lines are the results of a Gaussian deconvolution fitting. (F) SiO$_x$/NMC full cell performance with or without the Stabilized Lithium Metal Powder (SLMP) prelithiation capacity-enhancement additive, two cycles at C/20, two cycles at C/10, and then C/3. *(From (A, B, D–F) H. Zhao, Z.H. Wang, P. Lu, M. Jiang, F.F. Shi, X.Y. Song, Z.Y. Zheng, X. Zhou, Y.B. Fu, G. Abdelbast, X.C. Xiao, Z. Liu, V.S. Battaglia, K. Zaghib, G. Liu, Toward practical application of functional conductive polymer binder for a high-energy lithium-ion battery design, Nano Lett. 14 (11) (2014) 6704–6710. (C) A. Hirata, S. Kohara, T. Asada, M. Arao, C. Yogi, H. Imai, Y.W. Tan, T. Fujita, M.W. Chen, Atomic-scale disproportionation in amorphous silicon monoxide, Nat. Commun. 7 (2016).)*

during lithiation process. Large particle size reduces surface side-reactions toward electrolyte to improve both cycle and calendar life of the battery [5]. However, the major issues with SiO_x are its low electronic conductivities and large first lithiation irreversible capacity loss. Most of the battery grade SiO_x are carbon coated to increase electronic conductivity and to reduce surface reactions [5, 6]. However, the large first-cycle irreversible capacity loss is an intrinsic property of the SiO_x materials, and therefore, balancing cathode loading or introducing additional lithium is a necessary process for batteries with high SiO_x content anodes. The SiO_x material is one of the most successful classes of Si-based materials for the commercial Si-based anodes.

Carbon coating on Si particles is widely used to provide a stable interface between Si and electrolyte, and also to improve electronic conductivities for the Si-based materials. The most common and low-cost method for producing the carbon coating is through the use of organic precursor. The carbonization of the organic materials left a layer of the carbon materials on the surface of the Si particles. However, it is demonstrated that a plasma carbon coating process yields better battery performance for the carbon-coated Si materials [6]. This is because the plasma-deposited carbon is more graphitic in nature and is therefore more electronically and ionically conductive. More graphitic carbon also promotes better solid electrolyte interface (SEI) formation on its surface. These factors contribute to improved conductivity and reduced surface side-reactions to improve overall battery performance. The application of low-cost organic precursor coating followed by thermal decomposition is a widely used method for yielding carbon coating of the Si materials [7]. Although the decomposition left carbon on the Si particle surface, the carbon coatings are mostly amorphous with significant structure defects. Therefore the imperfect coating compromises electrochemical performance of the coated Si materials. The ideal choices of organic precursors are the structurally more organized molecules, such as π-π stack structure prior to carbonization. One example is the pyrene-type polymer materials that can form coating layers on the surface of Si particle. The pyrene moieties already form π-π stacking structures in the precursor so that carbonation temperature tends to be low to achieve significant graphitization [8] (Fig. 7.3). This type of coating provides superb protections to the Si surface. Carbon coating of Si materials has become a standard process to prevent the Si surface reactions and enhance conductivities. Most of the commercial Si particles for battery applications are carbon coated.

As carbon layer provides superb surface protection for Si materials, another major strategy to reduce Si surface reaction and accommodate Si volume expansion is through carbon composite with Si [7]. Carbon materials are both electrically conductive and able to provide lithium ion storage capabilities. Composite carbon with Si nanomaterials produces reduced surface area composite particles with high capacity and reduced overall volume expansion (Fig. 7.4). The petroleum pitch is one of the preferred candidates as it is readily

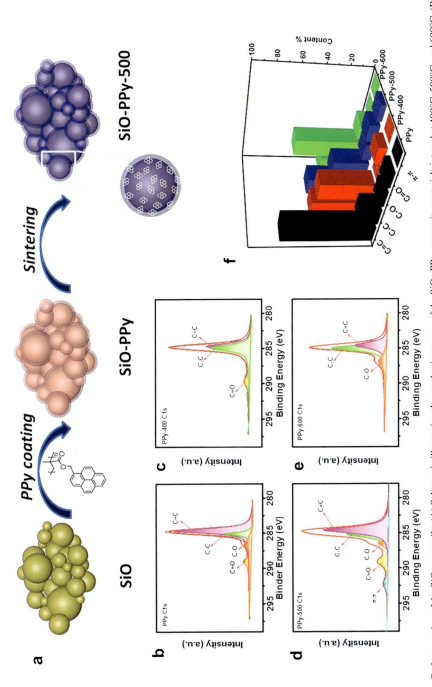

FIG. 7.3 Carbon coating of the SiO$_x$ materails. (A) Schematic illustrating the synthesis process of the SiO$_x$-PPy composite materails. (B–E) X-ray photoelectron spectroscopy (XPS) spectra of PPy coating and sintering at different temperatures. (F) The corresponding carbon bonding composition of untreated PPy and pyrolyzed PPy. High π-π bonding composition at 500°C corresponding to better electrochemical performance. *(From S. Fang, N. Li, T.Y. Zheng, Y.B. Fu, X.Y. Song, T. Zhang, S.P. Li, B. Wang, X.G. Zhang, G. Liu, Highly graphitized carbon coating on SiO with a π-π stacking precursor polymer for high performance lithium-ion batteries, Polymers 10 (6) (2018).)*

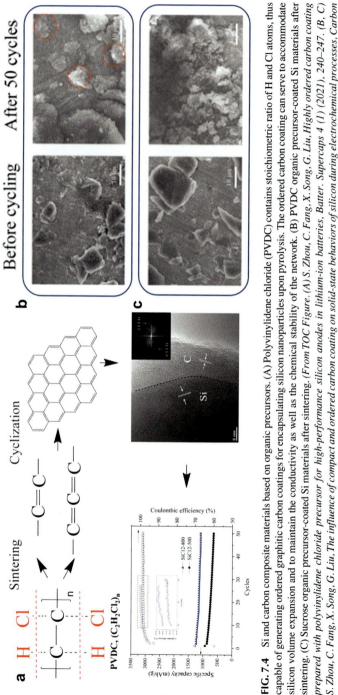

FIG. 7.4 Si and carbon composite materials based on organic precursors. (A) Polyvinylidene chloride (PVDC) contains stoichiometric ratio of H and Cl atoms, thus capable of generating ordered graphitic carbon coatings for encapsulating silicon nanoparticles upon pyrolysis. The ordered carbon coating can serve to accommodate silicon volume expansion and to maintain the conductivity as well as the chemical stability of the network. (B) PVDC organic precursor-coated Si materials after sintering. (C) Sucrose organic precursor-coated Si materials after sintering. (From TOC Figure. (A) S. Zhou, C. Fang, X. Song, G. Liu, Highly ordered carbon coating prepared with polyvinylidene chloride precursor for high-performance silicon anodes in lithium-ion batteries, Batter. Supercaps 4 (1) (2021), 240–247. (B, C) S. Zhou, C. Fang, X. Song, G. Liu, The influence of compact and ordered carbon coating on solid-state behaviors of silicon during electrochemical processes, Carbon Energy 2 (2020) 143–150.)

graphitized to provide superb solid-state ion transport property and readily passivates the surface. Most of the petroleum pitches have very high content of carbon element in the high 90%. The carbonization process tends to produce highly ordered carbon in this case. An alternative approach to demonstrate the importance of graphitization of carbon coating is through the use of 100% carbon retention precursors with a chemical preferred leaving group. In this case, a homogenous coating by the precursor can be retained, and graphitization can be achieved at much lower temperature [7,9].

Besides the SiO_x and Si/C composite approaches to form active and inactive composites, the Si nanoparticle can also be dispersed in an elastic organic matrix to form secondary particles [10] (Fig. 7.5). This approach requires the organic matrix to be both ionically and electronically conductive. Si electrodes based on this type of composite particles with very high areal capacity and loading have been fabricated with excellent rate performance. The conductive polymer in the composite also provides surface protections to the Si nanoparticles. Better control of Si surface reactions can make this method commercially successful.

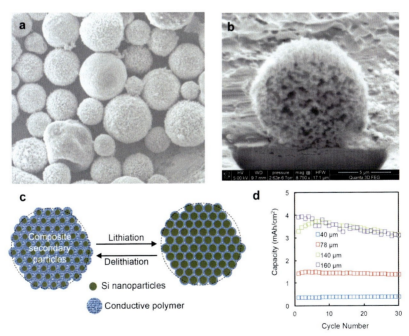

FIG. 7.5 (A) SEM images of the Si/conductive polymer composite secondary particles. (B) A single particle and its cross-section image. (C) Secondary composite particle has a stable dimension during Si lithiation and delithiation. The dotted lines are for visual guide. (D) Cycling performance of the electrodes with low to high loading. *(From S.D. Xun, B. Xiang, A. Minor, V. Battaglia, G. Liu, Conductive polymer and silicon composite secondary particles for a high area-loading negative electrode, J. Electrochem. Soc. 160 (9) (2013) A1380–A1383.)*

7.2 Silicon electrode binders

Si electrode binders were initially selected from the traditional lithium-ion electrode binders, such as polyvinylidene fluoride (PVDF), carboxymethyl cellulose (CMC), and (tyrene-butadiene rubber) SBR. It was quickly found that PVDF does not function well as Si electrode binders, due to the large volume changes of the Si material during the lithiation and delithiation process. Later, it was discovered that CMC is a better binder due to the formation of ester bonds between the carboxylate functional groups of the CMC molecules and hydroxide surface functional groups on the Si material [11]. This understanding has led to the further exploration of lithium polyacrylate (LiPAA) as Si electrode binders as the PAA has the highest density of carboxylate functional groups of any known polymers. Although both CMC and LiPAA binders have demonstrated superior performance than that of the PVDF binders, the battery life performance of the Si-based electrode is still far inferior to the carbon-based electrode. The drive to use high-capacity Si materials has led the research communities to use the bottom-up approach to design electrode binders for Si materials [12, 13]. Effective Si binders are considered as one of the enabling materials to bring Si-based battery to the market.

The critical requirements for electrode binders include strong adhesion to ensure good mechanical properties of the laminate, high and uniform electronic conductivities across the electrode, and excellent ion transport at the interface of active materials, binder, and electrolyte [14]. The binder also needs to have electrochemical stability in the voltage window at its operational potential. The large volume change of Si particles between charge and discharge has created a special challenge for adhesion and electric connection for the Si electrode. Majority of the bottom-up design of Si electrode binders is focused on solving these two major issues. We will use a class of specially designed functional conductive polymer adhesive binders to demonstrate the principles for the binder development for Si materials [12, 15, 16] (Fig. 7.6).

7.2.1 Adhesion

The adhesion between a polymer binder adhesive and particles can be achieved via three main mechanisms, including mechanical interlocking, Van der Waals force molecular adhesion, and chemical bonding. Both mechanical interlocking and Van der Waals force are physical adhesions between the particle substrate and the polymer adhesives. PVDF adhesion of electrode materials mainly uses the above-mentioned two types of forces. On the other hand, chemical bonding adhesion is to form permanent covalent or ionic bonds between the particle substrate and polymer adhesives. The chemical bonding energy is in the range of $\sim 100\,\text{kJ/mol}$, compared to that of the Van der Waals force of $\sim 10\,\text{kJ/mol}$, thus the chemical bonding adhesion is an order of magnitude stronger than the Van der Waals force adhesion. As for Si materials with large volume change during

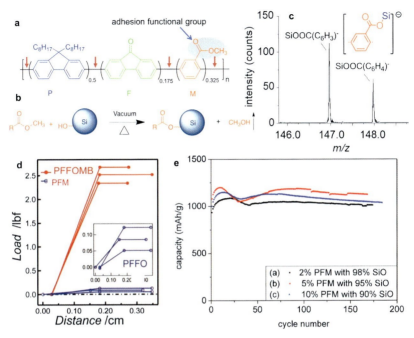

FIG. 7.6 (A) Chemical structure of the PFM functional conductive polymer binder. The red arrows indicate the most likely chemical bonds to be broken during the ionization process in the TOF-SIMS measurement. The blue shadow highlights the ester group that forms adhesion bonds with the Si-OH surface. (B) Transesterification reaction between the ester functional group of the PFM binder and the Si-OH surface group, which provides strong adhesion, during the electrode drying process. (C) TOF-SIMS result of the binder and active materials interface that shows the evidence of chemical bonding between the binder and the active materials. (D) Adhesion force plot of the Si/polymer binder electrode laminates, measured with scotch tape peel tests. Insert is the amplified PF data. The ester adhesion group containing PFM has very high adhesion strength. (E) Galvanostatic cycling performance of SiO anodes with 2%, 5%, and 10% by weight of a PFM binder using lithium counter electrodes at a C/10 rate (200 mA/g). The voltage is between 0.05 and 1 V. (From H. Zhao, Z.H. Wang, P. Lu, M. Jiang, F.F. Shi, X.Y. Song, Z.Y. Zheng, X. Zhou, Y.B. Fu, G. Abdelbast, X.C. Xiao, Z. Liu, V.S. Battaglia, K. Zaghib, G. Liu, Toward practical application of functional conductive polymer binder for a high-energy lithium-ion battery design, Nano Lett. 14 (11) (2014) 6704–6710. G. Liu, S.D. Xun, N. Vukmirovic, X.Y. Song, P. Olalde-Velasco, H.H. Zheng, V.S. Battaglia, L.W. Wang, W.L. Yang, Polymers with tailored electronic structure for high capacity lithium battery electrodes, Adv. Mater. 23 (40) (2011) 4679.)

cycling, the strong adhesion capability is a required property of a designed binder. One such example is the carboxylate-based adhesives such as CMC and PAA mentioned earlier. However, the density of adhesive groups, spatial factor as well as the activation energy of bond formation reaction need to be met to form chemical bonds between the Si particles and the adhesives. The carboxylate groups or carboxylate esters are incorporated in the binder to

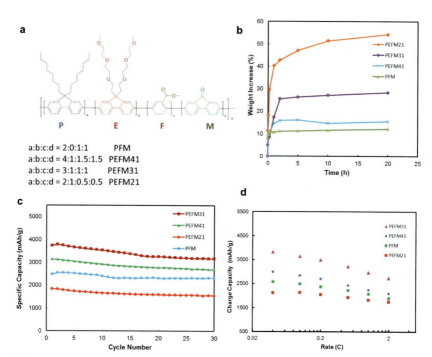

FIG. 7.7 (A) Molecular structure and the relative molar ratio of four functional polymer binders. (B) The swelling tests of polymer films in the EC/DEC (1:1) electrolyte. (C) Cycling performance of polymer/Si electrodes without any conductive additive. (D) The rate performance of the polymer/Si electrodes with four polymer binders at various charge rates. The discharge rate was kept at C/25. *(From M.Y. Wu, X.Y. Song, X.S. Liu, V. Battaglia, W.L. Yang, G. Liu, Manipulating the polarity of conductive polymer binders for Si-based anodes in lithium-ion batteries, J. Mater. Chem. A 3 (7) (2015) 3651–3658.)*

provide adhesion to the Si materials (Fig. 7.7). Other functional groups with strong adhesion force to Si surface such as dopamine are also explored to enhance the mechanical properties of the Si electrodes [17]. Fundamental studies of the adhesion force between the polymers with different adhesion groups have shown marked differences in the adhesion forces depending on both the chemistry of the adhesion groups as well as the density of the adhesion groups on the polymer chain [17, 18]. The formation of the chemical bonds between the Si particles and the functional adhesive binder normally requires dehydration at elevated temperatures. This dehydration step is achieved through the last step of the electrode coating and heat-drying process. Normally, the electrode will be dried at 150°C to remove water or NMP solvent during the electrode coating process. This high temperature drying condition also promotes dehydration, esterification, or transesterification of the binders with the Si surface hydroxide to form chemical bonds. This reaction has been confirmed in the composite electrode by time-of-flight secondary ion mass spectrometry (TOF-SIMS) method [5]. The extent of the formation of these bonds is still a subject to debate, because the formation of the bonds is at the particle surface, leading to stiffing of the polymer chains next to the surface, hence sterically hinders the further formation of the ester bonds between the polymers and the Si surfaces. However, the stable Si composite electrode performance based on those binders has demonstrated the effectiveness of this approach.

7.2.2 Ionic conductivity

Although the primary goal of binders is to hold all the electrode particles together and to withstand repeated volume changes of Si material during charge and discharge process, the binder also needs to ensure the mobility of lithium ions in the electrode, so the Si particles can be lithiated and delithiated [15]. Electrode binders need to have limited amount of electrolyte absorption capability when in contact with the electrolyte, e.g., PVDF has ~30% electrolyte intake and swelling. The electrolyte-swelled binder provides a lithium ion pass for the ions in the electrolyte to access the Si particles. However, the limited swelling of the binder could create significant charge transfer impedance of the electrode. Although higher electrolyte adsorption increases ion conductivity of the binder, excessive binder swelling normally reduces adhesion and lowers electrical conductivity of the electrode laminate [15]. The binder design, to a large extent, seeks a trade-off between adhesion and swelling, to maintain mechanical property, ion transport, and electrical conductivity of the electrode [19] (Fig. 7.7). The polymer binder molecule components allowing ion transport, among the most explored, include polar ethylene oxide segments and carboxylate function groups.

Polyethyleneoxide (PEO) is used as an ion-conducting polymer, which can dissolve lithium salt to form an ion-conducting amorphous PEO and lithium salt phase. The lithium ion is conducted through segmental motion of the PEO chains. Liquid electrolyte can significantly swell the PEO into a gel state, leading to enhance lithium ion transport but reduced mechanical strength. When PEO segments are incorporated in the designed polymer binders such as poly(9,9-dioctylfluorene-*co*-fluorenone-*co*-methylbenzoic ester) (PFM), they increase the electrolyte swelling of the binder. The ion transport is significantly enhanced through the swelled binder. However, the increase of PEO content in the electrolyte also leads to dramatic swelling of the binder, and significantly weakens mechanical properties. A series of studies of designed polymers with PFM backbone and PEO side-chain have demonstrated, when all other conditions are the same, higher PEO content increases electrolyte intake. The electrode performance peaks at electrolyte intake around 35%, and decreases when further PEO is incorporated into the binder structures [19]. As LiPAA binder is used for Si electrode for its superb adhesion, the lithium carboxylate functional groups also provide excellent electrolyte absorbent capabilities. The binders with lithium carboxylate groups enhance electrolyte absorption as well as provide additional lithium ion in the electrode.

7.2.3 Electrical conductivity

The long- and short-range electrical conductivity in the electrode is critical for proper operation of the battery. Traditional lithium ion battery electrode uses highly electrically conductive acetylene black with binder to ensure good electric interface conductivity in the electrode [14]. The electrical paths through current collector to the active material particles are formed via acetylene black nanosized particles and are maintained by electrode binders. Volume changes of the active materials over time tend to weaken these connections, leading to electrode failure. Electrically conductive polymer binders such as PFM for Si anode electrode are uniquely suitable for large volume change Si materials [5, 12, 20] (Fig. 7.8). As discussed, the strong adhesion of binder can be designed through the formation of covalent bond between the functional groups on polymer binders and the Si particles. Unlike nonconductive polymer binders and acetylene black conductive media, where the electrical connection is maintained by the physical adhesion of acetylene black nanoparticle with Si materials, conductive polymers with the designed functional groups can bind with the Si surface to provide molecular-level connection through strong covalent bonds. This proves to be extremely effective in accommodating the volume change of Si materials and provide electrical connections to the Si particles.

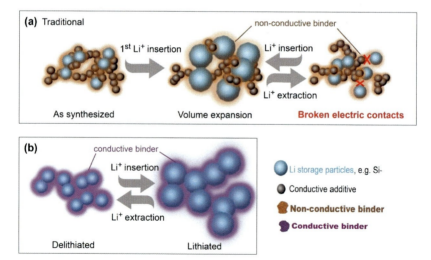

FIG. 7.8 Schematics of the technical approaches to address volume change issue in battery materials. (A) Traditional approaches use acetylene black (AB) as the conductive additive and PVDF polymer as mechanical binder. (B) Conductive polymer with dual functionality, as a conductor and binder, could keep both electrical and mechanical integrity of the electrode during the battery cycles. *(From G. Liu, S.D. Xun, N. Vukmirovic, X.Y. Song, P. Olalde-Velasco, H.H. Zheng, V.S. Battaglia, L.W. Wang, W.L. Yang, Polymers with tailored electronic structure for high capacity lithium battery electrodes. Adv. Mater. 23 (40) (2011) 4679.)*

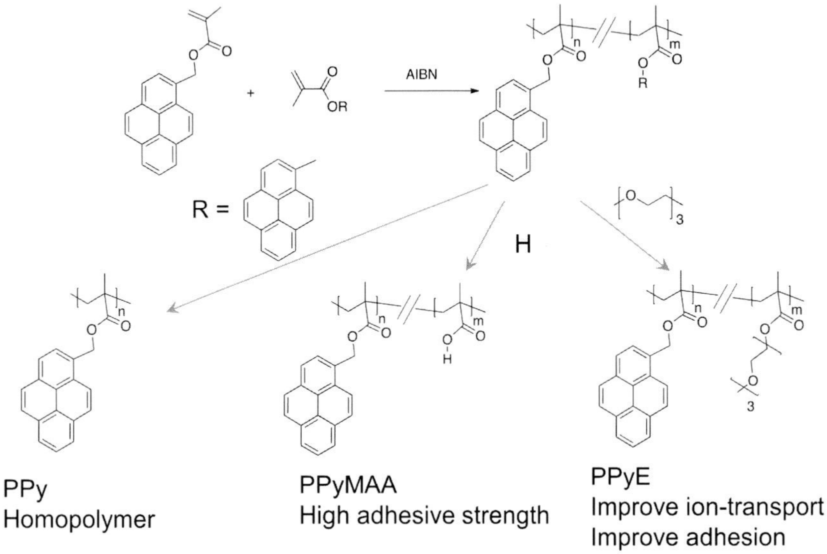

Active/inactive phases, binders, and impact of electrolyte **Chapter | 7 277**

PPy
Homopolymer

PPyMAA
High adhesive strength

PPyE
Improve ion-transport
Improve adhesion

FIG. 7.9 The functional polymer binders can be made with a radical polymerization process to incorporate functionalities at the side chains. The polymer backbone is methacrylate. The side chains can be electrically conducting pyrene moiety along with adhesion enhancement carboxylic acid groups and electrolyte-absorbing ethyleneoxide moiety. This generical polymer structure can be a fast route to screen binder functionalities for a variety of electrode materials.

7.2.4 A holistic approach

Combining all three critical properties of adhesion, ion transport, and electronical conductivity together provides a significant opportunity for bottom-up functional binder design for Si materials [5, 17, 18] (Fig. 7.9). Instead of the backbone electrically conducting polymers, a more flexible design of the side chain-conducting polymethacrylate pyrene polymer provides an easier maneuver to fast screening the binder functionalities. The backbone polymer binder structures are made from methacrylate radical polymerization processes, which are flexible to include functionalities in the monomers, to synthesize at lower cost, and to scale up. This side chain-conducting polymer approach provides a facile platform to screen functionalities for the binders for Si, and provides an avenue for low-cost mass production and commercialization.

7.3 Electrolytes and additives

The practical application of silicon anodes in lithium ion batteries (LIBs) is a difficult challenge due to the significant volume change of silicon (reversable volume expansion over three-fold upon full lithiation); this is a destructive factor for silicon anode as it undermines the electrode's integrity and also interrupts the stability of the SEI. These phenomena are responsible for rapid capacity loss, low Coulombic efficiency, and low cycling stability of Si-based LIBs [7,21,22]. A wide range of sophisticated silicon nanostructures have been proposed and produced, such as yolk-shell structures, hollow spheres, nanotubes, and porous composites, to mitigate the negative effects of silicon volume variation [23–27]. However, currently, the preparation of such nanostructures is rather complex and challenging in scale-up for actual industrial applications.

The electrolytes of LIBs make up a significant portion of the battery's materials and play important roles for battery's performance. The electrolytes used with silicon anodes are more than merely medium for lithium ion transportation because the electrolytes, commonly composed of organic species, participate in the electrochemical processes of the electrodes since the batteries commonly operate beyond the electrolyte's stable potential window. The electrolyte components have key impacts on formation, composition, and interfacial properties of SEIs, which dictate the long-term cycling behaviors of the cells [28–30] like that in the case of graphite anodes [31].

Generally, the electrolytes for silicon anodes in LIBs contain a lithium salt for Li^+ conductivity, a solvent medium that is composed of either small-molecule liquids or polymer composites, and optionally additives for tuning the properties of the bulk electrolyte. The development and application of new electrolyte components including organic solvents and additives are very convenient and practical with the current battery industry thanks to the wide application of graphite-based LIBs, which share similar needs of electrolytes as silicon anodes. Therefore it is strongly demanded to identify advanced electrolyte systems along with electrolyte composition optimization and new electrolyte development. This section covers the current progress on the selection and development of the electrolytes. In addition, analytical methods for reactions of electrolyte will be introduced as well.

7.3.1 Carbonate electrolytes and lithium salts

Small organic carbonate molecules are the most common choices of electrolytes for silicon anodes. These carbonate species include cyclic-shaped solvents such as ethylene carbonate (EC) and propylene carbonate (PC) and linear-shaped solvents like dimethyl carbonate (DMC) and ethyl methyl carbonate (EMC). The structures of solvents and additives described in this section are given in Fig. 7.10. Generally speaking, the cyclic solvents allow good solubility of lithium salts, while the linear solvents have low viscosity that realizes high ion conductivity of the cell [30]. Consequently, mixed carbonate solvents with both cyclic and linear molecules (e.g., EC/EMC electrolyte) are often employed to achieve optimum battery performance. The carbonate liquids are relatively stable, allowing satisfying compatibility with common electrochemical procedures and battery components. Therefore carbonate electrolytes have become the standard electrolytes for silicon-based batteries. On the other hand, the

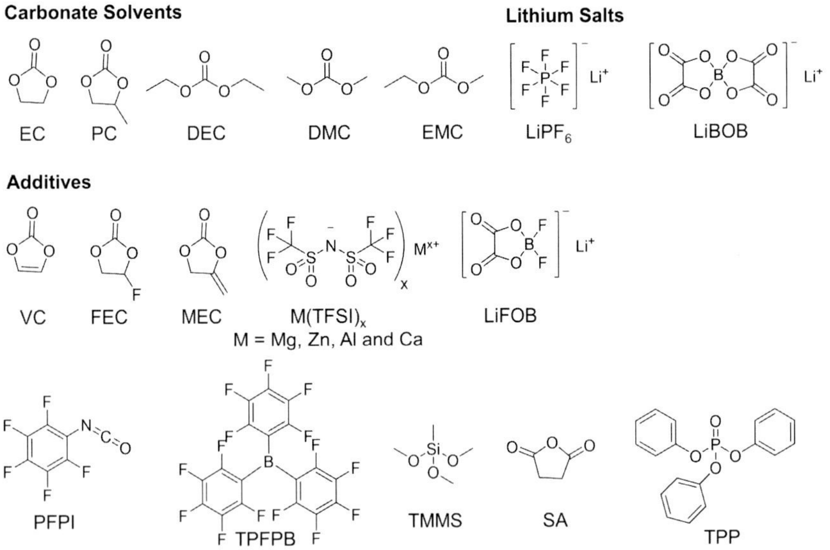

Carbonate Solvents

EC PC DEC DMC EMC

Lithium Salts

LiPF$_6$ LiBOB

Additives

VC FEC MEC M(TFSI)$_x$ (M = Mg, Zn, Al and Ca) LiFOB

PFPI TPFPB TMMS SA TPP

FIG. 7.10 Structures of carbonate solvents, lithium salts, and additives.

silicon-based batteries do operate below the stable potential window of the carbonate electrolytes, and as a result, the decomposition of the carbonate electrolytes is inevitable. The breakdown of EC-based electrolytes has been extensively studied [32, 33]. It has been widely acknowledged that EC produces lithium alkyl carbonate anion radical intermediate via one-electron ring-opening mechanism, which then generates lithium ethylene dicarbonate (LEDC), PEO, and other organic derivatives (Fig. 7.11A). The precise decomposition products of carbonate electrolytes are still under constant investigation. For example, the assignment of LEDC has proven skeptical, where a structurally similar molecule, lithium ethylene mono-carbonate (LEMC), has recently been proposed as the true EC decomposition product in the SEI layers [34].

It is worthwhile briefly noting here that other small-molecule organic liquids have also been explored as electrolytes for silicon anodes. For instance, ether-based solvents, which are widely applied in many other lithium battery systems [35,36], can also be used with silicon anodes. It has been demonstrated that 1,3-dioxolane (DOL) electrolytes led to improved high-rate cycling stability compared to that of EC/EMC electrolyte [37]. Fluorinated solvents that have been employed for lithium-sulfur batteries in recent years [38] were also applied to Si-based LIBs. A di-2,2,2-trifluoroethyl carbonate (TFEC) electrolyte system delivered a high initial reversible capacity of 2644 mAh/g and a fading rate of 0.064% per cycle capacity (over 300 cycles) for silicon nanoparticle anode at high lithium salt concentration [39]. Such an electrolyte system also has the advantage of nonflammability.

280 PART | III Electrolytes and SEI issues

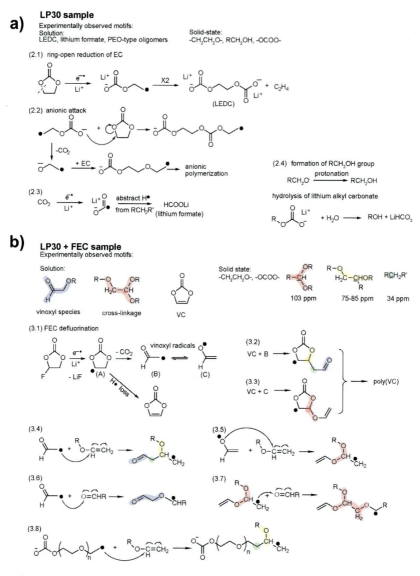

FIG. 7.11 (A) Decomposition of EC electrolyte. (B) Proposed reaction pathways of EC-based electrolyte containing FEC additive. *(From Y. Jin, N.-J.H. Kneusels, P.C.M.M. Magusin, G. Kim, E. Castillo-Martínez, L.E. Marbella, et al., Identifying the structural basis for the increased stability of the solid electrolyte interphase formed on silicon with the additive fluoroethylene carbonate, J. Am. Chem. Soc. 139 (42) (2017) 14992–15004.)*

The common choice of lithium salt for Si-based LIBs is lithium hexafluorophosphate (LiPF$_6$) [30]. LiPF$_6$ usually gives the best cell performance but its application is strongly limited by the safety concerns because of the sensitivity of LiPF$_6$ toward moisture and high temperature, which can readily lead to generation of fluoride compounds. Only a few alternative lithium salt options are available. Lithium bis(oxalato)borate (LiBOB) has been shown to be an effective cosalt with LiPF$_6$, where LiBOB helped to improve the reversable capacity of silicon/graphite/carbon composite electrodes [40]. Lithium bis(fluorosulfonyl)imide (LiFSI) has been examined as an independent lithium salt in a pouch cell system with silicon/graphite negative electrode and Li$_{1.03}$(Ni$_{0.5}$Co$_{0.2}$Mn$_{0.3}$)$_{0.97}$O$_2$ positive electrode [41]. It was demonstrated that LiFSI had comparable performance to that of LiPF$_6$ with the presence of fluoroethylene carbonate (FEC) in the electrolyte system, and that LiFSI showed best performance with vinylene carbonate (VC).

7.3.2 Electrolyte additives

An SEI film would be formed on the silicon anode with the precycles of the cell operation, which are often carried out at lower charge/discharge rates. The SEI layer helps to protect the silicon anode from parasite reactions with the electrolytes in later cycles, which contributes to the good long-term cycling performance of the cells [42]. Therefore the electrochemical decomposition of electrolytes for SEI formation is a key research topic. Due to the limited choices of electrolyte and lithium salt compounds for LIBs, additives have been a convenient and effective way for tuning the properties of the electrolyte systems [28,29]. Additives are foreign components commonly applied to the electrolyte bulk in small or minor quantities (e.g., 0.1–10 wt%) without altering the basic compositions of the solvents yet playing a critical role in achieving the desired functionalities of the electrolytes often by decomposition of the additive molecules during charge/discharge cycles. As the additives commonly serve as sacrificial components, there are no strict compatibility requirements for choice of additives, which allows a broad range of potential candidates. Additives also provide a cost-efficient approach to engineer the electrolytes because the additives are usually applied in small quantity, making up a rather small portion of the battery's total cost while contributing significantly to the battery's performance.

7.3.2.1 Ethylene carbonate derivative additives

Vinylene carbonate (VC) additive has been used to improve the performance of silicon anode for over a decade [43]. It has been demonstrated that VC can realize a few favored properties of the SEI layers, including smooth morphology, capability to isolate the electrode from the electrolyte, decreased LiF concentration, and invariant impedance during cell operation [43]. The VC additive does not change the major decomposition products of the base carbonate electrolytes in the SEI layers, which include polycarbonates and organic/inorganic lithium salts (e.g., ROCO$_2$Li, LiF). Due to these effects, the cycling performance of the silicon anode can be significantly increased by VC additive. The impact of VC on silicon anode is similar to that of graphite anode, where VC has been shown to promote passivation of carbon anode in carbonate electrolytes [44]. The main reduction product of VC has been proposed to be poly(VC) and CO$_2$, where poly(VC) is formed by radical-initiated polymerization of the double bond while CO$_2$ is produced by breakdown of the VC molecules in radical reactions and can further react to yield Li$_2$CO$_3$ [45].

Fluoroethylene carbonate (FEC) is the most frequently employed electrolyte additive that can significantly improve the long-term cycling performance of the silicon anodes, and it is also the most studied additive [45–48]. It should be noted that the precise electrochemical reaction pathways of FEC are still under continuing investigation, but it is generally believed that the involvement of FEC additive leads to formation of thin and dense SEI layers with high LiF content, which can suppress the decomposition of the electrolyte at the interphase and can improve the cycling performance of silicon electrodes. As FEC can generate VC in situ, the decomposition products of FEC also include poly(VC), CO$_2$, and LiCO$_3$, with addition of LiF that originates from the fluoride group of FEC [45]. It has also been demonstrated that the reaction of FEC is stoichiometric, and therefore, FEC is believed to produce low-mass poly(VC) products. A solid-state Nuclear Magnetic Resonance (ssNMR) study [32] of silicon nanowire electrodes confirmed that defluorination of FEC during battery operation yields VC and vinoxyl species, and also revealed that FEC promotes formation of crosslinked PEO species rather than linear soluble PEOs that are commonly observed with EC-based electrolytes. The possible reaction pathways and observed organic species fragments are shown in Fig. 7.11B.

Methylene ethylene carbonate (MEC), an additive with a structure resembling FEC, has been examined for silicon particle electrodes with 1.2M LiPF$_6$ EC/DEC (ethylene carbonate/diethyl carbonate 1:1) electrolyte [49]. MEC helped to improve the capacity retention of the silicon anode from 46% to 73% compared to the base electrolyte. The MEC-containing electrolyte was found to produce SEI layers with poly(MEC) content, which helped to prevent reduction reactions of the base electrolyte molecules.

7.3.2.2 Salt additives

A second salt could be employed in the electrolyte for tuning the electrolyte's properties. M(TFSI)$_x$ salts with different cations (M = Mg, Zn, Al, and Ca) have been examined as additives in the conventional 1.2M LiPF$_6$ 3:7 w/w EC:EMC electrolyte [50]. With the additive salts, Li-M-Si ternaries could be generated, which have been shown stable toward the electrolyte solvents. The substitution of M cations into the lithium silicide lattices helped to stabilize this highly reactive species at the interphase.

Lithium salts can also serve as additives. Lithium difluorooxalatoborate (LiFOB) and LiBOB at small doses (1–5 wt%) were found beneficial for cycling performance of the cells, which was rationalized by the rich content of oxalates and lithium fluorophosphate and the low concentration of LiF generated in the surface films with these additives on the electrode surfaces [51].

7.3.2.3 Additives based on reactive organic molecules

A highly reactive fluorine-rich molecule, pentafluorophenyl isocyanate (PFPI), was evaluated as additive for EC/DEC electrolyte in Si/NMC-111 full cells [52]. The addition of merely 2 wt% PFPI led to notably higher capacity retention and Coulombic efficiency (CE) after about 100 cycles. It was demonstrated that PFPI was a major component in the SEI layers, which reduced the immobilization of lithium ions at the interphase.

Another fluorine-rich additive, tris(pentafluorophenyl) borane (TPFPB), was reported for silicon thin-film anodes with 1M $LiClO_4$ EC/DEC (1/1) electrolyte [53]. The TPFPB additive realized doubled capacity retention after 100 cycles with higher Coulombic efficiencies. TPFPB improved the stability of the SEI layers and suppressed surface pulverization possibly by LiF formation and electrolyte decomposition mitigation.

Silane additives containing alkoxy groups such as trimethoxy methyl silane (TMMS) have been utilized to mediate the surface properties of silicon anodes [54]. The lithium ion in the electrolyte can react with the oxide and hydroxyl groups on the silicon anode surface, which is a common cause of the irreversible capacity loss of silicon anodes. Alkoxy silane additives can chemically react with the active groups on the silicon anode surfaces to passivate the electrode/electrolyte interface. Such a chemical process helps to improve the cycling performance of the cell. In addition, the silane additives can also prevent extensive mass accumulation on the silicon electrode that originates from irreversible reduction reactions of the bulk electrolyte.

Succinic anhydride (SA) was found favorable for tuning the properties of SEIs on amorphous silicon thin-film electrode [55]. A small amount of 3 wt% SA additive can effectively suppress the decomposition of $LiPF_6$ on the silicon electrode surface and promote the formation of SEI layers rich in hydrocarbons and Li_2CO_3, which realized good electrochemical performance of the silicon electrode with 80% capacity retention at 1800 mAh/g after 100 cycles.

For safety considerations of the electrolyte system, triphenyl phosphate (TPP) has been examined as an additive with a flame-retardant nature [56]. This additive was found to have satisfying compatibility with Si/C electrodes and has good electrochemical performance with FEC.

7.3.3 Polymer electrolytes

The common problems associated with the conventional small organic molecule electrolytes for application of LIBs include volatility, flammability, and limited adaptability to various temperatures. Therefore solid-state electrolytes have been widely investigated to replace the common organic liquids [30].

In addition, solid-state electrolytes can intrinsically reduce the repeated SEI formation reactions and thus result in higher cycling stability. For example, $Li_{6.4}La_3Zr_{1.4}Ta_{0.6}O_{12}$ (LLZTO) solid electrolytes have been examined for Si thin-film electrodes, which delivered 2200 mAh/g silicon capacity with retention of 72% after 100 cycles [57]. Recently, polymer-based solid electrolytes have attracted a lot of research interest as promising candidates for realizing solid electrolytes with high ionic conductivity and good performance at room temperature [58].

PEO is a widely investigated class of polymer matrix because the PEOs have high dielectric constant to allow good affinity to the dissolved lithium salts with sufficient flexibility of long-range segmental motion for lithium ion conductivity [59]. On the other hand, the ion conductivity of PEOs is generally very low at room temperature ($<10^{-6}$ S/cm) due to the semicrystalline properties of PEO molecules. Therefore the ion conductivity can be tuned by adjusting the structures of PEOs. A graft copolymer all-solid electrolyte based on copolymer of poly-(methoxy/hexadecal-poly(ethylene glycol) methacrylate) (termed 'PMH', see Fig. 7.12A) yielded a conductivity $10^{-3.9}$ S/cm at 30°C and $10^{-3.1}$ S/cm at 80°C [60]. This copolymer had a polymethacrylate backbone with mixed side chains of $-(OCH_2CH_2)_nOCH_3$ and $-(CH_2CH_2O)_nOC_{16}H_{33}$. It was demonstrated that the nonpolar hexadecal units interrupted the ordered

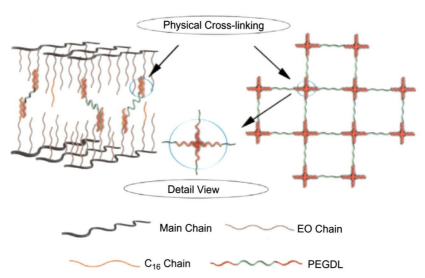

FIG. 7.12 Schematic demonstration of crosslinked PMH network. (A) PMH network with low concentration of C_{16} chains. (B) PMH network with high concentration of C_{16} chains. (C) PMH network physically crosslinked with PEGDL linker. (D) Self-crosslinked PEGDL network. *(From X. Zuo, X.-M. Liu, F. Cai, H. Yang, X.-D. Shen, G. Liu, Enhanced performance of a novel gel polymer electrolyte by dual plasticizers, J. Power Sources 239 (2013) 111–121.)*

alignment of the oligoethylene oxide chains and increased their mobility, which helped to increase the conductivity of this copolymer electrolyte.

To further increase the ion conductivity of polymer electrolytes, organic liquids can be applied as plasticizer into the solid polymer electrolyte to form gel polymer electrolytes (GPE) [59]. For the PMH system described earlier, a high concentration of the hexadecal chains in the network can aggregate to increase the mechanical stability of the network (Fig. 7.12B) yet still cannot satisfy the demand of high solvent content [60]. Therefore the PHM system was physically crosslinked with polyethylene glycol dilaurate (PEGDL) molecules (Fig. 7.12C), which then sustained 60 wt% nonvolatile polyethylene glycol dimethyl ether (PEGDME) and delivered a high conductivity of 8.2×10^{-3} S/cm at 30°C with $LiClO_4$ salt [61]. The PEGDL crosslinkers improved the mechanical properties of the PHM network by interactions with both the hexadecal chain units of PHM (Fig. 7.12C) and with themselves (Fig. 7.12D).

The GPE could also serve as a functional component in the battery for protection of the silicon materials. Recently, a cyanoresin-based organogel electrolyte has been investigated in high-loading silicon anode systems [62]. The cyanoresin organogel helped to reduce the amount of binder and conductive agents to allow 80 wt% loading of the active silicon content. A 75% capacity retention was obtained for this silicon anode after 150 cycles, with no capacity fading even raising to 5C rate. This system also realized a high rate capacity of about 1000 mAh/g at 10C rate, which was 10-fold higher than that of the conventional liquid electrolyte. It was proposed that cyanoresin organogel helped to maintain the integrity of the silicon electrode by providing additional cohesion between the silicon particles, which mitigated crack evolution and electrode thickness changes.

A different strategy for higher lithium ion transport efficiency is to chemically anchor the counter-anions onto the polymer backbones of the polymer electrolyte [63]. Silica nanoparticles can be grafted with polymer layers containing anion groups as host for lithium ions (Fig. 7.13A). The copolymerization of styrenesulfonate monomers and poly(ethylene oxide)-containing monomers realized nanocomposite electrolyte with an ion conductivity of about 10^{-6} S/cm (Fig. 7.13B) [64]. It was proposed that the PEO groups in the nanocomposite electrolyte assisted the dissociation of lithium ion from the sulfonate counter-anion on the polymer backbones. In comparison, silica particles grafted with only poly(lithium 4-styrenesulfonate) brushes exhibited much lower conductivity. Similarly, TFSI analogue groups have been grafted onto silica nanoparticles (Si-C5NTfLi, see Fig. 7.13C), which were mixed with 500 molecular mass polyethylene glycol dimethyl ether for lithium ion transportation [65]. The optimized ion conductivity for this system was 3×10^{-5} S/cm at 80°C. It was proposed that high ion concentration and flexibility of the polymer structures adjacent to the particles are crucial for good ion conductivity of this composite system.

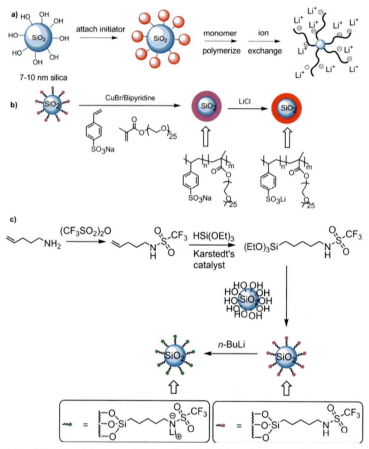

FIG. 7.13 Silicon particle composite electrolyte. (A) Schematic presentation of synthesis and structure of polyelectrolyte-decorated nanoparticles. (B) Copolymerization of sodium 4-styrenesulfonate and poly(ethylene oxide)methacrylate from silica nanoparticles and subsequent lithiation. (C) Synthetic route for Si-C5NHTf and lithiation to Si-C5NTfLi. *(From (A, B) H. Zhao, Z. Jia, W. Yuan, H. Hu, Y. Fu, G.L. Baker, G. Liu, Fumed silica-based single-ion nanocomposite electrolyte for lithium batteries, ACS Appl. Mater. Interfaces 7 (34) (2015) 19335–19341. (C) H. Zhao, F. Asfour, Y. Fu, Z. Jia, W. Yuan, Y. Bai, M. Ling, H. Hu, G. Baker, G. Liu, Plasticized polymer composite single-ion conductors for lithium batteries, ACS Appl. Mater. Interfaces 7 (34) (2015) 19494–19499.)*

7.3.4 Ionic liquid electrolytes

Ionic liquids (IL) are a class of organic salts commonly composed of a large organic cation and a small inorganic anion. The bulk size of the organic cations reduced the melting temperature of the compound to allow its existence in liquid form at room temperature. ILs have been shown compatible with the carbonaceous materials as anodes in LIBs [66, 67]. ILs as electrolytes for

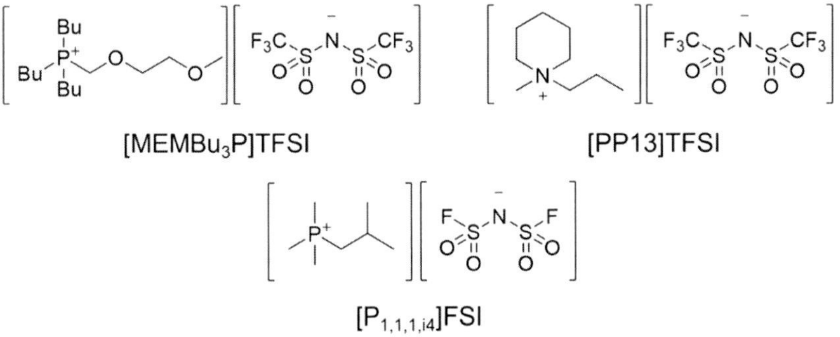

FIG. 7.14 Structures of some ILs.

silicon- or carbon-based batteries have the advantages of nonflammability, neglectable volatility, and high electrochemical stability, which address the common safety concerns of conventional organic electrolyte solvents. A few IL structures covered in this section are shown in Fig. 7.14.

One common anion used for ILs as electrolytes for Si-based LIBs is bis(trifluoromethylsulfonyl)imide (TFSI) [68,69]. A thick-film silicon electrode system has been examined with IL electrolyte that was composed of 1-methoxyethoxymethyl (tri-n-butyl)phosphonium [MEMBu$_3$P] cation and TFSI anion [70]. This IL system delivered an initial discharge capacity of 3450 mAh/g. In comparison, a commercially available IL system, 1-methyl-1-propylpiperidinium ([PP13])/TFSI, only realized an initial discharge capacity of 1900 mAh/g, although having similar retained capacity as that of [MEMBu$_3$P]/TFSI IL after 100 cycles. The conventional carbonate electrolyte propylene carbonate (PC) led to an initial discharge capacity of 3390 mAh/g but rather rapid capacity fade after the initial few cycles, which was because the electrode cycled in PC presented a pulverized surface with large particles.

Another IL anion is bis(fluorosulfonyl)imide (FSI) [71, 72]. For example, an IL electrolyte made of trimethyl isobutylphosphonium bis(fluorosulfonyl)imide (P$_{1,1,1,i4}$FSI) with high LiFSI salt concentration was examined for Si/graphene composite electrodes [73]. The (P$_{1,1,1,i4}$FSI) IL system delivered about 3.5 mAh/cm^2 capacity after 300 cycles at the current density of 1500 mA/g. In addition, this IL system also achieved satisfying capacity retention at 80°C after 60 cycles, which is beyond the operational temperature window of conventional carbonate electrolytes. The good performance was proposed to originate from the stable surface layers produced by the concentrated IL electrolyte that can effectively tolerate the volume expansion of the silicon materials.

7.3.5 Analytical methods of surface electrolyte interphases

Electrolyte components directly participate in the formation of SEIs, and therefore it is strongly demanded to reveal their precise reaction pathways, which, however, has turned out to be elusive. A wide variety of chemical characterization tools have been applied for retrieving the chemical information of the electrolyte systems and the SEI layers after battery operation [74]. X-ray Photoelectron Spectroscopy (XPS) can identify the bonding information of the elements of the electrode sample and imply the existence of organic species such as PEOs [75]. Nuclear magnetic resonance (NMR) is the most commonly used organic characterization method that can reveal the structure information of analytes, which has been used for analysis of electrolyte decomposition products [32, 76]. Gas chromatography mass spectrometry (GCMS) and liquid chromatography mass spectrometry (LCMS) are also standard organic analytical approaches that can utilize external chromatography to analyze complicated electrolyte or extracted SEI samples [77, 78]. Matrix-assisted laser desorption/ionization (MALDI) mass spectrometry can directly measure the electrolyte decomposition products on the electrode surfaces [79]. MALDI is principally an extremely sensitive technique that can observe the chemical transformations of single layers of molecules directly on the surfaces [80, 81], which also has the advantage of tolerating complicated chemical environments of the samples (residue inorganic species do not tend to interfere with MALDI measurement of organic species). This makes MALDI a promising analytical method for direct characterization of organic SEI components on the electrode surface (no extraction of the SEIs needed). Recently, MALDI has been successfully utilized to identify the PEO and polymethacrylate species in SEIs with exact structural assignments on a series of different electrode surfaces [87,88]. It is worth noting that the MALDI measurement was assisted by on-electrode chromatography technique, which proved critical for effective MALDI characterization of SEI components (details of on-electrode chromatography strategy provided in later section).

Fourier Transform Infrared Spectroscopy (FTIR) is a powerful analytical technique particularly for organic and polymeric molecules in battery systems [82–84]. A recent FTIR study examined the decomposition products of EC-based electrolytes. Diethyl 2,5-dioxahexane dicarboxylate (DEDOHC) and polyethylene were identified as electrolyte decomposition products on Sn and Ni electrodes, respectively. The identifications of these species were confirmed with synthetic reference compounds and quantum chemical (Hartree-Fock) calculations [85].

Chromatography is commonly applied to complicated organic and polymeric molecule samples to fractionate the analytes to facilitate their individual characterization. This is particularly important for SEI layers on electrode surfaces as the top layers of molecules could easily hinder the detection of buried layers underneath. Recently, an on-electrode chromatography method has been reported with the gradient polarity solvent wash (gradient wash) technique [86,87]. The gradient wash technique applies a sequence of elution solvents with rationally controlled and gradually increased polarities to verify the elution conditions of the electrode sample, which was tracked by FTIR measurement

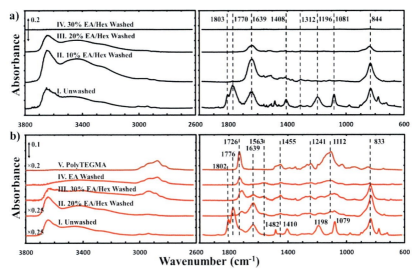

FIG. 7.15 On-electrode chromatography by gradient polarity solvent wash technique monitored by FTIR. (A) Electrode cycled with base electrolyte of 1.2M LiPF$_6$ EC/EMC. (B) Electrode cycled with TEGMA additive in the 1.2M LiPF$_6$ EC/EMC electrolyte. *(From C. Fang, Z. Liu, J. Lau, M. Elzouka, G. Zhang, P, Khomein, S. Lubner, P.N. Ross, G. Liu, Gradient polarity solvent wash for separation and analysis of electrolyte decomposition products on electrode surfaces. J. Electrochem. Soc. 167 (2) (2020) 020506.)*

after each elution [86]. For the electrode sample cycled in LiPF$_6$ EC/EMC electrolyte, 10%, 20%, and 30% ethyl acetate/hexane (EA/Hex) elution gradually rinsed off the EMC:LiPF$_6$ solvate, LEDC and LiHCO$_3$ species from the recovered electrode surfaces (Fig. 7.15A). When this on-electrode chromatography protocol was applied to electrode cycled with triethylene glycol methyl ether methacrylate (TEGMA) additive in the same electrolyte, where 10%–30% EA/Hex elution helped to remove the EC-electrolyte decomposition products, the presence of polyTEGMA that originated from the polymerization of the additive molecule was revealed and was confirmed with synthetic polyTEGMA sample (Fig. 7.15B). Obviously, this method could be coupled with various analytical techniques (such as MADLI described above) for fractionating the different organic species on the electrode surface to facilitate characterization of each component.

7.4 Conclusions

It is quite a journey to arrive where Si anode materials technology is today, although numerous challenges still exist to build a high capacity and long-life rechargeable Si-based battery. The SiO$_2$ layer on Si gives the Si materials a distinctive interfacial resistivity characteristic, compared to the carbon bases anode. It is still unclear whether the initial SiO_2 provides better passivation or not to the Si materials during the electrochemical process, although it is transformed into more ion-conducting forms of lithium silicates during lithiation. The Si volume expansion during lithiation creates new electrolyte reactive surfaces on Si. Both the SiO_x and carbon coating are used to lessen the volume expansion and to stabilize the Si surface, but only work to a certain extent to stabilize the interface. However, Si volume expansion still poses significant challenges to electrode integrity. The advancements of binder chemistry provide multiple functions to stabilize electrode mechanical structure via stronger adhesion, to improve lithium-ion mobility via ion transport channels, and to enhance electrical conductivity via conjugated molecular structures. The functional binder is an enabling factor for a robust Si-based electrode. The electrolytes and additives designed for Si-based materials are critical part of the puzzles to further stabilize the Si electrolyte interface. Both fluorinated electrolyte and additives are designed to provide improved passivation for the Si surface, which is modified by oxide, carbon, and binder, in the Si-based electrode. Although we remain curious about what happens at the Si and electrolyte interface, the battery engineering goal is to make the Si electrode good enough for the application in sight. Here, in-depth scientific discussions of materials of Si, binder, and electrolyte can provide insights to further improve Si battery performance.

References

[1] S. Xun, X. Song, L. Wang, M.E. Grass, Z. Liu, V.S. Battaglia, et al., The effects of native oxide surface layer on the electrochemical performance of Si nanoparticle-based electrodes, J. Electrochem. Soc. 158 (12) (2011) A1260–A1266.

[2] S. Xun, X. Song, M.E. Grass, D.K. Roseguo, Z. Liu, V.S. Battaglia, et al., Improved initial performance of Si nanoparticles by surface oxide reduction for lithium-ion battery application, Electrochem. Solid-State Lett. 14 (5) (2011) A61–A63.

[3] K. AlKaabi, D. Prasad, P. Kroll, N.W. Ashcroft, R. Hoffmann, Silicon monoxide at 1 atm and elevated pressures: crystalline or amorphous? J. Am. Chem. Soc. 136 (9) (2014) 3410–3423.

[4] A. Hirata, S. Kohara, T. Asada, M. Arao, C. Yogi, H. Imai, et al., Atomic-scale disproportionation in amorphous silicon monoxide, Nat. Commun. 7 (2016) 11591, https://doi.org/10.1038/ncomms11591.

[5] H. Zhao, Z.H. Wang, P. Lu, M. Jiang, F.F. Shi, X.Y. Song, et al., Toward practical application of functional conductive polymer binder for a high-energy lithium-ion battery design, Nano Lett. 14 (11) (2014) 6704–6710.

[6] I.H. Son, J.H. Park, S. Kwon, S. Park, M.H. Rummeli, A. Bachmatiuk, et al., Silicon carbide-free graphene growth on silicon for lithium-ion battery with high volumetric energy density, Nat. Commun. 6 (2015) 8.

[7] S. Zhou, C. Fang, X. Song, G. Liu, The influence of compact and ordered carbon coating on solid-state behaviors of silicon during electrochemical processes, Carbon Energy 2 (1) (2020) 143–150.

[8] S. Fang, N. Li, T.Y. Zheng, Y.B. Fu, X.Y. Song, T. Zhang, et al., Highly graphitized carbon coating on SiO with a π–π stacking precursor polymer for high performance lithium-ion batteries, Polymers 10 (6) (2018) 610.

[9] S. Zhou, C. Fang, X. Song, G. Liu, Highly ordered carbon coating prepared with polyvinylidene chloride precursor for high-performance silicon anodes in lithium-ion batteries, Batter. Supercaps 4 (2021) 240–247, https://doi.org/10.1002/batt.202000193.

[10] S.D. Xun, B. Xiang, A. Minor, V. Battaglia, G. Liu, Conductive polymer and silicon composite secondary particles for a high area-loading negative electrode, J. Electrochem. Soc. 160 (9) (2013) A1380–A1383.

[11] N.S. Hochgatterer, M.R. Schweiger, S. Koller, P.R. Raimann, T. Wohrle, C. Wurm, et al., Silicon/graphite composite electrodes for high-capacity anodes: influence of binder chemistry on cycling stability, Electrochem. Solid-State Lett. 11 (5) (2008) A76–A80.

[12] G. Liu, S.D. Xun, N. Vukmirovic, X.Y. Song, P. Olalde-Velasco, H.H. Zheng, et al., Polymers with tailored electronic structure for high capacity lithium battery electrodes, Adv. Mater. 23 (40) (2011) 4679–4683.

[13] H. Zhao, W. Yuan, G. Liu, Hierarchical electrode design of high-capacity alloy nanomaterials for lithium-ion batteries, Nano Today 10 (2) (2015) 193–212.

[14] G. Liu, H. Zheng, X. Song, V.S. Battaglia, Particles and polymer binder interaction: a controlling factor in lithium-ion electrode performance, J. Electrochem. Soc. 159 (3) (2012) A214–A221.

[15] M.Y. Wu, X.C. Xiao, N. Vukmirovic, S.D. Xun, P.K. Das, X.Y. Song, et al., Toward an ideal polymer binder design for high-capacity battery anodes, J. Am. Chem. Soc. 135 (32) (2013) 12048–12056.

[16] S.J. Park, H. Zhao, G. Ai, C. Wang, X.Y. Song, N. Yuca, et al., Side-chain conducting and phase-separated polymeric binders for high-performance silicon anodes in lithium-ion batteries, J. Am. Chem. Soc. 137 (7) (2015) 2565–2571.

[17] H. Zhao, Y. Wei, C. Wang, R.M. Qiao, W.L. Yang, P.B. Messersmith, et al., Mussel-inspired conductive polymer binder for Si-alloy anode in lithium-ion batteries, ACS Appl. Mater. Interfaces 10 (6) (2018) 5440–5446.

[18] H. Zhao, Y. Wei, R.M. Qiao, C.H. Zhu, Z.Y. Zheng, M. Ling, et al., Conductive polymer binder for high-tap-density nanosilicon material for lithium-ion battery negative electrode application, Nano Lett. 15 (12) (2015) 7927–7932.

[19] M.Y. Wu, X.Y. Song, X.S. Liu, V. Battaglia, W.L. Yang, G. Liu, Manipulating the polarity of conductive polymer binders for Si-based anodes in lithium-ion batteries, J. Mater. Chem. A 3 (7) (2015) 3651–3658.

[20] H. Zhao, Y.B. Fu, M. Ling, Z. Jia, X.Y. Song, Z.H. Chen, et al., Conductive polymer binder-enabled SiO-Sn$_x$Co$_y$C$_z$ anode for high-energy lithium-ion batteries, ACS Appl. Mater. Interfaces 8 (21) (2016) 13373–13377.

[21] Z. Lu, N. Liu, H.-W. Lee, J. Zhao, W. Li, Y. Li, et al., Nonfilling carbon coating of porous silicon micrometer-sized particles for high-performance lithium battery anodes, ACS Nano 9 (3) (2015) 2540–2547.

[22] J.R. Szczech, S. Jin, Nanostructured silicon for high capacity lithium battery anodes, Energy Environ. Sci. 4 (1) (2011) 56–72.

[23] H. Wu, G. Chan, J.W. Choi, I. Ryu, Y. Yao, M.T. McDowell, et al., Stable cycling of double-walled silicon nanotube battery anodes through solid–electrolyte interphase control, Nat. Nanotechnol. 7 (5) (2012) 310–315.

[24] Y. Son, Y. Son, M. Choi, M. Ko, S. Chae, N. Park, et al., Hollow silicon nanostructures via the Kirkendall effect, Nano Lett. 15 (10) (2015) 6914–6918.

[25] Y. Liu, Z. Tai, T. Zhou, V. Sencadas, J. Zhang, L. Zhang, et al., An all-integrated anode via interlinked chemical bonding between double-shelled–yolk-structured silicon and binder for lithium-ion batteries, Adv. Mater. 29 (44) (2017) 1703028.

[26] H. Zhao, Q. Yang, N. Yuca, M. Ling, K. Higa, V.S. Battaglia, et al., A convenient and versatile method to control the electrode microstructure toward high-energy lithium-ion batteries, Nano Lett. 16 (7) (2016) 4686–4690.

[27] X. Xiao, W. Zhou, Y. Kim, I. Ryu, M. Gu, C. Wang, et al., Regulated breathing effect of silicon negative electrode for dramatically enhanced performance of Li-ion battery, Adv. Funct. Mater. 25 (9) (2015) 1426–1433.

[28] A.M. Haregewoin, A.S. Wotango, B.-J. Hwang, Electrolyte additives for lithium ion battery electrodes: progress and perspectives, Energy Environ. Sci. 9 (6) (2016) 1955–1988.

[29] G.G. Eshetu, E. Figgemeier, Confronting the challenges of next-generation silicon anode-based lithium-ion batteries: role of designer electrolyte additives and polymeric binders, ChemSusChem 12 (12) (2019) 2515–2539.

[30] Z. Xu, J. Yang, H. Li, Y. Nuli, J. Wang, Electrolytes for advanced lithium ion batteries using silicon-based anodes, J. Mater. Chem. A 7 (16) (2019) 9432–9446.

[31] Q. Shi, W. Liu, Q. Qu, T. Gao, Y. Wang, G. Liu, et al., Robust solid/electrolyte interphase on graphite anode to suppress lithium inventory loss in lithium-ion batteries, Carbon 111 (2017) 291–298.

[32] Y. Jin, N.-J.H. Kneusels, P.C.M.M. Magusin, G. Kim, E. Castillo-Martínez, L.E. Marbella, et al., Identifying the structural basis for the increased stability of the solid electrolyte interphase formed on silicon with the additive fluoroethylene carbonate, J. Am. Chem. Soc. 139 (42) (2017) 14992–15004.

[33] G.V. Zhuang, K. Xu, H. Yang, T.R. Jow, P.N. Ross, Lithium ethylene dicarbonate identified as the primary product of chemical and electrochemical reduction of EC in 1.2 M LiPF6/EC:EMC electrolyte, J. Phys. Chem. B 109 (37) (2005) 17567–17573.

[34] L. Wang, A. Menakath, F. Han, Y. Wang, P.Y. Zavalij, K.J. Gaskell, et al., Identifying the components of the solid–electrolyte interphase in Li-ion batteries, Nat. Chem. 11 (9) (2019) 789–796.

[35] C. Fang, G. Zhang, J. Lau, G. Liu, Recent advances in polysulfide mediation of lithium-sulfur batteries via facile cathode and electrolyte modification, APL Mater. 7 (8) (2019), 080902.

[36] Z. Liu, X. He, C. Fang, L.E. Camacho-Forero, Y. Zhao, Y. Fu, et al., Reversible crosslinked polymer binder for recyclable lithium sulfur batteries with high performance, Adv. Funct. Mater. 30 (36) (2020) 2003605, https://doi.org/10.1002/adfm.202003605.

[37] V. Etacheri, U. Geiger, Y. Gofer, G.A. Roberts, I.C. Stefan, R. Fasching, et al., Exceptional electrochemical performance of Si-nanowires in 1,3-dioxolane solutions: a surface chemical investigation, Langmuir 28 (14) (2012) 6175–6184.

[38] Y. Zhao, C. Fang, G. Zhang, D. Hubble, A. Nallapaneni, C. Zhu, et al., A micelle electrolyte enabled by fluorinated ether additives for polysulfide suppression and Li metal stabilization in Li-S battery, Front. Chem. 8 (2020) 484.

[39] G. Zeng, Y. An, S. Xiong, J. Feng, Nonflammable fluorinated carbonate electrolyte with high salt-to-solvent ratios enables stable silicon-based anode for next-generation lithium-ion batteries, ACS Appl. Mater. Interfaces 11 (26) (2019) 23229–23235.

[40] M.-Q. Li, M.-Z. Qu, X.-Y. He, Z.-L. Yu, Electrochemical performance of Si/graphite/carbon composite electrode in mixed electrolytes containing LiBOB and LiPF[sub 6], J. Electrochem. Soc. 156 (4) (2009) A294.

[41] S.E. Trask, K.Z. Pupek, J.A. Gilbert, M. Klett, B.J. Polzin, A.N. Jansen, et al., Performance of full cells containing carbonate-based LiFSI electrolytes and silicon-graphite negative electrodes, J. Electrochem. Soc. 163 (3) (2015) A345–A350.

[42] K. Xu, Electrolytes and interphases in Li-ion batteries and beyond, Chem. Rev. 114 (23) (2014) 11503–11618.
[43] L. Chen, K. Wang, X. Xie, J. Xie, Effect of vinylene carbonate (VC) as electrolyte additive on electrochemical performance of Si film anode for lithium ion batteries, J. Power Sources 174 (2) (2007) 538–543.
[44] P. Ridgway, H. Zheng, G. Liu, X. Song, P. Ross, V. Battaglia, Effect of vinylene carbonate on graphite anode cycling efficiency, ECS Trans. 19 (25) (2009) 51–57, https://doi.org/10.1149/1.3247065.
[45] A.L. Michan, B.S. Parimalam, M. Leskes, R.N. Kerber, T. Yoon, C.P. Grey, et al., Fluoroethylene carbonate and vinylene carbonate reduction: understanding lithium-ion battery electrolyte additives and solid electrolyte interphase formation, Chem. Mater. 28 (22) (2016) 8149–8159.
[46] H. Nakai, T. Kubota, A. Kita, A. Kawashima, Investigation of the solid electrolyte interphase formed by fluoroethylene carbonate on Si electrodes, J. Electrochem. Soc. 158 (7) (2011) A798.
[47] N.-S. Choi, K.H. Yew, K.Y. Lee, M. Sung, H. Kim, S.-S. Kim, Effect of fluoroethylene carbonate additive on interfacial properties of silicon thin-film electrode, J. Power Sources 161 (2) (2006) 1254–1259.
[48] S.K. Heiskanen, J. Kim, B.L. Lucht, Generation and evolution of the solid electrolyte interphase of lithium-ion batteries, Joule 3 (10) (2019) 2322–2333.
[49] C.C. Nguyen, B.L. Lucht, Improved cycling performance of Si nanoparticle anodes via incorporation of methylene ethylene carbonate, Electrochem. Commun. 66 (2016) 71–74.
[50] B. Han, C. Liao, F. Dogan, S.E. Trask, S.H. Lapidus, J.T. Vaughey, et al., Using mixed salt electrolytes to stabilize silicon anodes for lithium-ion batteries via in situ formation of Li–M–Si ternaries (M = Mg, Zn, Al, Ca), ACS Appl. Mater. Interfaces 11 (33) (2019) 29780–29790.
[51] S. Dalavi, P. Guduru, B.L. Lucht, Performance enhancing electrolyte additives for lithium ion batteries with silicon anodes, J. Electrochem. Soc. 159 (5) (2012) A642–A646.
[52] R Nölle, A.J. Achazi, P. Kaghazchi, M. Winter, T. Placke, Pentafluorophenyl isocyanate as an effective electrolyte additive for improved performance of silicon-based lithium-ion full cells, ACS Appl. Mater. Interfaces 10 (33) (2018) 28187–28198, https://doi.org/10.1021/acsami.8b07683.
[53] G.-B. Han, J.-N. Lee, J.W. Choi, J.-K. Park, Tris(pentafluorophenyl) borane as an electrolyte additive for high performance silicon thin film electrodes in lithium ion batteries, Electrochim. Acta 56 (24) (2011) 8997–9003.
[54] Y.-G. Ryu, S. Lee, S. Mah, D.J. Lee, K. Kwon, S. Hwang, et al., Electrochemical behaviors of silicon electrode in lithium salt solution containing alkoxy silane additives, J. Electrochem. Soc. 155 (8) (2008) A583.
[55] G.-B. Han, M.-H. Ryou, K.Y. Cho, Y.M. Lee, J.-K. Park, Effect of succinic anhydride as an electrolyte additive on electrochemical characteristics of silicon thin-film electrode, J. Power Sources 195 (11) (2010) 3709–3714.
[56] M.C. Smart, F.C. Krause, C. Hwang, J. Soler, W.C. West, R.V. Bugga, et al., Electrolytes with improved safety developed for high specific energy Li-ion cells with Si-based anodes, ECS Trans. 50 (26) (2013) 365–374.
[57] C. Chen, Q. Li, Y. Li, Z. Cui, X. Guo, H. Li, Sustainable interfaces between Si anodes and garnet electrolytes for room-temperature solid-state batteries, ACS Appl. Mater. Interfaces 10 (2) (2018) 2185–2190.
[58] T. Bok, S.-J. Cho, S. Choi, K.-H. Choi, H. Park, S.-Y. Lee, et al., An effective coupling of nanostructured Si and gel polymer electrolytes for high-performance lithium-ion battery anodes, RSC Adv. 6 (9) (2016) 6960–6966.

[59] E. Quartarone, P. Mustarelli, Electrolytes for solid-state lithium rechargeable batteries: recent advances and perspectives, Chem. Soc. Rev. 40 (5) (2011) 2525–2540.
[60] X. Zuo, X.-M. Liu, F. Cai, H. Yang, X.-D. Shen, G. Liu, A novel all-solid electrolyte based on a co-polymer of poly-(methoxy/hexadecal-poly(ethylene glycol) methacrylate) for lithium-ion cell, J. Mater. Chem. 22 (41) (2012) 22265–22271.
[61] X. Zuo, X.-M. Liu, F. Cai, H. Yang, X.-D. Shen, G. Liu, Enhanced performance of a novel gel polymer electrolyte by dual plasticizers, J. Power Sources 239 (2013) 111–121.
[62] Y.-G. Cho, H. Park, J.-I. Lee, C. Hwang, Y. Jeon, S. Park, et al., Organogel electrolyte for high-loading silicon batteries, J. Mater. Chem. A 4 (21) (2016) 8005–8009.
[63] X.-G. Sun, J.B. Kerr, C.L. Reeder, G. Liu, Y. Han, Network single ion conductors based on comb-branched polyepoxide ethers and lithium bis(allylmalonato)borate, Macromolecules 37 (14) (2004) 5133–5135.
[64] H. Zhao, Z. Jia, W. Yuan, H. Hu, Y. Fu, G.L. Baker, et al., Fumed silica-based single-ion nanocomposite electrolyte for lithium batteries, ACS Appl. Mater. Interfaces 7 (34) (2015) 19335–19341.
[65] H. Zhao, F. Asfour, Y. Fu, Z. Jia, W. Yuan, Y. Bai, et al., Plasticized polymer composite single-ion conductors for lithium batteries, ACS Appl. Mater. Interfaces 7 (34) (2015) 19494–19499.
[66] H. Zheng, Q. Qu, L. Zhang, G. Liu, V.S. Battaglia, Hard carbon: a promising lithium-ion battery anode for high temperature applications with ionic electrolyte, RSC Adv. 2 (11) (2012) 4904–4912.
[67] H. Zheng, G. Liu, V. Battaglia, Film-forming properties of propylene carbonate in the presence of a quaternary ammonium ionic liquid on natural graphite anode, J. Phys. Chem. C 114 (13) (2010) 6182–6189.
[68] P.C. Howlett, D.R. MacFarlane, A.F. Hollenkamp, High lithium metal cycling efficiency in a room-temperature ionic liquid, Electrochem. Solid-State Lett. 7 (5) (2004) A97.
[69] J.-W. Song, C.C. Nguyen, S.-W. Song, Stabilized cycling performance of silicon oxide anode in ionic liquid electrolyte for rechargeable lithium batteries, RSC Adv. 2 (5) (2012) 2003–2009.
[70] H. Usui, Y. Yamamoto, K. Yoshiyama, T. Itoh, H. Sakaguchi, Application of electrolyte using novel ionic liquid to Si thick film anode of Li-ion battery, J. Power Sources 196 (8) (2011) 3911–3915.
[71] T. Sugimoto, Y. Atsumi, M. Kono, M. Kikuta, E. Ishiko, M. Yamagata, et al., Application of bis(fluorosulfonyl)imide-based ionic liquid electrolyte to silicon–nickel–carbon composite anode for lithium-ion batteries, J. Power Sources 195 (18) (2010) 6153–6156.
[72] D.M. Piper, T. Evans, K. Leung, T. Watkins, J. Olson, S.C. Kim, et al., Stable silicon-ionic liquid interface for next-generation lithium-ion batteries, Nat. Commun. 6 (1) (2015) 6230.
[73] R. Kerr, D. Mazouzi, M. Eftekharnia, B. Lestriez, N. Dupré, M. Forsyth, et al., High-capacity retention of Si anodes using a mixed lithium/phosphonium bis(fluorosulfonyl)imide ionic liquid electrolyte, ACS Energy Lett. 2 (8) (2017) 1804–1809.
[74] Y. Chu, Y. Shen, F. Guo, X. Zhao, Q. Dong, Q. Zhang, et al., Advanced characterizations of solid electrolyte interphases in lithium-ion batteries, Electrochem. Energy Rev. 3 (1) (2020) 187–219.
[75] C.K. Chan, R. Ruffo, S.S. Hong, Y. Cui, Surface chemistry and morphology of the solid electrolyte interphase on silicon nanowire lithium-ion battery anodes, J. Power Sources 189 (2) (2009) 1132–1140.
[76] Y. Jin, N.-J.H. Kneusels, C.P. Grey, NMR study of the degradation products of ethylene carbonate in silicon–lithium ion batteries, J. Phys. Chem. Lett. 10 (20) (2019) 6345–6350.

[77] C.L. Campion, W. Li, B.L. Lucht, Thermal decomposition of LiPF[sub 6]-based electrolytes for lithium-ion batteries, J. Electrochem. Soc. 152 (12) (2005) A2327.
[78] M. Tochihara, H. Nara, D. Mukoyama, T. Yokoshima, T. Momma, T. Osaka, Liquid chromatography-quadruple time of flight mass spectrometry analysis of products in degraded lithium-ion batteries, J. Electrochem. Soc. 162 (10) (2015) A2008–A2015.
[79] A. Abouimrane, S.A. Odom, H. Tavassol, M.V. Schulmerich, H. Wu, R. Bhargava, et al., 3-Hexylthiophene as a stabilizing additive for high voltage cathodes in lithium-ion batteries, J. Electrochem. Soc. 160 (2) (2012) A268–A271.
[80] J. He, C. Fang, R.A. Shelp, M.B. Zimmt, Tracking invisible transformations of physisorbed monolayers: LDI-TOF and MALDI-TOF mass spectrometry as complements to STM imaging, Langmuir 33 (2) (2017) 459–467.
[81] C. Fang, H. Zhu, O. Chen, M.B. Zimmt, Reactive two-component monolayers template bottom-up assembly of nanoparticle arrays on HOPG, Chem. Commun. 54 (58) (2018) 8056–8059.
[82] C. Korepp, H.J. Santner, T. Fujii, M. Ue, J.O. Besenhard, K.C. Möller, et al., 2-Cyanofuran—a novel vinylene electrolyte additive for PC-based electrolytes in lithium-ion batteries, J. Power Sources 158 (1) (2006) 578–582.
[83] H.J. Santner, C. Korepp, M. Winter, J.O. Besenhard, K.C. Möller, In-situ FTIR investigations on the reduction of vinylene electrolyte additives suitable for use in lithium-ion batteries, Anal. Bioanal. Chem. 379 (2) (2004) 266–271.
[84] Z. Li, C. Fang, C. Qian, S. Zhou, X. Song, M. Ling, et al., Polyisoprene captured sulfur nanocomposite materials for high-areal-capacity lithium sulfur battery, ACS Appl. Polym. Mater. 1 (8) (2019) 1965–1970.
[85] F. Shi, H. Zhao, G. Liu, P.N. Ross, G.A. Somorjai, K. Komvopoulos, Identification of diethyl 2,5-dioxahexane dicarboxylate and polyethylene carbonate as decomposition products of ethylene carbonate based electrolytes by Fourier transform infrared spectroscopy, J. Phys. Chem. C 118 (27) (2014) 14732–14738.
[86] C. Fang, Z. Liu, J. Lau, M. Elzouka, G. Zhang, P. Khomein, et al., Gradient polarity solvent wash for separation and analysis of electrolyte decomposition products on electrode surfaces, J. Electrochem. Soc. 167 (2) (2020), 020506.
[87] C. Fang, J. Lau, D. Hubble, P. Khomein, E.A. Dailing, Y. Liu, et al., Large-molecule decomposition products of electrolytes and additives revealed by on-electrode chromatography and MALDI, Joule 5 (2) (2021) 415–428, https://doi.org/10.1016/j.joule.2020.12.012.
[88] E.J. Hopkins, S. Frisco, R.T. Pekarek, C. Stetson, Z. Huey, S. Harvey, et al., Examining CO_2 as an additive for solid electrolyte interphase formation on silicon anodes, J. Electrochem. Soc. 168 (3) (2021) 030534, https://doi.org/10.1149/1945-7111/abec66.

Part IV

Achieving high performance

Chapter 8

Performance degradation modeling in silicon anodes

Partha P. Mukherjee and Ankit Verma
School of Mechanical Engineering, Purdue University, West Lafayette, IN, United States

8.1 Introduction

The development of next-generation high-energy density lithium-ion batteries (LIBs) comprised high-capacity anode and cathode materials has engendered considerable research on novel intercalation, alloying, and conversion material systems [1–3]. Silicon exhibits high theoretical specific capacity of 4200 mAh/g_{Si}, nearly ten times that of graphite, making it a desirable candidate as a next-generation anode for LIBs [4–7]. The silicon atom host can alloy with 4.4 atoms of lithium ($Li_{4.4}Si$), while graphite intercalates a maximum of 1/6 atom of Li per molecule (LiC_6), outlining the desirability of Si as anode. However, experimental cycling data sets have revealed inferior initial capacity (~75% of theoretical) and rapid capacity degradation, resulting in unacceptable cycle life with subsequent lithiation-delithiation.

The deleterious capacity loss in silicon anodes is attributable to the immense volumetric fluctuations of the intrinsic particles upon repeated alloying-dealloying with Li while cycling [8]. Graphite exhibits 10% volume expansion, while silicon particles can expand up to around 400% with full lithiation as its host architecture accommodates the 4.4 Li atoms per Si atom [9, 10]. The resulting volumetric strain initiates crack formation which propagate subsequently leading to fracture [11]. The brittle solid electrolyte interphase (SEI) layer also ruptures, which is unable to accommodate this large volumetric fluctuation. This leads to reformation of the SEI layer through contact between exposed particle surface and fresh electrolyte, which depletes the Li inventory, low coulombic efficiency, and exacerbating capacity fade [12, 13]. Crack coalescence contributes to hindered Li-ion diffusive transport inside the active material, and in extreme scenarios, it can also lead to complete rupture of the particles, resulting in particle isolation and exacerbated capacity fade.

In this chapter, we delineate the strategies used to model performance degradation interactions in silicon anodes cognizant of its large volume expansion.

The remaining part of the chapter is subdivided into two major sections. First, we outline the electrochemistry-transport interactions in silicon anodes for nano-sized particles to understand the performance-large volume expansion coupling devoid of mechanical damage [14]. Later, the mechano-electrochemical interaction dynamics are explored to delineate the performance-degradation complexations with fracture-based mechanical damage observed in micron-sized silicon particles [15–17]. Finally, we end with an outlook toward the need for the coupling of multiscale modeling paradigms for deep understanding of interfacial and bulk phenomena in silicon anodes via multiphysics simulations.

8.2 Morphology performance interactions in silicon anodes

The propensity of silicon particles to fracture during alloying/dealloying with lithium is drastically reduced in the nanometer limit due to lower values of diffusion-induced stress [18–21]. Stress-induced diffusion and pressure gradient effects also contribute to the minimization of fracture [22–24]. Consequently, nanospheres and nanowire configurations are being probed in detail to elucidate the morphological impact on performance [25–27]. Silicon nanospheres have been the focus of intensive investigations both experimentally and computationally. Chandrasekaran et al. examined lithium insertion/deinsertion in an evolving silicon particle at room temperature under both galvanostatic and potentiostatic control using a single-particle model, demonstrating the need for asymmetric transfer coefficients and sluggish kinetics resulting in kinetic hysteresis and potential gap in the Li—Si system [28]. Furthermore, they extended the work to a macrohomogeneous model incorporating porosity variations in the electrode due to the large molar volume changes [29]. The incorporation of volume changes in porous electrodes through porosity variation for anodes has been ably demonstrated [30–32]. Recently, Mai et al. reformulated the macrohomogeneous model coupling large deformations at the particle and electrode level with electrochemistry to simulate the performance of Si anode-NMC532 cathode cell [33]. Chan et al. grew silicon nanowires directly on current collector achieving theoretical charge capacity for silicon anodes with minimal capacity fading during cycling [34]. Although experimental studies on nanowires are available, computational studies are much sparse as compared to nanospheres. Charge transfer kinetics parameters have been reported by Swamy et al. using electrochemical impedance spectroscopy on single crystal silicon wafers [35].

In this section, the single-particle formalism with volume evolution will be discussed to analyze and contrast silicon lithiation in nanospheres and nanorods. The morphological variation requires differing formulations for the two configurations explored and provides insights into the impact of lithiation on the rate performance through diffusive transport and reaction kinetics considerations. Based on the rate performance results obtained, we delineate design guidelines for nanospheres and nanorods for superior performance.

8.2.1 Electrochemistry transport dynamics in silicon anodes

The problem formulation involves modeling of coupled solid-state diffusion of lithium inside the silicon nanospheres and nanorods/nanowires active material particles and lithiation reaction kinetics at the surface cognizant of the large volume expansion during lithium insertion in the silicon. The single-particle model is appropriately adjusted to reflect the dimensional changes for the silicon under lithiation. For both nanospheres and nanorods, only radial diffusion is considered. Although this approximation is particularly true for the nanospheres because of radial symmetry, its applicability is assumed for the nanorods predicated on theoretical reasoning and experimental data set. This is justified from the large length to radius ratio of the nanorods ($L/R \gg 1$) and the nature of the manufacturing process of the nanowires. The radial direction provides the shorter diffusion pathway for nanorods with large aspect ratio. Furthermore, the nanorods are grown on the current collectors thereby constraining the Li diffusion through the curved surface area (radial diffusion) and only one of the circular areas pointing toward the separator (axial diffusion). This is reflected in the radial expansion contributing to the bulk of the volume increase. Chan et al. reported a mean diameter increase of the nanowires during lithiation from 89 nm to 141 nm which contributes to around 70% of the final volume coming from radial diffusion with the remaining 30% coming from the length increase through axial diffusion [34]. The contribution of radial diffusion to volume expansion will increase further toward 100% as the aspect ratio is increased, consequently, radial diffusion is assumed to be the predominant transport pathway for nanorods and the nanorod length is assumed to be invariable throughout the lithiation.

At the single-particle level, the rate-determining transport mechanism is the solid-state diffusion inside the active material [36]. The electrolyte ionic diffusivity and conductivity is assumed to be high enough to neglect concentration and potential gradients of Li-ion within the electrolyte. Therefore, the active material is subjected to constant Li-ion flux from all directions irrespective of its position from the current collector under galvanostatic operation. Single phase diffusion inside the electrode particle is modeled using the Fick's law. In accordance with literature, in particular, Chandrasekaran et al., the final lithiated state of amorphous silicon for single-phase diffusion is considered to be $Li_{15}Si_4$ corresponding to a 280% increase in volume [28]. Thus, the final volume is 3.8 times the initial volume. The electrochemical reaction at the surface of the Si used to model the lithiation process is shown in Eq. (8.1).

$$Li^+ + e^- + Si_{4/15} \rightleftarrows LiSi_{4/15} \qquad (8.1)$$

This form has been chosen to keep the state of charge of Li between 0 and 1, also allowing for ease of applicability of the Butler-Volmer equation. The state of charge of Li inside the Si particle has a continuous profile between 0 and 1 as it has been considered the amorphous silicon particle with single phase diffusion as opposed to the crystalline silicon particle which shows segregation of the Li rich and the Li poor phases which can be treated through the Cahn-Hilliard formalism [15].

Furthermore, the voltage profile having multiple steps in it is seen for Li intercalation in graphite as well, with the plateaus coinciding with the coexistence of multiple phases. Consequently, the open circuit potential of graphite as a function of state of charge shows plateaus as the SOC varies from 0 to 1. In the literature, this can be adequately treated with single phase diffusion for the single-particle model or the macrohomogeneous pseudo 2D model [36, 37]. The same concept can be extended for lithiation in silicon adequately, and thus, we have a smooth transition in the concentration profiles throughout the silicon particle and a SOC varying from 0 to 1. We provide the governing differential equations and constitutive relations for the spherical nanoparticles and nanorods.

8.2.1.1 Nanospheres

Species balance in the nanosphere is given by the spherical diffusion equation in Eq. (8.2).

$$\frac{\partial c_s}{\partial t} = \frac{D_{Li}}{r^2} \frac{\partial}{\partial r}\left(r^2 \frac{\partial c_s}{\partial r}\right) \quad (8.2)$$

Here, c_s is the Li concentration inside the sphere, D_{Li} is the diffusion coefficient of Li into Si, and r, t represent the radial distance within the particle and time respectively. The subscript 's' is used to signify the sphere shape. Symmetry boundary condition is applied at the particle center (Eq. 8.3) and constant flux boundary condition is applied at the particle surface (Eq. 8.4) and the initial condition is given in Eq. (8.5).

$$r = 0, \frac{\partial c_s}{\partial r} = 0 \quad (8.3)$$

$$r = R_s, D_{Li}\frac{\partial c_s}{\partial r} = -\frac{\tilde{i}_s}{nF} \quad (8.4)$$

$$t = 0, c_s = 0 \,\mathrm{mol/m^3} \quad (8.5)$$

In the formulation, the apparent current density at the nanosphere surface, $\tilde{i}_s$, can be correlated to the actual current density i_s as shown in Eq. (8.6). Here R_0 is the initial radius of the particle prior to lithiation and $R_s(t)$ denotes the temporal evolution of the radius as the volume increases with lithiation, I is the applied current. For a detailed derivation of this formulation, the readers are referred to Chandrasekaran et al. [28]

$$\tilde{i}_s = i_s\left[\frac{R_0}{R_s(t)}\right]^3, i_s = \frac{I}{4\pi[R_s(t)]^2} \quad (8.6)$$

We use the Butler-Volmer formulation with symmetric transfer coefficients to describe the charge transfer kinetics at the surface of the silicon nanospheres (Eq. 8.7). Here i_0 is the exchange current density, F is the Faraday's constant, R is the universal gas constant, T is temperature and η is the overpotential for the electrochemical reaction.

$$i_s = i_0 \left[\exp\left(\frac{F}{2RT}\eta\right) - \exp\left(-\frac{F}{2RT}\eta\right) \right] \tag{8.7}$$

The exchange current density takes the form shown in Eq. (8.8). Here, k is the rate constant, c_e is the Li$^+$ ionic concentration in the electrolyte phase, c_{max} and $c_{max,0}$ are the maximum concentration of Li in the Si based on the final and initial volume ($V_{s,0}$) of Si respectively while the surface concentration is denoted by the subscript 'surf'. $\tilde{c}_s$ corresponds to the modified lithium concentration based on the volumetric change during lithiation. The corresponding forms for c_{max} and $c_{max,0}$ are specified in Eqs. (8.9) and (8.10) respectively. N_{tot} is the total number of moles of Li that can be inserted into the silicon electrode based on the mass of the pristine Si particle and a maximum molar Li:Si ratio of 3.75:1.

$$i_{s,0} = Fk(c_e)^{0.5}\left(c_{max,0} - \tilde{c}_{s,surf}\right)^{0.5}\tilde{c}_{s,surf}^{0.5}, \quad \tilde{c}_{s,surf} = c_{s,surf}\left[\frac{R_s(t)}{R_0}\right]^3 \tag{8.8}$$

$$c_{max} = \frac{V_{s,0}\dfrac{3.75}{\Omega_{Si}}}{V_{s,0}\left(1 + \dfrac{3.75\Omega_{Li}}{\Omega_{Si}}\right)} = 81967 \,\text{mol/m}^3 \tag{8.9}$$

$$N_{tot} = V_{s,max} c_{max} = \frac{4}{3}\pi(3.8 R_0^3) c_{max},$$
$$c_{max,0} = \frac{N_{tot}}{V_{s,0}} = \frac{N_{tot}}{\frac{4}{3}\pi R_0^3} = 3.8 c_{max} = 311474.6 \,\text{mol/m}^3 \tag{8.10}$$

The overpotential is correlated to the solid phase potential ϕ_1, electrolyte phase potential ϕ_2 and the open circuit potential (OCP) U of silicon lithiation based on the surface state of charge (SOC) $\theta_{s,surf}$ (Eq. 8.11). The dependence of the OCP on the SOC is obtained from GITT experiments in literature [28] and the surface SOC can be correlated to the current lithiated volume $V_s(t)$ and surface concentration (Eq. 8.12).

$$\eta = \phi_1 - \phi_2 - U(\theta_{s,surf}) \tag{8.11}$$

$$\theta_{s,surf} = \frac{V_s(t)}{N_{tot}} c_{s,surf} = \frac{\frac{4}{3}\pi[R_s(t)]^3}{N_{tot}} c_{s,surf} \tag{8.12}$$

Finally, the volume ratio of lithiated silicon, $V_s(t)$, and pristine spherical particle can be correlated to the molar volumes of lithium, Ω_{Li}, silicon, Ω_{Si}, and the average state of charge inside the particle, $\bar{\theta}_s$, to obtain a functional form for the temporal evolution of the nanosphere particle radius (Eq. 8.13).

$$\frac{V_s(t)}{V_{s,0}} = \frac{\frac{4}{3}\pi[R_s(t)]^3}{\frac{4}{3}\pi R_0^3} = \frac{\Omega_{Si} + 3.75\Omega_{Li}\bar{\theta}_s}{\Omega_{Si}} \Rightarrow \frac{R_s(t)}{R_0} = \left(\frac{\Omega_{Si} + 3.75\Omega_{Li}\bar{\theta}_s}{\Omega_{Si}}\right)^{1/3} \quad (8.13)$$

The average and local state of charge can be obtained from Eqs. (8.14) and (8.15) respectively. The average state of charge can be directly obtained from the constant current for galvanostatic discharge or can also be obtained from the average concentration field which is easier to evaluate during constant voltage simulations.

$$\bar{\theta}_s = \frac{It}{nFN_{tot}} = \frac{\frac{4}{3}\pi[R_s(t)]^3}{N_{tot}}\bar{c}_s, \quad \bar{c}_s = \frac{\int_0^{R_s(t)} c_s(4\pi r^2)\,dr}{\frac{4}{3}\pi[R_s(t)]^3} \quad (8.14)$$

$$\theta_s = \frac{\frac{4}{3}\pi[R_s(t)]^3}{N_{tot}}c_s = \frac{c_s}{c_{max}}\left[\frac{R_s(t)}{3.8^{1/3}R_0}\right]^3 \quad (8.15)$$

8.2.1.2 Nanorods

Species balance in the nanorods is given by the cylindrical diffusion equation. Here the subscript 'n' denotes the nanorod configuration.

$$\frac{\partial c_n}{\partial t} = \frac{D_{Li}}{r}\frac{\partial}{\partial r}\left(r\frac{\partial c_n}{\partial r}\right) \quad (8.16)$$

The boundary and initial conditions (Eq. 8.17) are similar to the nanosphere configuration differing only in the evaluation of the current densities (Eq. 8.18) based on the cylindrical shape of the nanorods. The curved surface area of the cylinder is the reaction area now (L_0 is the pristine length of the nanorod which is assumed to remain constant throughout lithiation). The volumetric dependence of a cylinder on the square of the radius is reflected in the equations for the nanorod as compared to the nanosphere which shows a third-degree dependence.

$$r = 0, \frac{\partial c_n}{\partial r} = 0; r = R_s, D_{Li}\frac{\partial c_n}{\partial r} = -\frac{\tilde{i}_n}{nF}; t = 0, c_n = 0 \,\text{mol/m}^3 \quad (8.17)$$

$$\tilde{i}_n = i_n\left[\frac{R_0}{R_n(t)}\right]^2, \quad i_n = \frac{I}{2\pi R_n(t)L_0} \quad (8.18)$$

Eqs. (8.19)–(8.21) give the Butler-Volmer and exchange current density correlation with the corresponding constitutive equations for maximum lithium concentrations. Again, we note the square dependence of the modified concentration on the nanorod radius ratio evolution. Another important point to note is the maximum concentrations based on initial and final volumes have the same value for both nanospheres and nanorods. This is an intrinsic property of lithiated silicon invariant of the morphology which also points to the consistency of the formulation.

$$i_n = i_0 \left[\exp\left(\frac{F}{2RT}\eta\right) - \exp\left(-\frac{F}{2RT}\eta\right) \right] \tag{8.19}$$

$$i_{n,0} = Fk(c_e)^{0.5}(c_{max,0} - \tilde{c}_{n,surf})^{0.5}\tilde{c}_{n,surf}^{0.5}, \tilde{c}_{n,surf} = c_{n,surf}\left[\frac{R_n(t)}{R_0}\right]^2 \tag{8.20}$$

$$N_{tot} = V_{n,max} c_{max} = \pi(3.8R_0^2)L_0 c_{max}, c_{max,0} = \frac{N_{tot}}{V_{n,0}} = \frac{N_{tot}}{\pi R_0^2 L_0} = 3.8 c_{max} \tag{8.21}$$

The overpotential equation (Eq. 8.22) is similar to the one for nanospheres differing only in the formulation of the surface state of charge (Eq. 8.23).

$$\eta = \phi_1 - \phi_2 - U(\theta_{n,surf}) \tag{8.22}$$

$$\theta_{n,surf} = \frac{V_n(t)}{N_{tot}} c_{n,surf} = \frac{\pi[R_s(t)]^2 L_0}{N_{tot}} c_{n,surf} \tag{8.23}$$

Again, the temporal evolution of the nanorod radius can be correlated to the volumetric evolution and consequently, the molar volumes and average state of charge inside the nanowire (Eqs. 8.24–8.26). It is to be noted that the nanowire radius increases by the power of 0.50 as compared to the nanosphere which increases by the power of 0.33. This indicates that, for the same initial radius, the fully lithiated nanorod radius will exceed the fully lithiated nanosphere radius.

$$\frac{V_n(t)}{V_{n,0}} = \frac{\pi[R_n(t)]^2 L_0}{\pi R_0^2 L_0} = \frac{\Omega_{Si} + 3.75\Omega_{Li}\bar{\theta}_n}{\Omega_{Si}} \Rightarrow \frac{R_s(t)}{R_0(t)} = \left(\frac{\Omega_{Si} + 3.75\Omega_{Li}\bar{\theta}_n}{\Omega_{Si}}\right)^{1/2} \tag{8.24}$$

$$\bar{\theta}_n = \frac{It}{nFN_{tot}} = \frac{\pi[R_n(t)]^2 L_0}{N_{tot}}\bar{c}_n, \bar{c}_n = \frac{\int_0^{R_n(t)} c_n(2\pi r L_0)dr}{\pi[R_n(t)]^2 L_0} \tag{8.25}$$

$$\bar{\theta}_n = \frac{\pi[R_n(t)]^2 L_0}{N_{tot}} c_n = \frac{c_n}{c_{max}}\left[\frac{R_s(t)}{3.8^{1/2}R_0}\right]^2 \tag{8.26}$$

The coupled equations are discretized using the finite volume framework utilizing central differencing spatial scheme with Euler implicit (backward) time stepping scheme. To obtain a good comparison between the nanospheres and nanorods, performance of equal volumes, (and hence mass) are contrasted between the two morphologies. Consequently, nanospheres of pristine radius R_0 are contrasted against nanorods of pristine radius R_0 and length, $L_0 = 4R_0/3$. Furthermore, equivalent volume high aspect ratio nanorods are investigated as well with $L_0/R_0 = 5, 10$ and contrasted with the cases to delineate the effect on the performance.

8.2.2 Representative highlights

Fig. 8.1A shows the schematic of the volume expansion of the silicon nanospheres and nanowires respectively. Radial diffusion in both cases leads to temporal increase of the radius with lithiation and the length remains constant for the nanorod. Material parameters for silicon, in particular, the transport, kinetic and thermodynamic properties are reported in Table 8.1 [28]. Furthermore, Table 8.2 lists the equivalent volume forms of the nanospheres and nanorods with $L_0/R_0 = 1.33, 5, 10$ respectively. The corresponding 1C current density for each of these configurations is also reported.

8.2.2.1 Nanospheres vs nanorods with same volume and same initial diameter

Fig. 8.1B and C shows the silicon local state of charge as a function of particle radius at 5C lithiation rate for the nanosphere and nanorods of same initial radius, $R_0 = 30$ nm. For equivalent volume comparison, the silicon nanorod has an initial length, $L_0 = 39.90$ nm. During lithiation as the Li enters the surface, the surface concentration is higher as compared to the concentration at the particle center. This results in a positive concentration gradient inside the particle as we go from the particle center to the surface. For 5C lithiation corresponding to a discharge of the Li—Si half-cell, the total nominal time of discharge is 720 s and correspondingly the concentration profiles are shown at 100 s, 200 s, 400 s, and 600 s. As the silicon particle gets alloyed, the concentration inside the particle increases resulting in the upwards shift of the concentration profiles with lithiation time. An important observation is that surface state of charge of the silicon nanorods increases at a faster pace as compared to the silicon nanosphere of the same diameter. This is directly correlated to the faster increase of the nanorod radius as compared to the nanosphere radius. Comparing Eqs. (8.13) and (8.24), we observe that the nanorod radius increases by the power of 0.50 as compared to the smaller power of 0.33 exhibited by the nanosphere radius. Diffusion limitations are exacerbated at larger particle radius as Li must diffuse through a longer distance in the nanorod as it reaches end of

Performance degradation modeling in silicon anodes **Chapter | 8** **307**

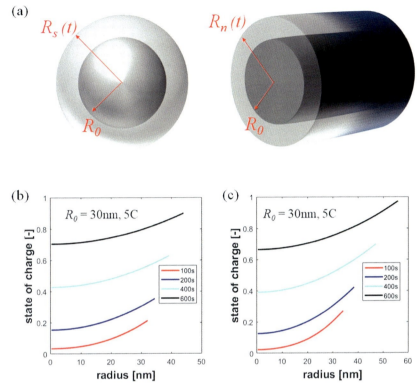

FIG. 8.1 (A) Schematic of radial lithiation of nanosphere (left) and nanorod (right). Dimensionless concentration profile as a function of radius for (B) nanosphere and (C) nanorod of equivalent volumes. Both the nanosphere and nanorod have initial radius 30 nm and the current rate of operation is 5C. Nanorods have steeper concentration gradient than nanospheres [14].

discharge. Consequently, the Li concentration shows a steeper concentration gradient in the nanorod as compared to the nanosphere. Although the larger surface area of the nanorod as compared to the nanosphere results in smaller effective current density (and hence lithium flux) at the nanorod surface as compared to the nanosphere, this advantage of the nanorod is overshadowed by the extra diffusion limitations.

The insights indicate that the nanowires are not always better as compared to nanospheres. In literature, experimental investigations of nanowire silicon have reported superior performance as compared to nanosphere silicon. This dichotomy can be explained by considering nanowires with larger aspect ratios. There is an optimum threshold of the nanorod length to radius ratio beyond which the nanorod rate performance exceeds the nanosphere rate performance for the same equivalent volume. In the following section, we investigate this phenomenon to delineate this threshold.

TABLE 8.1 List of baseline geometric, kinetic and transport parameters for silicon nanosphere and nanorods.

Parameter	Symbol	Value
Initial Radius	R_0 (nm)	30
Diffusivity	D_{Li} (m²/s)	10^{-18}
Rate constant	k [(m/s)(mol/m³)$^{-0.5}$]	2.5×10^{-9}
Molar volume of Li	Ω_{Li} (m³/mol)	9×10^{-6}
Molar volume of Si	Ω_{Si} (m³/mol)	1.2×10^{-5}
Maximum lithium concentration based on final radius	c_{max} (mol/m³)	81,967
Maximum lithium concentration based on initial radius	$c_{max,\,0}$ (mol/m³)	311,474.6
Density	ρ_{Si} (kg/m³)	2329
Temperature	T (K)	300

8.2.2.2 Nanospheres vs nanorods with same volume and varying diameter

Table 8.2 lists the nanorods of equivalent volume to nanospheres of radius 10 nm, 30 nm, 50 nm, 75 nm and 100 nm for aspect ratio $\frac{L_0}{R_0} = 1.33, 5, 10$. The corresponding 1C current density is also tabulated for all the scenarios. Fig. 8.2A–D displays contour plots of the obtained specific capacity in mAh/g as a function of the equivalent sphere diameter and C-rate for the nanospheres, nanorods with aspect ratio $L_0/R_0 = 1.33$, nanorods with aspect ratio $L_0/R_0 = 5$ and nanorods with aspect ratio $L_0/R_0 = 10$ respectively. As we go from Fig. 8.2A—nanospheres to Fig. 8.2B—nanorods with $L_0/R_0 = 1.33$, we see an increase in the low specific capacity region (red) at high C-rates and particle sizes. This is a direct consequence of the diffusion limitations overshadowing the active area increase. Consequently, nanorods with aspect ratio close to 1.33 are inferior to the nanospheres for all radii and C-rates. Moving on, if we compare Fig. 8.2A—nanospheres and Fig. 8.2C—nanorods with $L_0/R_0 = 5$, the specific capacity map is almost identical for all the C-rates and particle sizes. As we increase the aspect ratio, for the same equivalent volume, nanorod has a smaller initial radius as compared to the nanospheres. Consequently, the diffusion limitations start to diminish. In conjunction with the beneficial impact of additional surface area afforded by the nanowires, the nanowire performance begins to supersede that of the nanosphere. As the initial nanorod radius decreases with aspect ratio increase, the fully lithiated nanorod radii also remains smaller than the fully lithiated nanosphere radii. Consequently, throughout the entire lithiation process

TABLE 8.2 Dimensions of nanospheres and nanorods based on equivalent volume.

Nanosphere	Nanorod		Nanorod5		Nanorod10		
R_0 (nm)	R_0 (nm)	L_0 (nm)	R_0 (nm)	L_0 (nm)	R_0 (nm)	L_0 (nm)	1C Current (A)
10	10	13.33	6.44	32.20	5.11	51.09	3.4916×10^{-17}
30	30	40.00	19.31	96.55	15.33	153.26	9.4272×10^{-16}
50	50	66.67	32.18	160.90	25.54	255.43	4.3645×10^{-15}
75	75	100.0	48.27	241.35	38.31	383.15	1.4730×10^{-14}
100	100	133.33	64.37	321.85	51.09	510.87	3.4916×10^{-14}

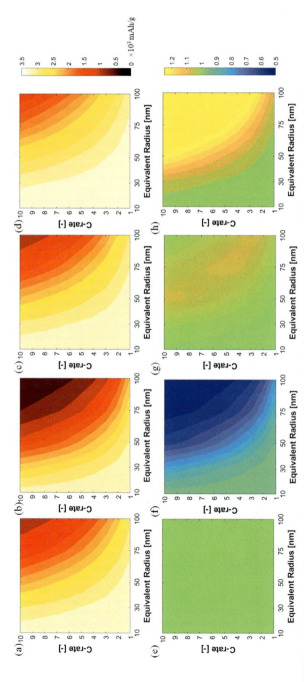

FIG. 8.2 Specific capacity contour plots as a function of equivalent spherical particle radius and C-rate for (A) nanosphere, (B) nanorods with aspect ratio $L_0/R_0 = 1.33$, (C) $L_0/R_0 = 5$ and (D) $L_0/R_0 = 10.0$. Ratio of specific capacity obtained to the specific capacity of nanospheres for (E) nanosphere, (F) nanorods with aspect ratio $L_0/R_0 = 1.33$, (G) $L_0/R_0 = 5$ and (H) $L_0/R_0 = 10.0$. Nanorods outperform nanospheres when its aspect ratio is greater than 5 [14].

the nanorod experiences inferior diffusion limitations as compared to the nanosphere. The alliance of inferior diffusion limitations and larger active area enabled minor kinetic limitations allows the nanorod performance to trump the nanosphere rate performance. This is further exemplified in Fig. 8.2D—nanorods with aspect ratio $L_0/R_0 = 10$ which has the least amount of low specific capacity (red) zones. Thus, it is imperative to go for nanorods with aspect ratio $L_0/R_0 \geq 5$ to achieve high-performance silicon anodes. Nanospheres can trump nanorods with aspect ratios less than 5 in performance based on diffusion limitations.

Fig. 8.2E–H further highlights this insight in a more compelling fashion. Here, we have shown the rate performance capacity contour maps for the morphologies normalized by the nanospheres specific capacity. Fig. 8.2E corresponds to the nanospheres, hence, normalized by its own capacity it exhibits a ratio of 1 throughout. Fig. 8.2F corresponds to the nanorods with $L_0/R_0 = 1.33$ and it displays blue zones with ratio < 1 at high C-rates and particle sizes, exemplifying its inferior performance. Fig. 8.2G shows the performance ratio for nanorods with $L_0/R_0 = 5$ and we see a performance ratio very close to or just greater than 1 for the entire range of particle radii and C-rates investigated. This further demonstrates $L_0/R_0 = 5$ as the threshold beyond which the nanorod performance starts to surpass that of the nanosphere. Finally, Fig. 8.2H compares the performance of nanorods with $L_0/R_0 = 10$ to that of the nanospheres, and it exhibits a ratio value greater than 1, sometimes as high as 1.2–1.4, for the entire span of C-rates and particle radius. This cements the fact that larger aspect ratio nanorods are viable candidates for silicon anodes.

The aforementioned results consider only radial diffusion in the nanowires. Short nanowires will have considerable areal contributions from the nanowire ends and axial diffusive transport can induce substantial concentration gradients in the axial direction. We perform a simple calculation to find the ratio of the curved surface ($A_{s,n} = 2\pi R L$) area to the circular end area $A_{c,n} = \pi R^2$. Note that we are still considering only one end, as in the experiments for the nanowires, the nanowires were grown directly from the current collector, so one end is not subject to any flux. This ratio comes out to be $\frac{A_s}{A_c} = 2\frac{L}{R}$. For L/R ratio changing from 1 to 10, the corresponding active area ratio changes from 2 to 20, thereby showing that the assumption of lithiation occurring primarily across the radial direction is strictly accurate for larger aspect ratios. Furthermore, for L/R ratio changing from 1 to 10, the corresponding diffusion lengths ratio that the Li$^+$ ions need to travel in the axial and radial directions go from 1 to 10 as well. As we go for larger aspect ratios, the diffusion length along the axial direction increases to much larger values as compared to the radial direction, thereby assuming radial diffusion being predominant very valid for long nanorods. The shortest nanorods that we consider have a L/R ratio 1.33, so we have an active area ratio 2.66 and diffusion length ratio 1.33 correspondingly. Allowing for axial diffusion as well and if diffusivity magnitudes in both the radial and axial direction are the same, we can predict that the nanorod will start outperforming spherical morphology at L/R ratios slightly less than 5 as well.

For short cylinder with both radial and axial diffusion contributions, concentration solve cognizant of both the diffusion modes is required (a 2D solve). The final expansion ratio in the radial direction (R_f/R_i) and axial direction (L_f/L_i) needs to be ascertained a priori as well which will be dependent on the magnitudes of the diffusivity magnitudes in the radial (D_r) and axial direction (D_z). Furthermore, the state of charge dependence of the radius ratio and length ratio needs to be ascertained. The assumption of unchanged length allowed for easier treatment of the radius ratio as a function of state of charge and is applicable for large aspect ratio nanorods.

8.3 Degradation phenomena in silicon anodes

Experimental and computational methodologies have been used to investigate lithium ion transport kinetics and associated stress generation inside high capacity anode active materials [38–41]. Lithium ion transport through two phase diffusion mechanism has been reported to be the contributing factor toward pulverization of silicon anodes [42]. Two-phase diffusional lithium transport dictates the first lithiation of the silicon active particle involving the conversion of crystalline silicon (Si) to amorphous lithiated silicon (Li_xSi) [43]. Subsequent delithiation-lithiation can be modeled as a single-phase diffusion process governed by the Fick's law of diffusion. The Cahn-Hilliard formulation based two phase transport model has been utilized by Chen *et al.* to model combined lithium diffusion and motion of the two-phase interface [44]. A similar formulation is adopted to capture two-phase diffusion in silicon in this chapter.

Silicon anodes have been extensively probed computationally for the large volume expansion based stress [45, 46]. Propagation of nucleated cracks in thin film silicon electrodes and silicon nanowires during the lithiation-delithiation process has been studied to identify the critical size [47]. Large elastic-plastic deformation of silicon has been incorporated in a few computational studies to model the transport and mechanics [48, 49]. The investigation of fracture tendencies inside silicon particle with surface film mimicking solid electrolyte interphase/secondary-phase layer is amenable through the lattice spring formalism [50, 51].

8.3.1 Mechano-electrochemical stochastics in silicon anodes

The problem formulation involves modeling of two coupled physical phenomenon, solid-state diffusion of lithium inside the active material particles and the resulting mechanical behavior involving large deformation, stress and fracture. The mechanical behavior is elucidated using a two-dimensional random lattice spring formulation with diffusion transport computed on a juxtaposed spherical coordinates computational grid. The complexity of the approach is in the accurate mapping of the field variables (like concentration) from the spherical grid to the spring network while accounting for large deformation of the spring network. The approach has been demonstrated for pristine low volumetric expansion active material like graphite [52] and high capacity active material particles like silicon and tin with two-phase diffusion [15]. In this section, we show the model and do a comparative analysis of the fracture map for two-phase versus single phase diffusion inside silicon while accounting for the presence of a surface film.

At the single-particle level, the rate-determining transport mechanism is the solid-state diffusion of Li-ion inside the active material. The electrolyte ionic diffusivity and conductivity is assumed to be high enough to neglect concentration and potential gradients of Li-ion within the electrolyte. Therefore, the active material is subjected to constant Li-ion flux from all directions irrespective of its position from the current collector under galvanostatic operation. Two-phase diffusion inside the active particle is modeled using the Cahn-Hilliard equation provided in Eq. (8.27). The corresponding single-phase diffusion formulation based on Fick's law is given in Eq. (8.28).

$$\frac{\partial \hat{c}_s}{\partial t} = \nabla . M_{Li}(\hat{c}_s) \nabla \left\{ c_{s,max} RT \left[\omega(1 - 2\hat{c}_s) + \ln \frac{\hat{c}_s}{1 - \hat{c}_s} \right] - \kappa \nabla^2 \hat{c}_s \right\} \quad (8.27)$$

$$\frac{\partial \hat{c}_s}{\partial t} = \nabla(D_s \nabla \hat{c}_s) \quad (8.28)$$

Symmetry boundary condition is applied at the particle center (Eq. 8.29) and constant flux boundary condition is applied at the particle surface (Eqs. 8.30 and 8.31). The particle surface boundary condition form manifests differently for the two-phase and single-phase models while the symmetry boundary condition form is the same for both models.

$$\text{Two} - \text{Phase and Single} - \text{Phase} : \left. \frac{\partial \hat{c}_s}{\partial r} \right|_{r=0} = 0 \quad (8.29)$$

$$\text{Two} - \text{Phase} : M_{Li}(\hat{c}_s) \nabla \left\{ c_{s,max} RT \left[\omega(1 - 2\hat{c}_s) + \ln \frac{\hat{c}_s}{1 - \hat{c}_s} \right] - \kappa \nabla^2 \hat{c}_s \right\} \bigg|_{r=R_p}$$

$$= \frac{I_{applied}}{SF} \quad (8.30)$$

$$\text{Single} - \text{Phase} : D_s \nabla \hat{c}_s |_{r=R_p} = \frac{I_{applied}}{SF} \quad (8.31)$$

Here $\hat{c}_s$ is the nondimensional concentration variable, defined as $\hat{c}_s = \frac{c_s}{c_{s,max}}$, where c_s denotes the molar concentration of lithium within the active material and $c_{s,max}$ corresponds to the maximum intercalated amount of lithium in the active material. M_{Li} is the mobility of lithium with functional dependence on concentration, defined as $M_{Li} = \frac{D_{Li}}{c_{s,max} RT}(\hat{c}_s(1 - \hat{c}_s))$, where D_{Li} corresponds to the diffusivity of lithium, R is the universal gas constant and T is temperature in Kelvin. The dimensionless parameter ω (Eq. 8.27) controls the shape of the double-well energy function characterizing the lithium-rich and lithium-depleted phases. The parameter κ modulates the contribution of the large concentration gradient at the two-phase interface to the free energy. Numerical solution of the Cahn-Hilliard model requires simplification of the fourth order governing differential equation into two coupled second order equations which are solved iteratively. Comparatively, the Fick's law formulation for single-phase diffusion is relatively straightforward as it leads directly to a single second order governing differential equation.

The concentration equations are further discretized using finite volume approach on a spherical grid. It is only solved for in the active material domain. The surface film domain is assumed devoid of concentration gradients and experiences stress from the active material swelling/shrinking during lithiation/delithiation. Consequently, a nominal concentration of zero is assigned to the surface film layer and only mechanics is solved for the film layer.

A modified lattice spring model capable of incorporating both large and small deformation is utilized to capture mechanical degradation inside the silicon anode with surface film. Detailed illustration of the lattice spring methodology amalgamating large deformation has been provided in Barai et al [15]. The salient features of this framework are summarized here along with the modifications required to incorporate surface film mechanics.

The governing virtual work expression utilized for this model is given in Eq. (8.32) [53].

$$\int_{t}{}^{t+\Delta t}_{t}V S_{ij} \delta_{t}^{t+\Delta t} \varepsilon_{ij} d^{t}V - \int_{t}{}^{t+\Delta t}_{t}A F_{i} \delta_{t}^{t+\Delta t} u_{i} d^{t}A = {}^{t+\Delta t}R \tag{8.32}$$

Here ^{t}V and ^{t}A signify the volume and area at the previous equilibrium configuration, $^{t+\Delta t}_{t}S_{ij}$ is the 2nd Piola-Kirchoff stress tensor, $^{t+\Delta t}_{t}\varepsilon_{ij}$ is the Green-Lagrange strain tensor, $^{t+\Delta t}_{t}F_{i}$ is the external force, $^{t+\Delta t}_{t}u_{i}$ is the displacement and $^{t+\Delta t}R$ is the residual. First term on the LHS represents the internal energy while the second term denotes the energy due to external loads. There are no external forces considered ($^{t+\Delta t}_{t}F_{i}=0$) in the present model, with the entire load coming from the internal diffusion and volumetric expansion-induced stress. Equilibrium configuration is solved for using residual minimization which signifies the balancing of the internal and external forces. The expanded form of the Green-Lagrange strain tensor is given in indicial notation by Eq. (8.33) [54].

$$^{t+\Delta t}_{t}\varepsilon_{ij} = \frac{1}{2}\left(\frac{\partial_{t} u_{i}}{\partial x_{j}} + \frac{\partial_{t} u_{i}}{\partial x_{j}} + \frac{\partial_{t} u_{k}}{\partial x_{i}}\frac{\partial_{t} u_{k}}{\partial x_{j}}\right) \tag{8.33}$$

Here, $_{t}u_{i}$ signifies the displacement from time step t to $t+\Delta t$ and x_{j} corresponds to the spatial coordinate. Linear elastic relation is considered between

Performance degradation modeling in silicon anodes **Chapter | 8 315**

the second Piola-Kirchoff stress and Green-Lagrange strain along the axial direction given by Eq. (8.34).

$$_{t}^{t+\Delta t}S_{11} = E \cdot {}_{t}^{t+\Delta t}\varepsilon_{11} \tag{8.34}$$

The Young's modulus, denoted by E, assumes different values for the active material and surface film. Other mechanical properties like Poisson's ratio and fracture threshold energy also assumes different values for silicon and film. Correspondingly, an indicator function is used to identify the different phases. The large volume expansion-based lattice spring model is applied to the silicon active material domain while small deformation-based lattice spring model mechanics is assigned to the surface film layer.

Fig. 8.3 shows a schematic representation of the initial, previous and final configuration of the pristine Si particle without surface film in the updated lagrangian lattice spring framework. An additive decomposition of the displacements instead of strains has been adopted here. The diffusion of Li atoms inside the active particle is a transient process resulting in concentration change which

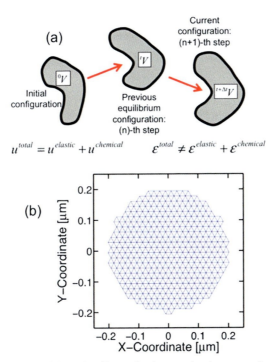

FIG. 8.3 (A) Schematic of the updated lagrangian framework to incorporate large volume expansion within the lattice spring model through additive decomposition of displacements. (B) Visualization of a 30 × 30 lattice spring mesh for a 400 nm diameter Si particle [15].

impacts the elastic length as the silicon particle expands/contracts upon lithiation/delithiation and is given by Eq. (8.35).

$$l_{elastic} = \frac{l_{total}}{1+(\Delta \hat{c}_s \Omega c_{s,max})} \tag{8.35}$$

Here, $l_{elastic}$ is the length of the spring element only due to elastic deformation, Ω is the volumetric expansion coefficient and $\Delta \hat{c}_s$ is the concentration change at a particular point inside the particle under a small time increment. The axial deformation giving rise to the internal stress can then be estimated as

$$\Delta l = l_{elastic} - L \tag{8.36}$$

where Δl signifies the change in length and L corresponds to the spring length at the previous equilibrium configuration. Since the propagation of mechanical stress waves is much faster than the diffusive transport of lithium, mechanical equilibrium inside the particle is achieved as soon as the lithium concentration change is established. Consequently, the diffusion behavior is time-dependent; however, a quasistatic stress solve is enough for elucidating the mechanics.

Microcrack formation is governed by the strain energy accumulation inside the spring element as the element deforms. If the element energy exceeds its fracture threshold, it is considered broken, and irreversibly removed from the lattice network. The strain energy of a spring element is estimated by Eq. (8.37).

$$\psi_e^{n+1} = \psi_e^n + \frac{1}{2}\Delta \vec{f} \cdot \Delta \vec{u} \tag{8.37}$$

The incremental internal force and displacement of the lattice spring element is denoted by $\Delta \vec{f}$ and $\Delta \vec{u}$ respectively. ψ_e^{n+1} and ψ_e^n represent the strain energy at the current and previous equilibrium configuration. Removal of the spring from the network is accomplished by neglecting its contribution to the stiffness matrix for subsequent force deformation analysis. The load carried by the broken spring is redistributed amongst the neighboring elements resulting in stress concentration effects. In this way, rupturing of the spring elements leads to the nucleation of microcracks which can further coalesce together to form spanning cracks.

8.3.2 Representative highlights

The investigation into the mechano-electrochemical interactions of active material and surface film requires accurate quantification of the mechanical properties of individual layers/phases. These parameters include Young's modulus, Poisson's ratio, mean fracture energy threshold per unit area and expansion coefficient. The mean fracture threshold energy per unit area is an indicator of the amount of strain energy that the material can absorb without fracture. If the strain energy exceeds this threshold, crack nucleation is initiated. The elastic modulus of the material directly relates to the amount of fracture damage

TABLE 8.3 List of transport and mechanical parameters for silicon and surface film.

Parameter	Symbol	Value
Diffusivity	D_{Li} (m²/s)	2×10^{-12}
Maximum lithium concentration	$c_{s,\,max}$ (mol/m³)	29.52×10^4
Density	ρ_{Si} (kg/m³)	2200
Temperature	T (K)	300
Non-dimensional enthalpy of mixing	ω (—)	2.6
Gradient energy coefficient	κ (J/m)	2.0×10^{-9}
Young's modulus of amorphous silicon	E_{aSi} (GPa)	45.0
Young's modulus of crystalline silicon	E_{cSi} (Gpa)	90.0
Young's modulus of surface film	E_{film} (GPa)	66.0
Fracture threshold energy of amorphous silicon	$\Psi_{t,\,aSi}$ (J/m²)	5.0
Fracture threshold energy of crystalline silicon	$\Psi_{t,\,cSi}$ (J/m²)	10.0
Fracture threshold energy of surface film	$\Psi_{t,\,film}$ (J/m²)	10.0

or crack propagation inside the material, hence material with high Young's modulus will have a greater propensity for fracture provided the fracture threshold energy for both the materials is the same. Material parameters for silicon active material and surface film (SEI/secondary phase) are reported in Table 8.3. In addition, the table also lists the bulk diffusion coefficient for Li in silicon and the nominal spherical silicon active material size [39, 44, 55, 56].

8.3.2.1 Concentration and damage profiles in silicon particle with surface film

Fig. 8.4 demonstrates the concentration contour plots for lithiation of a 500 nm crystalline silicon particle with 50 nm surface film through two-phase diffusion mechanism at 1C current rate of discharge. The areal current density at 1C rate is computed based on the theoretical gravimetric specific capacity of silicon (4200 mAh/g). Thus, the areal current density is calculated as (R_p is the particle diameter).

$$i = \frac{4200 \left(\frac{A}{kg}\right) \cdot \rho_{Si} \left(\frac{kg}{m^3}\right) \cdot \frac{4}{3}\pi R_p^3 \,(m^3)}{4\pi R_p^2 \,(m^2)} = 1400 \cdot \rho_{Si} \cdot R_p \,(A/m^2) \quad (8.38)$$

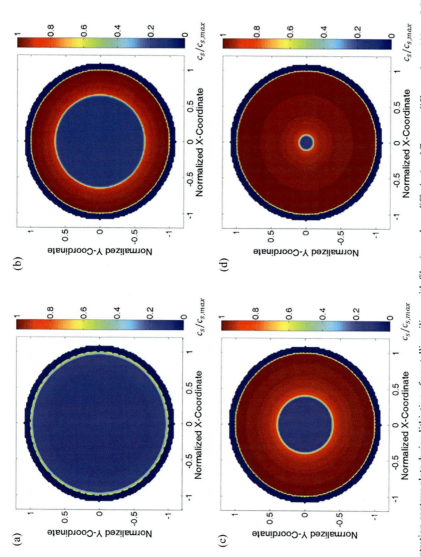

FIG. 8.4 Concentration contour plots during lithiation of crystalline silicon with film (two phase diffusion) at 1C rate at different times (A) $t=0.05$, (B) $t=0.35$, (C) $t=0.65$ and (D) $t=0.95$. The silicon particle boundary is delineated with a dashed yellow line. Presence of a sharp interface exhibits the two-phase diffusion mechanism. Film layer is devoid of concentration gradients and is assigned a nominal zero concentration [16].

Based on the model described earlier, the formation of a two-phase interface, across which sharp change in concentration occurs, is observed. As lithium ions diffuse into the silicon particle and alloyed with Si, lithium rich phase forms at the surface while the lithium deficient phase exists beyond the sharp two-phase interface close to the center of the particle. The two-phase interface moves inward as the lithiation progresses and reaches close to the center of the particle at the end of lithiation or discharge process. The lithiation profile reveals the two-phase diffusion mechanism governed by the Cahn-Hilliard formulation. The surface film (beyond the dashed yellow line in Fig. 8.4) is assigned a nominal Li-ion concentration of zero. Lithium diffusion inside the surface film is not considered with the focus aimed at understanding the mechanical response of the silicon and film system.

Alternatively, for the amorphous silicon particle, smooth transition of concentration with particle radius is observed, thereby revealing the single-phase diffusion mechanism governed by Fick's law. At the end of lithiation ($t^* = 1$), normalized concentration across the entire particle reaches close to unity, thereby indicating that the particle is full of lithium without any significant concentration gradient.

Fig. 8.5 demonstrates the damage profile within crystalline silicon with surface film during a single lithiation. The figures show the evolution of the lattice spring representation of the silicon active material (shown in blue) and surface film (shown in red) cluster. Fracture is interpreted as the missing bonds (white space) in the blue and red lattice system. Relatively large magnitude of fracture is observed in the crystalline silicon as compared to amorphous silicon thereby underlining the importance of the diffusion mechanism to fracture. In crystalline silicon, the existence of a sharp two-phase interface during two-phase diffusion leads to the formation of lithium rich and lithium poor phases with large discrepancy in strains. A sharp concentration gradient related stress exists only close to the two-phase interface creating diffusion-induced load close to the two-phase interface; however, its magnitude is small compared to the stress due to volumetric strain inhomogeneity. As lithiation progresses, the lithium rich phase is pushed outwards inducing tensile stress on the particle surface which leads to the initiation of surface crack fronts. Thus, the crystalline silicon particle disintegrates at the surface. Fragmentation of the brittle surface film is also observed as it is unable to accommodate the large volumetric expansion of the silicon particle.

In the single-phase amorphous silicon particle with surface film, the volumetric strain is homogeneously distributed throughout the particle dimension due to the smoother concentration variation. There is no lithium atom segregation of lithium rich and lithium poor zones. Thus, the lithium atoms can move to the center of the particle more freely as compared to the crystalline silicon particle where the lithium atoms are constrained by the two-phase interface. The concentration gradient-induced stress is always present throughout the amorphous silicon particle. Large volumetric strain is observed inside the amorphous

320 PART | IV Achieving high performance

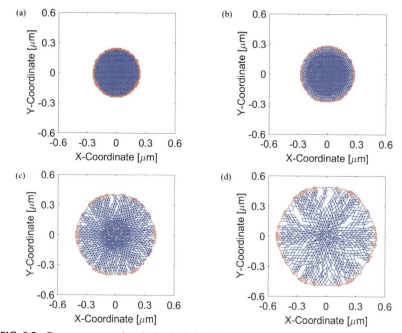

FIG. 8.5 Damage contour plots during lithiation of crystalline silicon with film (two phase diffusion) at 1C rate at different times (A) $t=0.05$, (B) $t=0.35$, (C) $t=0.65$ and (D) $t=0.95$. Silicon (shown in blue) shows fracture near particle surface due to two-phase lithiation and surface film (red) shows severe disintegration [16].

Si particle as well; however, the absence of an interface barrier (like the interface in two-phase diffusion process) allows for smoother expansion of the particle. Consequently, the total damage is much lower in amorphous silicon compared to the crystalline silicon because of smaller diffusion-induced stress and smaller disparity in volumetric strains throughout the radius of the amorphous Si particle.

Fragmentation of the surface film is observed as it is unable contain the large expansion of the silicon particle. Surface film property modification such that it exhibits elastomeric deformation is essential to mitigating surface film fracture in silicon. Elastomers can undergo large elastic deformation, stretching and returning to their original shape in a reversible manner. Thus, changing the nature of the diffusion process can help mitigate fracture inside the silicon particle; however, brittle surface film fracture continues unabated for both crystalline and amorphous silicon due to large inherent volumetric expansion of the Si particle.

An interesting feature can be observed from the fracture profile of crystalline silicon [16]. The crystalline silicon particle fractures close to the surface while fracture originates close to the center of the amorphous silicon particle during lithiation [16]. The amorphous silicon particle behavior resembles graphite behavior in the sense that tensile stresses originate at the center of the particle during lithiation with compressive stress at the surface. During delithiation, tensile stresses occur at the surface while compressive stresses form at the center. As silicon exhibits more strength in compression than tension, higher fracture is observed under tension. Consequently, the particle fractures in regions where tensile stresses originate. Thus, the differing fracture behavior of crystalline versus amorphous silicon is a direct consequence of the diffusion behavior inside the particle. Large volumetric expansion, by itself, cannot explain the fracture behavior as it occurs in both crystalline and amorphous silicon with lithiation. It is the nature of the diffusion process (single phase vs two-phase) that determines the fracture characteristics. Single phase diffusion materials (graphite, amorphous silicon) exhibit tensile fracture close to center of the particle during lithiation while two-phase diffusion materials (crystalline silicon) exhibit tensile fracture close to surface of the particle during lithiation and vice versa.

Quantification of the fracture inside the silicon particle with surface film is defined by defining a parameter named total damage density, DD_{total}. It is defined as the ratio of number of broken bonds to the total number of bonds in the silicon active material with surface film. Individual silicon/film damage density, DD_{Si} and DD_{film}, are also defined as the ratio of number of broken bonds in silicon/film ($n_{broken,total}$) to the total number of bonds in the silicon (n_{Si}) and film (n_{film}) agglomerate. The definitions have been kept the same mentioned earlier such that $DD_{total} = DD_{Si} + DD_{film}$.

$$DD_{total} = \frac{n_{broken,total}}{n_{Si} + n_{film}} \tag{8.39}$$

$$DD_{Si} = \frac{n_{broken,Si}}{n_{Si} + n_{film}}, DD_{film} = \frac{n_{broken,film}}{n_{Si} + n_{film}} \tag{8.40}$$

8.3.2.2 Effect of surface film on damage

As surface film thickness is increased, damage density inside the film decreases because there is no diffusion-induced stress inside the film, and hence, fracture inside the film will be mostly concentrated close to active material surface and portion of film far from the active material is relatively unaffected. In addition, as the film thickness increases there is more support structure for the film. The active material damage density is relatively unaffected by film thickness variation. Hence, a small thickness of surface film is desirable to reduce inert material quantity provided the integrity of the film can be maintained. Total damage density trend is dominated by damage inside film and hence mirrors the decreasing trend of film damage as film thickness increases.

As the film Young's modulus is increased, the fraction of broken bonds inside the film increases since more strain energy must be released. Increased Young's modulus implies increased strain energy accumulation which leads to higher number of broken bonds. A very interesting observation is that the fraction of broken bonds inside active material also shows a monotonically increasing trend with increased film's elasticity modulus [16]. Thus, the Young's modulus of the film has a direct impact on fracture inside the active material. As the elastic modulus of film increases, damage inside the film and the active material increases. This positive correlation provides us with a way to reduce fracture inside active material. Using a film of lower elastic modulus will help diminish fracture damage inside the active material. Mechanical damage hinders diffusion and causes particle isolation which leads to capacity fade, poor cycle life and reduced rate capabilities. Consequently, using a low elastic modulus film will improve the battery capacity and cycle life of Si based anode.

8.3.2.3 Fracture phase map and correlation to capacity fade

Fig. 8.6 depicts the fracture phase map quantified through the microcrack density for silicon anodes as a function of the partial molar volume, particle size and operating current rate [15,16]. The damage map for pristine Si particles without surface film exhibiting two-phase diffusion is displayed in Fig. 8.6A. It shows that smaller sized particles have little mechanical degradation even at large partial molar volumes while large sized particles can only sustain themselves at low partial molar volumes ratio. As the partial molar volumes are increased

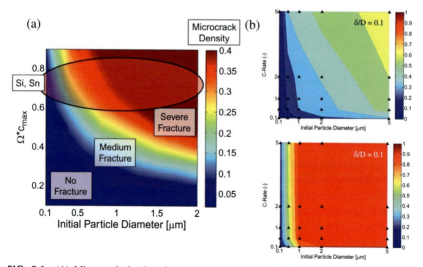

FIG. 8.6 (A) Microcrack density phase map between particle size and partial molar volume observed during a single lithiation-delithiation cycle for pristine Si particle without surface film [15]. (B) Fracture phase map for variation of particle size and C-rate with film thickness to particle size ratio, $\delta/D = 0.1$, for amorphous silicon with film (top) and crystalline silicon with film (bottom). Crystalline silicon exhibits much severe fracture as compared to amorphous silicon with increase in particle size and C-rate [16].

close to the ranges observed in Si anodes, fracture is imminent beyond a particle size of 100 nm. Fig. 8.6B give the fracture phase map for amorphous and crystalline silicon with film clearly demonstrating the need for going toward amorphous silicon to mitigate the fracture tendencies. Crystalline silicon particles beyond 100–200 nm show severe fracture at all C-rates while amorphous silicon particles can sustain reasonably well up to 1–2 μm. A direction toward improving the performance of silicon electrodes can be gleaned from the discussion presented earlier. Conversion of crystalline silicon to amorphous silicon is reported to occur over the first few lithiation-delithiation cycles. It is the opinion of the author to introduce extremely slow lithiation and delithiation (<C/100) of crystalline silicon particles with mid-ranged particle size (∼1–2 μm) for the first few cycles to enable complete conversion into amorphous silicon with minimal fracture. The decreased fracture tendency of amorphous silicon is a desirable characteristic and then can then be utilized to cycle the electrodes at higher C-rates. For amorphous silicon, the damage density for 1–2 μm particle lies within the reasonable limit (<0.2) for C-rates up to 2C. Hence, it is recommended an upper limit for C-rate of 2C for cycling of amorphous silicon particles. Thus, fracture inside the silicon active material can be minimized using the aforementioned cycling technique.

However, surface film rupture is inevitable for both amorphous and crystalline silicon owing to the high volumetric expansion of the silicon particle and brittle mechanics of the film. Consequently, capacity fade due to SEI film breakdown cannot be mitigated unless surface film modification is done to obtain elastomeric films. A small improvement in the fracture characteristics is obtained by decreasing the Young's modulus of the brittle film layer; however, it is not enough to resolve the capacity deterioration of silicon electrodes due to film breakage and resulting fresh SEI formation.

Fig. 8.7 shows the effects of C-rate and film to particle Young's moduli ratio on total damage density in a silicon particle of 500 nm covered with a surface film layer of 50 nm [17]. Both amorphous and crystalline Si particles with surface film are investigated. Fracture characteristics for amorphous and crystalline Si exhibit wide disparity owing to the nature of the underlying intercalated lithium diffusion process.

As the C-rate increases, damage density increases in both crystalline and amorphous Si. Similar trend is observed for increase of surface film to Si particle Young's moduli ratio. However, C-rate variation has a more pronounced effect on the total damage density as compared to Young's moduli change. Larger concentration gradients coupled with high volume expansion exacerbate the mechanical damage at high C-rates. Increase in film Young's modulus is reflected in the increases of strain energy accumulation rate. Fracture criterion is dictated by the fracture threshold energy which is kept fixed for all the variations explored. Faster rate of strain energy accumulation with higher Young's moduli results in larger number of bonds strain energy exceeding the fracture

324 PART | IV Achieving high performance

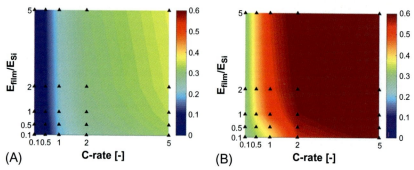

FIG. 8.7 Contour plots for total damage density as a function of C-rate and film to Si Young's moduli ratio (E_{film}/E_{Si}) for (A) amorphous Si with film and (B) crystalline Si with film. The Young's moduli ratio is varied by changing the film Young's modulus while keeping the Si Young's modulus constant. Crystalline Si with film exhibits higher damage density as compared to amorphous Si with film. [17].

threshold. This is demonstrated by the relatively higher magnitude of damage with increase in film to particle Young's modulus ratio.

Fracture in the silicon with surface film occurs all throughout the simulation time. As fracture progresses, the damaged bonds are removed implying there are no forces in these bonds. For computation of stress variation with time, we plot the stress in an undamaged surface film bond as the lithiation-delithiation progresses. Fig. 8.8A shows the evolution of stress in an undamaged surface film bond with lithiation-delithiation of an amorphous silicon particle with surface film at multiple C-rates. Here normalized time refers to the time for discharge/charge at a particular C rate divided by 1 h. For a 1C rate lithiation/delithiation, the total time for charge and discharge will be approximately 2 h and is normalized to 2. Correspondingly, for 2C current rate the normalized time goes to half of time for 1C rate and so on.

Stress in the surface film increases as lithiation progresses and is a maximum close to the end of lithiation. During delithiation, stress in the surface film relaxes as the particle contracts. There are no concentration gradient-induced stresses inside the surface film; the fracture is due to the film being unable to accommodate the high volumetric expansion of the Si particle. Consequently, we see minor increase in surface film stress with C-rate because a major contribution to stress is due to high volume expansion which is similar for all C-rates at the end of lithiation-delithiation. The C-rate change affects the concentration gradient-induced stress in the Si particle which in turn leads to small variation in stress in the surface film.

Fig. 8.8B shows the evolution of damage density with lithiation of an amorphous Si particle at 1C rate. The Si particle diameter simulated is 500 nm with 50 nm SEI thickness. Here, we try to characterize the damage that can lead to formation of new sites exposed to electrolyte that can cause irreversible Li loss

Performance degradation modeling in silicon anodes Chapter | 8 325

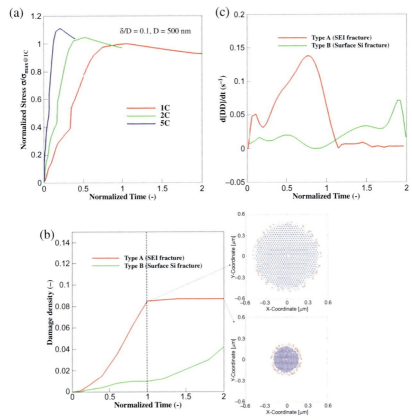

FIG. 8.8 (A) Evolution of stress in undamaged surface film bond with lithiation-delithiation of amorphous silicon particle with surface film at multiple C-rates. (B) Damage density vs time for lithiation-delithiation of 500 nm amorphous silicon particle with 50 nm surface film exhibiting single phase diffusion at 1C rate. (C) Evolution of rate of change of total damage density in silicon with surface film agglomerate at 1C rate lithiation-delithiation. The Si particle diameter simulated is 500 nm with 50 nm SEI thickness [17].

due to fresh SEI formation. Thus, two types of damage A and B have been defined that can contribute to Li inventory depletion. Type A damage is characterized by fracture of the surface film layer (as simulated by the atomistic model). Here the surface film is treated equivalent to chemical SEI film layer that forms on the Si particle surface. The SEI film layer reformation with cycling contributes to irreversible Li loss during battery operation. Type B damage is characterized by breaking up of the surface Si active material which can lead to fresh electrolyte flow in the silicon active material cracks and irreversible lithium loss due to SEI formation in these cracks.

It is evident from Fig. 8.8B that type A damage supersedes type B damage and thus, the dominant cause of capacity fading can be attributed to brittle SEI layer breakdown. In addition, type A damage is predominant during lithiation of the Si particle leading to high volumetric expansion of Si, which ruptures the surface film. Meanwhile, type B damage due to rupture of the Si surface is higher during delithiation as compared to lithiation. This is due to the formation of compressive stresses at the particle surface during lithiation and tensile stresses at the particle surface during delithiation. Since materials are more resilient to fracture in compression than in tension, we see higher amount of surface Si damage during delithiation.

Fig. 8.8C plots the rate of change of damage density corresponding to the two modes of damage defined previously (type A and type B) for the same active material and film structure and operating conditions as shown in Fig. 8.8B. Damage density rate is higher in surface film (SEI fracture) during lithiation due to high volumetric expansion and goes close to zero during delithiation owing to stress relaxation due to silicon particle contraction. There is more surface silicon fracture during delithiation as compared to lithiation due to the formation of tensile stresses at the surface during delithiation. This is evident from the higher rate of surface Si damage density increase during delithiation while it hovers close to zero during lithiation.

8.4 Summary and outlook

The complex interactions in silicon anodes present at a multitude of length and time scales necessitates a need for a coupled particle-scale electrochemistry-transport and large deformation mechanics solve, which can be upscaled to the cell level [57]. Furthermore, the immense volumetric fluctuations characteristic of silicon anodes contributes to a unique mode of capacity degradation through continuous breakdown and reformation of the SEI which serves the dual function of anode surface passivation and Li$^+$ ion-diffusion barrier. Coulombic efficiency and cycling performance of Si anodes is predominantly determined by its SEI state; however, SEI evolution, composition, and morphology understanding is lacking in the literature. Atomistic studies using density functional theory and *ab-initio* molecular dynamics can help bridge this knowledge gap and provide SEI kinetics and transport parameters [58]. These can be utilized in coupled reaction-diffusion-stress single-particle and macrohomogeneous models for prediction of performance and degradation-induced capacity fade of silicon-based lithium-ion batteries that exhibit good match with experimental cycling data sets. The upscaling of particle-scale models to cell level models can be accomplished through linkage of the particle volume change to porosity/active material fraction variations in the porous silicon electrode. It will also require appropriate mapping of particle level stress to cell level stress measures either through reduced order formulations [59] or a fully resolved direct numerical simulation (DNS) model [60].

Acknowledgments

The authors acknowledge the Electrochemical Society and Elsevier for figures reproduced in this chapter from the referenced publications of their journal, *Journal of The Electrochemical Society* and *Journal of Power Sources*, respectively.

References

[1] Y. Wang, G. Cao, Developments in nanostructured cathode materials for high-performance lithium-ion batteries, Adv. Mater. 20 (12) (2008) 2251–2269.
[2] M. Yoshio, H. Wang, K. Fukuda, Spherical carbon-coated natural graphite as a Lithium-ion battery-anode material, Angew. Chem. 115 (35) (2003) 4335–4338.
[3] J. Chen, Recent progress in advanced materials for lithium ion batteries, Materials 6 (1) (2013) 156–183.
[4] M. Kummer, et al., Silicon/polyaniline nanocomposites as anode material for lithium ion batteries, J. Electrochem. Soc. 161 (1) (2014) A40–A45.
[5] M. Green, et al., Structured silicon anodes for lithium battery applications, Electrochem. Solid St. 6 (5) (2003) A75–A79.
[6] P. Limthongkul, et al., Electrochemically-driven solid-state amorphization in lithium-silicon alloys and implications for lithium storage, Acta Mater. 51 (4) (2003) 1103–1113.
[7] S.R. Gowda, et al., Three-dimensionally engineered porous silicon electrodes for Li ion batteries, Nano Lett. 12 (12) (2012) 6060–6065.
[8] X.H. Liu, et al., Size-dependent fracture of silicon nanoparticles during lithiation, ACS Nano 6 (2) (2012) 1522–1531.
[9] X. Zhang, W. Shyy, A.M. Sastry, Numerical simulation of intercalation-induced stress in Li-ion battery electrode particles, J. Electrochem. Soc. 154 (10) (2007) A910–A916.
[10] J. Cho, Porous Si anode materials for lithium rechargeable batteries, J. Mater. Chem. 20 (20) (2010) 4009–4014.
[11] S.W. Lee, et al., Fracture of crystalline silicon nanopillars during electrochemical lithium insertion, Proc. Natl. Acad. Sci. 109 (11) (2012) 4080–4085.
[12] R. Deshpande, et al., Battery cycle life prediction with coupled chemical degradation and fatigue mechanics, J. Electrochem. Soc. 159 (10) (2012) A1730–A1738.
[13] J. Xu, et al., Electrode side reactions, capacity loss and mechanical degradation in lithium-ion batteries, J. Electrochem. Soc. 162 (10) (2015) A2026–A2035.
[14] A. Verma, A.A. Franco, P.P. Mukherjee, Mechanistic elucidation of Si particle morphology on electrode performance, J. Electrochem. Soc. 166 (15) (2019) A3852–A3860.
[15] P. Barai, P.P. Mukherjee, Mechano-electrochemical stochastics in high-capacity electrodes for energy storage, J. Electrochem. Soc. 163 (6) (2016) A1120–A1137.
[16] A. Verma, P.P. Mukherjee, Mechanistic analysis of mechano-electrochemical interaction in silicon electrodes with surface film, J. Electrochem. Soc. 164 (14) (2017) A3570–A3581.
[17] D.E. Galvez-Aranda, et al., Chemical and mechanical degradation and mitigation strategies for Si anodes, J. Power Sources 419 (2019) 208–218.
[18] Y.-T. Cheng, M.W. Verbrugge, Diffusion-induced stress, interfacial charge transfer, and criteria for avoiding crack initiation of electrode particles, J. Electrochem. Soc. 157 (4) (2010) A508–A516.
[19] R. Deshpande, Y.-T. Cheng, M.W. Verbrugge, Modeling diffusion-induced stress in nanowire electrode structures, J. Power Sources 195 (15) (2010) 5081–5088.

[20] J. Li, et al., Asymmetric rate behavior of Si anodes for lithium-ion batteries: ultrafast de-lithiation versus sluggish Lithiation at high current densities, Adv. Energy Mater. 5 (6) (2015) 1401627.
[21] X.H. Liu, et al., Self-limiting lithiation in silicon nanowires, ACS Nano 7 (2) (2013) 1495–1503.
[22] I. Ryu, et al., Size-dependent fracture of Si nanowire battery anodes, J. Mech. Phys. Solids 59 (9) (2011) 1717–1730.
[23] M.T. McDowell, et al., Novel size and surface oxide effects in silicon nanowires as lithium battery anodes, Nano Lett. 11 (9) (2011) 4018–4025.
[24] R. Grantab, V.B. Shenoy, Pressure-gradient dependent diffusion and crack propagation in lithiated silicon nanowires, J. Electrochem. Soc. 159 (5) (2012) A584–A591.
[25] M. Ashuri, Q. He, L.L. Shaw, Silicon as a potential anode material for Li-ion batteries: where size, geometry and structure matter, Nanoscale 8 (1) (2016) 74–103.
[26] R. Teki, et al., Nanostructured silicon anodes for lithium ion rechargeable batteries, Small 5 (20) (2009) 2236–2242.
[27] Z. Wen, et al., Silicon nanotube anode for lithium-ion batteries, Electrochem. Commun. 29 (2013) 67–70.
[28] R. Chandrasekaran, et al., Analysis of lithium insertion/deinsertion in a silicon electrode particle at room temperature, J. Electrochem. Soc. 157 (10) (2010) A1139–A1151.
[29] R. Chandrasekaran, T.F. Fuller, Analysis of the lithium-ion insertion silicon composite electrode/separator/lithium foil cell, J. Electrochem. Soc. 158 (8) (2011) A859–A871.
[30] T.R. Garrick, et al., Modeling volume change due to intercalation into porous electrodes, J. Electrochem. Soc. 161 (8) (2014) E3297–E3301.
[31] T.R. Garrick, et al., Modeling volume change in dual insertion electrodes, J. Electrochem. Soc. 164 (11) (2017) E3552–E3558.
[32] D.J. Pereira, J.W. Weidner, T.R. Garrick, The effect of volume change on the accessible capacities of porous silicon-graphite composite anodes, J. Electrochem. Soc. 166 (6) (2019) A1251–A1256.
[33] W. Mai, A. Colclasure, K. Smith, A reformulation of the Pseudo2D battery model coupling large electrochemical-mechanical deformations at particle and electrode levels, J. Electrochem. Soc. 166 (8) (2019) A1330–A1339.
[34] C.K. Chan, et al., High-performance lithium battery anodes using silicon nanowires, Nat. Nanotechnol. 3 (1) (2008) 31.
[35] T. Swamy, Y.-M. Chiang, Electrochemical charge transfer reaction kinetics at the silicon-liquid electrolyte interface, J. Electrochem. Soc. 162 (13) (2015) A7129–A7134.
[36] M. Guo, G. Sikha, R.E. White, Single-particle model for a lithium-ion cell: thermal behavior, J. Electrochem. Soc. 158 (2) (2011) A122–A132.
[37] Y. Ji, Y. Zhang, C.-Y. Wang, Li-ion cell operation at low temperatures, J. Electrochem. Soc. 160 (4) (2013) A636–A649.
[38] K. Zhao, et al., Large plastic deformation in high-capacity lithium-ion batteries caused by charge and discharge, J. Am. Ceram. Soc. 94 (s1) (2011) s226–s235.
[39] K. Zhao, et al., Inelastic Hosts as Electrodes for High-Capacity Lithium-Ion Batteries, AIP, 2011.
[40] S. Huang, et al., Stress generation during lithiation of high-capacity electrode particles in lithium ion batteries, Acta Mater. 61 (12) (2013) 4354–4364.
[41] G. Bucci, et al., Measurement and modeling of the mechanical and electrochemical response of amorphous Si thin film electrodes during cyclic lithiation, J. Mech. Phys. Solids 62 (2014) 276–294.

[42] T. Song, et al., Arrays of sealed silicon nanotubes as anodes for lithium ion batteries, Nano Lett. 10 (5) (2010) 1710–1716.
[43] S.K. Soni, et al., Diffusion mediated lithiation stresses in Si thin film electrodes, J. Electrochem. Soc. 159 (9) (2012) A1520–A1527.
[44] L. Chen, et al., A phase-field model coupled with large elasto-plastic deformation: application to lithiated silicon electrodes, J. Electrochem. Soc. 161 (11) (2014) F3164–F3172.
[45] S.K. Soni, et al., Stress mitigation during the lithiation of patterned amorphous Si islands, J. Electrochem. Soc. 159 (1) (2011) A38–A43.
[46] K. Higa, V. Srinivasan, Stress and strain in silicon electrode models, J. Electrochem. Soc. 162 (6) (2015) A1111–A1122.
[47] T.K. Bhandakkar, H. Gao, Cohesive modeling of crack nucleation under diffusion induced stresses in a thin strip: implications on the critical size for flaw tolerant battery electrodes, Int. J. Solids Struct. 47 (10) (2010) 1424–1434.
[48] C.V. Di Leo, E. Rejovitzky, L. Anand, Diffusion–deformation theory for amorphous silicon anodes: the role of plastic deformation on electrochemical performance, Int. J. Solids Struct. 67 (2015) 283–296.
[49] X. Zhang, A. Krischok, C. Linder, A variational framework to model diffusion induced large plastic deformation and phase field fracture during initial two-phase lithiation of silicon electrodes, Comput. Methods Appl. Mech. Eng. 312 (2016) 51–77.
[50] C.-F. Chen, P. Barai, P.P. Mukherjee, An overview of degradation phenomena modeling in Lithium-ion battery electrodes, Curr. Opin. Chem. Eng. 13 (2016) 82–90.
[51] P. Barai, P.P. Mukherjee, Stochastics of diffusion induced damage in intercalation materials, Mater. Res. Exp. 3 (10) (2016) 104001.
[52] P. Barai, P.P. Mukherjee, Stochastic analysis of diffusion induced damage in lithium-ion battery electrodes, J. Electrochem. Soc. 160 (6) (2013) A955–A967.
[53] K.-J. Bathe, Finite Element Procedures, Klaus-Jurgen Bathe, 2006.
[54] J.N. Reddy, An Introduction to Nonlinear Finite Element Analysis: With Applications to Heat Transfer, Fluid Mechanics, and Solid Mechanics, OUP Oxford, 2014.
[55] Y. Qi, et al., Lithium concentration dependent elastic properties of battery electrode materials from first principles calculations, J. Electrochem. Soc. 161 (11) (2014) F3010–F3018.
[56] P. Johari, Y. Qi, V.B. Shenoy, The mixing mechanism during lithiation of Si negative electrode in Li-ion batteries: an ab initio molecular dynamics study, Nano Lett. 11 (12) (2011) 5494–5500.
[57] A.A. Franco, Multiscale modelling and numerical simulation of rechargeable lithium ion batteries: concepts, methods and challenges, RSC Adv. 3 (32) (2013) 13027–13058.
[58] F. Soto, et al., Modeling solid-electrolyte interfacial phenomena in silicon anodes, Curr. Opin. Chem. Eng. 13 (2016) 179–185.
[59] P. Barai, et al., Reduced order modeling of mechanical degradation induced performance decay in lithium-ion battery porous electrodes, J. Electrochem. Soc. 162 (9) (2015) A1751–A1771.
[60] M.E. Ferraro, et al., Electrode mesoscale as a collection of particles: coupled electrochemical and mechanical analysis of NMC cathodes, J. Electrochem. Soc. 167 (1) (2020) 013543.

Chapter 9

Nanostructured 3D (three dimensional) electrode architectures of silicon for high-performance Li-ion batteries

Surendra K. Martha, Liju Elias, and Sourav Ghosh
Department of Chemistry, Indian Institute of Technology Hyderabad, Sangareddy, Telangana, India

9.1 Introduction

Lithium-ion batteries (LIBs) have emerged as part of day to day life by fulfilling myriads of end-user needs. The explosive growth of LIBs is due to attractive characteristics such as high energy density (150–200 Wh kg^{-1}), high operating voltage (∼3.7 V), limited self-discharge and low maintenance requirements compared to other leading rechargeable batteries like lead-acid, nickel-based batteries (nickel-cadmium and nickel metal-hydride batteries), etc. [1–6]. However, the ever-increasing demand for clean and renewable energy greatly increases the need to develop LIBs with improved energy density to meet the needs of existing and anticipated applications in consumer electronics, electric vehicles, etc. [3, 4]. As the electrode materials (anode and cathode) play a pivotal role in the performance of LIBs, especially on determining the output voltage and specific capacity, enormous research efforts have taken place globally to develop new electrode materials with high energy density and cycling stability than the state of the art electrode materials [6–13].

Among the numerous anode materials reported during the last several decades as a potential alternative to graphite [14], silicon is considered as the most promising material with high theoretical capacity (4200 mAh g^{-1}, corresponds to fully lithiated state of Li$_{4.4}$Si) and moderate operation voltage (<0.5 V vs. Li/Li$^+$) [15–27]. Silicon, being the second most abundant element on earth's crust (28% of the crust's mass) can serve as more-effective and environmentally benign anode material than the other reported ones. Silicon is an anode material, which can electrochemically alloy and de-alloy with lithium at room temperature, and each silicon atom can alloy with 4.4 lithium leading to the formation of $Li_{22}Si_5$ alloy. Hence, the theoretical specific capacity of silicon anode is ~10 times and ~20 times higher than that of graphite and $Li_4Ti_5O_{12}$ (LTO) anodes, respectively [27, 28]. However, silicon undergoes huge volume change (~400%) during the charge-discharge processes. The volume change causes a series of issues like cracking and fracturing of the anode, degradation, and regeneration of solid electrolyte interphase (SEI), loss of electronic conductivity, etc., which results in pulverization and thereby leading to capacity loss, poor cycling stability and low coulombic efficiency. Therefore, as the demerits outweighed the merits of silicon, the commercial application of silicon as anode material for LIBs have been hindered.

9.2 Structure and electrochemistry of silicon

Silicon (Si) is a p-block element in the periodic table with atomic number 14, having four electrons in the valence shell. Metallic silicon is gray in color and acts as an intrinsic semiconductor. Silicon dioxide (SiO_2) is the naturally occurring oxide of silicon and is the major component of sand and rocks. The crystalline Si possesses a diamond cubic close-packed structure with lattice constants of 5.431 Å. As it is a semiconductor, the conductivity of silicon increases with temperature. Silicon-based materials have been widely explored for various industrial applications like electronics, semiconductors, silicone, cosmetics, health care, etc. [29–32] In the case of energy storage applications, the use of Si becomes more facile as the infrastructure for processing Si in the pure form is already optimized by other industries.

Silicon can electrochemically alloy and de-alloy with lithium under room temperature. Hence, it can be used as an anode material for LIBs. Other than silicon, the elements such as Sn, Sb, Al, Mg, Bi, In, Zn, Pb, Ag, Pt, Au, Cd, As, Ga, and Ge are also electrochemically active towards lithium [10, 13, 27]. Owing to the high specific capacities (Fig. 9.1A) of alloy type anodes, the alloying (lithiation) reactions of metallic or semi-metallic elements and their compounds with lithium have been attained much interest during the past few decades.

The possible Si-Li alloy phases that can be formed in crystalline and amorphous silicon during heat treatment and their corresponding calculated formation energies are shown in Fig. 9.1B [33]. Although the alloy phases shown in Fig. 9.1B obtained through heat treatment may not be exactly reproduced during electrochemical alloying, but it gives an overall idea of the possible alloy phases that can be formed during lithiation.

The data in Fig. 9.1B shows crystalline Si-Li phases as the much-expected phases with lower Gibbs free energy than the amorphous phases during lithiation. However, during electrochemical lithiation at room temperature, the crystalline silicon undergoes a phase transformation from crystalline to amorphous in the first lithiation itself and remains amorphous [34–36]. Hence, electrochemical lithiation leads to the formation of amorphous Li_xSi (with

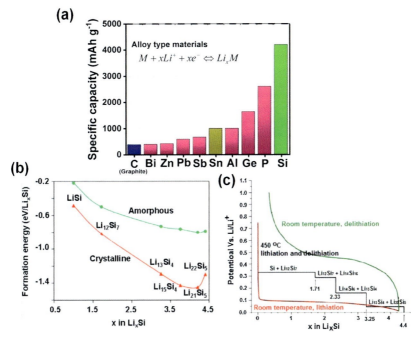

FIG. 9.1 (A) A comparison of the specific capacities of different types of alloy type anode materials; (B) Si-Li alloy phases of crystalline and amorphous silicon during lithiation and their calculated formation energies; (C) the electrochemical lithiation (red) and de-lithiation (green) curves in comparison with the theoretical high temperature (at 450°C) voltage curves (black) of crystalline silicon. *((B) Reproduced with permission from C.-M. Wang, X, Li, Z. Wang, W. Xu, J. Liu, F. Gao, L. Kovarik, J.-G. Zhang, J. Howe, D.J. Burton, In situ TEM investigation of congruent phase transition and structural evolution of nanostructured silicon/carbon anode for lithium ion batteries, Nano Lett. 12 (3) (2012) 1624–1632. (C) Reproduced with permission from H. Wu, Y. Cui, Designing nanostructured Si anodes for high energy lithium ion batteries, Nano Today 7 (5) (2012) 414–429. Copyright Elsevier 2012.)*

$0 < x < 3.75$) alloys rather than the expected crystalline alloys. The mechanism of formation of amorphous Li_xSi during lithiation (discharge) is given in Eq. (9.1) [16, 36, 37].

$$Si_{(crystalline)} + xLi^+ + xe^- \rightarrow Li_xSi_{(amorphous)} + (3.75-x)Li^+ + (3.75-x)e^- \quad (9.1)$$

But when the value of "x" reaches 3.75, a crystalline $Li_{15}Si_4$ alloy forms. This transformation of the highly lithiated amorphous phase to a metastable crystalline phase occurs at lower potentials (<50 mV vs. Li/Li$^+$) [36–39]. The subsequent de-lithiation (charge) results in the formation of amorphous silicon as the final product along with some residual $Li_{15}Si_4$ phase as given in Eq. (9.2). The formation of residual $Li_{15}Si_4$ phase after the first de-lithiation can be avoided by adjusting the operating voltage window.

$$\text{Li}_{15}\text{Si}_4 \text{ (crystalline)} \rightarrow \text{Si}_{\text{(amorphous)}} + y\text{Li}^+ + ye^- + \text{Li}_{15}\text{Si}_4 \text{ (residual)} \qquad (9.2)$$

The electrochemical lithiation and de-lithiation curves in comparison with the theoretical high temperature (at 450°C) voltage curves of crystalline silicon are shown in Fig. 9.1C [40–42]. The *red* and *green* lines in Fig. 9.1C shows the electrochemical lithiation and de-lithiation, respectively at room temperature. The *black* line shows the theoretical high-temperature lithiation and de-lithiation at 450°C.

9.3 The failure mechanism of silicon anodes

Despite the attractive theoretical capacity of silicon, there are many complications in its practical application due to mechanical issues. The fundamental material challenges to use silicon as an effective anode material for LIBs are represented schematically in Fig. 9.2 [41]. The major failure modes of silicon-based anodes are due to; (i) material pulverization [43–45], (ii) morphology and volume change of the entire electrode [15, 46], and (iii) continuous solid electrolyte interphase (SEI) formation [47–52].

The lithiation/de-lithiation results in large volume expansion/contraction of silicon, leading to developing large internal stress. As the silicon particles grow larger and larger to ~400% of their original size during lithiation, induce

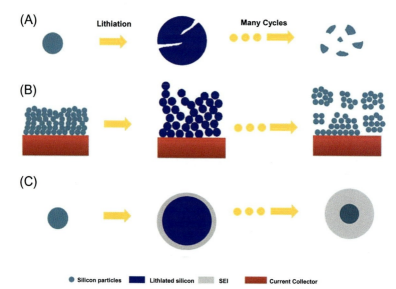

FIG. 9.2 The schematic of the mechanism of major failure modes of silicon-based anodes; (A) material pulverization, (B) morphology and volume change of the entire electrode, and (C) continuous SEI formation. *(Reproduced with permission from H. Wu, Y. Cui, Designing nanostructured Si anodes for high energy lithium ion batteries, Nano Today 7 (5) (2012) 414–429.)*

extremely large internal stress, results in cracking and pulverization [53, 54]. The volume change, cracking and pulverization of silicon particles deteriorate the electrode structure integrity, which leads to loss of electrical contact with the current collector, increase in internal resistance, and thereby capacity fade [55, 56]. The unstable SEI formation during cycling is another reason for the failure of silicon anodes [49, 51, 57, 58]. The stable SEI, formed by the decomposition of electrolyte at lower potentials on the anode surface during discharge, is favorable for the stability of anode. The stable SEI formed during the first lithiation can act as a barrier, which allows lithium ion passage and prevents electron flow, and thereby reduces further electrolyte decomposition. Hence, it can enhance the C-rate performance, cycle-life and thereby the overall performance of the cell. However, in the case of silicon anode, the SEI layer is formed during lithiation state deteriorates during de-lithiation or shrinking of the silicon particles. The large volume changes in silicon particles during lithiation/de-lithiation (expansion/contraction) leads to break down of SEI into separate pieces, and results in the formation of new SEI on the exposed surface of silicon in the subsequent cycles. The continuous SEI formation-breaking-reformation process in silicon anode during cycling leads to excessive consumption of lithium ions and electrolyte, and thereby the formation of extremely thick SEI layer. Hence, the electrochemical performance in terms of capacity and cycle-life decays due to poor electronic conductivity effected from increased internal resistance and exhaustion of the electrolyte.

9.4 Mitigation strategies

Several approaches have been reported in the last few decades to overcome the limitations of silicon-based anode materials for LIBs based on the fundamental understanding of alloying and failure mechanisms of Si anode. The major efforts are based on the modification of the size, structure, and composition of the silicon anodes. The strategy of downsizing the Si particles to the nanoscale is strategically beneficial due to the superior electrochemical performance of nanomaterials over bulk materials by facilitating fast Li-ion diffusion. The nanostructured silicon particles provide enough space to accommodate the volume change during lithiation and reduce the issue of pulverization by limiting the electrode strain. The Silicon nanoparticles can offer good electronic contact, reduce electrode impedance and provides enhanced rate performance and cycling stability.

In structural modifications, the nano-engineering by controlling the size and shape of silicon particles to achieve zero-dimensional (0D), one-dimensional (1D), two dimensional (2D) and three dimensional (3D) materials are reported to be effective in overcoming the challenges of silicon-based anodes and the corresponding structural and morphological changes with cycling are represented in the Fig. 9.3 [46]. The different types of silicon nanostructures are synthesized and studied as the electrode material to overcome the issue of volume

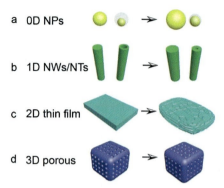

FIG. 9.3 Schematic of Si nanostructures and morphology change after cycling: (A) 0D Si nanoparticles and yolk-shell nanostructure; (B) 1D Si nanowires and nanotubes; (C) 2D thin films; (D) 3D porous structure. *(Reproduced with permission from J.-Y. Li, Q. Xu, G. Li, Y.-X. Yin, L.-J. Wan, Y.-G. Guo, Research progress regarding Si-based anode materials towards practical application in high energy density Li-ion batteries, Mater. Chem. Front. 1 (9) (2017) 1691–1708.)*

change and pulverization, which includes nanoparticles [55, 59–68], nanowires [23, 69–76], nanorods [77–80], nanotubes [81–83], interconnected hollow nanospheres [84–86], nanosheets, and thin films [87–96]. Further, the efforts to incorporate pores or voids to the silicon nanostructures to accommodate the large volume change during lithiation also attained much interest to circumvent the inherent problems of silicon anodes [97–105].

Although the large volume change and (nano)material pulverization can be addressed by the careful design of nanostructured electrode architectures, the formation of stable SEI on these nano-architectures is difficult to accomplish due to their high surface-to-volume ratio. Hence, to attain long cycling stability, the strategy of developing composites of silicon with carbonaceous materials have attracted much interest. As the carbonaceous materials form stable SEI on their surface, the stress of volume change during cycling reduces and enhances the electrical conductivity. Therefore, carbon has emerged as an indispensable part of anodes for LIB applications.

Gu et al. reported the method of developing a silicon-carbon composite by embedding the silicon nanoparticles in the carbon fiber matrix to address the issue of volume expansion [106]. However, the in-situ TEM study on the lithiation/de-lithiation mechanism of the developed silicon-carbon fiber composite showed the issue of carbon fiber cracking during cycling. The continuous cracking of the carbon fiber during lithiation leads to large electrolyte consumption due to the formation of the new SEI layer on the fractured carbon fiber surface. Moreover, the cracking of the carbon fiber matrix can lead to the failure of the anode material due to loss of electrical contact. Similarly, silicon nanoparticles embedded in board-like porous carbon matrix reported by Wu et al. shows enhanced electrochemical performance with a reversible capacity of

1249 mAh g^{-1} even after 100 cycles at a current density of 0.5 A g^{-1}. [107] Although, by careful design of the carbon matrix, the issue of pulverization is addressed to a certain extent in silicon nanoparticles embedded composites, the poor cycling stability remained a serious issue. Hence, carbon coating on nanostructured silicon particles has emerged as an attractive strategy.

Considering the 0D Si nanoparticles and carbon coating, Si nanoparticles with the various particle size of 5, 10, and 20 nm using reverse micelles at high pressure and temperature in a bomb is synthesized and investigated as the anode for LIB [55]. It is observed that 10 nm-sized Si particles showed the highest charge capacity, with poor coulombic efficiency and capacity retention than 20 nm sized particles. However, capacity retention and coulombic efficiency were improved significantly when the carbon coating was applied to 10 nm-sized Si. In the carbon coating approach, the design of yolk-shell silicon-carbon composite comprises of the silicon yolk and thin layer of carbon coating as shell offers excellent electrochemical performance [68, 108, 109]. In this core-shell design, as the silicon is completely covered with the thin carbon layer, the carbon shell provides enough mechanical strength for the silicon particles to expand freely without pulverization (Fig. 9.4A) [68]. In short, the carbon coating can function as the conductive network, provide mechanical strength against volume change, maintain consistency in SEI formation and thereby enhances the electrochemical performance of the anode material. Liu et al. reported a yolk-shell shell design with 100% capacity retention over 300 cycles, and long term cycling stability over 1000 cycles with a capacity retention of 74% [68]. In another report, a novel interconnected Si hollow nanosphere electrode with the discharge capacity of 1420 mAh g^{-1} at the end of 700 cycles, at C/2 current rate is reported [84]. The capability of accommodating large volume changes without pulverization is an effective way of mitigating the volume expansion/contraction during cycling. The 0D silicon nanoparticles are attractive due to their simple and easy synthesis routes and the possibility of developing composite electrodes with different forms of carbon by simple methods. As the 0D hollow silicon nanostructures are less prone to a stress fracture, they show better electrochemical performance than the solid nanospheres.

The 1D architectures (nanowires, nanotubes, etc.) also show enhanced performance by providing enough space to accommodate volume change and proper electrical contact with the current collector. The 1D silicon nanowires grown directly on the current collector, as shown in Fig. 9.4B, delivered a charge capacity (4277 mAh g^{-1}) almost similar to the theoretical charge capacity (4200 mAh g^{-1}) and a discharge capacity of 3124 mAh g^{-1} during the first cycle at C/20 rate. The efficient charge transport through the 1D electronic pathways of the silicon nanowires enhances the electrochemical performance and long term cyclability (>200 cycles) of the electrodes without pulverization [23]. Similarly, 1D silicon nanotubes are also reported to be promising with less diffusion induced stress and good electronic conductivity. Another strategy of making a sealed array of silicon nanotubes by combining the design advantages

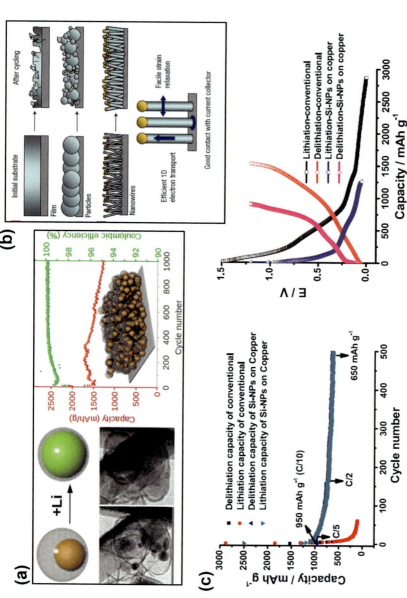

FIG. 9.4 (A) Schematic of the yolk-shell design with void space between core silicon and thin carbon layer before and after lithiation, in-situ TEM results showing the volume change during lithiation, and cycling stability of the electrode for 1000 cycles. (B) Schematic of morphological changes that occur in Si substrate, film, and nanowires during electrochemical cycling. (C) Galvanostatic charge-discharge voltage profile and cycling performance for the conventional electrode (Si 50wt%, Binder 10wt%, Carbon 40wt%) and pressure embodied Si-NPs on copper foil current collector electrode during the first cycle at C/10 rate. ((A) Reproduced with permission from N. Liu, H. Wu, M.T. McDowell, Y. Yao, C. Wang, Y. Cui, A yolk-shell design for stabilized and scalable Li-ion battery alloy anodes, Nano Lett. 12 (6) (2012) 3315–3321. (B) Reproduced with permission from C.K. Chan, H. Peng, G. Liu, K. Mcllwrath, X.F. Zhang, R.A. Huggins, Y. Cui, High-performance lithium battery anodes using silicon nanowires, Nat. Nanotechnol. 3 (1) (2008) 31. (C) Reproduced with permission from K.K. Sarode, R. Choudhury, S.K. Martha, Binder and conductive additive free silicon electrode architectures for advanced lithium-ion batteries, J. Energy Storage 17 (2018) 417–422.)

of both nanowires and hollow structures show good capacity retention (>80% after 50 cycles) and high intial Coulombic efficiency (>85%) [110]. Further, the carbon-coated free-standing 1D silicon nanowire electrodes show a capacity of ~800 mAh g^{-1} even after 659 cycles with a Coulombic efficiency of 99.9%.

The nanostructured 2D silicon thin films are reported to be promising with enhanced cycling stability and rate capabilities due to minimal volume expansion and easy accessibility to lithium ions. However, the performance of the electrode depends highly on the characteristics of the thin film thickness, surface morphology, and the degree of adhesion with the current collector. The vacuum-deposited amorphous n-type silicon film on nickel substrate with a thickness of 50 nm shows an initial capacity of ~3750 mAh g^{-1} and cycling stability without significant capacity fade at 1C rate for over 200 cycles. But with an increase in thickness to 275 nm, the electrode performance decreased considerably to an initial capacity of 2200 mAh g^{-1} [111]. Despite the promising electrochemical performance of silicon thin films, the practical application of 2D silicon electrodes is hindered due to cost of thin-film fabrication.

Apart from all these approaches, the development of binder-free electrode architectures to facilitate efficient charge transport by avoiding the extra weight and resistance of the binder is also a popular method to enhance the performance of silicon anode. An anode architecture prepared by pressure-embedding the silicon nanoparticles on copper foil current collector with an even distribution of the nanoparticles and having enough space between the particles to accommodate the volume change shows better performance than the conventional silicon electrode [92]. The electrode fabrication approach also minimizes the contact resistance, facilitates the electron transport and thus improves cycle stability. The pressure-embedded silicon nanoparticles electrode shows an initial discharge capacity of ~1250 mAh g^{-1} at the C/10 rate (Fig. 9.4C) and retains more than 650 mAh g^{-1} at C/2 rate even after 500 cycles.

Nanostructured Si materials could address the issues of electrode pulverization and promote fast Li-ion diffusion. However, nanostructured materials suffer several disadvantages, including poor tap density resulting from the high surface area, uncontrolled electrolytic reaction at the electrode-electrolyte interface, and higher inter-particle resistance. Most of the nanostructured Si anodes are investigated in the laboratory scale with minimal mass loading of few milligrams per cm^2. To meet the commercial requirement, the active material loadings need to be high enough on a typical conventional 2D current collector. But maintaining high energy density and rate capability at the same time is highly challenging for the nanostructured electrodes because of ion transport limitations and increased electrode resistance for electron transport. It is not a simple engineering to attain high electrochemical performance of nanostructured materials at high mass loadings. The scaling is, in fact, a fundamental scientific challenge. 3D electrode architectures having a 3D conductive framework which acts as a 3D current collector and a 3D porous network is another effective way for fast ion transport. The 3D electrode architecture capable of delivering efficient charge by utilizing all the electrode materials even from the bulk of a thick electrode, which is desired for the realization of high capacity, high rate electrode.

9.5 Nanostructured 3D electrode architectures

The design of nanostructured 3D electrode architectures can offer the large effective surface area of the active material, mechanical stability, easy electrolyte access, effective ionic transport, etc. In silicon-based anodes, the 3D electrode architectures can address the issue of pulverization and thereby enhance the overall electrochemical performances of the electrode. Additionally, the fabrication of 3D structure could be inexpensive in comparison to 0D/1D/2D Si nanostructure, because it involves mainly a template method and a chemical pore-forming method. The 3D electrode architectures of silicon can be developed either by using 3D porous silicon nanostructures and their composites, self-standing silicon-based electrodes by employing various types of carbon/polymer scaffolds (such as graphene, CNT, CNF materials) and 3D-porous metallic current collectors (Cu and Ni foam). The recent advancement in the development of 3D electrode architecture of Si-based anode material are categorized broadly and discussed below.

9.5.1 3D porous silicon-carbon composite

The attractive electrochemical performance of the silicon nanowires based electrodes lead to the idea of developing 3D porous silicon architectures with enough free space to accommodate the volume change. The 3D porous silicon electrode can overcome the limitation of very low mass loading density of the silicon nanowires based anodes, and the presence of interconnected pores in the 3D design helps to reduce the issue of pulverization during cycling.

Novel 3D porous carbon-silicon frameworks have been designed by self-assembly of the phenol-formaldehyde resin by the assistance of triblock poly(propylene oxide)-poly(ethylene oxide)-poly(propylene oxide) (PEO-PPO-PEO) copolymers. The overall fabrication of porous carbon-Si nanocomposite is shown in Fig. 9.5A. Further, the electrochemical performance of porous Carbon-Si hybrid exhibits an initial discharge capacity of 1868 mAh g^{-1}, with a capacity retention of 1000 mAh g^{-1} for 100 cycles, as shown in Fig. 9.5B. The excellent electrochemical performance is attributed to the formation of continuous mesoporous structures in the exclusive 3D conductive frameworks [112]. Further, the performance of the 3D macroporous silicon electrode is reported to be enhanced by developing a silver coating through a silver mirror reaction [104]. The overall schematic illustration of the formation of Ag nanoparticles interconnect network over the entire surface area to provide a better electron pathway from the current collector is shown in Fig. 9.5C. The silver-coated macroporous 3D electrode architecture of silicon shows an initial capacity of ~3585 mAh g^{-1} at a current density of 500 mA g^{-1} and cycling stability over

Nanostructured 3D electrode architectures of silicon **Chapter | 9** 341

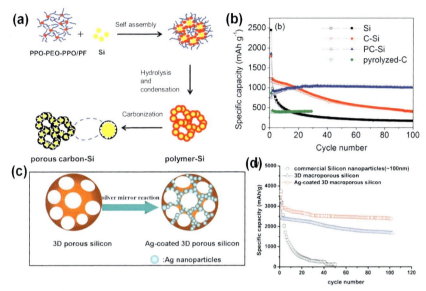

FIG. 9.5 (A) Synthesis route of porous Carbon-Si nanocomposite; (B) Capacity vs. cycle number of bare Si, Carbon-Si, and porous Carbon-Si hybrid. (C) Schematic illustration of the formation of Ag nanoparticles interconnect network over the entire surface area providing better electron pathway from the current collector; (D) The cycle stability of 3D macroporous Si, Ag-coated 3D macroporous Si, and commercial Si nanoparticles. *((A) and (B) Reproduced with permission from M.S. Wang, Y. Song, W.L. Song, L.Z. Fan, Three-dimensional porous carbon–silicon frameworks as high-performance anodes for lithium-ion batteries, ChemElectroChem 1 (12) (2014) 2124–2130. (C) and (D) Reproduced with permission from Y. Yu, L. Gu, C. Zhu, S. Tsukimoto, P.A. van Aken, J. Maier, Reversible storage of lithium in silver-coated three-dimensional macroporous silicon, Adv. Mater. 22 (20) (2010) 2247–2250.)*

100 cycles with 85% capacity retention at 100 mA g^{-1} (Fig. 9.5D). The silver coating improves the conductivity of the 3D porous silicon anode and enhances the overall electrochemical performances. Although the electrodes based on 3D silicon nanostructures are interesting, the issues of mass loading and structural stability are prevailing. Hence, in 3D electrode architectures, nanostructured silicon loaded 3D frameworks of carbonaceous materials or other current collectors have been attained much interest to effectively address the issue of pulverization and cyclability than 3D silicon structures.

Among different carbonaceous materials, graphene has been widely used as the framework to support the silicon nanoparticles and thereby to make 3D electrode architectures. Pyrolytic carbon-coated silicon nanoparticles embedded in the 3D graphene framework is reported as an efficient anode material for LIBs (Fig. 9.6A) [113]. The double protection of silicon particles by graphene framework and amorphous carbon coating is achieved through the solvothermal reaction (180°C for 12 h) of graphene oxide and silicon nanoparticles followed by

342 PART | IV Achieving high performance

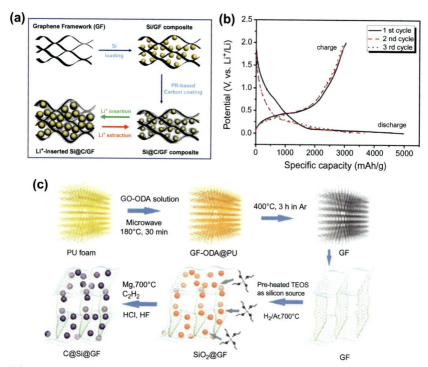

FIG. 9.6 (A) Schematic of the synthesis procedure for the development of carbon-coated silicon nanoparticles embedded on 3D graphene framework electrode and its lithiation/de-lithiation mechanism, (B) voltage profiles of the electrode during the first three cycles at a current density of 0.1 A g^{-1}; (B). (C) Schematic illustration of the synthesis process for mesoporous C@Si@graphene foam nano architectures. *((A) and (B) Reproduced with permission from F. Zhang, X. Yang, Y. Xie, N. Yi, Y. Huang, Y. Chen, Pyrolytic carbon-coated Si nanoparticles on elastic graphene framework as anode materials for high-performance lithium-ion batteries, Carbon 82 (2015) 161–167. (C) Reproduced with permission from S. Chen, P. Bao, X. Huang, B. Sun, G. Wang, Hierarchical 3D mesoporous silicon@ graphene nanoarchitectures for lithium ion batteries with superior performance, Nano Res. 7 (1) (2014) 85–94.)*

amorphous carbon coating. Phenolic resins are used as carbon precursors to obtain the amorphous carbon coating. The charge-discharge profile of the electrodes are shown in Fig. 9.6B and the electrode shows good cycle stability over 200 cycles with a capacity retention of ~85% and ~100% coulombic efficiency. The excellent 3D architecture of the electrode with a double protection mechanism provides mechanical and structural stability along with electrical conductivity and thereby gives enhanced electrochemical performance.

The nanostructured electrode of hierarchical 3D carbon-coated mesoporous silicon nanospheres on graphene foam delivers attractive electrochemical performances [114]. The electrode architecture is developed by combining a thermal bubble ejection assisted chemical-vapor-deposition and magnesiothermic reduction methods (Fig. 9.6C). The electrode shows a specific capacity of 1200 mAh g^{-1} at a current density of 1 A g^{-1}, and good cycling stability over 200 cycles without any structural deformation. The enhanced performance of the electrode is attributed to the synergistic effects on pulverization, achieved from the combination of mesoporous silicon nanospheres and graphene foam.

The 3D hierarchical macro/mesoporous silicon network encapsulated within a carbon matrix obtained through magnesiothermic reduction of SiO$_2$/carbon composite is also reported to show enhanced performance than the porous silicon and conventional silicon-carbon composite [115]. Further, a silicon-graphene composite (1:4 ratio) with spherical morphology and 3D conductive network prepared by a facile spray-drying method followed by low-temperature reduction serves as a good electrode material for LIBs. The electrode design shows a high initial capacity of 1298.1 mAh g^{-1} at 100 mA g^{-1} and good rate capability.

Further, the 3D nitrogen-doped graphene/silicon composite electrode prepared by dispersing silicon nanoparticles uniformly on the surface of the nitrogen-doped graphene matrix gives mesoporous structure with a large surface area. The composite electrode material is synthesized through electrostatic self-assembly followed by hydrothermal processing and heat treatment (Fig. 9.7A) [116]. The developed 3D electrode architecture delivers a capacity of 1132 mAh g^{-1} even after 100 cycles at a current density of 5 A g^{-1}. The enhanced performance of the electrode is mainly attributed to the porous design and improved intrinsic electrical conductivity resulted from the presence of nitrogen-doped graphene in the composite.

In another strategy, the 3D electrode architecture developed by dispersing silicon nanoparticles on nitrogen-doped graphitized carbon shows attractive electrochemical performance. The N-doped graphitized carbon acts as a robust 3D conductive network and provides structural stability. The 3D support was prepared from ZIF-67 (metal-organic framework) precursor by magnesiothermic reduction, as shown in Fig. 9.7B [117]. The electrode gives a capacity of 900 mAh g^{-1} at 0.2 A g^{-1} over 300 cycles with a capacity retention of 85.5% (Fig. 9.7C) and also gives a high discharge capacity of 880 mAh g^{-1} at a current density of 1 A g^{-1}. The enhanced conductivity due to the 3D network and the efficiency of the design to accommodate the volume change during cycling helps the electrode to attain good stability and electrochemical performance.

An electrode architecture developed by depositing Si and alumina (Al$_2$O$_3$) on 3D hollow carbon nanospheres interconnected films is another attractive nanostructured anode design of silicon. A template-directed carbon segregation method is used for the synthesis of hollow carbon nanospheres, and the electrophoretic deposition technique is employed to make them as an interconnected 3D carbon nanospheres substrate film [118]. The Si and Al$_2$O$_3$ are deposited on the substrate film through plasma-enhanced chemical vapor deposition (PECVD) and atomic layer deposition (ALD), respectively. The overall schematic of hollow CNS/Si/Al$_2$O$_3$ core-shell film fabrication processes is shown in Fig. 9.8. The

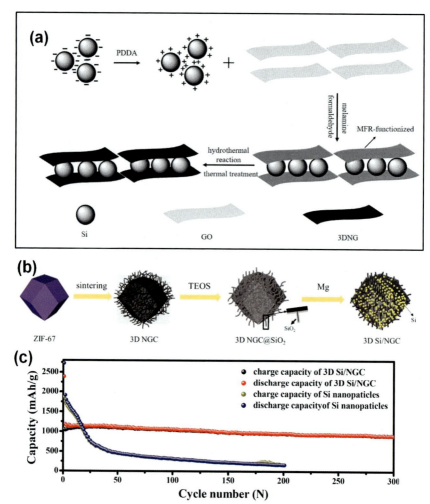

FIG. 9.7 (A) Schematic of the stepwise synthesis procedure for the development of a 3D nitrogen-doped graphene/silicon composite electrode. (B) Schematic of the development of the 3D silicon/N-doped graphitized carbon composite electrode; (C) the cycle performance of 3D Si/NGC silicon nanoparticles at 0.2 A g^{-1}. ((A) Reproduced with permission from X. Tang, G. Wen, Y. Song, Stable silicon/3D porous N-doped graphene composite for lithium-ion battery anodes with self-assembly, Appl. Surf. Sci. 436 (2018) 398–404. (B) and (C) Reproduced with permission from T. Mu, P. Zuo, S. Lou, Q. Pan, H. Zhang, C. Du, Y. Gao, X. Cheng, Y. Ma, H. Huo, A three-dimensional silicon/nitrogen-doped graphitized carbon composite as high-performance anode material for lithium ion batteries, J. Alloys Compd. 777 (2019) 190–197.)

electrode shows remarkable capacity retention of 1560 mAh g^{-1} at a current density of 1 A g^{-1}, and the presence of Al$_2$O$_3$ coating helps to achieve cycling stability over 100 cycles. As the 3D network of hollow carbon nanospheres film acts as the conductive support and provides voids for the stress-free volume change of silicon during cycling, the electrode shows enhanced electrochemical performances.

Nanostructured 3D electrode architectures of silicon **Chapter | 9** 345

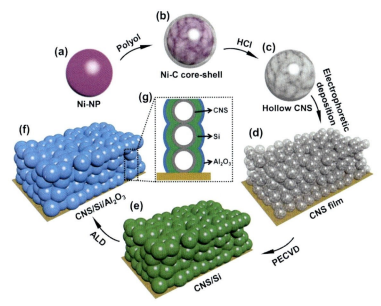

FIG. 9.8 Schematic of hollow CNS/Si/Al$_2$O$_3$ core-shell film fabrication processes. (A) Ni-NPs prepared by solution reaction. (B) Ni-C core-shell structure fabricated by template-directed carbon segregation method. (C) Hollow CNSs obtained after HCl etching. (D) CNS film deposited on stainless steel by the electrophoretic deposition technique. (E) CNS/Si obtained after silicon deposition with PECVD. (F) CNS/Si/Al$_2$O$_3$ obtained after Al$_2$O$_3$ deposition with ALD. (G) The inner structure of CNS/Si/Al$_2$O$_3$. *(Reproduced with permission from B. Li, F. Yao, J.J. Bae, Chang, M.R. Zamfir, D.T. Le, D.T. Pham, H. Yue, Y.H. Lee, Hollow carbon nanospheres/silicon/alumina core-shell film as an anode for lithium-ion batteries. Sci. Rep. 5 (1) (2015) 1–9.)*

The hierarchical arrangement of carbon-coated silicon nanoparticles on a 3D CNT network achieved through electrostatic self-assembly acts as a promising electrode material for LIBs [119]. The 3D electrode architecture contains large silicon content (72.4%), and the 3D CNT network offers structural stability and enhanced conductivity. A schematic of the synthesis of 3D electrode architecture is given in Fig. 9.9A. The electrode shows excellent performance in terms of reversible capacity (989.5 mAh g^{-1} at 0.5C) even after 1000 cycles with a capacity retention of ~86% (Fig. 9.9B). The novel structural design of the electrode offers good electrochemical performance.

Other than following the conventional ideas of electrode design, many innovative ideas have been reported in developing the nanostructured 3D electrode architectures of silicon for LIBs. A 3D silicon/carbon core-shell electrode architecture developed by combining the ideas of the 3D electrode and silicon/carbon core-shell design shows better performance than the individual designs [120]. The core-shell silicon/carbon is prepared by the thermal decomposition method and the 3D electrode with enough free space between the core-shell structures are obtained through laser processing. The good rate capability

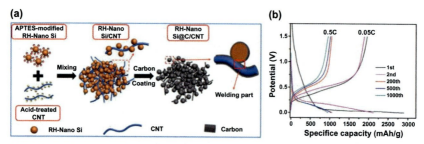

FIG. 9.9 (A) Schematic of the synthesis procedure involved in the development of carbon-coated silicon nanoparticles on 3D CNT network, and (B) characteristic voltage profile of the material at different C rates. *(Reproduced with permission from W. An, B. Xiang, J. Fu, S. Mei, S. Guo, K. Huo, X. Zhang, B. Gao, P.K. Chu, Three-dimensional carbon-coating silicon nanoparticles welded on carbon nanotubes composites for high-stability lithium-ion battery anodes, Appl. Surf. Sci. 479 (2019) 896–902.)*

(1170 mAh g^{-1} at 8 A g^{-1}) and cycling stability (>300 cycles) of the 3D core-shell electrode design is attributed to the large effective surface area to reduce the polarization effect and enough free space to release the physical stress during cycling (Fig. 9.10).

In another novel idea, nanosized silicon particles were developed from the commercial micro-sized particles through a low-cost fluid-induced fracture technique followed by an electrode level thermolysis for the cyclization of the polyacrylonitrile binder to form a 3D conductive anode matrix serves as a promising anode for LIB [121]. A detailed step-wise synthesis of the electrode architecture is schematically represented in Fig. 9.11A. The electrode architecture gives a capacity of 3081 mAh g^{-1} at 0.1 A g^{-1} (Fig. 9.11B) and good cycling stability at higher current density (2 A g^{-1}) over 500 cycles with a capacity of 1423 mAh g^{-1}. In this design, the binder is converting to a 3D structure and covers the silicon nanoparticles, and the presence of microchannels helps to accommodate volume change and provides mechanical stability.

9.5.2 3D porous metallic current collector as a scaffold

Other than current carbon-based collectors, copper and nickel-based nanostructured 3D frameworks are also widely used to develop 3D electrode architectures for LIBs to replace conventional Cu current collector with the objective of high material loading of silicon. The traditional battery electrodes use conventional metal foil current which restricts the higher mass loading due to the tendency of delaminating of the thick electrode from the flat current collector, and the kinetic limitation in lithium-ion diffusion through the thick electrode. So this 3D metal scaffold could help to enhance the energy density as well as the power density of the cell by enhancing the active material loading as well as better-interconnected network for faster Li-ion and electronic diffusion.

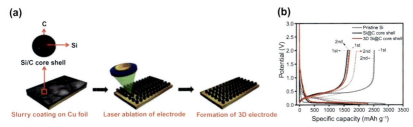

FIG. 9.10 (A) Schematic of the synthesis procedure of 3D silicon/carbon core-shell electrode architecture, and (B) voltage profiles of the 3D electrode in comparison with silicon/carbon core-shell and pristine silicon electrodes. *(Reproduced with permission from J.S. Kim, W. Pfleging, R. Kohler, H.J. Seifert, T.Y. Kim, D. Byun, H.-G. Jung, W. Choi, J.K. Lee, Three-dimensional silicon/carbon core–shell electrode as an anode material for lithium-ion batteries, J. Power Sources 279 (2015) 13–20.)*

Silicon particles with the size around 4 μm have been sealed into a compressed 3D Cu foam matrix as shown in Fig. 9.12A [122]. The electrode was fabricated by dipping the Cu foam inside the Si suspension, followed by pressing the electrode to seal the active material inside the foam by the simple calendaring process. The conventional electrode delivered an initial discharge capacity of 2000 mAh g^{-1}, with 10% capacity retention at the end of 20 cycles. However, a 3D integrated electrode, delivered an initial discharge capacity of 2500 mAh g^{-1}, with a capacity retention of 2000 and 1200 mAh g^{-1} at the end of 100 and 200 cycles, respectively. Even though the proposed design could not prevent the issue of unstable SEI formation, but this architecture could be able to trap the active material inside the 3D porous metal current collector [122]. The methodology proposed here proves its compatibility with roll-to-roll coating methods, which could be scaled up for large area electrodes.

A strategy of ultrafast laser-induced macro/nano surface patterning of copper current collector to form nano-cavities and pores on the surface followed by silicon nanoparticle deposition results in the formation of 3D electrode architecture. The direct ultrafast laser ablation allows the proper removal of the material to deep into the copper foil without melt formation as in thermal processing [123]. The electrode material shows an initial capacity of >1000 mAh g^{-1} at 3C rate and cycling stability over 400 cycles with a capacity retention of 55% at C/5 rate.

A 3D electrode architecture developed by depositing silicon-aluminum (Si-Al) film through magnetron sputtering on copper foam current collector shows enhanced electrochemical performance than the conventional Si-Al film [124]. The electrode delivers an initial capacity of 3348 mAh g^{-1} with a coulombic efficiency of 93.6%. The 3D macroporous structure of the electrode ensures good electrical/ionic conductivity and reduced pulverization issues and thereby improved electrochemical performances. The silicon-based 3D cellular electrode developed by casting the ball-milled silicon powders into the 3D copper

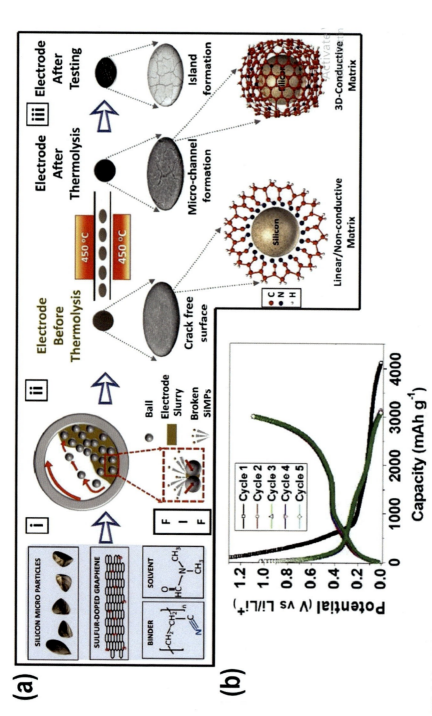

FIG. 9.11 (A) Schematic of the development of the 3D electrode architecture, and (B) characteristic voltage profile of the material at 0.1 A g^{-1}. *(Reproduced with permission from R. Batmaz, F.M. Hassan, D. Higgins, Z.P. Cano, X. Xiao, Z. Chen, Highly durable 3D conductive matrixed silicon anode for lithium-ion batteries, J. Power Sources 407 (2018) 84–91.)*

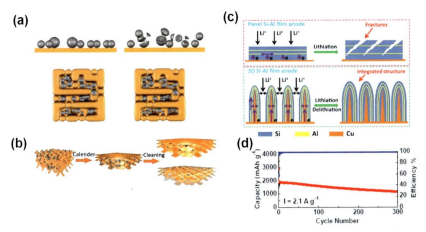

FIG. 9.12 (A) Schematic representation of conventional and 3D copper current collector, Si particle can come out when expansion happens in convention current collector but stay stuck in 3D electrode; (B) The steps of the integrated electrode for trapping silicon particles. (C) Schematic illustration of the mechanism of the 3D-Si-Al electrode for improving the electrochemical performance; (D) Cycle performance of 3D-MSAF anode in an FEC-containing electrolyte at the current density of 2.1 A g^{-1}. ((A, B) Reproduced with permission from C. Zhao, S. Li, X. Luo, B. Li, W. Pan, H. Wu, Integration of Si in a metal foam current collector for stable electrochemical cycling in Li-ion batteries, J. Mater. Chem. A 3 (18) (2015) 10114–10118. (C, D) Reproduced with permission from Q. Zhang, J. Liu, Z.-Y. Wu, J.-T. Li, L. Huang, S.-G. Sun, 3D nanostructured multilayer Si/Al film with excellent cycle performance as anode material for lithium-ion battery, J. Alloys Compd. 657 (2016) 559–564.)

cellular architectures obtained by the multistep electrodeposition process is also reported to have improved electrochemical performance [125]. The reduction in the strain-induced structural deformation of the material resulted from the electrode design improves the overall performance.

A novel type of 3D interconnected network of graphene wrapped porous silicon sphere was fabricated using layer-by-layer assembly, followed by magnesiothermic-reduction. The composite slurry of 3D interconnected Si @ Graphene composite (70 wt%), 15 wt% Super P carbon black, and 15 wt% poly(vinylidene fluoride) (PVDF) in N-methyl-2-pyrrolidone was coated on the Copper foam [126]. The schematic illustration of the Si@Graphene network is shown in Fig. 9.13A. The Si@Graphene network delivered a high reversible capacity of 1299.6 mAh g^{-1} at the end of the 25th cycle; in comparison, bare Si spheres could deliver only 431.5 mAh g^{-1}, in the potential range of 0.01–1.2 V at a current density of 0.05C (1C = 4200 mA g^{-1}). The unique porous, interconnected structural features of this anode are beneficial for faster ionic transport, which leads to higher rate capability and long cycle life.

Si-based composite electrode coated on conventional Cu foil current collector display a very short cycle life. The conventional electrode with active material loading between 1 and 4 mg cm^{-2} could limit the capacity to 1200 mAh g^{-1}

350 PART | IV Achieving high performance

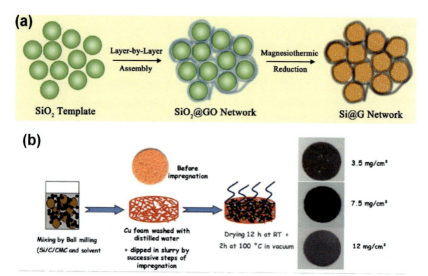

FIG. 9.13 (A) Schematic illustration for the synthesis of the Si@Graphene network. (B) Schematic illustration of the fabrication procedure and images of the prepared electrode. *((A) Reproduced with permission from P. Wu, H. Wang, Y. Tang, Y, Zhou, T. Lu, Three-dimensional interconnected network of graphene-wrapped porous silicon spheres: in situ magnesiothermic-reduction synthesis and enhanced lithium-storage capabilities, ACS Appl. Mater. Interfaces 6 (5) (2014) 3546–3552. (B) Reproduced with permission from D. Mazouzi, D. Reyter, M. Gauthier, P. Moreau, D. Guyomard, L. Roué, B. Lestriez, Very high surface capacity observed using Si negative electrodes embedded in copper foam as 3D current collectors, Adv. Energy Mater. 4 (8) (2014) 1301718.)*

with very limited cycle life. However, much better cycling stability could be achieved, when the copper foam is employed as a current collector by maintaining a similar electrode formulation (Fig. 9.13B) [127]. It has been observed that the Si loading could reach up to 10 mg cm^{-2} with surface capacity around 10 mAh cm^{-2}, having cycling stability of around 400 cycles with commercial micrometric particles. Again it has been observed that the use of carbon nanofiber (CNF) instead of carbon black (CB) as conducting additive in slurry could lead to better coulombic efficiency and capacity retention as a consequence of the better electronic wiring provide by CNF, which allows homogeneous functioning of the composite electrode. Further, CNF composite mass provides better C rate performance due to higher lithium-ion diffusion through the conductive framework and low electrode polarization. The volumetric and gravimetric energy densities are increased by 23% and 19% for the LiFePO$_4$/Si@Cu-foam full cells in relation to conventional LiFePO$_4$//Graphite lithium-ion cell.

The 3D bicontinuous silicon anode developed by depositing silicon film on a porous nickel framework shows improved performance than the conventional foil-supported silicon film [128]. The porous 3D nickel framework is obtained by electroplating nickel on an opal colloidal crystal template and subsequent removal of the template. The electrode shows an initial capacity of 3568 mAh g^{-1} at 0.5C rate, and cycling stability over 100 cycles with 85% of capacity retention. As the 3D porous nickel scaffold serves as both a conductive network and a solution to accommodate the volume change of silicon during cycling, the electrode shows improved performance than the conventional silicon film anodes. The silicon thin film developed on nickel form (3D conductive framework) through electrodeposition technique shows improved capacity retention (2800 mAh g^{-1} after 80 cycles at 0.36 A g^{-1}) and good cycling stability (>100 cycles). The electrode design attracted attention in terms of the simple synthesis procedure adopted for the development of 3D architecture.

A novel 3D multilayered assembly of Si/RGO nanohybrid on a porous Ni foam substrate is designed as illustrated in Fig. 9.14A [129]. In this approach GO was initially deposited on the porous Ni foam by immersing Ni foam in aqueous GO followed by deposition of Si NPs onto the GO-coated Ni foam using the dip-coating approach. The multilayered structure of alternating Si NP layers and GO layers was obtained through layer-by-layer deposition followed by rapid thermal reduction of GO to RGO at 600°C for 2 h in a flow

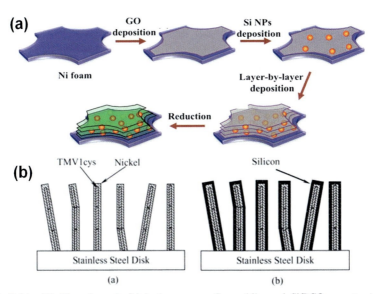

FIG. 9.14 (A) The schematic fabrication process for multilayered Si/RGO nanostructures. (B) Diagram of a patterned 3D- (tobacco mosaic virus) TMV1cys/Ni current collector and 3D silicon anode. *((A) Reproduced with permission from J. Chang, X. Huang, G. Zhou, S. Cui, P.B. Hallac, J. Jiang, P.T. Hurley, J. Chen, Multilayered Si nanoparticle/reduced graphene oxide hybrid as a high-performance lithium-ion battery anode, Adv. Mater. 26 (5) (2014) 758–764. (B) Reproduced with permission from X. Chen, K. Gerasopoulos, J. Guo, A. Brown, C. Wang, R. Ghodssi, J.N. Culver, A patterned 3D silicon anode fabricated by electrodeposition on a virus-structured current collector, Adv. Funct. Mater. 21 (2) (2011) 380–387.)*

of Ar and H$_2$ mixture. The outstanding Li-ion storage capacity of 2300 mAh g^{-1} at 0.05C, 700 mAh g^{-1} at 10C of the bulk electrode was obtained due to conductive pathways provided by the porous Ni scaffold, the well-controlled layered alternating Si/RGO conducting nanostructures, and the excellent mechanical stability. The electrode showed superior capacity retention of 87% (630 mAh g^{-1}) at a current rate of 10C after 152 cycles and 780 mAh g^{-1} at 3C after 300 cycles. These aforementioned results reveal that the multilayered Si/RGO electrode is effective in mitigating the volume expansion/contraction issue during lithiation/delithaiation, and enhancing the charge transfer and charge diffusion during the cycling process. The fabrication of 3D nanostructures electrode architecture by dip-coating is low-cost, binder-free, and easy to scale up, and holds great potential for application of high-performance Si-based materials in LIBs.

Another electrode fabricated by electrodepositing silicon nanoparticles on self-assembled 3D virus-structured of nickel as a current collector (Fig. 9.14B) acts as good anode material for LIBs [130]. This current collector is composed of self-assembled nanowire-like rods of genetically modified tobacco mosaic virus (TMV1cys), chemically coated on to nickel current collector to create a high surface area conductive substrate. The method allows making a highly porous silicon electrode with the high surface area, and the complex network of current nickel collector ensures conductive support to the electrode architecture. The developed anode material shows a good initial capacity of 2300 mAh g^{-1} and retain >1200 mAh g^{-1} capacity after >170-cycles. Further, the issues of large irreversible capacity and cycling stability are addressed by developing a standard carbon coating on the electrode surface.

9.5.3 3D carbon-based scaffold

To tackle the volume expansion and structural disintegration during cycling, a 3D electrode fabrication is demonstrated, in which Si nanoparticles are dispersed within 3D electrode graphene coated porous stainless steel current collector followed by deposition of second graphene layer above the silicon layer [131]. Perfluorohexane is used as a precursor for graphene which is depositied onto the current collector by CVD followed by heating at 950°C under argon/hydrogen atmosphere. This electrode architecture contains the Si nanoparticles within the porous graphene network, which further supports the charge transfer process throughout the Si nanoparticles surface (schematically shown in Fig. 9.15). As a result, these electrodes were able to cycle up to 1000 cycles with 100% coulombic efficiency, while maintaining 88% capacity retention. The electrode delivered a capacity of 0.365 mAh cm^{-2}, at a higher current rate of 0.5 mA cm^{-2}.

Graphene sheets can be stacked to form graphene papers, which are electrically conducting and mechanically robust, and could be prepared easily from the exfoliated graphite [59, 132]. The potential low-cost starting material and

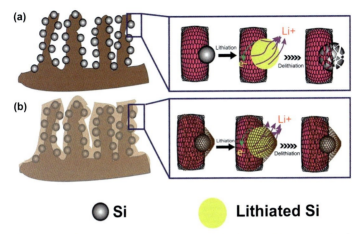

FIG. 9.15 Schematically representation of lithiation and delithiation of Si nanoparticles in (A) 3D graphene-coated porous stainless steel current collector (pSS-Gr/Si), (B) SiNPs sealed inside pores-under graphene sheets (pSS-Gr/Si/Gr). *(Reproduced with permission from M.V. Shelke, H. Gullapalli, K. Kalaga, M.T.F. Rodrigues, R.R. Devarapalli, R. Vajtai, P.M. Ajayan, Facile synthesis of 3D anode assembly with Si nanoparticles sealed in highly pure few layer graphene deposited on porous current collector for long life Li-ion battery, Adv. Mater. Interfaces 4 (10) (2017) 1601043.)*

fabrication process of this graphene papers makes them attractive current collectors for energy storage materials. Si-graphene paper composite is prepared by dispersing the Si nanoparticles and GO suspension, followed by suction filtration and reduction under Ar—H$_2$ mixture gas at 700°C for 1 h. The composite electrode delivered a capacity >2200 and 1500 mAh g^{-1} after 50 and 200 cycles, respectively.

In another report, ultrathin-graphite foam (UGF) with a mass density of 10–20 mg cm^{-3}, and a specific surface area of 20% of the projected area of UGF sheet with the interconnected porous 3D structure having 450-μm pore diameter make it appealing as the current collector [133, 134]. The Si/Graphene/UGF 3D electrode is fabricated by drop-casting the graphene-coated Si nanoparticles on the UGF sheets, and the corresponding SEM and TEM images are shown in Fig. 9.16A–F [133]. The electrode delivered a gravimetric capacity of 983 and 513 mAh g^{-1}, when the Si loading on UGF was controlled to 1.5 and 0.4 mg cm^{-2}, which further correspond to 25.2 and 8.6 wt% Si mass loading, respectively. The good electrochemical performance of this 3D structure is because of the electron-conducting property of UGF and graphene network, faster Li-ion diffusion due to higher surface area of UGF to anchor Si nanoparticles, and graphene sheets could act as buffers to the stress originates during alloying-dealloying.

In another novel nanostructure design, simultaneously Si nanoparticles encapsulated in graphene sheets are anchored on vertically aligned graphene

354 PART | IV Achieving high performance

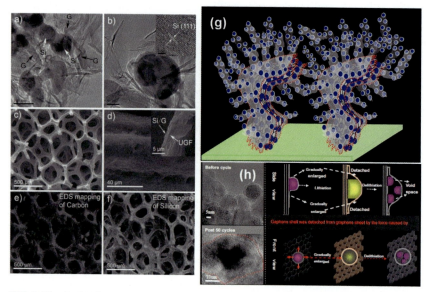

FIG. 9.16 (A, B) TEM images of graphene encapsulated Si, (B inset) HRTEM of the Si nanoparticle, (C, D) SEM images of the structure of Si/graphene dropped on UGF, (D inset) cross-sectional view of the composite, and corresponding EDS element mapping images of (E) C and (F) Si acquired at the same region with Figure C. (G) graphic model of SiNPs@GNS–GrTr; (H) Schemes for the volume changes of the SiNPs@GNS-GrTr anode during the cycling process. *((A–F) Reproduced with permission from J. Ji, H. Ji, L.L. Zhang, X. Zhao, X. Bai, Fan, F. Zhang, R.S. Ruoff, Graphene-encapsulated Si on ultrathin-graphite foam as anode for high capacity lithium-ion batteries, Adv. Mater. 25 (33) (2013) 4673–4677. (G, H) Reproduced with permission from N. Li, S. Jin, Q. Liao, H. Cui, C. Wang, Encapsulated within graphene shell silicon nanoparticles anchored on vertically aligned graphene trees as lithium ion battery anodes, Nano Energy 5 (2014) 105–115.)*

trees (represented as SiNPs@GNS-GrTr) [135]. The designed tree-like 3D electrode is mechanically robust, flexible, and lose internal structure could easily accommodate the volume changes i.e., expansion/contraction during lithiation / delithalition. The electrode exhibits a reversible capacity of 1528 mAhg^{-1} at a current density of 150 mA g^{-1}, with 88.6% capacity retention for 50 cycles. A graphic model of SiNPs@GNS-GrTr is shown in Fig. 9.16G. The graphene trees are directly grown on the current collector, and the uniform dispersion of this three-dimensional nanostructure provides a short diffusion path for the Li-ion and electrons, which results in the excellent electrochemical performance as an anode in LIBs.

The continued search for better anodes has fascinated to design Silicon oxycarbide glass-graphene composite paper derived from molecular precursor 1,3,5,7-tetramethyl-1,3,5,7-tetravinylcyclotetrasiloxane polymeric precursor and graphene oxide as graphene precursor [136]. The digital picture of a large area (approximately 15 cm × 2.5 cm) self-standing anode material and the

Nanostructured 3D electrode architectures of silicon **Chapter | 9** 355

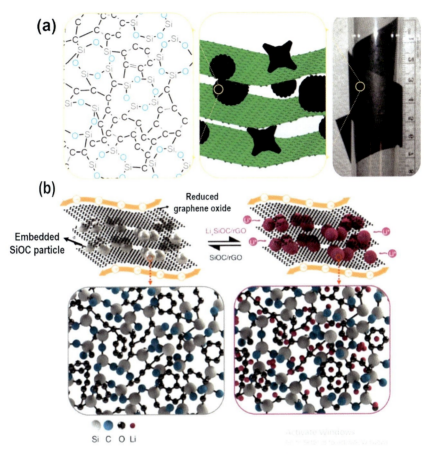

FIG. 9.17 (A) Digital camera picture and schematic illustration of the proposed hybrid structure of the freestanding paper along with the atomic structure of pyrolyzed SiOC particle; (B) Schematic representing the mechanism of lithiation/delithiation in SiOC particles. *(Reproduced with permission from L. David, R. Bhandavat, U. Barrera, G. Singh, Silicon oxycarbide glass-graphene composite paper electrode for long-cycle lithium-ion batteries, Nat. Commun. 7 (1) (2016) 1–10.)*

schematic illustration of a proposed hybrid structure of the freestanding paper along with the atomic structure of pyrolyzed SiOC particle is shown in Fig. 9.17A. The paper electrode delivered a charge capacity of 588 mAh g^{-1} at the end of the 1000th cycle. Fig. 9.17B represents the mechanism of lithiation and delithiation in SiOC particles, which further confirms that the majority of lithiation occurs via adsorption of ions at the disordered carbon phase.

Silicon is incorporated into the 3D-CNT sponge structure to achieve a high capacity of 40 mAh cm^{-2}, which is approximately 10 times higher than current Li-ion battery technology [137]. The resulting Si-CNT sponge is flexible and

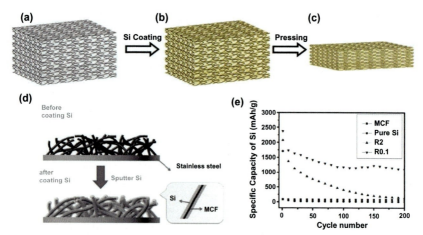

FIG. 9.18 (A) CNT sponge used in CVD; (B) Conformal CVD deposition of amorphous Si onto CNT surface to form Si-CNT coaxial nanostructure; (C) Pressed Si-CNT. (D) Schematic representation of Si/MCFs electrode with three-dimensional structure; (E) Cycling tests for cells at a current density of 750 μA cm^{-2}. ((A–C) Reproduced with permission from L. Hu, H. Wu, Y. Gao, A. Cao, H. Li, J. McDough, X. Xie, M. Zhou, Y. Cui, Silicon–carbon nanotube coaxial sponge as Li-ion anodes with high areal capacity, Adv. Energy Mater. 1 (4) (2011) 523–527. (D, E) Reproduced with permission from K.-F. Chiu, S.-H. Su, H.-J. Leu, C.-Y. Wu, Silicon thin film anodes coated on micron carbon-fiber current collectors for lithium ion batteries, Surf. Coat. Technol. 267 (2015) 70–74.)

mechanically withstanding in a solvent medium. The preparation of Si-CNT 3D-film through CVD of Si on CNT sponge is represented in Fig. 9.18A–C. The electrode delivered the initial charge and discharge capacities of 3200 and 2750 mAh g^{-1}, with a coulombic efficiency of 86% at C/5 rate between the voltage ranges of 0.05–1 V, respectively. However, the capacity of the electrode drops to 1300 mAh g^{-1} from 2800 mAh g^{-1}, when the lower cut-off potential is limited from 0.05 to 0.17 V, while the cycling retention improves dramatically. The excellent electrochemical behavior of this 3D electrode is realized based on the highly conductive porous CNT backbone with uniform deposition of Si over the entire 3D framework.

Another 3D carbon-silicon electrode prepared by using micron carbon-fibers obtained using thermal chemical vapor deposition (TCVD) as a current collector, followed by the deposition of silicon through radio frequency magnetron sputtering shows good performance as anode material for LIBs [138]. The electrode design is reported to have good capacity retention and cycling stability (~1087 mAh g^{-1} even after 200 cycles). The efficiency of the electrode design to buffer the internal stress developed during cycling resulted in the improved electrochemical performance.

A flexible freestanding 3D silicon-carbon nanofiber (500–600 nm length) composite prepared through electrospinning shows good stability and electrochemical performance [139]. A schematic of the development of the 3D

Nanostructured 3D electrode architectures of silicon **Chapter | 9** 357

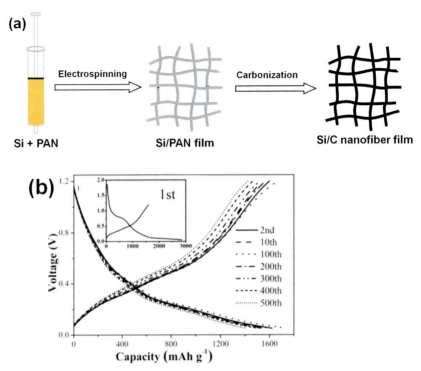

FIG. 9.19 (A) Schematic of the development of the 3D silicon-carbon nanofiber (500–600 nm length) composite; (B) characteristic voltage profile. *(Reproduced with permission from E. Qu, T. Chen, Q. Xiao, G. Lei, Z. Li, Flexible freestanding 3D Si/C composite nanofiber film fabricated using the electrospinning technique for lithium-ion batteries anode, Solid State Ionics 337 (2019) 70–75.)*

electrode architecture is given in Fig. 9.19A. The material gives a stable capacity of 1620 mAh g^{-1} over 500 cycles Fig. 9.19B. The flexible freestanding design with well dispersed and uniformly encapsulated silicon nanoparticles inside the carbon nanofiber gives enhanced electrochemical performance with reduced volume change during cycling.

A new strategy to develop 3D silicon-carbon (Si-C) free-standing electrodes by using carbonized P-pitch as high-temperature binder and carbon fiber as current collector instead of the conventional organic binder and copper current collector attained much interest [140]. A schematic of the 3D electrode fabrication is shown in Fig. 9.20. The organic binder and conducting diluent free Si-C 3D electrode architecture developed at an annealing temperature of 1000°C shows enhanced electrochemical performance with capacities over 2000 and 1000 mAh g^{-1} at C/10 and 5C rates (Fig. 9.20C and D), respectively for more than 100 cycles and 250 cycles. 3D electrode architecture of Si-NPs on CF allows the continuous conducting framework having good conductivity and

358 PART | IV Achieving high performance

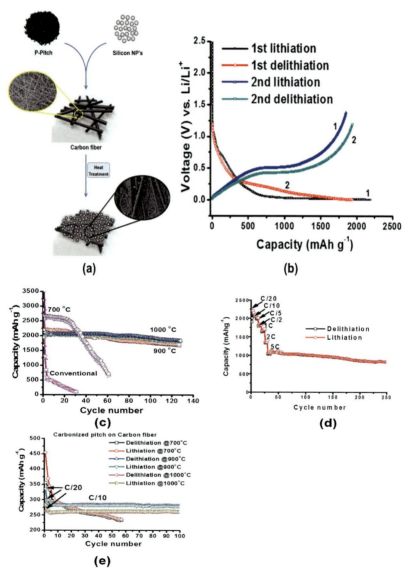

FIG. 9.20 (A) Schematic of the steps involved in the fabrication of Si-C 3D electrode architectures, and (B) the charge/discharge profiles of the Si-C free-standing 3D electrode fabricated at an annealing temperature of 1000°C. (C) Comparison of cycle life of 3D Si-C composite electrodes prepared at 700°C, 900°C, 1000°C and conventional electrode (as indicated), (D) C rate performance for the Si-C composite free-standing electrode annealed at 1000°C and (E) Capacity vs. cycle number for the Pitch coated carbon fibers (No Si-NPs) annealed at 700, 900 and 1000°C (as indicated). *(Reproduced with the permission from S.K. Kumar, S. Ghosh, S.K. Malladi, J. Nanda, S.K. Martha, Nanostructured silicon–carbon 3D electrode architectures for high-performance lithium-ion batteries, ACS Omega 3 (8) (2018) 9598–9606.)*

flexible network of carbon formed from the pitch and carbon fiber which accommodates the volume change of silicon during charge and discharge. Pitch forms a conducting carbon coating along with the fiber and coats along with the particles and increases contact strength. Increasing temperature from 700°C to 1000°C, increases the structural ordering of carbon and silicon which allows the coating of carbon layers throughout Si-NPs. The integrity of the Si-C composite electrode is well-maintained due to the good mechanical strength of the carbon fiber and pitch. In the present electrode, approach pitch improves the connectivity between fiber-fiber and fiber-Si active material contacts, thereby reducing the internal impedance of the cell. The 3D electrode CF architecture with pitch controls the degradation of Si anode. The enhanced performance of the Si-C 3D electrode architectures is attributed to the synergistic effect of the carbon-coated silicon nanoparticles and the 3D carbon fiber current collector. In the current electrode approach when carbon fibers are used as current collectors, they contribute 10% capacity to the Si-C composite electrodes (Fig. 9.20E). Besides, the carbonized pitch delivers capacity (260–280 mAh g^{-1}) to the total capacity of the electrode. Besides, the full cells assembled using 3D Si-carbon composite electrode and Mg-F doped LMR-NMC cathode show the high open-circuit voltage of >4V, the high energy density of >500 Wh kg^{-1} in the voltage range between 4.6 and 2.0 V.

In another report, 3D Si-based nanostructure is fabricated by encapsulating Si nanoparticles in carbon nanofiber, followed by wrapping with the graphene nanosheets [141]. The overall fabrication process of Si/CNF-Graphene composite is illustrated in Fig. 9.21A–D. Further, Fig. 9.21E–G represents the cross-sectional FESEM image of the composite, clearly signify the multilayer structure of Si/CNF-Graphene composite. The 3D carbon nanofiber network not only acts as a current collector but also promotes the charge transfer and maintains stable electrical contact of the Si nanoparticles. The composite electrode exhibits a stable reversible capacity of 878 mAh g^{-1} at a current density of 100 mA g^{-1} for 100 cycles. In addition to the conducting network, the 2D graphene sheet was serving as a protective layer for Si particles, which can prevent the continuous SEI formation, resulting in good cycling stability even at the higher current rate of 5 A g^{-1}.

In another strategy a silicon/graphene/carbon nano fiber ((FSiGCNFs)) 3D hybrid electrode architecture with atomic control of the expansion space designed by ALD is reported as a space as the binder-free anode for flexible LIBs, shown in Fig. 9.22 [142]. The electrode architecture provides porous structure throughout the matrix and limits the SEI formation and increases its stability, provides highly efficient channels for the fast transport of both electrons and lithium ions during cycling, The Si nanoparticle in FSiGCNFs have controllable void spaces ensure excellent mechanical strength and provides sufficient space to overcome the issues of volume expansion of Si nanoparticles during lithiation and delithiation processes. The 3D FSiGCNF electrode, shows remarkable cycling stability of 2002 mAh g^{-1} at current density of 700 mA g^{-1}

360 PART | IV Achieving high performance

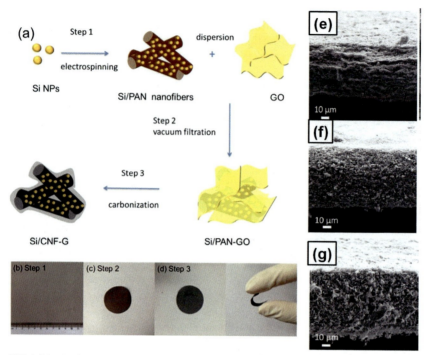

FIG. 9.21 (A) Schematic illustration of the preparation procedures for the Si/CNF-G nanostructure. Photographs of (B) Si/PAN nanofibers, (C) Si/polymer nanofibers/GO sheets, and (D) Si/CNF-G sheets, (E)–(G) Cross-sectional FESEM images of the different composite. *(Reproduced with permission from M.S. Wang, W.L. Song, L.Z. Fan, Three-dimensional interconnected network of graphene-wrapped silicon/carbon nanofiber hybrids for binder-free anodes in lithium-ion batteries, ChemElectroChem 2 (11) (2015) 1699–1706.)*

after 1050 cycles, 1201 mAh g^{-1} at 1400 mA g^{-1} after 2800 cycles and 582 mAh g^{-1} at high currents of 28,000 mA g^{-1}.

All these features offer ample opportunities for modifying towards better Si-anode materials for optimal cell performance. Most of the works presented in carbon fiber-based 3D electrode architecture are limited mostly to coin-tape or swage-lock type cells. For tabbing large format pouch cells, the carbon felt can be metalized by using sputtering) or metal coating followed by spot welding with metal foil or by using conducting epoxy with metal foil.

9.5.4 3D porous conductive polymer framework

In another strategy, an electrode architecture is developed by encapsulating silicon nanoparticles in a 3D porous conductive polymer framework to surpass the inherent problem of pulverization in silicon-based anodes [143]. The hydrogel composite of silicon nanoparticles/conductive polymer (polyaniline) is

Nanostructured 3D electrode architectures of silicon **Chapter | 9** **361**

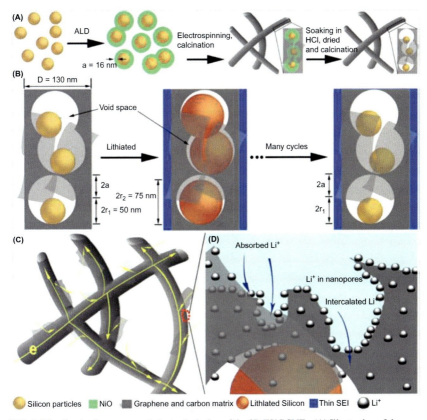

FIG. 9.22 Synthesis process and electrode design of the 3D FSiGCNFs. (A) Illustration of the synthesis process of the 3D FSiGCNFs. The 3D FSiGCNFs were first synthesized by ALD, followed by electrospinning, annealed at 450°C in Ar atmosphere to obtain the Si/NiO/graphene/carbon composite, and soaked in hydrochloric acid solution to remove the NiO to form precise and controllable expansion space and annealed at 800°C in Ar atmosphere for 2h. (B) Schematic diagram of the 3D FSiGCNF electrode design. Illustration of (C) electron transmission and (D) Li$^+$ storage in the 3D FSiGCNF film. *(Reproduced with permission from J. Zhu, T. Wang, F. Fan, L. Mei, B. Lu, Atomic-scale control of silicon expansion space as ultrastable battery anodes, ACS Nano 10 (9) (2016) 8243–8251.)*

prepared by the solution-phase synthesis method through mixing the silicon nanoparticles with phytic acid and aniline in water. The 3D porous structure is obtained through rapid polymerization and cross-linking of aniline by the addition of an oxidizer (ammonium persulfate) into the hydrogel. The silicon nanoparticles/conductive polymer hydrogel composite electrode is designed in such a way that each silicon nanoparticle is perfectly encapsulated inside the conductive polymer coating and well connected to the highly porous hydrogel framework. The 3D porous conductive polymer matrix allows fast electronic and ionic transport, and electrode design offers enough space to lodge

362 PART | IV Achieving high performance

the volume change during cycling. The electrode delivers excellent deep cycling stability of 1600 mAh g^{-1} for 1000 deep cycles at 1 A g^{-1}, and 550 mAh g^{-1} after 5000 cycles, which corresponds to 91% capacity at 6 A g^{-1} (Fig. 9.23).

An attractive strategy to develop a flexible 3D silicon/carbon fiber paper electrode attained much attention due to its attractive design and electrochemical performance. The electrode architecture is prepared by combining electrospraying and electrospinning methods followed by carbonization. The technique of simultaneous electrospraying and electrospinning of nanostructured silicon-polyacrylonitrile clusters and polyacrylonitrile fibers, respectively, followed by carbonization resulted in the formation of the flexible 3D silicon/carbon fiber paper electrode (Illustrated in Fig. 9.24) [144]. The electrode delivers a capacity of ~1600 mAh g^{-1} and shows cycling stability over 600 cycles. The electrode architecture is a unique design having flexibility, mechanical stability, and enhanced electrochemical performance. The combined effect of the carbon fiber network and uniformly distributed

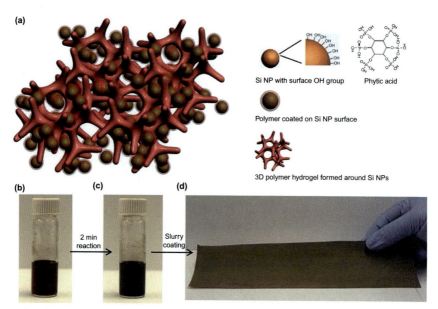

FIG. 9.23 (A) Schematic illustration of 3D porous SiNP/conductive polymer hydrogel composite electrodes, (B) First, SiNPs were dispersed in the hydrogel precursor solution containing the crosslinker (phytic acid), the monomer aniline and the initiator ammonium persulphate. (C) After several minutes of chemical reaction, the aniline monomer was crosslinked, forming a viscous gel with a dark green color. (D) The viscous gel was then bladed onto a 5 × 20 cm^2 copper foil current collector and dried. *(Reproduced with permission from H. Wu, G. Yu, L. Pan, N. Liu, M.T. McDowell, Z. Bao, Y. Cui, Stable Li-ion battery anodes by in-situ polymerization of conducting hydrogel to conformally coat silicon nanoparticles, Nat. Commun. 4 (1) (2013) 1–6.)*

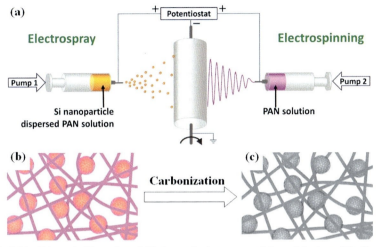

FIG. 9.24 Schematic illustration of (A) the synthesis process and the architecture of the flexible 3D Si/C fiber paper electrode (B) before, and (C) after carbonization. *(Reproduced with permission from Y. Xu, Y. Zhu, F. Han, C. Luo, C. Wang, 3D Si/C fiber paper electrodes fabricated using a combined electrospray/electrospinning technique for Li-ion batteries, Adv. Energy Mater. 5 (1) (2015) 1400753.)*

carbon-coated silicon nanoparticle clusters resulted in the improved performance of the electrode.

9.6 Conclusions

Silicon is a promising anode material for rechargeable LIBs and can replace conventional carbon-based anodes due to high theoretical capacity. In this chapter, the main development of silicon-based anode is described, beginning with nano-sized Si material to composite materials, and finally 3D structured materials. The combination of these three approaches to fabricate Si-based composites results in enhanced cyclic performance. Three dimensional electrode architecture-based porous Si has attracted significant attention due to beneficial structural features and great promise for practical applications. The porous electrode structure accommodates volume change during cycling without losing the structural integrity of the electrode. However, the overall properties of Si-based anodes in 3D electrode architectures are not very satisfactory for practical applications when we consider their overall gravimetric and volumetric energy densities, C-rate performance, and cycling stability in large format cells. There are still several technological hurdles that must be overcome, such as higher active area mass loading of silicon nanoparticles to reach >3 mAhcm^{-2} surface capacity. Developing low-cost processing methods and exploiting nonexpensive materials are essential for the practical applications of the Si-based anodes. Although there is still a long way to go for cost-effective large-scale applications of Si-anodes, it is envisioned that the electrode architecture will help to develop better Si-C composite electrodes in future.

References

[1] J.R. Owen, Rechargeable lithium batteries, Chem. Soc. Rev. 26 (4) (1997) 259–267.
[2] J.M. Tarascon, M. Armand, Issues and challenges facing rechargeable lithium batteries, Nature 414 (6861) (2001) 359–367.
[3] M. Armand, J.-M. Tarascon, Building better batteries, Nature 451 (7179) (2008) 652–657.
[4] B. Scrosati, J. Garche, Lithium batteries: status, prospects and future, J. Power Sources 195 (9) (2010) 2419–2430.
[5] M.S. Whittingham, Lithium batteries and cathode materials, Chem. Rev. 104 (10) (2004) 4271–4302.
[6] P.G. Bruce, B. Scrosati, J.M. Tarascon, Nanomaterials for rechargeable lithium batteries, Angew. Chem. Int. Ed. 47 (16) (2008) 2930–2946.
[7] V. Etacheri, R. Marom, R. Elazari, G. Salitra, D. Aurbach, Challenges in the development of advanced Li-ion batteries: a review, Energy Environ. Sci. 4 (9) (2011) 3243–3262.
[8] P. Roy, S.K. Srivastava, Nanostructured anode materials for lithium ion batteries, J. Mater. Chem. A 3 (6) (2015) 2454–2484.
[9] M.A. Rahman, G. Song, A.I. Bhatt, Y.C. Wong, C. Wen, Nanostructured silicon anodes for high-performance lithium-ion batteries, Adv. Funct. Mater. 26 (5) (2016) 647–678.
[10] M.-S. Balogun, Y. Luo, W. Qiu, P. Liu, Y. Tong, A review of carbon materials and their composites with alloy metals for sodium ion battery anodes, Carbon 98 (2016) 162–178.
[11] W.-J. Zhang, A review of the electrochemical performance of alloy anodes for lithium-ion batteries, J. Power Sources 196 (1) (2011) 13–24.
[12] A.S. Arico, P. Bruce, B. Scrosati, J.-M. Tarascon, W. Van Schalkwijk, Nanostructured materials for advanced energy conversion and storage devices, in: Materials for Sustainable Energy: A Collection of Peer-Reviewed Research and Review Articles From Nature Publishing Group, World Scientific, 2011, pp. 148–159.
[13] M. Obrovac, V. Chevrier, Alloy negative electrodes for Li-ion batteries, Chem. Rev. 114 (23) (2014) 11444–11502.
[14] R. Yazami, P. Touzain, A reversible graphite-lithium negative electrode for electrochemical generators, J. Power Sources 9 (3) (1983) 365–371.
[15] J.W. Choi, D. Aurbach, Promise and reality of post-lithium-ion batteries with high energy densities, Nat. Rev. Mater. 1 (4) (2016) 1–16.
[16] D. Ma, Z. Cao, A. Hu, Si-based anode materials for Li-ion batteries: a mini review, Nano-Micro Lett. 6 (4) (2014) 347–358.
[17] B. Liang, Y. Liu, Y. Xu, Silicon-based materials as high capacity anodes for next generation lithium ion batteries, J. Power Sources 267 (2014) 469–490.
[18] Y. Liu, K. Hanai, J. Yang, N. Imanishi, A. Hirano, Y. Takeda, Silicon/carbon composites as anode materials for Li-ion batteries, Electrochem. Solid-State Lett. 7 (10) (2004) A369–A372.
[19] I.H. Son, J.H. Park, S. Kwon, S. Park, M.H. Rümmeli, A. Bachmatiuk, H.J. Song, J. Ku, J.W. Choi, Choi, J.-m., Silicon carbide-free graphene growth on silicon for lithium-ion battery with high volumetric energy density, Nat. Commun. 6 (2015) 7393.
[20] M. Ashuri, Q. He, L.L. Shaw, Silicon as a potential anode material for Li-ion batteries: where size, geometry and structure matter, Nanoscale 8 (1) (2016) 74–103.

[21] D. Larcher, C. Mudalige, A. George, V. Porter, M. Gharghouri, J. Dahn, Si-containing disordered carbons prepared by pyrolysis of pitch/polysilane blends: effect of oxygen and sulfur, Solid State Ionics 122 (1–4) (1999) 71–83.
[22] C. Wang, G. Wu, X. Zhang, Z. Qi, W. Li, Lithium insertion in carbon-silicon composite materials produced by mechanical milling, J. Electrochem. Soc. 145 (8) (1998) 2751.
[23] C.K. Chan, H. Peng, G. Liu, K. McIlwrath, X.F. Zhang, R.A. Huggins, Y. Cui, High-performance lithium battery anodes using silicon nanowires, Nat. Nanotechnol. 3 (1) (2008) 31.
[24] J.K. Lee, C. Oh, N. Kim, J.-Y. Hwang, Y.-K. Sun, Rational design of silicon-based composites for high-energy storage devices, J. Mater. Chem. A 4 (15) (2016) 5366–5384.
[25] L. Liu, J. Lyu, T. Li, T. Zhao, Well-constructed silicon-based materials as high-performance lithium-ion battery anodes, Nanoscale 8 (2) (2016) 701–722.
[26] H. Wu, G. Zheng, N. Liu, T.J. Carney, Y. Yang, Y. Cui, Engineering empty space between Si nanoparticles for lithium-ion battery anodes, Nano Lett. 12 (2) (2012) 904–909.
[27] C.-M. Park, J.-H. Kim, H. Kim, H.-J. Sohn, Li-alloy based anode materials for Li secondary batteries, Chem. Soc. Rev. 39 (8) (2010) 3115–3141.
[28] C. Sandhya, B. John, C. Gouri, Lithium titanate as anode material for lithium-ion cells: a review, Ionics 20 (5) (2014) 601–620.
[29] M. Tilli, M. Paulasto-Krockel, M. Petzold, H. Theuss, T. Motooka, V. Lindroos, Handbook of Silicon Based MEMS Materials and Technologies, Elsevier, 2020.
[30] T.D. Lee, A.U. Ebong, A review of thin film solar cell technologies and challenges, Renew. Sust. Energ. Rev. 70 (2017) 1286–1297.
[31] M.-J. Bañuls, R. Puchades, Á. Maquieira, Chemical surface modifications for the development of silicon-based label-free integrated optical (IO) biosensors: a review, Anal. Chim. Acta 777 (2013) 1–16.
[32] J. Casady, R.W. Johnson, Status of silicon carbide (SiC) as a wide-bandgap semiconductor for high-temperature applications: a review, Solid State Electron. 39 (10) (1996) 1409–1422.
[33] C.-M. Wang, X. Li, Z. Wang, W. Xu, J. Liu, F. Gao, L. Kovarik, J.-G. Zhang, J. Howe, D.J. Burton, In situ TEM investigation of congruent phase transition and structural evolution of nanostructured silicon/carbon anode for lithium ion batteries, Nano Lett. 12 (3) (2012) 1624–1632.
[34] M. Obrovac, L. Christensen, Structural changes in silicon anodes during lithium insertion/extraction, Electrochem. Solid-State Lett. 7 (5) (2004) A93.
[35] M.T. McDowell, Y. Cui, Single nanostructure electrochemical devices for studying electronic properties and structural changes in lithiated Si nanowires, Adv. Energy Mater. 1 (5) (2011) 894–900.
[36] T. Hatchard, J. Dahn, In situ XRD and electrochemical study of the reaction of lithium with amorphous silicon, J. Electrochem. Soc. 151 (6) (2004) A838–A842.
[37] J. Li, J. Dahn, An in situ X-ray diffraction study of the reaction of Li with crystalline Si, J. Electrochem. Soc. 154 (3) (2007) A156–A161.
[38] L. Beaulieu, T. Hatchard, A. Bonakdarpour, M. Fleischauer, J. Dahn, Reaction of Li with alloy thin films studied by in situ AFM, J. Electrochem. Soc. 150 (11) (2003) A1457–A1464.
[39] V. Chevrier, J. Zwanziger, J. Dahn, First principles study of Li-Si crystalline phases: charge transfer, electronic structure, and lattice vibrations, J. Alloys Compd. 496 (1–2) (2010) 25–36.
[40] R.A. Huggins, Lithium alloy negative electrodes, J. Power Sources 81 (1999) 13–19.
[41] H. Wu, Y. Cui, Designing nanostructured Si anodes for high energy lithium ion batteries, Nano Today 7 (5) (2012) 414–429.
[42] R.A. Sharma, R.N. Seefurth, Thermodynamic properties of the lithium-silicon system, J. Electrochem. Soc. 123 (12) (1976) 1763–1768.

[43] I. Ryu, J.W. Choi, Y. Cui, W.D. Nix, Size-dependent fracture of Si nanowire battery anodes, J. Mech. Phys. Solids 59 (9) (2011) 1717–1730.
[44] X.H. Liu, L. Zhong, S. Huang, S.X. Mao, T. Zhu, J.Y. Huang, Size-dependent fracture of silicon nanoparticles during lithiation, ACS Nano 6 (2) (2012) 1522–1531.
[45] X.H. Liu, H. Zheng, L. Zhong, S. Huang, K. Karki, L.Q. Zhang, Y. Liu, A. Kushima, W.T. Liang, J.W. Wang, Anisotropic swelling and fracture of silicon nanowires during lithiation, Nano Lett. 11 (8) (2011) 3312–3318.
[46] J.-Y. Li, Q. Xu, G. Li, Y.-X. Yin, L.-J. Wan, Y.-G. Guo, Research progress regarding Si-based anode materials towards practical application in high energy density Li-ion batteries, Mater. Chem. Front. 1 (9) (2017) 1691–1708.
[47] X. Wu, Z. Wang, L. Chen, X. Huang, Ag-enhanced SEI formation on Si particles for lithium batteries, Electrochem. Commun. 5 (11) (2003) 935–939.
[48] G. Li, H. Li, Y. Mo, L. Chen, X. Huang, Further identification to the SEI film on Ag electrode in lithium batteries by surface enhanced Raman scattering (SERS), J. Power Sources 104 (2) (2002) 190–194.
[49] A. Tokranov, R. Kumar, C. Li, S. Minne, X. Xiao, B.W. Sheldon, Control and optimization of the electrochemical and mechanical properties of the solid electrolyte interphase on silicon electrodes in lithium ion batteries, Adv. Energy Mater. 6 (8) (2016) 1502302.
[50] S.-P. Kim, A.C. Van Duin, V.B. Shenoy, Effect of electrolytes on the structure and evolution of the solid electrolyte interphase (SEI) in Li-ion batteries: a molecular dynamics study, J. Power Sources 196 (20) (2011) 8590–8597.
[51] C.K. Chan, R. Ruffo, S.S. Hong, Y. Cui, Surface chemistry and morphology of the solid electrolyte interphase on silicon nanowire lithium-ion battery anodes, J. Power Sources 189 (2) (2009) 1132–1140.
[52] P. Verma, P. Maire, P. Novák, A review of the features and analyses of the solid electrolyte interphase in Li-ion batteries, Electrochim. Acta 55 (22) (2010) 6332–6341.
[53] M.W. Verbrugge, D.R. Baker, X. Xiao, Q. Zhang, Y.-T. Cheng, Experimental and theoretical characterization of electrode materials that undergo large volume changes and application to the lithium–silicon system, J. Phys. Chem. C 119 (10) (2015) 5341–5349.
[54] L. Beaulieu, K. Eberman, R. Turner, L. Krause, J. Dahn, Colossal reversible volume changes in lithium alloys, Electrochem. Solid-State Lett. 4 (9) (2001) A137–A140.
[55] H. Kim, M. Seo, M.H. Park, J. Cho, A critical size of silicon nano-anodes for lithium rechargeable batteries, Angew. Chem. Int. Ed. 49 (12) (2010) 2146–2149.
[56] J. Graetz, C. Ahn, R. Yazami, B. Fultz, Highly reversible lithium storage in nanostructured silicon, Electrochem. Solid-State Lett. 6 (9) (2003) A194–A197.
[57] Y. Oumellal, N. Delpuech, D. Mazouzi, N. Dupre, J. Gaubicher, P. Moreau, P. Soudan, B. Lestriez, D. Guyomard, The failure mechanism of nano-sized Si-based negative electrodes for lithium ion batteries, J. Mater. Chem. 21 (17) (2011) 6201–6208.
[58] I.A. Shkrob, J.F. Wishart, D.P. Abraham, What makes fluoroethylene carbonate different? J. Phys. Chem. C 119 (27) (2015) 14954–14964.
[59] J.K. Lee, K.B. Smith, C.M. Hayner, H.H. Kung, Silicon nanoparticles–graphene paper composites for Li ion battery anodes, Chem. Commun. 46 (12) (2010) 2025–2027.
[60] R.D. Cakan, M.-M. Titirici, M. Antonietti, G. Cui, J. Maier, Y.-S. Hu, Hydrothermal carbon spheres containing silicon nanoparticles: synthesis and lithium storage performance, Chem. Commun. 32 (2008) 3759–3761.
[61] S. Chen, M.L. Gordin, R. Yi, G. Howlett, H. Sohn, D. Wang, Silicon core–hollow carbon shell nanocomposites with tunable buffer voids for high capacity anodes of lithium-ion batteries, Phys. Chem. Chem. Phys. 14 (37) (2012) 12741–12745.

[62] X. Li, P. Meduri, X. Chen, W. Qi, M.H. Engelhard, W. Xu, F. Ding, J. Xiao, W. Wang, C. Wang, Hollow core–shell structured porous Si–C nanocomposites for Li-ion battery anodes, J. Mater. Chem. 22 (22) (2012) 11014–11017.
[63] H.-H. Li, J.-W. Wang, X.-L. Wu, H.-Z. Sun, F.-M. Yang, K. Wang, L.-L. Zhang, C.-Y. Fan, J.-P. Zhang, A novel approach to prepare Si/C nanocomposites with yolk–shell structures for lithium ion batteries, RSC Adv. 4 (68) (2014) 36218–36225.
[64] S.H. Ng, J. Wang, D. Wexler, K. Konstantinov, Z.P. Guo, H.K. Liu, Highly reversible lithium storage in spheroidal carbon-coated silicon nanocomposites as anodes for lithium-ion batteries, Angew. Chem. Int. Ed. 45 (41) (2006) 6896–6899.
[65] A. Magasinski, P. Dixon, B. Hertzberg, A. Kvit, J. Ayala, G. Yushin, High-performance lithium-ion anodes using a hierarchical bottom-up approach, Nat. Mater. 9 (4) (2010) 353–358.
[66] X.-Y. Zhou, J.-J. Tang, J. Yang, J. Xie, L.-L. Ma, Silicon@ carbon hollow core–shell heterostructures novel anode materials for lithium ion batteries, Electrochim. Acta 87 (2013) 663–668.
[67] Y. Park, N.S. Choi, S. Park, S.H. Woo, S. Sim, B.Y. Jang, S.M. Oh, S. Park, J. Cho, K.T. Lee, Si-encapsulating hollow carbon electrodes via electroless etching for lithium-ion batteries, Adv. Energy Mater. 3 (2) (2013) 206–212.
[68] N. Liu, H. Wu, M.T. McDowell, Y. Yao, C. Wang, Y. Cui, A yolk-shell design for stabilized and scalable Li-ion battery alloy anodes, Nano Lett. 12 (6) (2012) 3315–3321.
[69] K. Peng, J. Jie, W. Zhang, S.-T. Lee, Silicon nanowires for rechargeable lithium-ion battery anodes, Appl. Phys. Lett. 93 (3) (2008), 033105.
[70] C.K. Chan, R. Ruffo, S.S. Hong, R.A. Huggins, Y. Cui, Structural and electrochemical study of the reaction of lithium with silicon nanowires, J. Power Sources 189 (1) (2009) 34–39.
[71] R. Teki, M.K. Datta, R. Krishnan, T.C. Parker, T.M. Lu, P.N. Kumta, N. Koratkar, Nanostructured silicon anodes for lithium ion rechargeable batteries, Small 5 (20) (2009) 2236–2242.
[72] H.T. Nguyen, F. Yao, M.R. Zamfir, C. Biswas, K.P. So, Y.H. Lee, S.M. Kim, S.N. Cha, J.M. Kim, D. Pribat, Highly interconnected Si nanowires for improved stability Li-ion battery anodes, Adv. Energy Mater. 1 (6) (2011) 1154–1161.
[73] M.T. McDowell, S.W. Lee, I. Ryu, H. Wu, W.D. Nix, J.W. Choi, Y. Cui, Novel size and surface oxide effects in silicon nanowires as lithium battery anodes, Nano Lett. 11 (9) (2011) 4018–4025.
[74] X.H. Liu, L.Q. Zhang, L. Zhong, Y. Liu, H. Zheng, J.W. Wang, J.-H. Cho, S.A. Dayeh, S.T. Picraux, J.P. Sullivan, Ultrafast electrochemical lithiation of individual Si nanowire anodes, Nano Lett. 11 (6) (2011) 2251–2258.
[75] A.M. Chockla, K.C. Klavetter, C.B. Mullins, B.A. Korgel, Tin-seeded silicon nanowires for high capacity Li-ion batteries, Chem. Mater. 24 (19) (2012) 3738–3745.
[76] K.-Q. Peng, X. Wang, L. Li, Y. Hu, S.-T. Lee, Silicon nanowires for advanced energy conversion and storage, Nano Today 8 (1) (2013) 75–97.
[77] Y. Zhou, X. Jiang, L. Chen, J. Yue, H. Xu, J. Yang, Y. Qian, Novel mesoporous silicon nanorod as an anode material for lithium ion batteries, Electrochim. Acta 127 (2014) 252–258.
[78] S. Soleimani-Amiri, S.A. Safiabadi Tali, S. Azimi, Z. Sanaee, S. Mohajerzadeh, Highly featured amorphous silicon nanorod arrays for high-performance lithium-ion batteries, Appl. Phys. Lett. 105 (19) (2014) 193903.
[79] F.F. Cao, J.W. Deng, S. Xin, H.X. Ji, O.G. Schmidt, L.J. Wan, Y.G. Guo, Cu-Si nanocable arrays as high-rate anode materials for lithium-ion batteries, Adv. Mater. 23 (38) (2011) 4415–4420.
[80] S.H. Nguyen, J.C. Lim, J.K. Lee, Electrochemical characteristics of bundle-type silicon nanorods as an anode material for lithium ion batteries, Electrochim. Acta 74 (2012) 53–58.

[81] M.-H. Park, M.G. Kim, J. Joo, K. Kim, J. Kim, S. Ahn, Y. Cui, J. Cho, Silicon nanotube battery anodes, Nano Lett. 9 (11) (2009) 3844–3847.
[82] H. Wu, G. Chan, J.W. Choi, I. Ryu, Y. Yao, M.T. McDowell, S.W. Lee, A. Jackson, Y. Yang, L. Hu, Stable cycling of double-walled silicon nanotube battery anodes through solid–electrolyte interphase control, Nat. Nanotechnol. 7 (5) (2012) 310.
[83] J.K. Yoo, J. Kim, Y.S. Jung, K. Kang, Scalable fabrication of silicon nanotubes and their application to energy storage, Adv. Mater. 24 (40) (2012) 5452–5456.
[84] Y. Yao, M.T. McDowell, I. Ryu, H. Wu, N. Liu, L. Hu, W.D. Nix, Y. Cui, Interconnected silicon hollow nanospheres for lithium-ion battery anodes with long cycle life, Nano Lett. 11 (7) (2011) 2949–2954.
[85] W.-H. Lee, D.-Y. Kang, J.S. Kim, J.K. Lee, J.H. Moon, Uniformly dispersed silicon nanoparticle/carbon nanosphere composites as highly stable lithium-ion battery electrodes, RSC Adv. 5 (23) (2015) 17424–17428.
[86] P. Gao, J. Fu, J. Yang, R. Lv, J. Wang, Y. Nuli, X. Tang, Microporous carbon coated silicon core/shell nanocomposite via in situ polymerization for advanced Li-ion battery anode material, Phys. Chem. Chem. Phys. 11 (47) (2009) 11101–11105.
[87] J. Yin, M. Wada, K. Yamamoto, Y. Kitano, S. Tanase, T. Sakai, Micrometer-scale amorphous Si thin-film electrodes fabricated by electron-beam deposition for Li-ion batteries, J. Electrochem. Soc. 153 (3) (2006) A472–A477.
[88] J.W. Wang, Y. He, F. Fan, X.H. Liu, S. Xia, Y. Liu, C.T. Harris, H. Li, J.Y. Huang, S.X. Mao, Two-phase electrochemical lithiation in amorphous silicon, Nano Lett. 13 (2) (2013) 709–715.
[89] E. Pollak, G. Salitra, V. Baranchugov, D. Aurbach, In situ conductivity, impedance spectroscopy, and ex situ Raman spectra of amorphous silicon during the insertion/extraction of lithium, J. Phys. Chem. C 111 (30) (2007) 11437–11444.
[90] A. Netz, R.A. Huggins, W. Weppner, The formation and properties of amorphous silicon as negative electrode reactant in lithium systems, J. Power Sources 119 (2003) 95–100.
[91] H. Jung, M. Park, Y.-G. Yoon, G.-B. Kim, S.-K. Joo, Amorphous silicon anode for lithium-ion rechargeable batteries, J. Power Sources 115 (2) (2003) 346–351.
[92] K.K. Sarode, R. Choudhury, S.K. Martha, Binder and conductive additive free silicon electrode architectures for advanced lithium-ion batteries, J. Energy Storage 17 (2018) 417–422.
[93] X. Yu, F. Xue, H. Huang, C. Liu, J. Yu, Y. Sun, X. Dong, G. Cao, Y. Jung, Synthesis and electrochemical properties of silicon nanosheets by DC arc discharge for lithium-ion batteries, Nanoscale 6 (12) (2014) 6860–6865.
[94] C. Pereira-Nabais, J. Swiatowska, M. Rosso, F.O. Ozanam, A. Seyeux, A.L. Gohier, P. Tran-Van, M. Cassir, P. Marcus, Effect of lithiation potential and cycling on chemical and morphological evolution of Si thin film electrode studied by ToF-SIMS, ACS Appl. Mater. Interfaces 6 (15) (2014) 13023–13033.
[95] M. Park, G. Wang, H.-K. Liu, S. Dou, Electrochemical properties of Si thin film prepared by pulsed laser deposition for lithium ion micro-batteries, Electrochim. Acta 51 (25) (2006) 5246–5249.
[96] Z. Lu, J. Zhu, D. Sim, W. Zhou, W. Shi, H.H. Hng, Q. Yan, Synthesis of ultrathin silicon nanosheets by using graphene oxide as template, Chem. Mater. 23 (24) (2011) 5293–5295.
[97] D.P. Wong, R. Suriyaprabha, R. Yuvakumar, V. Rajendran, Y.-T. Chen, B.-J. Hwang, L.-C. Chen, K.-H. Chen, Binder-free rice husk-based silicon–graphene composite as energy efficient Li-ion battery anodes, J. Mater. Chem. A 2 (33) (2014) 13437–13441.
[98] L. Shen, X. Guo, X. Fang, Z. Wang, L. Chen, Magnesiothermically reduced diatomaceous earth as a porous silicon anode material for lithium ion batteries, J. Power Sources 213 (2012) 229–232.

[99] Y. Liu, B. Chen, F. Cao, H.L. Chan, X. Zhao, J. Yuan, One-pot synthesis of three-dimensional silver-embedded porous silicon microparticles for lithium-ion batteries, J. Mater. Chem. 21 (43) (2011) 17083–17086.
[100] M.-S. Wang, L.-Z. Fan, M. Huang, J. Li, X. Qu, Conversion of diatomite to porous Si/C composites as promising anode materials for lithium-ion batteries, J. Power Sources 219 (2012) 29–35.
[101] N. Liu, K. Huo, M.T. McDowell, J. Zhao, Y. Cui, Rice husks as a sustainable source of nanostructured silicon for high performance Li-ion battery anodes, Sci. Rep. 3 (1) (2013) 1–7.
[102] M. Ge, J. Rong, X. Fang, A. Zhang, Y. Lu, C. Zhou, Scalable preparation of porous silicon nanoparticles and their application for lithium-ion battery anodes, Nano Res. 6 (3) (2013) 174–181.
[103] J. Cho, Porous Si anode materials for lithium rechargeable batteries, J. Mater. Chem. 20 (20) (2010) 4009–4014.
[104] Y. Yu, L. Gu, C. Zhu, S. Tsukimoto, P.A. van Aken, J. Maier, Reversible storage of lithium in silver-coated three-dimensional macroporous silicon, Adv. Mater. 22 (20) (2010) 2247–2250.
[105] B.M. Bang, J.I. Lee, H. Kim, J. Cho, S. Park, High-performance macroporous bulk silicon anodes synthesized by template-free chemical etching, Adv. Energy Mater. 2 (7) (2012) 878–883.
[106] M. Gu, Y. Li, X. Li, S. Hu, X. Zhang, W. Xu, S. Thevuthasan, D.R. Baer, J.-G. Zhang, J. Liu, In situ TEM study of lithiation behavior of silicon nanoparticles attached to and embedded in a carbon matrix, ACS Nano 6 (9) (2012) 8439–8447.
[107] L. Wu, J. Yang, X. Zhou, M. Zhang, Y. Ren, Y. Nie, Silicon nanoparticles embedded in a porous carbon matrix as a high-performance anode for lithium-ion batteries, J. Mater. Chem. A 4 (29) (2016) 11381–11387.
[108] H. Su, A.A. Barragan, L. Geng, D. Long, L. Ling, K.N. Bozhilov, L. Mangolini, J. Guo, Colloidal synthesis of silicon–carbon composite material for lithium-ion batteries, Angew. Chem. Int. Ed. 56 (36) (2017) 10780–10785.
[109] S. Guo, X. Hu, Y. Hou, Z. Wen, Tunable synthesis of yolk–shell porous silicon@ carbon for optimizing Si/C-based anode of lithium-ion batteries, ACS Appl. Mater. Interfaces 9 (48) (2017) 42084–42092.
[110] T. Song, J. Xia, J.-H. Lee, D.H. Lee, M.-S. Kwon, J.-M. Choi, J. Wu, S.K. Doo, H. Chang, W. I. Park, Arrays of sealed silicon nanotubes as anodes for lithium ion batteries, Nano Lett. 10 (5) (2010) 1710–1716.
[111] S. Ohara, J. Suzuki, K. Sekine, T. Takamura, A thin film silicon anode for Li-ion batteries having a very large specific capacity and long cycle life, J. Power Sources 136 (2) (2004) 303–306.
[112] M.S. Wang, Y. Song, W.L. Song, L.Z. Fan, Three-dimensional porous carbon–silicon frameworks as high-performance anodes for lithium-ion batteries, ChemElectroChem 1 (12) (2014) 2124–2130.
[113] F. Zhang, X. Yang, Y. Xie, N. Yi, Y. Huang, Y. Chen, Pyrolytic carbon-coated Si nanoparticles on elastic graphene framework as anode materials for high-performance lithium-ion batteries, Carbon 82 (2015) 161–167.
[114] S. Chen, P. Bao, X. Huang, B. Sun, G. Wang, Hierarchical 3D mesoporous silicon@ graphene nanoarchitectures for lithium ion batteries with superior performance, Nano Res. 7 (1) (2014) 85–94.
[115] X. Zuo, Y. Xia, Q. Ji, X. Gao, S. Yin, M. Wang, X. Wang, B. Qiu, A. Wei, Z. Sun, Self-templating construction of 3D hierarchical macro-/mesoporous silicon from 0D silica nanoparticles, ACS Nano 11 (1) (2017) 889–899.

[116] X. Tang, G. Wen, Y. Song, Stable silicon/3D porous N-doped graphene composite for lithium-ion battery anodes with self-assembly, Appl. Surf. Sci. 436 (2018) 398–404.

[117] T. Mu, P. Zuo, S. Lou, Q. Pan, H. Zhang, C. Du, Y. Gao, X. Cheng, Y. Ma, H. Huo, A three-dimensional silicon/nitrogen-doped graphitized carbon composite as high-performance anode material for lithium ion batteries, J. Alloys Compd. 777 (2019) 190–197.

[118] B. Li, F. Yao, J.J. Bae, J. Chang, M.R. Zamfir, D.T. Le, D.T. Pham, H. Yue, Y.H. Lee, Hollow carbon nanospheres/silicon/alumina core-shell film as an anode for lithium-ion batteries, Sci. Rep. 5 (1) (2015) 1–9.

[119] W. An, B. Xiang, J. Fu, S. Mei, S. Guo, K. Huo, X. Zhang, B. Gao, P.K. Chu, Three-dimensional carbon-coating silicon nanoparticles welded on carbon nanotubes composites for high-stability lithium-ion battery anodes, Appl. Surf. Sci. 479 (2019) 896–902.

[120] J.S. Kim, W. Pfleging, R. Kohler, H.J. Seifert, T.Y. Kim, D. Byun, H.-G. Jung, W. Choi, J.K. Lee, Three-dimensional silicon/carbon core–shell electrode as an anode material for lithium-ion batteries, J. Power Sources 279 (2015) 13–20.

[121] R. Batmaz, F.M. Hassan, D. Higgins, Z.P. Cano, X. Xiao, Z. Chen, Highly durable 3D conductive matrixed silicon anode for lithium-ion batteries, J. Power Sources 407 (2018) 84–91.

[122] C. Zhao, S. Li, X. Luo, B. Li, W. Pan, H. Wu, Integration of Si in a metal foam current collector for stable electrochemical cycling in Li-ion batteries, J. Mater. Chem. A 3 (18) (2015) 10114–10118.

[123] Y. Zheng, P. Smyrek, J.-H. Rakebrandt, H.J. Seifert, W. Pfleging, C. Kübel, Silicon-based 3D electrodes for high power lithium-ion battery, in: 2017 IEEE International Conference on Manipulation, Manufacturing and Measurement on the Nanoscale (3M-NANO), IEEE, 2017, pp. 61–64.

[124] Q. Zhang, J. Liu, Z.-Y. Wu, J.-T. Li, L. Huang, S.-G. Sun, 3D nanostructured multilayer Si/Al film with excellent cycle performance as anode material for lithium-ion battery, J. Alloys Compd. 657 (2016) 559–564.

[125] T. Jiang, S. Zhang, X. Qiu, W. Zhu, L. Chen, Preparation and characterization of silicon-based three-dimensional cellular anode for lithium ion battery, Electrochem. Commun. 9 (5) (2007) 930–934.

[126] P. Wu, H. Wang, Y. Tang, Y. Zhou, T. Lu, Three-dimensional interconnected network of graphene-wrapped porous silicon spheres: in situ magnesiothermic-reduction synthesis and enhanced lithium-storage capabilities, ACS Appl. Mater. Interfaces 6 (5) (2014) 3546–3552.

[127] D. Mazouzi, D. Reyter, M. Gauthier, P. Moreau, D. Guyomard, L. Roué, B. Lestriez, Very high surface capacity observed using Si negative electrodes embedded in copper foam as 3D current collectors, Adv. Energy Mater. 4 (8) (2014) 1301718.

[128] H. Zhang, P.V. Braun, Three-dimensional metal scaffold supported bicontinuous silicon battery anodes, Nano Lett. 12 (6) (2012) 2778–2783.

[129] J. Chang, X. Huang, G. Zhou, S. Cui, P.B. Hallac, J. Jiang, P.T. Hurley, J. Chen, Multilayered Si nanoparticle/reduced graphene oxide hybrid as a high-performance lithium-ion battery anode, Adv. Mater. 26 (5) (2014) 758–764.

[130] X. Chen, K. Gerasopoulos, J. Guo, A. Brown, C. Wang, R. Ghodssi, J.N. Culver, A patterned 3D silicon anode fabricated by electrodeposition on a virus-structured current collector, Adv. Funct. Mater. 21 (2) (2011) 380–387.

[131] M.V. Shelke, H. Gullapalli, K. Kalaga, M.T.F. Rodrigues, R.R. Devarapalli, R. Vajtai, P.M. Ajayan, Facile synthesis of 3D anode assembly with Si nanoparticles sealed in highly pure few layer graphene deposited on porous current collector for long life Li-ion battery, Adv. Mater. Interfaces 4 (10) (2017) 1601043.

[132] M.J. Allen, V.C. Tung, R.B. Kaner, Honeycomb carbon: a review of graphene, Chem. Rev. 110 (1) (2010) 132–145.
[133] J. Ji, H. Ji, L.L. Zhang, X. Zhao, X. Bai, X. Fan, F. Zhang, R.S. Ruoff, Graphene-encapsulated Si on ultrathin-graphite foam as anode for high capacity lithium-ion batteries, Adv. Mater. 25 (33) (2013) 4673–4677.
[134] H. Ji, L. Zhang, M.T. Pettes, H. Li, S. Chen, L. Shi, R. Piner, R.S. Ruoff, Ultrathin graphite foam: a three-dimensional conductive network for battery electrodes, Nano Lett. 12 (5) (2012) 2446–2451.
[135] N. Li, S. Jin, Q. Liao, H. Cui, C. Wang, Encapsulated within graphene shell silicon nanoparticles anchored on vertically aligned graphene trees as lithium ion battery anodes, Nano Energy 5 (2014) 105–115.
[136] L. David, R. Bhandavat, U. Barrera, G. Singh, Silicon oxycarbide glass-graphene composite paper electrode for long-cycle lithium-ion batteries, Nat. Commun. 7 (1) (2016) 1–10.
[137] L. Hu, H. Wu, Y. Gao, A. Cao, H. Li, J. McDough, X. Xie, M. Zhou, Y. Cui, Silicon–carbon nanotube coaxial sponge as Li-ion anodes with high areal capacity, Adv. Energy Mater. 1 (4) (2011) 523–527.
[138] K.-F. Chiu, S.-H. Su, H.-J. Leu, C.-Y. Wu, Silicon thin film anodes coated on micron carbon-fiber current collectors for lithium ion batteries, Surf. Coat. Technol. 267 (2015) 70–74.
[139] E. Qu, T. Chen, Q. Xiao, G. Lei, Z. Li, Flexible freestanding 3D Si/C composite nanofiber film fabricated using the electrospinning technique for lithium-ion batteries anode, Solid State Ionics 337 (2019) 70–75.
[140] S.K. Kumar, S. Ghosh, S.K. Malladi, J. Nanda, S.K. Martha, Nanostructured silicon–carbon 3D electrode architectures for high-performance lithium-ion batteries, ACS Omega 3 (8) (2018) 9598–9606.
[141] M.S. Wang, W.L. Song, L.Z. Fan, Three-dimensional interconnected network of graphene-wrapped silicon/carbon nanofiber hybrids for binder-free anodes in lithium-ion batteries, ChemElectroChem 2 (11) (2015) 1699–1706.
[142] J. Zhu, T. Wang, F. Fan, L. Mei, B. Lu, Atomic-scale control of silicon expansion space as ultrastable battery anodes, ACS Nano 10 (9) (2016) 8243–8251.
[143] H. Wu, G. Yu, L. Pan, N. Liu, M.T. McDowell, Z. Bao, Y. Cui, Stable Li-ion battery anodes by in-situ polymerization of conducting hydrogel to conformally coat silicon nanoparticles, Nat. Commun. 4 (1) (2013) 1–6.
[144] Y. Xu, Y. Zhu, F. Han, C. Luo, C. Wang, 3D Si/C fiber paper electrodes fabricated using a combined electrospray/electrospinning technique for Li-ion batteries, Adv. Energy Mater. 5 (1) (2015) 1400753.

Chapter 10

Processing and properties of silicon anode materials

Raj N. Singh

School of Materials Science and Engineering, Oklahoma State University, Tulsa, OK, United States

10.1 Introduction

Silicon is a promising anode material for Li-ion batteries due to its high theoretical specific capacity of more than 10 times (∼4200 mAh/g) that of the commercially used graphite [1]. Silicon undergoes an enormous (∼400%) volume expansion forming a $Li_{4.4}Si$-like phase during lithiation [2, 3]. It is also found that $Li_{15}Si_4$ ($Li_{3.75}Si$) is the highest possible lithiated phase instead of $Li_{22}Si_5$ corresponding to $Li_{4.4}Si$ [4]. Nevertheless, the $Li_{15}Si_4$ phase undergoes 370% volume expansion [5], which leads to significant stress, fracture and electrical isolation of silicon-based anode causing capacity loss in batteries. In addition, the uncontrolled growth of solid-electrolyte-interphase (SEI) occurs on newly exposed/fractured silicon surfaces [6]. This results in capacity fade and poor cycling performance. The stress associated with the volume expansion plays a key role in fracture of the silicon-based electrodes as evidenced from the stress evolution verified using in-situ and ex-situ studies [7–11] Chon et al. demonstrated in-situ measurement of stress of 0.5 GPA by means of change in stress-induced substrate curvature in amorphous Li_xSi layer on a thick silicon wafer [7]. This indirect method relies on the accuracy of the measurement of thickness. Zeng et al. measured the lithiation-induced stress using in-situ Raman spectroscopy on silicon nano-particles mixed with many other materials forming the anode and found tensile stress of 0.2 GPa during the beginning of lithiation, followed by a transition to a compressive stress of up to 0.3 GPa upon further lithiation [8]. However, these studies were limited in scope, they did not report on the nature of stress evolution after subsequent lithiation and de-lithiation cycles. During lithiation, crystalline silicon (c-Si) undergoes a phase transition forming an amorphous Li_xSi (a-Li_xSi) on a core of unreacted c-Si. The a-Li_xSi can further transform to a crystalline lithiated phase corresponding to $Li_{15}Si_4$, depending on the lithiation potential [8, 9]. So, there exists a boundary between a-Li_xSi and unreacted c-Si and c-$Li_{15}Si_4$. More recently, it is reported that the lithiated zone in silicon itself has two sub-zones depending on the lithium content (x). One is at the initial surface of lithiation (where $x \sim 2.5$ in Li$_x$Si) and a second is a much less lithiated region deep into the crystal ($x \sim 0.1$–0.17) [10, 11]. These phase transitions can lead to the evolution of residual stress from volume mismatch or differential strain between different sub-zones.

Therefore, direct knowledge of the evolution of residual stress is critical for a better understanding of stress, deformation, and finally fracture mechanism of silicon-based anodes as described later from our own research. A fundamental study was performed to identify the nature, magnitude, and evolution of the residual stress after full de-lithiation of c-Si anodes subjected to many lithiation-delithiation cycles. This is done with the help of strain-dependent frequency shift of the Raman spectra as a probe for measuring strain/stress and using a mechanics-based approach to decipher the magnitude of the residual stress [12–16]. Subsequently, the evolution of the residual stress and its consequences were supported by microstructural studies [16].

Many processing strategies have been employed to overcome the capacity fade as a result of the volume change and stress buildup leading to failure of Si electrodes. Besides the theoretical understanding of volume change and stress calculation, most of the research was confined to reducing the dimension of the particle size, so that the increased surface area can minimize or accommodate the stress developed in the Si material and associated volume expansion. Thus, over the last few years, various kinds of nanostructures of silicon such as nanowires, nanoparticles, hollow spheres, core-shell structures [17–26] have been employed. However, they are not easy to scale up or cost-effective as they employ single-crystalline nanostructured silicon and require high-precision fabrication technologies, which are neither the common practice for Li-ion battery industry nor desirable. A majority of these studies were focused on highly advanced architectures created by means of 3D patterning and electrochemical etching of wafer, which are also not desirable approaches for state-of-the-art manufacturing practices used for high capacity Li-ion batteries, especially for electric vehicles. Also, there are reports that the alloying of silicon by lithiation depends on the crystalline orientation [27, 28]. So, defining the orientation and aligning the particular favored orientation for easy lithiation direction is also a challenging task for single-crystalline-based silicon electrode materials.

Many researchers attempted to fabricate composite electrodes with different allotropes of carbon including graphene and carbon nanotubes (CNTs) [11, 20, 29–39]. Also, there are few reports available on the use of porous current collectors [40], addition of some additives of metal nanoparticles or films [36, 41–46] or uses of some other electrolytes [47] or binders [35], to improve performance. All these studies add cost to the final product due to either extra processing steps or the addition of an extra amount/number of advanced nanomaterials, but the capacity fading problem remains. Currently, the most significant challenge is getting high reversible capacity from Si-based electrodes (especially at high Si loading), which can be robust, easy to scale, and compete with the graphite-based electrode materials in terms of performance, manufacturing practice, and cost.

To address these challenges, our group investigated the processing of Si anodes. The objective of one approach was to develop a silicon-based anode using a robust and innovative process for realizing a high reversible capacity, from easily available metallurgical grade polycrystalline silicon powder. It employed a modified metal-assisted chemical etching process to create nanostructured pores and nanofibers on micron size polycrystalline silicon that was coated with carbon using a solution-based approach. The battery performance of the processed anodes was then evaluated in a half cell (vs. Li) configuration displaying improved capacity retention as described later [48].

In the second approach, the etching of Si to create nanofibers was eliminated to further simplify the process for a lower cost. In this second processing approach, milled low-grade polycrystalline silicon powder was coated with carbon using an innovative approach of solution-assisted coating using furfuryl alcohol as a source of carbon. In this newly developed approach, a mixture of milled silicon and superconducting carbon particles is dispersed in the furfuryl alcohol, which is rapidly gelled and polymerized to arrest the dispersed phases. This is then dried and fired in an inert atmosphere at 950°C to convert the furfuryl alcohol to carbon coating on particles. The nanostructured superconducting carbon helped in separating the silicon particles while coating with furfuryl-derived carbon. The carbon layer outside silicon particles along with the superconducting carbon enhanced the conductivity of the electrode and likely limited the uncontrolled growth of SEI, if any. These attributes also led to a better capacity retention of Si-based anodes in Li-ion cells both at room and higher temperatures and at higher current density/charging rates as described in our recent publication [49].

10.2 Experimental methods

10.2.1 Single crystal silicon anode preparation and electrochemical testing for residual stress

The electrochemical tests were performed in a custom-made electrochemical cell containing single-sided polished Si (100) wafers (Wafer World Inc.; p-type, thickness 475–575 μm, resistivity ~1–20 Ω-cm) as an anode (after washing in acetone and dilute HF solution to remove any grease/dirt and native oxide layer) and Li metal (MTI Corporation) as the counter electrode inside an Ar-filled glovebox. The electrolyte was 1.0 M LiPF$_6$ in 1:1 (w/w) ethylene carbonate/diethyl carbonate (Sigma-Aldrich). Battery cycling was performed using a MACCOR eight-channel battery analyzer (4300M). The cells were cycled between 1.5 V and 1 mV versus Li/Li$^+$ at a constant current of 100 μA. The residual stress was measured after cycling and complete de-lithiation.

Raman measurements were carried out using a Nicolet Almega XR Dispersive Raman 960 spectrometer with 532 nm laser over the range of 140–2000 cm^{-1} with power varied between 10% and 80% of incident laser power (Max power: 150 mW; Magnification 50×). The Raman instrument was optimized by varying the laser power at a single point to minimize the sample heating effect. The Raman spectra were taken both before and after the test under similar conditions to further minimize the thermal effects. A collection time of the 30s was used for each spectrum and 20 spectra were taken to reduce noise and avoid CCD saturation. PeakFit software after linear background subtraction was used to analyze the Raman spectra for peak position and full width at half maxima (FWHM). All the measurements were done on samples after cleaning with diethyl carbonate (DEC) to remove electrolyte then vacuum drying. Field emission scanning electron microscopy (Hitachi S-4800) was used to observe the images before and after electrochemical tests.

10.2.2 Processing of etched anode from Si powder

10.2.2.1 Synthesis and characterizations

Micron size polycrystalline Silicon powder (5–100 μm; metallurgical grade from global metallurgical Inc.) was first rinsed by acetone. After drying, the silicon powder (2.5 g) was dispersed in an etchant solution containing, 0.15 g AgNO$_3$ (Sigma Aldrich) in 25 mL of 4.6 M hydrofluoric acid solution (40%, Sigma-Aldrich). Another solution comprised of 0.12 M hydrogen peroxide (Sigma-Aldrich) was added drop-wise into the previous rotating and uniformly premixed solution at an interval of 2–3 min for about 30 min. The whole mixture was then left for etching the micron size silicon particles at room temperature for 2 h to create porous nanostructures on the surfaces of the silicon particles. The etching process was stopped by spraying deionized water and powder was separated by filtering from the solution. The filtered powder was soaked in nitric acid solution (65%, Sigma-Aldrich) for 20 min to remove the silver deposits on the etched particle surfaces and washed thoroughly using deionized water, and separated successively using a centrifugation (VWR 13,000 rpm, 15 min). This step was repeated several times until the pH of the solution became neutral. Finally, the powder was dried on a hot plate before moving on to the next step of carbon coating.

After etching, the etched silicon particles having nanofibers were mixed with superconducting carbon nanopowder (US Research Nanomaterials, Inc.; size: 5–100 nm; conductivity: 2–4 × 10^5 S m^{-1}; purity: 97.5%) in 90:10 wt. ratio in furfuryl alcohol (FFA; 98% Across organic) followed by de-agglomeration and separation of the two phases. After homogeneous and uniform mixing, the solution was quickly gelled by adding HCl while rotating on a magnetic stirrer. The gelled sample was then dried at ∼100°C for about 2 h. The polymer/gel coated sample thus created was kept inside a tube furnace (Thermolyne Type

59,300 High-temperature Tube furnace) and the temperature was raised to 600°C and held for 1 h, the temperature was further increased to 750°C at a constant heating rate of 5°C min^{-1} and held there for 2 h under continuous N$_2$ atmosphere to form a uniform carbon-coating on the etched silicon particles, which were networked and separated by the superconducting carbon nanostructures. The thickness of the carbon coating was optimized by controlling the concentration of furfuryl alcohol in the polymer gel.

Powder X-ray diffraction patterns were obtained using a Bruker AXS D8 X-ray diffractometer with CuK$_\alpha$ radiation between 10 and 90 degree (2θ values). Field emission scanning electron microscopy (Hitachi S-4800) was used to analyze the images before and after coating the etched silicon particles. Raman measurements were carried out using a Nicolet Almega XR Dispersive Raman 960 spectrometer with 532 nm laser excitation over the range 140–2000 cm^{-1} at 80% of incident laser power (max power: 150 mW). A collection time of 30s was used for each spectrum and for each acquisition 20 spectra were accumulated to reduce noise and avoid CCD saturation. To analyze the Raman spectra, the peaks were fitted using "PeakFit" software after linear background subtraction.

10.2.2.2 Electrode preparation, cell assembly and electrochemical characterizations

The coated material was ground and added with a small amount (2–5 wt%) of superconducting carbon black (97.5%, US Research Nanomaterials, Inc.) and 10 wt% of polyvinylidene fluoride (PVDF) in an N-Methylpyrrolidine (97%, Sigma-Aldrich) solution and mixed overnight to make the slurry, which then was coated on an unpolished surface of a single-sided polished copper foil (9 μm thick 99.99%, MTI Corp.) using tape casting. After drying the organic solvent at 120°C using a hot plate for 10–15 min the coated foil was then transferred to a vacuum oven for storage. The coated foil was pressed using a laminating press to make a dense and uniformly thick electrode. The pressed electrode was cut into 1 cm^2 pieces to function as an anode for a half-cell. The electrochemical tests were performed in a 2032 type coil cell with the laboratory-made anode and Li metal (MTI Corp.) as the counter electrode inside an Ar-filled glovebox. The electrolyte used was 1.0 M LiPF$_6$ in 1:1 (w/w) ethylene carbonate/diethyl carbonate (Sigma-Aldrich). "Celgard" separator was used for coin cells after soaking them in the electrolyte. Cyclic voltammetry measurements were conducted using a potentiostat (Princeton Applied Research, *V*ersaSTAT 4) between 2 V and 1 mV at a scan rate of 1 mV s^{-1}. Electrochemical impedance spectroscopy measurements were also carried out using the same potentiostat used for cyclic voltammetry measurements, between 1 MHz and 10 mHz. Battery cycling was performed using a MACCOR battery tester (4300 M). The voltage cutoff was set to 1.5 V and 1 mV versus Li/Li$^+$, and the cycling current was set to 500 μA cm^{-2} (unless stated otherwise).

10.2.3 Processing of anode from Si powder (unetched)

10.2.3.1 Materials synthesis and characterizations

Raw polycrystalline silicon powders (5–100 μm; metallurgical grade, Global Metallurgical Inc.) were ball milled in a medium of ethyl alcohol (40% by vol), 10% of silicon (by volume), and 50% by volume of the jar was filled with the zirconia balls. The final parameters chosen for milling were optimized after several runs. After drying, the milled silicon particles were mixed with superconducting carbon (US Research Nanomaterials, Inc.; size: 5–100 nm; conductivity: 2–4×10^5 S m^{-1}; purity: 97.5%) in 90:10 wt. ratio in furfuryl alcohol (FFA 98%—Acros Organics) followed by de-agglomeration using a probe sonicator. Separation and uniform mixing of both components (superconducting carbon and milled silicon) are very important in this processing approach to improve on mixing. For uniform and homogeneous mixing both the components were sonicated separately using ethyl alcohol and then added together for further sonication. After homogeneous and uniform mixing, the solution was rapidly gelled by the addition of HNO_3 while rotating on a magnetic stirrer. The gelled sample was then dried at ~100°C for ~~about~~ 2 h. The polymer/gel coated sample was placed inside a tubular furnace (Thermolyne Type 59,300 high-temperature tube furnace) and the temperature was raised to two intermediate temperatures of 550°C and 750°C for 1 h of holding time for each step before taking it to the final temperature of 950°C at a constant heating rate of 10°C min^{-1} and held there for 2 h under a continuous flow of N_2 atmosphere. The particle size of milled Si powder was measured at a regular interval of milling time using Malvern Zeta Sizer ZS90 with 633 nm laser excitation. Raman spectroscopy was carried out using a Nicolet Almega XR Dispersive Raman 960 spectrometer at 532 nm laser excitation over the range 200–4000 cm^{-1} at 80% of incident laser power (Max power: 150 mW).

10.2.3.2 Electrode preparation, cell assembly and electrochemical characterizations

The carbon coated silicon particles (85 wt%) were mixed with a small amount (5 wt%) of superconducting carbon (97.5%, US Research Nanomaterials, Inc.) and 10 wt% of PVDF in an N-Methylpyrrolidine (97%, Sigma-Aldrich) solution for overnight to make the slurry, which then was coated on copper foil (9 μm thick 99.99%, MTI Corp.). After drying the organic solvent at 120°C using a hot plate the coated foil was then transferred to a vacuum oven for storage. The coated foil was pressed using a laminating press to make uniform thickness. The pressed electrode was cut into 1 cm^2 pieces to serve as an anode. The electrochemical tests were performed in the 2032 type coin cell with both half- and full-cell arrangements. For the half cell, Li metal (MTI Corp.) was used as the counter electrode and for full cell, $LiNi_{0.5}Co_{0.2}Mn_{0.3}O_2$ (Toda NCM 523, provided by Argonne National Lab) was used as cathode vs. the silicon-based anode.

The cathode loading used was 80:10:10 (wt%); 80 wt% was NCM 523, 10 wt% was superconducting carbon black and the remaining 10 wt% PVDF. The electrolyte used was 1.0 M LiPF$_6$ in 1:1 (w/w) ethylene carbonate/diethyl carbonate (BASF). "Celgard" separator was used for coin cells after soaking them in the electrolyte. Electrochemical impedance spectroscopy measurements were carried out using *V*ersaSTAT 4 (Princeton Applied Research) potentiostat between 1 MHz and 10 mHz. Battery cycling was performed using a MACCOR battery tester (4300 M). The voltage cutoff was set to 1.5 V and 1 mV versus Li/Li$^+$, and the cycling current was set to 0.8 A/g (unless stated otherwise) for half cells. Full cells were tested between the voltage window of 2.7–4.2 V at a current density of 200 mA/g.

10.3 Results and discussion

10.3.1 Evolution of residual stress in single crystal Si and fracture

Raman spectroscopy was used to study the evolution of residual stresses after complete de-lithiation by collecting spectra after 1, 10, and 50 cycles using a beam size of ∼1 μm. Fig. 10.1 shows the Raman data after different cycles performed on the same single-crystalline silicon (c-Si) anode along with the same c-Si before any lithiation, and a standard amorphous silicon (a-Si; Strem Chemicals Inc.) for comparison. It shows only the first-order Raman peak from silicon and not from any other materials or electrolyte because the electrodes were cleaned before taking the Raman spectra. The Raman spectrum was taken on silicon before the test shows a sharp and intense peak appearing at ∼522.34 cm^{-1} corresponding to c-Si. The peak intensity decreases with the increase in cycle number and the intensity of the peak after 50 cycles is similar to the standard reference a-Si. In addition, the peak also shifts and broadens towards lower wavenumbers. The data presented here is the average of spectra taken on four different spots within each sample and is representative of the general trend. The details of peak shift and broadening calculated from FWHM are summarized in Table 10.1. The FWHM increases with an increase in the number of lithiation/de-lithiation cycles. To simulate the data from zero to an infinite number of cycles (complete amorphization of silicon) the data for c-Si before any lithiation and standard a-Si were put together with the cycled electrodes as well (Fig. 10.1A). The FWHM found in our study varies from 3.9 cm^{-1} (corresponding to c-Si before any lithiation) to 12.84 cm^{-1} when it is fully amorphous.

Fig. 10.1B shows the deconvolution of each Raman spectrum taken on the silicon electrode surface after different cycle number. It shows one sharp peak (at ∼520 cm^{-1}) and one broad peak (at ∼510 cm^{-1}). The solid line represents the experimental data whereas the dashed and dotted lines represent the deconvoluted peaks. The broad peak, which appears after the 1st cycle, becomes a broad hump as the cycle number is increased. It is interesting to see the

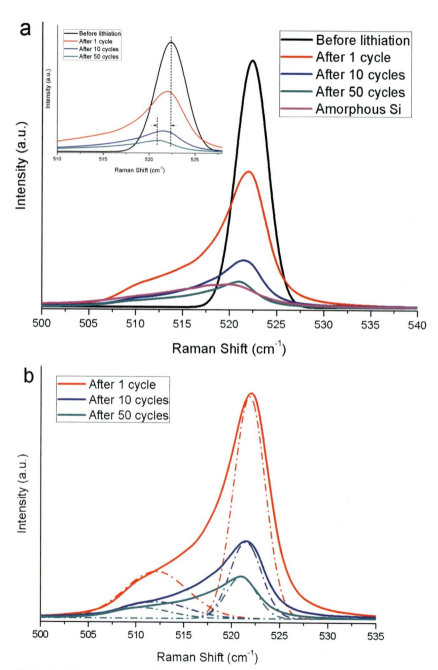

FIG. 10.1 (A) Raman spectra of the silicon anode before and after cycling along with the standard a-Si and (B) deconvolution *(dashed lines)* of first-order Raman peaks of the silicon anode cycled at different number of cycles [16].

TABLE 10.1 Calculated peak position, FWHM, peak shift and the corresponding tensile residual stress found in silicon electrodes after different cycle number [16].

Sample	Peak position (cm^{-1})	FWHM (cm^{-1})	Peak shift (cm^{-1})[a]	Residual stress (MPa)
c-Si before lithiation	522.34	3.90	–	–
After 1st cycle	522.01	5.93	0.33 ± 0.05	69 ± 11
After 10 cycles	521.45	6.83	0.89 ± 0.12	186 ± 26
After 50 cycles	520.95	7.01	1.39 ± 0.26	291 ± 56
a-Si (standard)	519	12.84	–	–

[a]Peak shift was calculated taking c-Si as a reference.

existence of the first-order Raman peak close to ~520 cm^{-1}, which corresponds to c-Si, and remains even after 50 cycles, although shifted with reduced intensity when compared to c-Si before lithiation. This is due to the unreacted or unlithiated silicon core underneath.

It should be mentioned that the laser wavelength (532 nm) used in our study is expected to penetrate beyond the subsurface region up to a micron depending upon the crystallinity [50] and the large thickness (~500 μm) of the unreacted silicon wafer still remaining after limited initial cycles.

These broad peaks at lower wavenumber along with the presence of a relatively sharp peak have been previously observed and attributed to amorphous silicon from a mixed-phase silicon film [51–53]. The presence of a-Si in those studies was not as a result of electrochemical lithiation/de-lithiation but was prepared by conventional alloying approaches. The increase in FWHM with decreasing wavenumber indicates the overall amorphous nature of silicon due to lithiation and de-lithiation at the surface or sub-surface level because most of the response in Raman spectra comes from the near-surface region. It is reasonable to propose that an adequate amount of unreacted c-Si beneath the a-Si [54] is also sampled by Raman and the shift of the c-Si peak towards the lower wavenumber could be due to the lattice distortion or presence of residual stress. The frequency/wavenumber associated with the first-order Raman peak is sensitive to stress/strain even at the nanoscale, which makes Raman spectroscopy a promising tool to measure both the nature and magnitude of stress [8, 55].

Anastassakis et al. found that with applied uniaxial stress there is a splitting of optical phonon into a singlet with eigenvector parallel to the stress and a doublet with eigenvectors perpendicular to the stress [15]. Besides the splitting, due to the hydrostatic component of the applied stress, they could also find a shift in the frequency of the Raman peak. In other words, if there is an evolution or existence of stress, it would shift the first-order Raman peak and from the shift one can determine the stress associated with it using a relation by Kim et al. [56]. Equation 10.1 was derived by assuming that stress is hydrostatic, each component of strain can be treated similarly. This approach was successfully used to determine stress in Si nanoparticles embedded in a SiC matrix under hydrostatic stress created by the expansion of the Si upon freezing below the melting point by one of the authors of the current paper [56].

Following the analysis of Anastassakis et al., one can obtain a single equation from the dynamical matrix, which was given by Kim et al. [56] as,

$$\lambda = p\epsilon + 2q\epsilon \tag{10.1}$$

where $\epsilon = \epsilon_{xx} = \epsilon_{yy} = \epsilon_{zz}$ is the strain component by assuming the stress is hydrostatic and each component of the strain can be taken as equal in all three directions, p and q are material dependent coefficients, the value of which for silicon was given by Chandrasekhar et al. as $p = -1.43 \times 10^{28}\, s^{-2}$ and $q = -1.89 \times 10^{28}\, s^{-2}$ [57]. $\lambda = \omega^2 - \omega_0^2$, and $\omega - \omega_0 \cong \lambda/2\omega_0$, the strain-dependent frequency shift (from the unstressed phonon peak ω_0). By relating the strain to stress and compliance constants C_{11} and C_{12}, Eq. (10.1) can be modified as shown in Eq. (10.2),

$$\frac{\lambda}{\sigma} = (C_{11} + 2C_{12})(p + 2q) \tag{10.2}$$

where $C_{11} = 76.8 \times 10^{-13}\, m^2/N$, $C_{12} = -21.4 \times 10^{-13}\, m^2/N$ for silicon [15, 57, 58].

The final expression for the residual stress can be given by Eq. (10.3) as,

$$\sigma = \frac{(\omega - \omega_0)2\omega_0}{(C_{11} + 2C_{12})(p + 2q)} = 209\, \text{MPa} \Delta\omega\, (cm^{-1}) \tag{10.3}$$

where $\Delta\omega$ is the shift in wavenumber from the equilibrium or stress-free state.

Table 10.1 gives the position of the peak, FWHM, peak shift, and corresponding data for the residual stress. The residual stress becomes 291 ± 56 MPa (tensile) for a shift of $1.39 \pm 0.26\, cm^{-1}$ after 50 cycles. But the increase in residual stress is non-linear up to 50 cycles. After 1 cycle, it is 69 ± 11 MPa, but at 10 cycles, it reaches a value of 186 ± 26 MPa. Therefore, the residual stress buildup more rapidly up to 10 initial cycles and then slows down as shown in Table 10.1. This approach is more direct and no calibration or thickness measurement is needed to quantify the residual stress; the sources of errors in our approach and data are much less as compared to the previous studies on the measurement of lithiation/de-lithiation induced stress [7, 8]. It should also be mentioned that Shi et al. using finite element analysis modeled maximum principal stress through the electrode thickness across the lithiated/unlithiated interface following the crack deflection and propagation paths and found tensile stress in the range of 332 MPa just below the crack surface, which could be the residual stress [50], but there was no direct experimental verification of the modeling results. However, this value is within 15% of the stress observed after 50 cycles.

The highest Li-containing phase has the stoichiometry of $Li_{22}Si_5$, which was made (metallurgically) by alloying lithium with silicon. From the crystallographic analysis, it was found that the unit cell volume (160.2 Å^3 vs. 6592 Å^3) or the volume per Si (20 Å^3 vs. 82.4 Å^3) atom in a $Li_{22}Si_5$ phase could reach more than 4 times unalloyed Si [59]. However, the existence of this Li-rich phase is controversial for electrochemically lithiated silicon. Theoretical studies have led to the conclusion that lithium atoms reside at the interstitial site inside bulk Si; at higher concentrations, lithium clusters promote the breaking of Si-Si bonds [60, 61]. If this is the situation, then during lithiation, the measured stress in partially-lithiated silicon is compressive, as it should be because the size of Li atom (1.23 Å, 1.55 Å, 0.60 Å(+1)), respectively as covalent, atomic, and ionic radii is larger than the size of the Si atom (1.11 Å, 1.32 Å, 0.41 Å(+4)), respectively as covalent, atomic and ionic radii. This is consistent with the observations that doping Si with B (smaller atom) leads to tensile stress and doping Si with Ge or Sb (larger atoms) leads to compressive stress as suggested by Narayan [62]. Chon et al. found that the measured stress (compressive) could reach up to 0.5 ± 0.1 GPa upon lithiation in crystalline Si [7]. But during delithiation, the stress becomes tensile and cracks form at ~2 GPa.

Upon delithiation the top layer of silicon will try to contract, whereas the underlying material, which was still unreacted will constrain/restrict it. Therefore, delithiation will create tensile stress, which was observed in this chapter and the literature [7]. It should be mentioned that our measurement of stresses is after delithiation based on the fact that the final cell potential was at 1.5 V, the highest extraction potential used in our study. Therefore, the measured tensile stress in our study by the shift in the Raman peak for Si is consistent with this explanation as well. Therefore, both explanations, first based on atomic volume considerations during lithiation and delithiation and second constraints placed by the underlying unreacted Si, are consistent with the experimental findings of tensile stress upon delithiation and as measured by Raman spectroscopy of our study.

To explain the effect of residual stress and its consequence to failure, we characterized the evolution of the surface morphology of the underlying silicon electrode after successive cycles of operation as shown in Fig. 10.2. All the images were taken after cleaning the electrolyte residue and removing loose cracked or delaminated particles if any. Just after the first cycle, when there was no delamination, we observed the initiation of some tiny cracks at nanoscale (Fig. 10.2B). The cracks become denser and wider with increasing number

384 PART | IV Achieving high performance

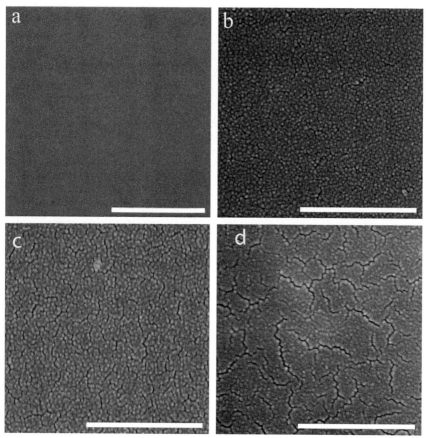

FIG. 10.2 SEM showing surface morphology of the silicon electrode obtained (A) before and after (B) 1, (C) 10, and (D) 50 cycles. Scale bars are for 500 nm.

of cycles shown in Fig. 10.2C and D. In contrast, the SEM image of the same electrode shows a very smooth surface even at higher magnification before cycling (Fig. 10.2A). The surface also becomes rough even after 1 cycle and this could probably be due to the amorphous nature of the thin electrode layer formed upon cycling, the existence of which was already verified by the Raman spectrum (Fig. 10.1). Shi et al. [63] explained the fracture mechanism of silicon electrode based on finite element modeling. For the first cycle, they argued that the residual stress was the primary reason for the crack initiation and propagation, but there was no experimental proof of it. Once the cracks initiate and grow in a vertical direction a region will come where there will a boundary region between reacted a-lithiated Si and unreacted c-Si. The vertical cracks can be deflected energetically at the interface between the a-lithiated Si and c-Si because of the low interfacial fracture toughness [7, 63]. Once the crack grows vertically and delaminates, the unreacted c-Si continues to undergo lithiation/delithiation in successive cycles. The underlying silicon after delamination is shown in Fig. 10.2C and D. The visible cracks are likely a result of the tensile residual stress, which can again act as a crack initiation site for the delamination of the next layer. This is the reason for observing crystalline silicon peak after initial cycles, which diminished with time and number of cycles.

These observations suggest that the failure of silicon anodes is not catastrophic but gradual. This layer-by-layer lithiation and de-lithiation builds up the residual stress gradually and leads to the failure of the electrode. The cracks initiated by residual stresses can grow further by fatigue and hence can lead to premature failure of the silicon-based electrodes [59]. Thus, these findings on the existence of tensile residual stress and its measurement should be a step forward to design next-generation silicon-based electrodes for Li-ion batteries.

10.3.2 Processing and properties of hierarchical Si anode from etched silicon powder

10.3.2.1 Coating on hierarchical nanostructures

The pristine micron size polycrystalline particles of silicon were etched using a modified metal-assisted chemical etching method as described in the experimental section. The initial size of the particles ranged from few microns to about 100 µm. Etching results in an overall porous structure with nanofibers on it are shown in Fig. 10.3A and B.

After processing, the etched silicon particles with porous regions surrounding the nanofibers were mixed vigorously with superconducting carbon nanoparticles in acetone followed by their de-agglomeration. The mixture after drying was then uniformly dispersed in a stirred solution of furfuryl alcohol (FFA). The process is intended to separate the etched silicon particles using the conductive carbon particles before final and outer carbon over-coating on less agglomerated or individual etched particles using a solution-coating approach. It is to be noted that the conductive carbon particles used here are much more electrically-conducting than conventional carbon black and also of different sizes in the nano regime, which has the advantage of fitting inside the pores easily. The other important aspect of our processing approach is rapid polymerization of the entire solution containing etched-silicon particles, conductive carbon and FFA using a polymerizing agent (i.e., HCl). This processing step is designed to dramatically reduce motion of the dispersed solution by which one can obtain coating either on individual particles or particles networked with some conductive carbon. Here in this study, we used HCl in addition to heating the solution to polymerize FFA as quickly as possible. This first reported use of FFA to process a silicon-based anode can be attributed to two important but different considerations: (i) very low viscosity at room

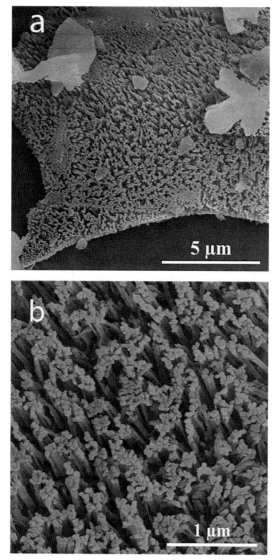

FIG. 10.3 Typical morphology of silicon particles (A, B) after etching at two different magnifications [48].

temperature, enabling modification of coating thickness by changing viscosity, either by heating or by extra polymerizing agent (i.e., HCl), and (ii) low cost.

XRD patterns were collected from the as-received polycrystalline silicon powder and compared to a pattern taken from a carbon-coated etched silicon sample. The majority of the peaks can be assigned to silicon along with a carbon signature. It should be noted that the peaks coming from graphitic planes of carbon from superconducting carbon black (SCB), which was mixed/stuffed within the etched silicon before carbon coating, are suppressed after coating.

To further confirm the coating as well as to determine the quality of carbon a more surface sensitive Raman spectroscopy was used, which can be performed on a very localized region of interest with a spot size of 1.1 µm. To check and confirm the carbon coating after pyrolyzing, one silicon sample was also just coated using FFA-derived carbon without any SCB added before polymerization. Fig. 10.4 shows Raman spectra taken on the carbon-coated etched silicon particles along with a spectral response from carbon-coated etched silicon without any SCB added as well as only SCB as a reference. The overall higher intensity of the peak coming from silicon at $\sim$522 cm^{-1} indicates that most of the underlying material is still silicon with lesser amounts of carbonaceous material as a thin coating. The strong presence of the peaks associated with the carbonaceous phase indicates that the quality and uniform coverage of a carbon coating is achieved. It should be noted that the peak intensity coming from the carbonaceous phase is strong enough although less than the peak intensity coming from the underlying silicon. The presence of D and G bands can be distinguished after deconvolution of the observed peaks. The red line curve represents

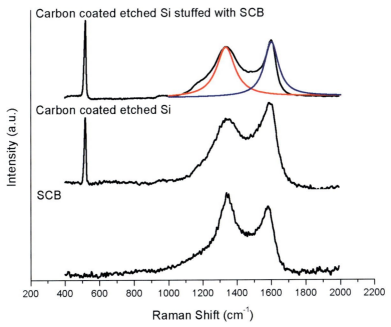

FIG. 10.4 Raman spectra of carbon-coated etched silicon sample along with a spectral response from carbon-coated etched silicon without any SCB added before coating as well as only SCB as a reference [48].

the D band located at ~1334 cm^{-1} and the blue line curve is associated with the G band, which appears at ~1598 cm^{-1}. Interestingly the ratio of intensity coming from the D band (I_d) to the G band intensity (I_g) denoted by R (I_d/I_g) is in the same range (~0.8) for silicon coated with carbon, no matter whether it was mixed with SCB before coating or not. Whereas the R-value increased above 1 (~1.1) for SCB alone. For the carbon-coated sample the lower value of R (I_d/I_g ~0.8) <1, indicates the coating with less disordered carbon.

10.3.2.2 Electrochemical performance of etched Si anode

The electrochemical behavior of the electrode was first analyzed by cyclic voltammetry. The broad peak at ~0.5 V during the first cycle of lithiation of pristine polycrystalline silicon is related to the formation of the solid electrolyte interphase (SEI) due to reaction with the electrolyte and active silicon material (Fig. 10.5A). It should be noted that the peak due to SEI formation is also present at almost the same potential value for the etched silicon particles (Fig. 10.5B). However, it suddenly disappeared from the sample when it is filled and coated with the carbonaceous material (Fig. 10.5C). It should also be noted that by filling the void spaces between the silicon nanofibers and coating with carbon reduces the overall surface area and hence the amount of SEI because SEI formation is a surface area-dependent reaction. In this case, reducing the surface area will only reduce the formation of SEI but will not interfere or hinder the number of lithium ions to intercalate. Since the graphitic layer itself should be a stable host for lithium ions, it should be able to intercalate and de-intercalate lithium ions reversibly. Thus, all the carbon structures inside and outside will allow lithium to diffuse inside to enter into the silicon lattice and the high surface areas created by etching and from the silicon nanofibers will be available for intercalation.

The appearance of a broad shoulder at ~0.25 V in Fig. 10.5A is probably due to the formation of the metastable Li-rich clusters, which forms within the a-Li$_x$Si matrix [64]. These Li-rich clusters with distinct short-range order have been observed for bulk silicon previously [65]. It is worth noting that this peak is absent in both Fig. 10.5B and Fig. 10.5C, which corresponds to the presence of nanostructured silicon. Further, it should be mentioned that any metastable phase would have a detrimental effect on capacity retention. Thus, avoiding this metastable phase during cycling leads to better cyclability as verified later. The absence of any peak at 0.5 or 0.25 V in etched and coated samples is due to the modifications done by the carbon coating as well as the presence of silicon nanofibers as discussed earlier. However, for all the samples and for all the cycles of lithiation the sharp intense peak centered at ~0.01 V corresponds to the complete alloying reaction between silicon and lithium. The anodic reactions in both Fig. 10.5A and B, show one broad shoulder centered at 0.43 V and one peak centered at 0.6 V ascribed to the de-lithiation from Li$_x$Si.

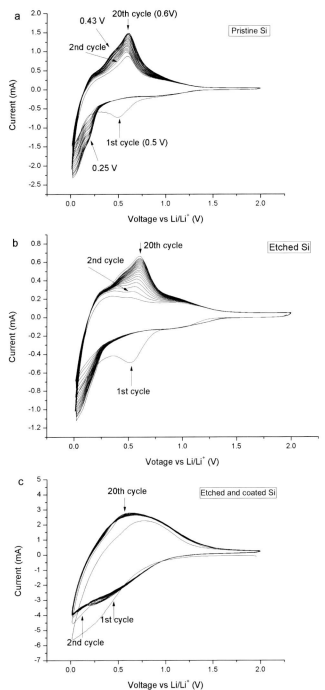

FIG. 10.5 Cyclic voltammetry curves from electrodes made using (A) pristine, (B) etched and (C) carbon-coated etched silicon [48].

The reproducibility of the coated sample can be related to the overlapping of the de-alloying curves as is evident from Fig. 10.5C.

The electrochemical impedance spectroscopy (EIS) curves for the electrodes made from etched and coated silicon powders before cycling and after 200 cycles are shown in Fig. 10.6 along with the curves for pristine silicon and etched silicon electrodes. To extract the numerical values for the resistance, the experimental curves were fitted using an Excel solver, where the equivalent resistance was calculated as per the RC equivalent circuit modeled by Ruffo et al. [66]. The resistive components (R_C, R_S, and R_E) were then calculated from the intercepts of the fitted semicircle as given in Table 10.2. Where R_C is the contact resistance between electrode and copper substrate/current collector, and R_S is the total resistance of the electrode and copper substrate/current collector, and $R_E = R_S - R_C$ is the effective electrode resistance. The increase in resistance of the etched silicon electrode in comparison to the electrode based on pristine silicon is probably due to more SEI layer formation on the silicon nanofibers created by etching. But modification of the similar etched sample by carbon coating results in much lower resistance (R_E of 54.8 Vs 134.1 Ω) by a factor of almost 2.5. This lowering of resistance could be due to the uniform SEI formation from outer shell carbon coating or the conductive carbon particles used for stuffing and separating the etched silicon particles before coating. However, 200 cycles of the same coated sample resulted in a larger semicircle

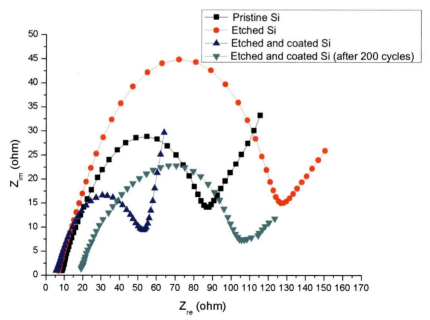

FIG. 10.6 Nyquist plots of pristine, etched, etched, and carbon-coated (both before and after 200 cycles) of silicon samples [48].

TABLE 10.2 Calculated electrode resistance from the Nyquist plots [48].

Sample	Contact resistance R_C (ohm)	Total resistance R_S (ohm)	Net electrode resistance R_E (ohm)
Polycrystalline Si	8.51	102.29	93.78
Etched Si	6.57	140.66	134.09
Etched and coated sample before cycling	5.32	60.18	54.86
Etched and coated sample after 200 cycles	19.50	111.10	91.60

than before cycling. Both R_C and R_S values increased after 200 cycles indicating higher contact resistance from the substrate as well as from the bulk of the electrode. This could be due to the fracture of core silicon particles, especially large size particles, where even after etching the material was not converted to fully nanostructured porous silicon. However, the resistance is still less than the resistance of the etched silicon electrode alone, even after 200 cycles as can be seen from Table 10.2 and Fig. 10.6.

Charge-discharge curves for the cell using an electrode of etched and coated silicon are shown in Fig. 10.7A along with the curves for cells using pristine and etched silicon electrodes. The highest lithiation capacity of ~3214 mAh/g was observed for the anodes based on pristine silicon in the first cycle. However, for the anodes based on etched and coated silicon, the first lithiation cycle (discharge) specific capacity of ~2750 mAh/g is obtained. The high initial specific capacity is an indication of the major participation of silicon within the carbonaceous network. The subsequent charge/discharge cycles maintained improved coulombic efficiency (99.6%) with a reversible capacity of ~778 mAh/g even after 200 cycles. This high reversible capacity with very high coulombic efficiency is among the best values reported for silicon-based composite anodes [11, 29–32, 34–37]. The etched and coated electrode shows improved cyclability when compared to electrodes based on pristine silicon or even etched silicon, which have lost all their capacity even after 15–20 cycles. The worse cycle life in terms of capacity fading for the unetched and uncoated samples (made conventionally) is probably due to the fracture and detachment from each other and from the current collector as well. It should be remembered that in a conventional approach, we have not followed the steps for avoiding agglomeration and coating. The coated sample was processed by freezing (rapidly) the mixture of conductive carbon and etched silicon in a polymeric solution, which created an electrically well-connected electrode. Upon firing, the frozen polymeric gel produced an outer layer of carbon coating on

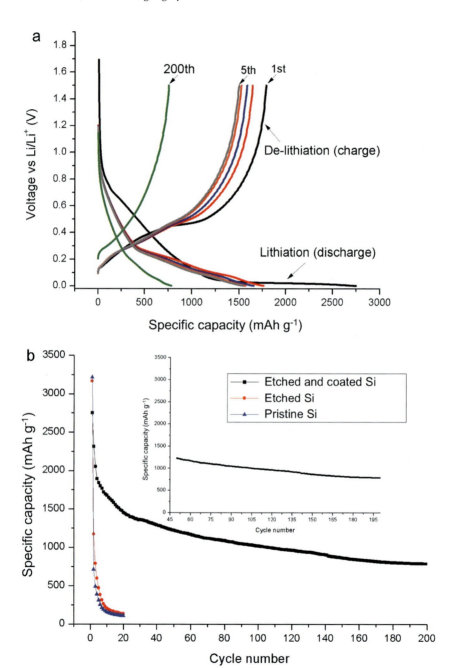

FIG. 10.7 (A) Charge-discharge curves for the cell using etched and coated silicon anode up to 200 cycles and (B) specific capacity vs. cycle number of the same anode along with the anodes based on pristine and etched silicon. Inset shows stable capacity retention of the etched and coated silicon anode between 50 and 200 cycles [48].

individually etched particles. The failure of pristine polycrystalline silicon electrode is apparent, due to its fracture, which is unavoidable for the large micron size silicon particles. Even after etching, the electrode failed probably due to the isolation caused by the fracture from the un-etched core silicon.

The improved performance and capacity retention of the etched and coated silicon can be attributed to the newly adopted processing of the electrode. In this approach, uniform electrical connectivity was achieved by the conducting phase of SCB and carbon coating during the processing step in which rapid freezing by gelation of the FFA created coating on individual etched particles. The outer shell coating with carbon improved the stability of SEI layer formation, which could also decrease the capacity loss. The absence of an obvious peak due to SEI formation can be inferred from the cyclic voltammetry (Fig. 10.5C). This may be an indication of a more uniform SEI layer formation. The controlled and uniform SEI layer can also contribute to improved reversible capacity. A long and gradual increase in the slope starting at $\sim$0.1 V (Fig. 10.7A, first lithiation cycle) is characteristic of alloying reaction between lithium and host silicon matrix. Subsequent lithiation curves show a slope starting at $\sim$0.3 V (Fig. 10.7A, 2nd cycle onwards), probably as a result of lithiation of the amorphous silicon [21]. Fig. 10.7B shows improved reversible capacity with high coulombic efficiency, especially after a few initial cycles. The improvement in capacity retention can be observed from the relatively flat nature of the curve when plotted between 50 and 200 cycles (inset of Fig. 10.7B).

From Fig. 10.8 it can be seen that the first cycle coulombic efficiency of the etched and coated silicon anode is nearly 65%, which is much higher than the value of 23% for pristine polycrystalline silicon, and 38% for the etched silicon. However, the coulombic efficiency of the coated sample has reached to nearly 99% only after a few cycles. This improvement in coulombic efficiency can be observed from Fig. 10.8.

Our approach ensures that the pores or voids available between the silicon fibers are filled with different shapes and sizes of the conductive carbon nanostructures. This may provide the conducting path even after fracturing of silicon, if any, thereby enhancing the conductivity of the active anode materials, which is required for getting high capacity even after numerous cycles. Furthermore, the conformal and uniform carbon coating on individual particles controls the amount of SEI formation without which there would have been even more successive loss in capacity as silicon particles fracture and create new surfaces. It also provides a stable and flexible outer shell, which could be beneficial for the larger particles especially after fracture to keep the fractured particles in contact and act as an electrical bridge between the fractured particles and the outside particles/current collector. Therefore, more silicon could be actively available for lithiation. The presence of an enhanced number of reversible sites for intercalation in carbon-based anode materials also depends on the level of graphitization. It has been reported that increased defects and/or disordered carbon result in increased irreversible capacity by providing more irreversible sites

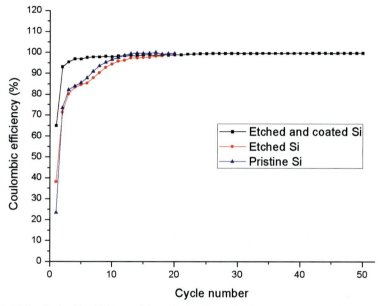

FIG. 10.8 Coulombic efficiency of the etched and coated silicon anode along with pristine and etched silicon anodes for reference [48].

for lithium ions [67]. The lower value of the ratio I_d/I_g in Raman spectra of the coated carbon also indicates the presence of less disordered carbon, thereby facilitating easy path for lithium to go inside silicon and come back again during lithiation and de-lithiation steps.

10.3.3 Processing and properties of Si anode from unetched-silicon powder

10.3.3.1 Coating on silicon powder and electrode performance

The polycrystalline micron size silicon particles were ball milled using zirconia balls, ethanol and 30 g of silicon particles in a single batch. The milling duration was set to 24 h. The size of the particles was measured at various times during milling. The particle size was about 350 nm for 24 h milling time [49].

The electrode made from 24 h (350 nm) milled silicon particles without any coating failed after only a few cycles with a starting capacity of ~1300 mAh/g. Although the mean size is several hundred nm there still exist a number of relatively large particles close to sub-micron to few microns. This is typical behavior of all uncoated particles because lithiation leads to fracture and loss of electrical interconnection among Si particles [49]. The average particle size for milled powder of 350 nm is more than the critical size of 150 nm for inducing fracture assuming the shape of the particles is spherical [68, 69]. It has been proposed that below this threshold size there should be no strain-induced fractures during lithiation/delithiation cycles. Interestingly, on the other hand, the same particles after carbon coating as described in the experimental section show impressive performance as described in the following section.

Cycling performances of cells made using both coated and uncoated 24 h milled silicon is shown in Fig. 10.9. The electrode made using milled silicon without any carbon coating failed after only two cycles. On the other hand, the electrode made using carbon-coated milled silicon shows a reversible capacity of 1050 mAh/g after 200 cycles with an initial capacity of 1658 mAh/g. One can also observe a steady increase in the overall capacity as the cycle number increases. The reason for such an increase is currently under further study. Also, the coulombic efficiencies for the coated samples were more than 99% after few initial cycles.

We propose that a thin layer of oxide over the milled silicon particles may act as a buffer and could help in other ways such as by preventing the formation of SiC during carbon coating. The formation of SiC phase(s) (such as beta-SiC) is detrimental to electrode performance. SiC formation starts at the Si-C interface; formation depends upon the temperature and reaction time. It is reported that formation of SiC on (100) silicon substrate requires annealing at 800°C for a period of time $(t) > 100$ min and at 900°C for $t > 25$ min [70]. The temperature and duration of exposure for carbon coating are suitable for the formation of SiC. The reduction of oxides in general by carbon is well documented in the literature [71, 72]. In this study, the carbonization process involves formation of carbon from polymer. Carbon (along with hydrogen) is a reducing agent, which prevails during the process of carbonization of FFA coated milled silicon. It has been observed that Si in reaction with decomposed carbon forms SiC especially at higher reaction temperature and longer time [73]. In such instances, if there exists a thin layer of SiO_x, instead of producing SiC, the reducing agents would react with the oxide layer first and reduce the SiO_x into Si. As a result, one could simultaneously avoid the formation of SiC as well as reduce the oxides present over silicon particles. So, to avoid the formation of

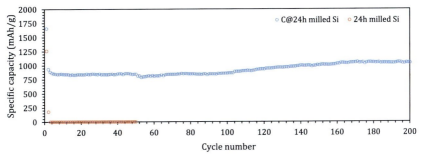

FIG. 10.9 Comparison of cycling performance of cells comprised of both the carbon-coated and uncoated 24 h milled silicon electrodes [49].

SiC, an optimized layer of SiO$_x$ could be very useful by reducing the oxide layer and preventing carbonization of silicon. It is feasible to tailor the oxide layer thickness in such a way that it would only be able to prevent the formation of carbide by reducing the oxide layer. In that way, one can achieve multiple benefits from a thin layer of oxide with optimum thickness.

The electrochemical impedance spectroscopy (EIS) of the electrode was investigated [49]. The electrodes made using uncoated milled (24 h) silicon particles display very high resistance ($\geq 200\,\Omega$) only after 50 cycles; the electrodes subsequently failed and the tests were terminated. In contrast, when the same Si powder was coated with carbon and tested for 200 cycles it resulted in lower ($\leq 100\,\Omega$) resistance. It bears repeating that the uncoated silicon electrodes were only tested for 50 cycles and the coated electrodes were tested for 200 cycles. Even after that, the uncoated electrodes displayed at least three times higher resistance (230.1 vs 68.4 Ω). Among all the electrodes, the one with carbon-coated 24 h-milled silicon showed the lowest electrode resistance for both pre- and post-cycling. It should be mentioned that the 24 h-milled electrode demonstrated better electrochemical performance among the electrodes. The effectiveness of the carbon coating (along with freezing the solution and extra carbon particles) in increasing the overall conductivity can be observed from the data summarizes in Table 10.3.

Cycling resulted in an increased amount of contact resistance as compared to as-prepared cells. For carbon-coated 24 h-milled silicon electrodes, cycling up to 200 cycles resulted in resistance of ~68 Ω, compared to the initial resistance of ~30 Ω. After only 50 cycles of uncoated silicon electrodes, the electrode

TABLE 10.3 Electrochemical impedance parameters obtained from milled silicon and carbon-coated milled silicon electrodes [49].

Electrode material		Contact resistance R_C (Ω)	Total resistance R_S (Ω)	Net electrode resistance R_E (Ω)
24 h milled Si	Before cycling	11.4	75.6	64.2
	After 50 cycles (failed)	22.3	212.6	190.3
C@24 h milled Si[a]	Before cycling	5.8	36.0	30.2
	After 200 cycles	31.2	99.6	68.4

[a] C@24 h milled Si stands for carbon coated over 24 h milled Si.

resistance nearly tripled to 190 Ω (from initial resistance of 64 Ω) for the 24 h-milled Si. Such an increase in the electrode resistance resulted in a significant capacity loss (Fig. 10.9) and could be attributed to fracture and hence isolation of particles and/or uncontrolled formation of SEI layer.

10.3.3.2 Effect of charging current on the electrode performance

Besides high capacity, cost, and safety, the performance of any electrode at higher current densities or charging rates is another consideration for its successful application. Thus, the electrode made using carbon-coated 24 h-milled silicon was tested at various current densities as shown in Fig. 10.10.

To get a true picture of the effect of charging current, the cell was cycled for 80 cycles as per normal test procedure at 0.6 A/g before the higher rate test. A stable capacity of over 1400 mAh/g was obtained at 0.6 A/g. It is decreased to about 1150 when the current density was doubled (1.2 A/g). The test results exhibit excellent cyclability with a high capacity of over 700 mAh/g at the current density of 6 A/g after 60 cycles of rate test, which is 10 times the initial current density. However, further increase in current densities to 12 A/g for charging is found to be detrimental to the performance of the electrode with capacity decreasing over 70% to 200 mAh/g. These results demonstrate excellent performance of the Si-based anodes at higher charging rates or current density.

10.3.3.3 Effect of ambient temperature on the electrode performance

The performance and safety of any battery will depend on the performance of electrode materials at higher temperatures—up to 40°C above room temperature (RT) of 22°C. It is important to know how the electrode behaves at a typical temperature (≥ 20°C). Thus, one of the electrodes based on carbon-coated 24 h-milled silicon particles was tested at various temperatures from 20°C to 60°C at

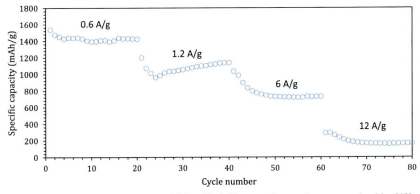

FIG. 10.10 Rate test of carbon-coated 24 h milled silicon anode at various current densities [49].

an interval of 10°C and compared with the similar electrode tested throughout at room temperature. All the tests were carried out in a temperature-controlled chamber, as described earlier; the cell was kept in the middle of the chamber to maintain cell temperature uniformity. The temperature was monitored with the help of a thermocouple kept close to the cell. All the cycling data were recorded after keeping the cell at the desired temperature for more than an hour to achieve a reliable and homogenous temperature; results are shown in Fig. 10.11.

Initially, the temperature was set to 40°C, similar to a very warm summer temperature. The initial jump in capacity could be ascribed to accelerated reaction due to higher temperature or enhanced conductivity of the cell. After 20 cycles of test at 40°C, the capacity reached a value of about 1370 mAh/g. When the same electrode was tested at room temperature it could attain only about 840 mAh/g after 20 cycles. After 20 cycles at 40°C when the cell was tested at room temperature for another 10 cycles one could observe the matching capacity of both the cells at the end of 10 cycles. It also shows the reproducibility of our results. Next, this cell was tested at 50°C and showed an enhanced capacity of 1470 mAh/g and after additional 20 cycles of testing at same 50°C it could still retain ~1200 mAh/g, which is still higher than the capacity at room temperature after similar number of cycles. There is a small jump from this capacity to ~1260 mAh/g when the temperature was further increased to 60°C. After running it at this temperature for additional 20 cycles the reversible capacity reached a value close to 1100 mAh/g. But a decrease in temperature from 60°C to 30°C and then to 20°C lowered the capacity. When the temperature was increased again to 25°C, the capacity reached a value close to the room temperature data. Throughout the test, it can be seen that the cell shows better performance when exposed to higher (up to 60°C) than room temperature.

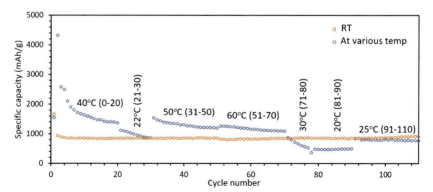

FIG. 10.11 Comparison of cycling performances of the cells at different temperatures comprised of carbon-coated 24 h milled silicon particles. Data mentioned in brackets are starting and ending cycle numbers. Data for two different cells are shown, one tested throughout RT and the other at various temperatures including RT [49].

It should be remembered that the cell has gone through different levels of alternating high and low temperature during testing. But still, it could retain its original value even after 100 cycles. In addition, higher capacities are recorded at higher temperatures than at room temperature. It is well known that the state of charge changes with a number of cycles. Therefore, the loss in capacity due to cycling is actually overcome by the performance of the battery at temperatures higher than the room temperature.

Our tests show better performance at higher ambient temperatures than at room temperature. This could be considered as a compromise between capacity and cycle life because it has been observed that higher temperatures, especially greater than 50°C, reduces cycle life [74]. However, this could be specific to an electrode and may not be a general phenomenon. The cell tested at different temperatures was also investigated at regular intervals using electrochemical impedance spectroscopy. The impedance data at different temperatures and number of testing cycles are given in Table 10.4. The impedance spectra were taken twice when the cell was tested at higher temperatures ($\geq$RT). The first at the higher temperature at which the test was performed and the second at the room temperature. Whenever the impedance spectra were taken after a certain number of cycles at the higher as well as the room temperature it is interesting to observe that there is up to 2.5-fold increase in net electrode resistance when tested at room temperature after a fixed number of cycles at higher temperatures. Usually, after a certain number of cycles, it is expected to get similar resistance values if not the same. But the results show two different values far apart from each other. These lower resistance values at higher temperature are in fact in good agreement with the higher capacity values that we observed from our earlier discussions on electrochemical performance. It is reasonable to imagine that a lower resistance electrode would be expected to have higher capacity. A higher temperature is expected to decrease the resistance for materials showing semiconducting (Si) and ionic conduction (electrolyte).

10.3.3.4 Performance in a full cell

The anode made using carbon-coated 24 h-milled silicon was also used in a full cell arrangement against NCM cathode (Toda NCM 523). Full cell performance depends on both the cathode and anode materials. A half-cell was used in order to study the performance of the cathode alone as shown in Fig. 10.12A. The NCM cathode shows a reversible capacity of $\sim$122 mAh/g after 100 cycles with a capacity retention of more than 92%. Therefore, NCM cathodes are an excellent candidate for use in a full cell.

Fig. 10.12B demonstrates the cycling performance of the full cell with NCM cathode employed in concert with a carbon-coated 24 h-milled silicon as anode; the specific capacities are calculated according to the mass of NCM based cathode. The full cell capacity decreases gradually and attained a very stable

TABLE 10.4 Electrochemical impedance parameters obtained from carbon-coated 24 h milled silicon electrode tested at different temperatures [49].

Cycle number[a]	Temperature (°C)	Contact resistance R_C (Ω)	Total resistance R_S (Ω)	Net electrode resistance R_E (Ω)
Before cycling	22 (RT)	5.2	48.6	43.4
After 20 cycles	40	7.0	158.4	151.4
	RT	22.4	292.6	270.2
After 30 cycles (20+10)	RT	9.3	162.4	153.1
After 50 cycles (30+20)	50	7.2	197.5	190.3
	RT	13.3	333.4	320.1
After 70 cycles (50+20)	60	13.9	156.9	143.0
	RT	12.7	372.3	359.6
After 90 cycles (70+20)	30 and 20	9.6	260.3	250.7
After 110 cycles (90+20)	25	26.0	241.6	215.6

[a] The numbers mentioned in the brackets are total cycle numbers. A sum of cycle numbers tested in the earlier temperature plus the cycle numbers in the current temperature profile. For example, in 3rd row the cell was tested at RT for 10 cycles after testing for 20 cycles at 40°C. So, the sum is (20+10=30).

reversible capacity of ~76 mAh/g after 100th cycles, which is comparable to those of reported full cells based on silicon anodes [75–77].

The carbon coating on the surface of silicon particles increased the overall conductivity and also avoided direct contact of silicon with the electrolyte, this contact has been shown to result in uncontrolled SEI formation with anode fracture [78, 79]. The increase in overall conductivity of the electrode by the superconducting carbon as well as coated carbon provide channels for electron transport, whereas the direct contact between the carbon and electrolyte prevents uncontrolled SEI growth, which could be detrimental if there is any agglomerated or larger-sized (>10 μm) silicon particles. The porous electrode structure created by the superconducting carbon particles and the coated carbon buffer on the Si anode material helped accommodate the volume change of silicon and provided access to electrolyte, resulting in attaining higher capacity, especially at higher charging rates.

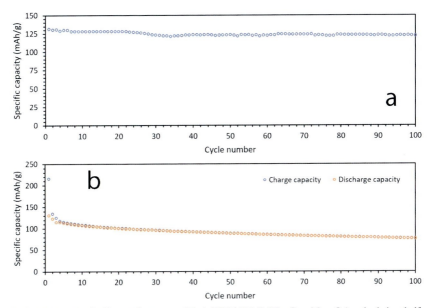

FIG. 10.12 (A) Cycling performance of Toda NCM 523 (LiNi$_{0.5}$Co$_{0.2}$Mn$_{0.3}$O$_2$) cathode in a half cell, and (B) Cycling performance of full cell comprised of carbon coated 24 h milled silicon anode and Toda NCM 523 (LiNi$_{0.5}$Co$_{0.2}$Mn$_{0.3}$O$_2$) cathode. Specific capacities were calculated according to the mass of NCM cathode [49].

10.4 Conclusions

The evolution of tensile residual stress and its consequences were investigated experimentally by shift in first order Raman spectrum and by morphological evolution of cracks in the single crystal silicon electrode. The residual stress builds up gradually leading to crack initiation and propagation. The residual stress was tensile and increased from 69 ± 11 MPa (after 1st cycle) to 186 ± 26 MPa after 10th cycles of lithiation/de-lithiation. After 50 cycles of operation, it reached 291 ± 56 MPa. These findings also demonstrate the role of tensile residual stress on the failure mechanism of crystalline silicon electrodes, and confirm the progressive nature of the failure of silicon electrodes. The fundamental knowledge gained through such studies is expected to further advance the current understanding of failure mechanisms of silicon-based electrodes and help in designing reliable high-capacity electrodes for the next-generation Li-ion batteries.

An innovative and robust method of processing high reversible capacity anode using low grade polycrystalline silicon was successfully demonstrated. This approach ensures uniform carbon coating on the individual etched silicon particles with nanofibers and stuffed with and separated by conductive carbon for achieving superior performance. The outer shell provides structural stability and controls the amount of SEI formation, whereas the carbon black inside the porous nanostructure of silicon provided a conductive path. Key results and conclusions of our processing studies are:

- A new approach to processing, including the creation of silicon nanofibers, providing channels for fast electronic and ionic transfer, as well as free space for accommodating volume expansion.
- Separating etched silicon particles and rapid gelation during solution coating enabling avoiding agglomeration of particles and facilitating coating of individual particles by carbon.
- Controlled SEI formation kinetics in the anode with etched and coated silicon was established as indicated by the absence of obvious SEI peak for silicon at 0.5 V.
- The etched and coated silicon anode showed huge improvement (2.5× times) in the overall conductivity leading to superior capacity retention.
- The synergistic effect of etching and coating on silicon exhibited an impressive specific capacity of approximately 1000 mAh/g after 100 cycles while the capacity remained at ~780 mAh/g even after 200 cycles with a very high coulombic efficiency of 99.6%.

Another processing route for fabricating silicon-based anode from a more cost-effective unetched Si powder material is demonstrated in a coin cell. The excellent electrochemical performance of carbon-coated 24 h-milled silicon anodes can be mainly attributed to the optimum size and improved conductivity resulting from the superconductive carbon and the conformal carbon coating on the active anode particles. A compromise between particle size, oxide layer thickness, and reduction during the coating process in a reducing atmosphere is found to play an important role in the overall performance of the electrode. The electrode possesses superior rate performance (~750 mAh/g) even at a current density of 6 A/g. The performance and safety of the silicon-based anode are demonstrated at various temperatures (up to 60°C) higher than the room temperature. A full cell with good cycling stability using carbon-coated milled silicon as the anode and $LiNi_{0.5}Co_{0.2}Mn_{0.3}O_2$ as the cathode is also demonstrated. Further optimization of the processing can lead to the excellent performance of Si-based anodes for high capacity and performance Li-ion cells.

Acknowledgments

The authors gratefully acknowledge the Oklahoma State University for the financial support, Argonne National Laboratory for providing the cathode materials, and OCAST (Oklahoma Center for Advancement of Science and Technology) award # AR17-016-2.

References

[1] B.A. Boukamp, G.C. Lesh, R.A. Huggins, All-solid lithium electrodes with mixed-conductor matrix, J. Electrochem. Soc. 128 (1981) 725–729.

[2] L.Y. Beaulieu, K.W. Eberman, R.L. Turner, L.J. Krause, J.R. Dahn, Colossal reversible volume changes in lithium alloys, Electrochem. Solid-State Lett. 4 (2001) A137–A140.

[3] U. Kasavajjula, C. Wang, A.J. Appleby, Nano- and bulk-silicon-based insertion anodes for lithium-ion secondary cells, J. Power Sources 163 (2007) 1003–1039.

[4] M.N. Obrovac, L. Christensen, Structural changes in silicon anodes during lithium insertion/extraction, Electrochem. Solid-State Lett. 7 (2004) A93–A96.

[5] C.Y. Chen, T. Sano, T. Tsuda, K. Ui, Y. Oshima, M. Yamagata, M. Ishikawa, M. Haruta, T. Doi, M. Inaba, S. Kuwabata, In situ scanning electron microscopy of silicon anode reactions in lithium-ion batteries during charge/discharge processes, Sci. Rep. 6 (2016) 36153.

[6] H. Wu, G. Chan, J.W. Choi, I. Ryu, Y. Yao, M.T. McDowell, S.W. Lee, A. Jackson, Y. Yang, L. Hu, Y. Cui, Stable cycling of double-walled silicon nanotube battery anodes through solid-electrolyte interphase control, Nat. Nanotechnol. 7 (2012) 310–315.

[7] M.J. Chon, V.A. Sethuraman, A. McCormick, V. Srinivasan, P.R. Guduru, Real-time measurement of stress and damage evolution during initial lithiation of crystalline silicon, Phys. Rev. Lett. 107 (2011), 045503.

[8] Z. Zeng, N. Liu, Q. Zeng, S.W. Lee, W.L. Mao, Y. Cui, In situ measurement of lithiation-induced stress in silicon nanoparticles using micro-Raman spectroscopy, Nano Energy 22 (2016) 105–110.

[9] L. Luo, J. Wu, J. Luo, J. Huang, V.P. Dravid, Dynamics of electrochemical lithiation/delithiation of graphene-encapsulated silicon nanoparticles studied by in-situ TEM, Sc. Rep. 4 (2014) 3863.

[10] B.K. Seidlhofer, B. Jerliu, M. Trapp, E. Hüger, S. Risse, R. Cubitt, H. Schmidt, R. Steitz, M. Ballauff, Lithiation of crystalline silicon as analyzed by operando neutron reflectivity, ACS Nano 10 (2016) 7458–7466.

[11] C. Botas, D. Carriazo, W. Zhang, T. Rojo, G. Singh, Silicon-reduced graphene oxide self-standing composites suitable as binder-free anodes for lithium-ion batteries, ACS Appl. Mater. Interfaces 8 (2016) 28800–28808.

[12] W.E. Jahsman, F.A. Field, The effect of residual stresses on the critical crack length predicted by the griffith theory, J. Appl. Mech. 30 (1963) 613.

[13] S.S. Mitra, O. Brafman, W.B. Daniels, R.K. Crawford, Pressure-induced phonon frequency shifts measured by Raman scattering, Phys. Rev. 186 (1969) 942.

[14] S. Ganesan, A.A. Maradudin, J. Oitmaa, A lattice theory of morphic effects in crystals of the diamond structure, Ann. Phys. 56 (556) (1970).

[15] E. Anastassakis, A. Pinczuk, E. Burstein, F.H. Pollak, M. Cardona, Effect of static uniaxial stress on the Raman spectrum of silicon, Solid State Commun. 8 (1970) 133.

[16] M. Jana, R.N. Singh, A study of evolution of residual stress in single crystal silicon electrode using Raman spectroscopy, Appl. Phys. Lett. 111 (6) (2017) 1–5.

[17] E.V. Astrova, G.V. Li, A.M. Rumyantsev, V.V. Zhdanov, Electrochemical characteristics of nanostructured silicon anodes for lithium-ion batteries, Semiconductors 50 (2016) 276–283.

[18] D.V. Novikov, E.Y. Evschik, V.I. Berestenko, et al., Electrochemical performance and surface chemistry of nanoparticle Si@SiO$_2$ Li-ion battery anode in LiPF$_6$-based electrolyte, Electrochim. Acta 208 (2016) 109–119.

[19] M. Ashuri, Q. He, Y. Liu, K. Zhang, S. Emani, M.S. Sawicki, J.S. Shamie, L.L. Shaw, Hollow silicon nanospheres encapsulated with a thin carbon shell: an electrochemical study, Electrochim. Acta 215 (2016) 126–141.
[20] H. Zhang, X. Li, H. Guo, Z. Wang, Y. Zhou, Hollow Si/C composite as anode material for high performance lithium-ion battery, Powder Technol. 299 (2016) 178–184.
[21] J. Chen, X. Huang, Hollow Silicon Structures for Use as Anode Active Materials in Lithium-Ion Batteries, 2015. Patent No. WO2015153637.
[22] M. Green, Structured Silicon Anode, 2013. Patent No. US8384058.
[23] J.K. Lee, B.W. Cho, J.M. Woo, H.S. Kim, K.Y. Chung, W.Y. Chang, S.O. Kim, S.E. Park, Method of Preparing Bundle Type Silicon Nanorod Composite Through Electroless Etching Process Using Metal Ions and Anode Active Material for Lithium Secondary Cells Comprising the Same, 2014. Patent No. US8741254.
[24] M. Green, F.M. Liu, Method of Fabricating Fibres Composed of Silicon or a Silicon-Based Material and Their Use in Lithium Rechargeable Batteries, 2007. Patent No. WO2007083155.
[25] M. Green, Silicon Anode for a Rechargeable Battery, 2016. Patent No. US9252426.
[26] C. Zhou, M. Ge, J. Rong, X. Fang, Nanoporous Silicon and Lithium Ion Battery Anodes Formed Therefrom, 2013. Patent No. US20130252101.
[27] S.W. Lee, H.W. Lee, I. Ryu, W.D. Nix, H. Gao, Y. Cui, Kinetics and fracture resistance of lithiated silicon nanostructure pairs controlled by their mechanical interaction, Nat. Commun. 6 (2015) 7533.
[28] S.W. Lee, M.T. McDowell, L.A. Berla, W.D. Nix, Y. Cui, Fracture of crystalline silicon nanopillars during electrochemical lithium insertion, Proc. Natl. Acad. Sci. U. S. A. 109 (2012) 4080–4085.
[29] Y. Li, B. Chang, T. Li, L. Kang, S. Xu, D. Zhang, L. Xie, W. Liang, One-step synthesis of hollow structured Si/C composites based on expandable microspheres as anodes for lithium ion batteries, Electrochem. Commun. 72 (2016) 69–73.
[30] H. Yu, X. Liu, Y. Chen, H. Liu, Carbon-coated Si/graphite composites with combined electrochemical properties for high-energy-density lithium-ion batteries, Ionics 22 (2016) 1847–1853.
[31] U. Toçoğlu, G. Hatipoğlu, M. Alaf, F. Kayış, H. Akbulut, Electrochemical characterization of silicon/graphene/MWCNT hybrid lithium-ion battery anodes produced via RF magnetron sputtering, Appl. Surf. Sci. 389 (2016) 507–513.
[32] Z.D. Huang, K. Zhang, T.T. Zhang, et al., Binder-free graphene/carbon nanotube/silicon hybrid grid as freestanding anode for high capacity lithium ion batteries, Comp. Part A: Appl. Sci. Manuf. 84 (2016) 386–392.
[33] H. Wu, G. Zheng, N. Liu, T.J. Carney, Y. Yang, Y. Cui, Engineering empty space between Si nanoparticles for lithium-ion battery anodes, Nano Lett. 12 (2012) 904–909.
[34] J. Li, J. Wang, J. Yang, X. Ma, S. Lu, Scalable synthesis of a novel structured graphite/silicon/pyrolyzed-carbon composite as anode material for high-performance lithium-ion batteries, J. Alloys Compd. 688 (2016) 1072–1079.
[35] D.L. Schulz, J. Hoey, J. Smith, et al., Si_6H_{12}/polymer inks for electrospinning a-Si nanowire lithium ion battery anodes, Electrochem. Solid-State Lett. 13 (2010) A143–A145.
[36] H.A. Lopez, Y.K. Anguchamy, H. Deng, Y. Han, C. Masarapu, S. Venkatachalam, S. Kumar, High Capacity Anode Materials for Lithium Ion Batteries, 2015. Patent No. US9190694.
[37] L. Zhang, L. Yue, S. Wang, X. Zhao, Nano-Silicon Composite Lithium Ion Battery Anode Material Coated With Poly (3,4-Ethylenedioxythiophene) as Carbon Source and Preparation Method Thereof, 2016. Patent No. US9437870.

[38] I. Do, H. Wang, H.M. Bambhania, L. Wang, Silicon-Graphene Nanocomposites for Electrochemical Applications, 2014. Patent No. US20140255785.
[39] M. Tatsuhiro, C.J. Chen, T.F. Hung, et al., Multilayer Si/Graphene Composite Anode Structure, 2015. Patent No. US20150004494.
[40] E.M. Berdichevsky, S. Han, Y. Cui, R.J. Fasching, G.E. Loveness, W.S. DelHagen, M.C. Platshon, Open Structures in Substrates for Electrodes, 2014. Patent No. US8637185.
[41] Q. Wang, L. Han, X. Zhang, J. Li, X. Zhou, Z. Lei, Ag-deposited 3D porous Si anodes with enhanced cycling stability for lithium-ion batteries, Mater. Lett. 185 (2016) 558–560.
[42] S.H. Kim, S.H. Yook, A.G. Kannan, S.K. Kim, C. Park, D.W. Kim, Enhancement of the electrochemical performance of silicon anodes through alloying with inert metals and encapsulation by graphene nanosheets, Electrochim. Acta 2019 (2016) 278–284.
[43] N. Fukata, M. Mitome, Y. Bando, W. Wu, Z.L. Wang, Lithium ion battery anodes using Si-Fe based nanocomposite structures, Nano Energy 26 (2016) 37–42.
[44] M. Green, Electrode Comprising Structured Silicon-Based Material, 2015. Patent No. US9012079.
[45] P.J. Rayner, A Method of Making Silicon Anode Material for Rechargeable Cells, 2015. Patent No. US8962183.
[46] W. Xu, Silicon and Lithium Silicate Composite Anodes for Lithium Rechargeable Batteries, 2013. Patent No. US20130230769.
[47] M. Nie, D.P. Abraham, Y. Chen, A. Bose, B.L. Lucht, Silicon solid electrolyte interphase (SEI) of lithium ion battery characterized by microscopy and spectroscopy, J. Phys. Chem. C 117 (2013) 13403–13412.
[48] M. Jana, T. Ning, R.N. Singh, Hierarchical nanostructured silicon-based anodes for lithium-ion battery: processing and performance, Mater. Sci. Eng. B 232–235 (2018) 61–67.
[49] M. Jana, R.N. Singh, A facile route for processing of silicon-based anode with high capacity and performance, Materialia 6 (2019) 100314.
[50] J. Song, C. Yang, H. Hu, X. Dai, C. Wang, H. Zhang, Penetration depth at various Raman excitation wavelengths and stress model for Raman spectrum in biaxially-strained Si, Sci. China Phys., Mech. Astronomy 56 (2013) 2065–2070.
[51] Z. Iqbal, S. Veprek, Raman scattering from hydrogenated microcrystalline and amorphous silicon, J. Phys. C 15 (1982) 377–392.
[52] A.T. Voutsas, M.K. Hatalis, J. Boyce, A. Chiang, Raman spectroscopy of amorphous and microcrystalline silicon films deposited by low-pressure chemical vapor deposition, J. Appl. Phys. 78 (1995) 6999–7006.
[53] D. Levi, B. Nelson, J. Perkins, In-Situ studies of the growth of amorphous and microcrystalline silicon using real-time spectroscopic ellipsometry, in: Natl. Center for Photovoltaics and Solar Prog. Rev. Meeting, 2003, pp. 1–4.
[54] C.K. Chan, H. Peng, G. Liu, K. McIlwrath, X.F. Zhang, R.A. Huggins, Y. Cui, High-performance lithium battery anodes using silicon nanowires, Nat. Nanotechnol. 3 (2008) 31–35.
[55] M. Hanbücken, P. Müller, R.B. Wehrspohn, Mechanical Stress on the Nanoscale: Simulation, Material Systems and Characterization Techniques, John Wiley & Sons, 2011.
[56] J. Kim, S. Tlali, H.E. Jackson, J.E. Webb, R.N. Singh, A micro-Raman investigation of the SCS-6 SiC fiber, J. Appl. Phys. 82 (1997) 407–412.
[57] M. Chandrasekhar, J.B. Renucci, M. Cardona, Effects of interband excitations on Raman phonons in heavily doped n-Si, Phys. Rev. B 17 (1978) 1623–1633.
[58] D.E. Gray, American Institute of Physics Handbook, McGraw-Hill, 1972.

[59] T.I. Ramjaun, H.J. Stone, L. Karlsson, M.A. Gharghouri, K. Dalaei, R.J. Moat, H.K.D.H. Bhadeshia, Surface residual stresses in multipass welds produced using low transformation temperature filler alloys, Sci. Technol. Weld. Join. 19 (2014) 623–630.

[60] H. Kim, K.E. Kweon, C.-Y. Chou, J.G. Ekerdt, G.S. Hwang, A comparative first-principles study of the structural and electronic properties of the liquid Li-Si and Li-Ge alloys, J. Phys. Chem. C 114 (2010) 17,942.

[61] G.D. Watkins, F.S. Ham, Electron paramagnetic resonance studies of a system with orbital degeneracy: the lithium donor in silicon, Phys. Rev. B 1 (1970) 4071.

[62] J. Narayan, Recent progress in thin film epitaxy across the misfit scale, Acta Materialia 61 (2013) 2703.

[63] F. Shi, Z. Song, P.N. Ross, G.A. Somorjai, R.O. Ritchie, K. Komvopoulos, Failure mechanisms of single-crystal silicon electrodes in lithium-ion batteries, Nat. Commun. 7 (2016) 11886.

[64] M. Sternad, M. Forster, M. Wilkening, The microstructure matters: breaking down the barriers with single crystalline silicon as negative electrode in Li-ion batteries, Sci. Rep. 6 (2016) 1.

[65] P. Limthongkul, Y.I. Jang, N.J. Dudney, Y.M. Chiang, Electrochemically-driven solid-state amorphization in lithium-silicon alloys and implications for lithium storage, Acta Mater. 51 (2003) 1103.

[66] R. Ruffo, S.S. Hong, C.K. Chan, R.A. Huggins, Y. Cui, Impedance analysis of silicon nanowire lithium ion battery anodes, J. Phys. Chem. C 113 (2009) 11390–11398.

[67] M. Jana, A. Sil, S. Ray, Morphology of carbon nanostructures and their electrochemical performance for lithium ion battery, J. Phys. Chem. Solids 75 (2014) 60–67.

[68] X.H. Liu, L. Zhong, S. Huang, S.X. Mao, T. Zhu, J.Y. Huang, Size-dependent fracture of silicon nanoparticles during lithiation, ACS Nano 6 (2012) 1522–1531.

[69] Y.H. Cho, S. Booh, E. Cho, H. Lee, J. Shin, Theoretical prediction of fracture conditions for delithiation in silicon anode of lithium ion battery, APL Mater. 5 (2017) 1–8.

[70] L. Moro, A. Paul, D.C. Lorents, R. Malhotra, R.S. Ruoff, P. Lazzeri, L. Vanzetti, A. Lui, S. Subramoney, Silicon carbide formation by annealing C60 films on silicon, J. Appl. Phys. 81 (9) (1997) 6141–6146.

[71] M. Ksiazek, M. Tangstad, E. Dalaker, E. Ringdalen, Reduction of SiO_2 to SiC using natural gas, Metall. Mater. Trans. A 1 (2014) 272–279.

[72] M.V. Vlasova, N.G. Kakazei, V.N. Minakov, G.A. Puchkovskaya, V.S. Sinel'nikova, T.V. Tomila, V.I. Shcherbina, Formation of silicon carbide in the reaction of reduction of silica by carbon, Sov. Powder Metall. Met. Ceram. 28 (1989) 718–723.

[73] K.V. Emtsev, A. Bostwick, K. Horn, J. Jobst, G.L. Kellogg, L. Ley, J.L. McChesney, T. Ohta, S.A. Reshanov, J. Röhrl, E. Rotenberg, A.K. Schmid, D. Waldmann, H.B. Weber, T. Seyller, Towards wafer-size graphene layers by atmospheric pressure graphitization of silicon carbide, Nat. Mater. 8 (2009) 203–207.

[74] F. Leng, C.M. Tan, M. Pecht, Effect of temperature on the aging rate of Li ion battery operating above room temperature, Sc. Rep. 5 (2015) 1–12.

[75] K.S. Eom, T. Joshi, A. Bordes, I. Do, T.F. Fuller, The design of a Li-ion full cell battery using a nano silicon and nano multi-layer graphene composite anode, J. Power Sources 249 (2014) 118.

[76] L.W. Ji, H.H. Zheng, A. Ismach, Z.K. Tan, S.D. Xun, E. Lin, V. Battaglia, V. Srinivasan, Y.G. Zhang, Graphene/Si multilayer structure anodes for advanced half and full lithium-ion cells, Nano Energy 1 (2012) 164–171.

[77] C.L. Li, C. Li, W. Wang, Z.F. Mutlu, J. Bell, K. Ahmed, R. Ye, M. Ozkan, C.S. Ozkan, Silicon derived from glass bottles as anode materials for lithium ion full cell batteries, Sci. Rep. 7 (2017) 1–11.
[78] N. Lin, J. Zhou, L. Wang, Y. Zhu, Y. Qian, Polyaniline-assisted synthesis of Si@C/RGO as anode material for rechargeable lithium-ion batteries, ACS Appl. Mater. Interfaces 7 (2015) 409–414.
[79] N. Lin, T. Xu, T. Li, Y. Han, Y. Qian, Controllable self-assembly of micro-nanostructured Si-embedded graphite/graphene composite anode for high-performance Li-ion batteries, ACS Appl. Mater. Interfaces 9 (2017) 39318–39,325.

Part V

Applications and future directions

Chapter 11

Advanced silicon-based electrodes for high-energy lithium-ion batteries

Dominic Leblanc[a], Abdelbast Guerfi[a], Myunghun Cho[a], Andrea Paolella[a], Yuesheng Wang[a], Alain Mauger[b], Christian Julien[b], and Karim Zaghib[a]

[a]Center of Excellence in Transportation Electrification and Energy Storage (CETEES), Hydro-Québec, Varennes, QC, Canada, [b]Institut de Minéralogie, de Physique des Matériaux et de Cosmochimie (IMPMC), Sorbonne Université, UMR-CNRS 7590, Paris, France

11.1 Introduction

The industry of lithium-ion batteries (LIBs) is expanding rapidly because of their numerous applications, including the market penetration for electric vehicles (EVs) and the regulation of the intermittence of electricity produced by windmills or photoelectric plants before integration with smart grids. However, the development of EVs is still hampered by their limited range. The autonomy of an EV powered by a LIB is determined by the number of lithium ions that can transit between the two electrodes of the battery. In commercial LIBs, the negative electrode (conventionally called the anode) is generally fabricated from graphite. The capacity of the graphite anode is considerably higher than that of the material of the positive electrode (cathode) currently used in LIBs. Therefore, the energy density of LIBs is primarily limited by the cathode. However, the capacity of the anode is an important parameter because the ratio of the negative electrode capacity to that of the positive electrode (N/P ratio), also known as the "balance of the battery," must be positive. The reason is that the positive electrode with a high capacity would send an excess of Li$^+$ ions to the negative electrode under deep charging conditions. As a result, the lithium in excess would accumulate at the surface of the negative electrode, forming a lithium metal film. This lithium plating poses a severe threat to safety because of the formation of lithium oxides and the reactivity of the lithium metal with the electrolyte. Therefore, the batteries must be constructed such that the capacity of the negative electrode is higher than that of the positive electrode. This condition imposed by safety concerns implies that substituting graphite with a material that has a higher specific capacity is desirable to increase the energy density of LIBs.

For an understanding of the interest in silicon (Si) as an anode material for LIBs, consider the binary phase diagram for Li and Si shown in Fig. 11.1. Various stable compounds can be formed during the lithiation of silicon ($Li_{12}Si_7$, Li_7Si_3, $Li_{13}Si_4$, and $Li_{22}Si_5$). The corresponding redox potentials vs. Li^+/Li are listed in Table 11.1.

Thus, the working potential of amorphous Si (*a*-Si) anode is ≤ 0.4 V vs. Li^+/Li, which is the lowest potential of anode materials for LIBs, except for graphite (0.15–0.25 V) and Li^0 itself. The effect of this slightly higher working potential of Si compared with graphite on the energy density is negligible concerning the beneficial effect of the capacity. Silicon has a theoretical capacity of

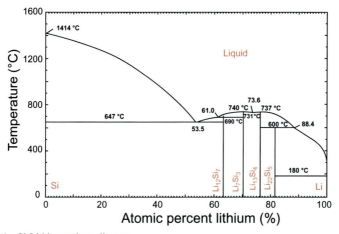

FIG. 11.1 Si-Li binary phase diagram.

TABLE 11.1 Plateau potentials and composition ranges of Li-Si systems at 25°C.

Biphasic reaction	Potential vs. Li^+/Li (mV)	x in Li_xSi
Si/$Li_{12}Si_7$	374	0.00–1.71
$Li_{12}Si_7$/Li_7Si_3	336	1.71–2.33
Li_7Si_3/$Li_{13}Si_4$	193	2.33–3.25
$Li_{13}Si_4$/$Li_{22}Si_5$	100	3.25–4.40

TABLE 11.2 Crystal structure data of the Li-Si system.

Phase	Composition (mol% Si)	Pearson symbol	Volume per Si (Å³)	Expansion (%)	C_{th} (mAh/g)
(βLi)	0.0	cI2	–	–	–
$Li_{22}Si_5$	18.5	cF432	82.4	320%	4199
$Li_{13}Si_4$	23.5	oP24	67.3	243%	3101
Li_7Si_3	30.0	hR7	51.5	163%	2227
$Li_{12}Si_7$	36.8	oP152	43.5	122%	1636
(Si)	100.0	cF8	19.6	0%	0

4200 mAh/g when it is cycled between Si and $Li_{22}Si_5$, which is approximately an order of magnitude higher than that of graphite (372 mAh/g). Nevertheless, the problem inhibiting the commercialization of Si as an anode material is the large volume variations in Si during charge-discharge cycles (Table 11.2). The expansion from Si to $Li_{22}Si_5$ in the lithiation process exceeds 300%. Such a large change in volume results in the cracking or even pulverization of the particles. In addition, the contraction in volume by the same amount in a reversible process during delithiation (discharge) results in a disconnection of the particles from the carbon additive of the anode and the current collector. This, in turn, results in a significantly poor cycling ability of the Si-anodes. Therefore, the challenge is to build Si-based electrodes that can alleviate the volume change of Si particles during cycling without any damage. This implies the synthesis of well-structured Si particles at a low cost.

Nano-sized Si particles have the advantage in that they alleviate the volume change and mitigate fractures more easily than larger particles; this increases the cycling ability [1–5]. However, the volumetric capacity of nano Si-based anodes is smaller because of their intrinsically low tap density [6]; in addition, their industrial fabrication is expensive. In contrast, microscopic particles enable an increase in Si loading, but avoiding their fracture during cycling is more difficult; this motivated the research on using Si nanoscale building blocks for micro-sized materials [7]. Therefore, the performance of Si-based anodes largely depends on their size, geometry, and structure [8]. The purpose of this paper is to review the progress on Si-based anodes in recent years under these different aspects. A review that focused on the engineering of silicon architectures and the construction of silicon-based composites is available in [9], while a review that focused on the preparation and characterization of nanostructured Si electrodes is available in [10].

11.2 Silicon as a low-cost anode material

Silicon can be obtained by the carbothermal reduction of quartz (SiO_2) in an arc furnace [11] or by the thermal decomposition of gaseous precursors such as trichlorosilane ($SiHCl_3$) or silane (SiH_4) in the Siemens process [12].

11.2.1 Metallurgical-grade silicon produced by the carbothermal process

Metallurgical-grade silicon can be obtained by the carbothermal reduction of quartz (SiO_2) in an arc furnace according to the following overall reaction:

$$SiO_2 + 2C \rightarrow Si + 2CO \qquad (11.1)$$

A typical arc furnace has a crucible with a diameter of approximately 10 m. Its crucible is continuously filled with the charge of raw materials (quartz, charcoal, and wood chips). The furnace has three carbon electrodes, which are submerged in the load and provide a three-phase current to heat the contents of the furnace. The temperature of the charge increases to approximately 2000°C in the hottest zone. At this temperature, the silicon dioxide is reduced to liquid silicon by the carbon. The liquid silicon is collected in a pocket from the tap hole at the base of the oven. The silicon is purified in the ladle by injecting gas ($O_2 + N_2$) through a nozzle or by adding a synthetic slag ($SiO_2 + CaO$) to reduce its calcium and aluminum content. After purification, the liquid silicon is poured into molds to solidify it; subsequently, it is crushed according to the customer's specification.

The typical purity of metallurgical silicon is approximately 97%–99% Si. The main impurities are iron, aluminum, calcium, and titanium. The major industrial companies in the metallurgical silicon sector are Globe Specialty Metals (USA), Simcoa (Australia), Elkem (Norway), Ferro-Atlantica (Spain), and Dow Corning (USA). China also has several small producers who, when combined, are one of the most important sources of silicon supply. The typical price for metallurgical silicon is currently approximately 3 USD/kg. For high-purity silicon, another process is available: the Siemens process [13].

11.2.2 High-purity silicon produced via the Siemens process

The silicon used in semiconductor applications must be extremely pure (99.9999999% Si or 9 N). This purity is obtained by the chemical purification of the metallurgical silicon [14]. In a first step, the metallurgical silicon must be transformed into a gaseous precursor (at 300°C):

$$Si\,(s) + 3\,HCl\,(g) \rightarrow HSiCl_3\,(g) + H_2\,(g) \qquad (11.2)$$

Advanced Si-based electrodes for high-energy Li-ion batteries Chapter | 11   415

The trichlorosilane (HSiCl$_3$) is then purified by distillation and reconverted into the metallic form using thermal decomposition (CVD) in the presence of hydrogen in a Siemens reactor [15]:

$$4HSiCl_3\,(g) + 2H_2\,(g) \rightarrow 3Si\,(s) + SiCl_4\,(g) + 8HCl\,(g) \qquad (11.3)$$

The major companies in this field are Hemlock Semiconductor (USA), Wacker Chemie (Germany), Tokuyama (Japan), and MEMC (USA). The company REC (USA) employs a variant of the Siemens process that uses the thermal decomposition of silane [16]:

$$SiH_4\,(g) \rightarrow Si\,(s) + 2H_2\,(g) \qquad (11.4)$$

The direct supply from polysilicon manufacturers under long-term contracts is often preferable as it guarantees regular supply at a firm price. The financial clauses of these contracts are confidential; however, the prices are deemed to be in the range between 40 and 80 USD/kg.

11.3 Mechanically milled silicon nanopowder

The use of nanomaterials is mandatory for the manufacture of silicon anodes [17]. The major advantage of the use of nanometric powders (10–100 nm) of silicon over coarser powders (1–10 μm) is that the silicon nanoparticles can more easily accommodate the volume change during the lithiation-delithiation process without cracking.

Metallurgical-grade silicon is available in a coarse size, normally in the form of lumps with a diameter of less than 100 mm. Therefore, they must be ground using a series of equipment to avoid contaminating the powder generated. The main source of contamination is the wear and tear of the parts of the grinding equipment. Silicon is a hard material (hardness: 6.5 Mohs) and has sharp edges. Therefore, abrasion-resistant materials must be selected for the parts to be in contact with silicon.

To perform the task, we used jaws, rollers, and conduits with an alumina coating (hardness: 9 Mohs). Dry grinding was conducted using three pieces of equipment: (i) jaw crusher (100 mm → 10 mm), (ii) roller crusher (10 mm → 1 mm), and (iii) jet mill (1 mm → 1–10 μm). Wet grinding was conducted in isopropanol. To reduce the risk of contamination, we used grinding balls composed of stabilized zirconium oxide (ZrO$_2$ 93%, Y$_2$O$_3$ 5%; hardness >9 Mohs).

The nanoscale silicon powder was characterized chemically and physically and was then used to fabricate a silicon anode. The cycling characteristics of the nano-silicon anode were compared with an anode of the same composition but composed of micrometric particles of silicon.

11.3.1 Fracture mechanics: Critical defect size

Most materials are composed of crystal domains in which atoms are regularly arranged in a three-dimensional network. The configuration of atoms is determined by their size and type of chemical bonds. In the crystal lattice, these interatomic bonds only act over short distances and can be broken if they are subjected to a tensile or compressive force.

Internal stresses are not uniformly distributed in the materials because they are dispersed in the form of grains of different sizes. The stress distribution depends on the mechanical properties of the materials and, above all, on the presence of cracks or defects, which serve as stress concentration sites. Griffith demonstrated that the increase in stress at a crack head (σ_f) is inversely proportional to the square root of the crack length (a) perpendicular to the direction of the stress [18] (Fig. 11.2):

$$\sigma_f \, a^{1/2} \approx C^{te}. \tag{11.5}$$

Because a "perfect" material has maximum stress at rupture of a finite value (σ_Y), a critical value for the length of crack (a_{crit}) exists at which the material can no longer break through crack propagation (Fig. 11.3).

In theory, although mechanical grinding assumes that the material is brittle, in practice, the crystals can store energy without breaking and release this energy when mechanical stress is eliminated. Such behavior is known as elasticity. When a fracture occurs, part of the stored energy is transformed into free surface energy, which is the potential energy of the atoms on the newly produced surfaces. The Griffith criterion designates the maximum stress (σ) beyond which a brittle material breaks. The value of this threshold stress is expressed as a function of the elastic modulus of the material (E_M) and of the energy (γ) required to generate a new surface [19]:

$$\sigma \geq \sqrt{\frac{2E_M \gamma}{\pi a}}. \tag{11.6}$$

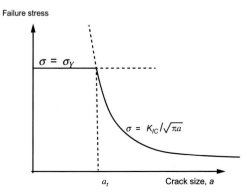

FIG. 11.2 Failure constraint as a function of crack size.

Advanced Si-based electrodes for high-energy Li-ion batteries Chapter | 11 417

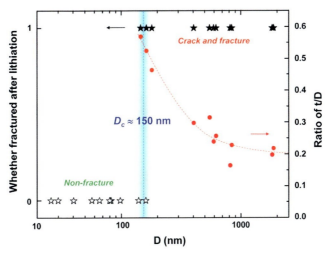

FIG. 11.3 First lithiation test results: The presence of cracking according to the thickness of the Li$_x$Si shell (t) as a function of the size (D) of the silicon particles; the graph indicates that under a critical size (D_c) of approximately 150 nm, the particles do not crack.

The value of the surface energy of a crystal is a value that is difficult to access [19]. For silicon, it has been evaluated at $\gamma = 1240$ erg/cm (1.240 J/m) for the plane (111) [20]. Using the elastic limit at break $\sigma_Y = 2.8$ GPa and the elastic modulus $E_M = 170$ GPa [12], we can calculate the critical defect size (a_{crit}):

$$a_{crit} = \frac{2E_M\gamma}{\pi\sigma_Y^2} = 1.7 \times 10^{-8} \text{m} \approx 20 \text{nm} \quad (11.7)$$

Thus, the calculated critical defect size is approximately 20 nm. This value is indicative of the order of magnitude of the particle size sought. Thus, if the size of the critical defect is equal to or higher than the size of the particles of silicon, these should not crack via crack propagation (fragile behavior).

Griffith's theory highlights the existence of a critical defect size at which a change occurs in the stress relaxation mechanism: The fragile behavior of material disappears to facilitate ductile behavior. Thus, if the particle size is lower than the dimension of the critical defect size, the brittle behavior of the silicon particles should disappear. Thus, experiments have established that the size of the silicon particles must be lower than 150 nm so that their cracking is no longer observed during their lithiation (Fig. 11.3) [21].

11.3.2 Decrease in the mechanical size of silicon

Silicon nanopowders were prepared from commercial metallurgical-grade (MG-Si) lumps via a conventional mechanical attrition process using stepwise dry milling (jaw crusher, roll mill, and jet mill), followed by a wet milling step. The mechanical attrition process is one of the most widely used processes to produce fine particles [22]. This technique is particularly efficient for silicon because of its hard and brittle nature. The flowchart to prepare the MG-Si nanopowders with the images of the typical equipment used for each process is available in [23].

The main characteristics to consider when selecting a dry grinding method are the particle size distribution obtained, the nature of the grinding media that can contaminate the material, and the capacity to increase the production scale of the grinding technology. The small size that can be achieved using dry grinding methods is on the micron scale [24].

Typically, the MG-Si lumps are first crushed into centimeter-sized particles using a jaw crusher with a zirconia ceramic liner to avoid metal contamination. Subsequently, these particles are processed in a zirconia roll mill (Makino, MRCA-1) to obtain millimeter-size particles. Finally, the millimeter-sized particles are ground using a jet mill to achieve micrometer-sized particles. After jet milling, the particle size distribution (PSD) ranges between 2 and 20 µm.

Wet grinding uses a reservoir containing the suspension to be ground (powder and solvent), which feeds a grinding chamber filled with small beads that are set in motion by a high-speed agitator. The function of the agitator is to create shear and impact forces between the balls and material to be ground. Wet grinding has several advantages including the following: it produces a uniform distribution of particle sizes, and the addition of a liquid decreases the tendency of the material to agglomerate. Industrial wet nano-grinding bead mill equipment is readily available commercially and can be used to decrease MG-Si to particle sizes lower than 100 nm [23]. The wet milling of silicon must be performed in organic solvents (i.e., isopropanol, heptane, or toluene) with high boiling points (>80°C) and low vapor pressure. Water is not a good solvent because silicon nanopowders are oxidized by water accompanied by hydrogen gas evolution.

11.3.3 Size effects on electrochemical properties

To illustrate the effect of the size of the Si particles on the electrochemical properties, we prepared an anode using the aforementioned nanometer-sized Si powder obtained by wet grinding and the micrometer-sized Si particles obtained by jet milling for comparison. In both scenarios, the silicon powder was mixed with acetylene carbon black (Denka Black) and sodium alginate (Aldrich) at a ratio of 50:25:25 using water as a solvent to achieve a viscosity of ~8500 cP for coating. A high-energy mixer (SPEX Certiprep) was used to de-agglomerate and mix the nanopowder. The slurry was coated on a copper foil to achieve loadings of approximately 0.6 mg/cm. The electrode was pre-dried at 75°C in a convection oven and then carefully dried at 110°C under a mild vacuum for 12 h.

CR2032 coin cells (Hohsen) were assembled in a He-filled glove box using a Celgard 3501 separator and 200-µm lithium foil anode (FMC Lithium).

Advanced Si-based electrodes for high-energy Li-ion batteries **Chapter | 11** **419**

The electrolyte comprises 1 mol L^{-1} LiPF$_6$ in a mixture of ethylene carbonate (EC) and diethyl carbonate (DEC) (7:3 by volume) with the addition of 10 vol% of fluoroethylene carbonate (FEC) (Ube). The cells were galvanostatically charged and discharged at 25°C using a VMP3 cycler (Bio-Logic) at a C/24 rate for formation cycles and a C/6 rate for life cycles over the voltage range of 0.005–1.0 V vs. Li$^+$/Li.

The two composite electrodes fabricated using the micrometer-sized ($d_{50} \sim 8\,\mu m$) and nanometer-sized powder ($d_{50} \sim 100\,nm$) were electrochemically tested in half-cell using lithium metal as a counter electrode. Fig. 11.4A and B show the first two cycles (formation cycles) of the half-cells at a C/24 rate. The first discharge curve (insertion of lithium) exhibited an outstanding specific capacity that included the irreversible capacity loss (ICL), which is typically related to the formation of a solid electrolyte interphase (SEI) on the electrode surface [25, 26]. We observed that nanometer-sized MG-Si powder exhibited a lower specific capacity (3510 vs. 4980 mAh/g) and a higher ICL (31% vs. 11%) compared with the micrometer-sized powder. Such behavior is attributed to a higher surface area and a higher amount of native oxide initially covering the silicon particles [27–29]. This oxide layer

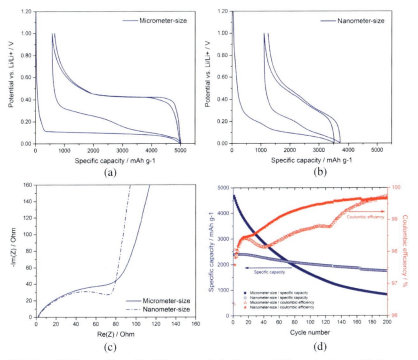

FIG. 11.4 Stability cycling of the Si/Li coin cells for 8-μm and 100-nm powders and their corresponding coulombic efficiencies.

is electrochemically inactive and considered to impede charge transport at the particle interface [27, 30] such that it inhibits the complete lithiation of the Si core to yield a lower reversible specific capacity. However, the electrochemical impedance measured on both electrodes before formation cycles (Fig. 11.4C) are in a similar resistance range (approx. 100 Ω). Fig. 11.4D shows the cycle performance at a rate of C/6 for micrometer- and nanometer-sized silicon electrodes. Micrometer-sized silicon had a higher initial reversible capacity (4685 mAh/g) but exhibited a faster capacity decay over cycles (15% of initial capacity at 140 cycles). On the contrary, the nanometer-sized silicon exhibited a smaller initial capacity (2400 mAh/g) but had a higher capacity retention over cycles (more than 80% of initial capacity at 140 cycles), which was evidence of the better structural integrity of the nanoparticles. In addition, the coulombic efficiency was observed to remain low and unstable during the whole cycle life, which indicated that the loss of the active material continued over the cycles, probably due to the lack of a stable SEI along with the generation of the new Si surface or loss of electric contact caused by the pulverized particles, particularly for the micrometer-sized powder.

After the cycle test, a post mortem analysis using scanning electron microscopy (SEM) indicated that the electrode thickness increased significantly; the micrometer-sized electrode had an even higher expansion than that of the nanometer-sized one (500% vs. 250%). Both electrodes were highly deformed and cracked at the end of their lives. Cracks were observed inside the composite electrode and at the interface with the current collector, which implied that a higher binding strength was required to maintain the cohesion and adhesion of the electrode. Note the cracks were observed despite the mid-size of the particles being 100 nm, i.e., smaller than the 150 nm suggested in Fig. 11.3 as the size limit below which cracks should not occur. The reason is that this size limit depends on different factors, including the geometry of the particles; thus, even Si-structures ≤150 nm thick may crack upon cycling [31].

11.3.4 Wet grinding optimization

The micrometer-sized silicon powder was again wet-milled using 1.0-mm diameter beads and isopropyl alcohol (IPA). The solid content was 20% and the samples were collected at different milling times (0, 3, 9, and 24 h) for physical characterization (Fig. 11.5) and evaluation of the electrochemical performance.

Fig. 11.6 shows the voltage profiles during the formation cycles along with the first coulombic efficiency and discharge capacity. The initial (0-h) sample (Fig. 11.6A) exhibited a typical discharge profile of the micrometer-sized silicon powder with a plateau at approximately 0.4 V; the 3-h milled sample (Fig. 11.6B) still maintained approximately 40% of the plateau at 0.4 V. However, after 9 h of milling and higher (Figs. 11.6C and D), the plateau at 0.4 V disappeared and the voltage profiles were nearly linear beginning at

FIG. 11.5 SEM image of Si powder obtained at different milling times (A) 0h, (B) 3h, (C) 9h, and (D) 24h of wet-milling with 1 mm diameter beads and 2200 rpm agitating speed in the 20 wt% solid contents of IPA solution.

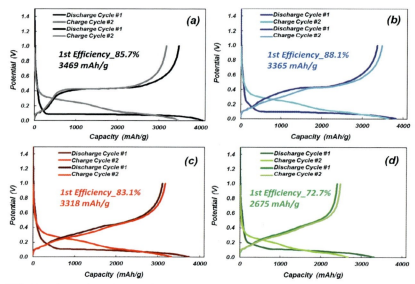

FIG. 11.6 Formation voltage profile of Si powder wet-milled at different times: (A) 0 h, (B) 3 h, (C) 9 h, and (D) 24 h.

approximately 0.1 to approximately 0.6 V. We observed that the first coulombic efficiency (CE) and discharge capacity decreased with milling time.

11.4 Silicon nanospheres using induced plasma

An inductively coupled plasma (ICP) is a heat source free of contamination, which is particularly suitable for high-purity processes. An ICP has a relatively large volume in a high-temperature zone (5000–10,000 K) and a relatively low gas velocity, making it an ideal tool for high-temperature material processing, where the melting or evaporation of the material is required. In an ICP, virtually any type of gas (e.g., H_2, O_2, N_2, Ar, NH_3, CH_4, or various mixtures) can be used as the plasma working gas, making the induction plasma both a heat and chemical reaction source [32].

The process begins with the vaporization of precursor materials using the high enthalpy of the ICP; subsequently, the material vapor is transported to the tail or fringe of the plasma where the temperature decreases drastically. This temperature gradient facilitates the formation of a highly supersaturated vapor, which results in the rapid production of numerous nanoparticles via homogeneous nucleation, heterogeneous condensation, and coalescence. Moreover, thermal plasmas are generated in the presence of any gas or gas mixture, enabling many chemical reactions to occur.

In thermal plasma synthesis, the precursors can be gases, liquids, or solids before being injected into the plasma. The availability of gas-phase precursors for pure metals is severely limited, so the most commonly used reactants for plasma synthesis are solid materials with the same compositions as the nanopowders. The in-flight vaporization of the feed material is a prerequisite for the homogeneous nucleation of the nanoparticles and the growth of monodispersed nanopowders without aggregation.

The PSD of nanopowders can be controlled during the plasma process using different means. The most important one is the cooling rate of the gaseous product. Therefore, to obtain ultrafine nanometric particles, the cooling rate of the vapor has to be extremely rapid, typically in the range of 10^4–10^6 K s^{-1}. Such a high quench rate is frequently achieved through the proper design of the reactor and the injection of a large amount of cooling gas. In addition, the plasma operational conditions (pressure, power level, etc.) have, to a certain extent, an effect on the size and distribution of the powders that are produced.

In general, nanometric materials in induction plasma can be synthesized through one of two approaches: the first is through a physical process (i.e., evaporation-condensation), and the second is through a chemical process (i.e., evaporation-reaction-condensation). Silicon nanopowder can be realized through both techniques using precursors such as micrometer-sized silicon powder (physical process) or derived from silane gas (chemical process).

11.4.1 Synthesis of nano-silicon using the physical process

In this process, the precursor material consists of micrometer-sized solid silicon particles that are prepared through mechanical milling and sizing [33]. The plasma is merely the heat source used to evaporate the materials. The rapid quenching of the vapor is necessary to nucleate particles before the vapor impinges on the cooled walls of the plasma reactor. The supersaturation of the vapor species due to high quench rates provides the driving force for particle nucleation. High quench rates lead to the production of ultrafine particles (down to an nm size) via homogeneous nucleation. The typical size of the nanoparticles ranges from 20 to 200 nm depending on the operating conditions.

The chemical composition of nanopowders produced by the physical process is directly related to the starting materials. However, depending on the difference in the boiling points of impurities in the starting material, the final nanopowder products may be purer than the starting powder because of the refining effect caused by volatility differences [25].

11.4.2 Synthesis of nano-silicon using the chemical process

In the chemical process, one or more gases react with the vapor generated from the precursor material introduced into the plasma reactor, and the resultant product condenses to form nanoparticles. Any material exposed to thermal plasmas at these high temperatures ($T \geq 10^4$ K) will be decomposed, possibly into its elemental constituents. The gaseous reactant can be a component in the plasma or separately injected at appropriate locations (into the plasma zone or in the reactor).

The composition of nanopowders produced by a chemical process may be complicated due to unexpected intermediate reactions in the plasma. The operational conditions must be carefully controlled to ensure the completeness of the desired reaction or synthesis. Note that for the evaporation-condensation process with a chemical reaction involved, the focus should be placed on controlling the quench rate. A considerably high quench rate may cause negative effects as some chemical reactions are lowered by the abrupt drop in temperature.

Silicon nanopowders are prepared by the decomposition of silane gas in a radio-frequency (RF) thermal plasma. In contrast to the chloride precursors ($SiCl_4$, $SiHCl_3$), silane does not produce corrosive chlorine gas in the plasma. The thermal plasma offers a high-temperature and contamination-free environment to produce pure silicon powder from SiH_4:

$$SiH_4\,(g) \rightarrow Si(s) + 2H_2\,(g), \tag{11.8}$$

with the evolution of hydrogen gas as a by-product. Because the gas-phase precursor feed rate is easier to control precisely, more homogeneous and uniform nucleation of the nanoparticles is expected to occur, leading to nanoparticles with a narrower size distribution. By properly controlling the coalescence of the nanoparticles in the quenching stage, the silane gas also offers the possibility to produce particles with a smaller mean diameter as compared with the evaporation-condensation process. Silicon nanospheres synthesized by silane decomposition in an RF induction plasma exhibit the same structure and morphology as the evaporation-condensation process [26].

11.4.3 In situ TEM lithiation of silicon nanospheres

The silicon nanospheres have a regular surface that can easily be observed using a transmission electron microscope (TEM). The nanoparticles will be observed under a TEM during their in situ lithiation to better understand the mechanism of silicon lithiation. The method consists of mounting a sample of silicon nanopowders on a sample holder specially designed in an electrochemical cell (open cell). The cathode, fabricated from a platinum wire, is fixed (top) and the anode, fabricated from a tungsten wire, is mobile because it is actuated by a piezoelectric element (bottom).

We performed an in situ TEM technique for the observation of a typical 200-nm pristine spherical silicon particle synthesized using an induced plasma. When a bias potential of -2 to -5 V was used to start lithiation of the silicon particle, a rapid reaction from the interior of the particle was observed, forming a core-shell structure. The crystalline core was gradually transformed into an amorphous LixSi alloy. The thickness of the amorphous phase was not uniform around the particle and resulted in an anisotropic lithiation [22]. The crack formation was not observed; however, we could not fully lithiate the larger particles (200 nm) in our experiment. Because the particles were clustered together and no conductive binder was used, the conductivity might have been extremely low to achieve full lithiation; however, some smaller particles (~50 nm in size) fully reacted with lithium without fracturing.

11.4.4 Si thin films

When Si is crystallized, its expansion during lithiation is anisotropic, which increases the stress and strain fields in the particles. Hence, the thin Si films are fabricated using a-Si, which decreases the irreversible capacity loss in the first cycle and improves the cycling ability. The performance largely depends on the adhesion of the film to the support, film thickness, and deposition rate. When these parameters are optimized, the results are remarkable: long cycle lives up to 3000 cycles were obtained with Si films that were prepared using physical vapor deposition (PVD) [34, 35]. For instance, by optimizing the synthesis parameters via PVD and n-type-doping Si with phosphorous to increase the electrical conductivity, the capacity of a 50 nm-thick n-Si later delivered 2000 mAh/g after 3000 cycles at a heavy load of a 30-C rate [34] (Fig. 11.7). Magnetron sputtering is also a powerful synthesis process with good results. A 275 nm-thick Si film synthesized using this process exhibited a high capacity of 3134 mAh/g at a 0.025-C rate. The capacity retention was 61.3% at a

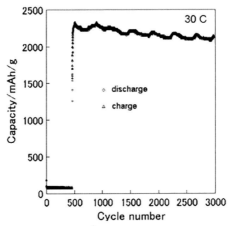

FIG. 11.7 The capacity retention of 500 Å thick n-Si during charge-discharge cycling at a 30-C charge-discharge rate in propylene carbonate containing 1 mol L^{-1} LiClO$_4$. *(Reproduced with permission from T. Takamura, S. Ohara, M. Uehara, J. Suzuki, K. Sekine, A vacuum deposited Si film having a Li extraction capacity of over 2000 mAh g^{-1} with a long cycle life. J. Power Sources 129 (2004) 96–100. Copyright 2004 Elsevier.)*

0.5-C rate for 500 cycles [36]. The results seemed less remarkable than those obtained using the aforementioned PVD, but note that the results in [36] were obtained on a much thicker film. However, note that magnetron sputtering is an expensive synthesis process because the growth rate of the film using this technique is typically 14 nm min^{-1}. Zhu et al. proposed a sandwich-structure Si/d-Ti$_3$C$_2$ exhibiting a capacity of 130 mAh g^{-1} at a current density of 500 mA g^{-1} after 200 cycles [37]. Anodes obtained by protecting Si films with graphene drapes exhibited long cycle lives (>1000 charge-discharge steps) with an average specific capacity of ~806 mAh/g. The volumetric capacity averaged over 1000 charge-discharge cycles was ~2821 mAh cm^{-3}, which was 2–5 times higher than what is reported in the literature for Si nanoparticle-based electrodes [38].

11.4.5 Si nanowires

Thin films or nanoparticles can be used as anodes with good electrochemical properties, but at low loadings only. The advantage of nanowires (NWs) is that they provide sufficient space between them to accommodate their change in volume during cycling. Each NW can be directly connected to the current collector, making conductive carbon additive and polymer binders unnecessary; thus, the loading is increased [39]. These features explain the possibility to increase the loading with the NW morphology. In addition, the entanglement of the NWs prevents them from detaching from the substrate, enhancing the cycle life and producing a quantity of Si of 1.2 mg cm^{-2} [40]. Therefore, the problem of a low volumetric energy density, which is the Achilles' heel of any Li-ion battery with nano-sized active particles, is overcome. The growth methods have been reviewed in another study [41]. The most popular one is the vapor-liquid-solid (VLS) growth in a chemical vapor deposition reactor under a flowing Si-bearing gas. In the first step, the growth of the NWs in a planar furnace is vertical. Entanglement can be obtained, and they become entangled when the temperature in the gas phase above the furnace is decreased rapidly. We have already mentioned that the Si films must be amorphous. For the same reason, the best result is obtained when the crystalline core of the NWs is embedded in an *a*-Si shell. This shell prevents cracking in the surface layer during cycling, which increases the cycle life. For this purpose, an *a*-Si layer is deposited using CVD over the crystalline Si NWs grown using VLS [40]. However, note that the yield of CVD is low, typically 0.75 mg/h, such that the process is not scalable. In addition, note that the best results were demonstrated with double-walled Si NWs.

11.4.6 Porous silicon

The pores of several nanometers provide sufficient free space to accommodate volume expansion during the lithiation of porous silicon. Thus, Si porous structures have a much better cycling ability than their non-porous counterparts [42–46]. Li et al. demonstrated that a large mesoporous silicon sponge (>20 μm), prepared using the anodization method, can limit particle volume expansion at full lithiation to ~30% and prevent pulverization in bulk silicon particles [47]. This sponge as the anode delivered a capacity of ~750 mAh g^{-1} based on the total electrode weight with >80% capacity retention over 1000 cycles at 1 A/g; this was a remarkable result if we consider that the area-specific-capacity was high (~1.5 mAh/cm). A metal-assisted wet chemical etching was used to obtain porous Si NWs of 5–8 μm length and a fore size of 10 nm [48]. When a Si NW was used as an anode, a capacity of 2400 mAh/g was demonstrated, with an outstanding initial coulombic efficiency of 91% and stable cycling performance. Another example is the high performance of porous doped NWs obtained by the direct etching of boron-doped Si wafers. The primary role of boron-doping was to increase the electrical conductivity and thus the C-rate; however, it generated defective sites that facilitated the etching and formation of pores. A capacity above 1000 mAh/g was demonstrated after 2000 cycles, at the current of 18 A/g (4.5 C). The only problem with NWs is the slow growth of these structures such that more scalable syntheses of other porous nano-structures are required. These have been reviewed in another study [49]. A solution is to begin with commercial silicon nanoparticles and treat them in a process involving boron-doping and electroless etching [50]. Boron-doping is obtained by mixing the commercial particles with boric acid in solution, then drying and annealing at 900°C in an argon atmosphere. Electroless etching is obtained in the presence of AgNO$_3$ in hydrofluoric acid (HF) solution. The two reactions responsible for the etching are simultaneous:

$$4\,Ag^+ + 4\,e^- \rightarrow 4\,Ag, \tag{11.9}$$

$$Si + 6F^- \rightarrow [SiF_6]^{2-} + 4\,e^-, \tag{11.10}$$

i.e., silicon donates electrons to reduce Ag$^+$ to Ag and is etched by F$^-$. Ag$^+$ reacts preferentially at the defective dopant sites where pores are generated at the surface [51]. Different porosities can be obtained simply by adjusting the ratio of boric acid over silicon. To optimize the rate capability, Ge et al. wrapped porous Si nanoparticles in graphene. The corresponding anode delivered a capacity of 1400 and 1000 mAh/g after 200 cycles at C/4 and C/2 rates, respectively [50]. Porous Si particles were also obtained from crushed boron-doped Si wafers and the addition of an H$_2$O$_2$ etchant [52]. With a pore size of several hundreds of nanometers, the particles thus obtained delivered 2000 mAh/g for the 50 cycles tested [53].

A nanostructured Si@C@carbon nanotube composite with interpenetrating voids was designed to enhance the rate capability [54]. The anode with this composite and additional porous graphene at the weight ratio of 2:8 as the active material delivered a capacity of 600 mAh/g at 4 A/g, with a capacity retention of 80% at 800 mAh/g after 200 cycles (Fig. 11.8). Similarly, three-dimensional

428 PART | V Applications and future directions

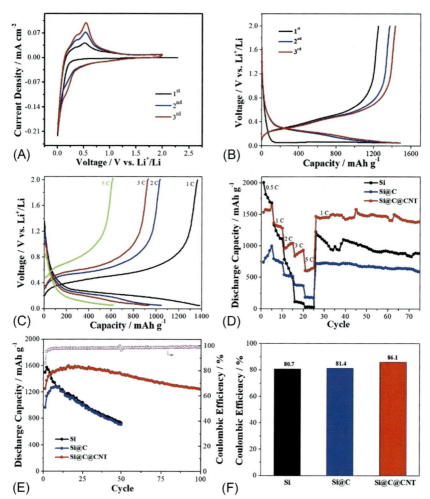

FIG. 11.8 (A) CV curves of the Si@C@CNT electrode were obtained at a scan rate of 0.1 mV s^{-1} between 0.01 and 2V vs Li$^+$/Li. (B) Galvanostatic charge-discharge curves of the Si@C@CNT electrode at the rate of 0.2C (1C = 1500 mA g^{-1}). (C) Galvanostatic charge-discharge curves of the Si@C@CNT electrode at different current rates of 1, 2, 3, and 5C. (D) Rate performance of the pure Si, Si@C, and Si@C@CNT electrodes. (E) Cycling performance of the pure Si, Si@C, and Si@C@CNT electrodes with the capacity retention of 80% at 0.2C. (F) Initial coulombic efficiency of the pure Si, Si@C, and Si@C@CNT electrodes. *(Reproduced with permission from X. Zhu, S. H. Cho, R. Tao, X. Jia, Y. Lu, Building high-rate silicon anodes based on hierarchical Si@C@CNT nanocomposite. J. Alloys Compd. 791 (2019) 1105–1113. Copyright 2019 Elsevier.)*

silicon/carbon nanotube capsule composites delivered a capacity of 547 mAh/g at 10 A/g and retained 1226 mAh/g after 100 cycles at 0.5 A/g [55].

Another possibility is an electrochemical etching with HF to obtain a porous Si film; this exhibited a capacity of 1260 mAh/g when it was combined with polyacrylonitrile (PAN) [56], and increased to 2000 mAh/g after Au-coating [57]. Another solution is Si deposition into a porous template. Here, porous Si is obtained in the form of an inverse structure by filling the voids of the template. Frequently, SiO_2 is used to form an opal structure. The synthesis can be obtained using the CVD process [43], or by filling the void space of the template with a gel-like silicon precursor followed by annealing to solidify the gel and obtain rigid porous Si [42]. The gel was obtained by reducing $SiCl_4$ with sodium naphthalene in glyme and capping this product with an n-butyl group by reaction with LiC_4H_9 to protect Si from reaction with SiO_2 to form SiO_{2-x} during annealing. The same gel was used to obtain porous Si-carbon core-shell NWs that were 6.5 nm in diameter using an SBA-15 template [58]. A capacity of 2738 mAh/g was retained after 80 cycles.

Instead of filling the void space, the porous SiO_2 template can be converted into Si directly by the reduction of magnesium operating at 650°C according to the following reaction:

$$2Mg\,(g) + SiO_2\,(s) \rightarrow 2MgO\,(s) + Si\,(s) \qquad (11.11)$$

Subsequently, the MgO and unreacted SiO_2 are removed by washing with HCl and HF. A 3D-mesoporous Si with an effective surface of $74.2\,m^2/g$ obtained by this process delivered, after carbon coating via a CVD process, a capacity of 1500 mAh/g after 100 cycles [59].

A different approach was used by Cen et al. who constructed a Si anode with a microsized-branched structure obtained from acid-etching of cast Al-Si alloy scraps [60]. Compared with the typical, low-loading ratio of significantly lower than 1 mg/cm in nano-sized Si, this micro-sized Si anode has a typical loading ratio of 2 mg/cm. The branched Si with carbon coating demonstrated an initial discharge capacity of 3153 mAh/g at the current rate of 1/16C and maintained a capacity of 1133 mAh/g at the 100th cycle, under a current rate of 1/4C. In addition, the rate performance was promising, with a discharge capacity of 488 mAh/g at the current rate of 1C. To our knowledge, this is the best result obtained with micro-sized particles.

11.4.7 Nano-Si/pyrolysis carbon microparticles

Owing to their voids or pores and uniform carbon coating, such particles often have high capacities exceeding 1000 mAh/g and a good cycle life [42, 61–75]. Si nanoparticles encapsulated in porous carbon matrix built using carbon nanotubes (CNTs) and nitrogen-doped carbon delivered a reversible specific capacity of 1380 mAh/g at a current density of 0.5 A/g, maintained at 1031 mAh/g after 100 cycles [75]. The Si particles prepared from rice husks were porous with a specific surface area of $288.4\,m^2/g$. Lin et al. used a scalable mechanical pressing approach with remarkable industrial compatibility to fabricate nanostructured Si secondary clusters with denser packing ($1.38\,g\,cm^{-3}$, pellet form), exhibiting a higher tap density ($0.91\,g\,cm^{-3}$, powder form) [66]. Over 95%

of initial capacity was retained after 1400 cycles at 1C, with an average specific capacity of 1250 mAh/g. After uniformly integrating CNTs into the clusters, the rate capability was increased, with capacities of 1140 and 880 mAh/g at 2 and 4C, respectively. The mass loading of these Si/C anodes was 2.02 mg/cm. Another advantage is the remarkable industrial compatibility of the synthesis process, contrary to the electrospun and micro-emulsion methods often used. Another industrially established process is the spray drying process used by Jung et al. [64] to produce Si/C materials with porous structures. However, despite their high performance, the Si/C materials suffer from disadvantages that hinder their application [76]. HF etching is a dangerous process inappropriate to large-scale fabrication. The low coulombic efficiency is not a problem when the Si anode is tested using a Li-metal counter electrode because the Li-metal anode is a reservoir of lithium; however, in a battery, the initial irreversible loss is a clear disadvantage. With the few exceptions outlined in this section, the mass loading of the microparticles is minimal.

11.4.8 Nano-Si/graphite microparticles

Si/C composites are composed of a graphitic carbon/nano-Si core and an amorphous carbon shell. The function of the graphitic carbon is to increase the electrical conductivity, buffer the variation of volume during cycling, and lower the specific surface area. The carbon shell protects the particles against side reaction with the electrolyte and will be analyzed in a subsequent section devoted to the stabilization of the solid electrolyte interface (SEI). Attention is focused in this section on the synthesis of the graphitic carbon/nano-Si core.

Spherical graphite is a commercial product, on which Si was deposited to form active particles of Si anodes. Si nanoparticles were attached to the surface of graphite by the carbonization of coal-tar pitch [77]. The corresponding anode delivered a capacity of 712 mAh g^{-1} and exhibited capacity retention of 88% at a higher rate of 5C (3250 mA/g). Ko et al. used a CVD process to synthesize Si-nanolayer-embedded graphite/carbon hybrids [78]. The advantage of the CVD is the uniform deposition of Si nanolayers on the surface of graphite, which explains the excellent results obtained with this composite after carbon coating: an initial coulombic efficiency of ~92% and a capacity of ~517 mAh/g for a high areal capacity loading >3.3 mAh/cm. A full-cell using this anode and a $LiCoO_2$ cathode achieved a high energy density of 1043 Wh L^{-1} and 92% capacity retention after 100 cycles.

Owing to the small specific surface area (SSA) of spherical graphite, attention was focused on flake graphite, which is also a commercial product but has a much larger SSA. The flake graphite was used as the building block to synthesize composites where nano-Si@C particles were uniformly dispersed in micrographite 3D conductive matrices. Such systems have been explored by Guo et al. [79–87]. The results are shown in Fig. 11.9 for the scenario in which the samples were prepared using a spray drying process [81]. Through

Advanced Si-based electrodes for high-energy Li-ion batteries Chapter | 11 431

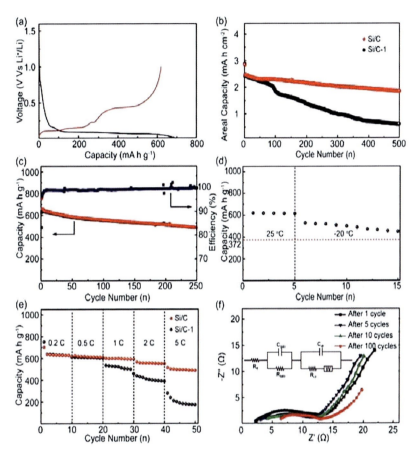

FIG. 11.9 Electrochemical properties of two Si/C microspheres prepared using a spray drying process from a slurry composed of 3-g nano-Si and 20-g flake graphite with (Si/C) and without (Si/C-1) carbon deposition. (A) Initial charge-discharge profiles of Si/C anodes at 0.1C. The cycling performance of densely compacted Si/C anodes at (B) 25, (C) 55, and (D) −20°C. All batteries were measured at 0.1C for the first cycle and 0.5C for later cycles. (E) Rate capabilities of Si/C and Si/C-1 anodes measured under various current densities from 0.2 to 5C. (F) Nyquist plots of Si/C anodes after different cycles and the equivalent circuit (insert). *(Reproduced with permission from Q. Xu, J. Y. Li, J. K. Sun, Y. X. Yin, L. J. Wan, Y. G. Guo, Watermelon-inspired Si/C microspheres with hierarchical buffer structures for densely compacted lithium-ion battery anodes. Adv. Energy Mater. 7 (2017) 1601481. Copyright 2017 Wiley.)*

optimizing the size of the particles and the size distribution of the flakes to obtain an efficient occupation of space and filling of the interstitial space, a coulombic efficiency of 99.8% was obtained, with a capacity of 620 mAh/g stable over 500 cycles under a high pressing density of 1.1 mg/cm and reversible areal capacity of 2.54 mAh/cm. Consequently, pouch cells fabricated with these Si/C composites and 3D conductive networks are an attractive alternative for use in LIBs. Luo et al. fabricated walnut-structured Si/C composites with an average particle size of 14 μm using nano-Si dispersed in a graphitic matrix [88]. The full-cell with this anode and a $LiCoO_2$ cathode exhibited capacity retention of 88% after 100 cycles, confirming the promising alternatives of these Si/C composites for high-energy-density LIBs. In a recent review, Chae et al. highlighted the necessity for the co-utilization of graphite and Si in terms of commercialization and reviewed the research progress on the graphite/Si anode [89].

Expanded graphite (EG) was also considered and led to interesting results. In particular, a carbon nanotube intertwined with an expanded graphite/porous Si composite using the in situ magnesiothermic reduction method, where porous Si nanoparticles (NPs) were dispersed in the interspaces constructed by EG sheets, delivered capacities of 2618 mAh/g at 0.2 A/g and 1390 mAh/g at 4 A/g, maintaining a capacity of 2152 mAh/g after 100 cycles at 0.4 A/g [90].

11.4.9 Nano-Si/graphene composite

Owing to its flexibility and mechanical strength, graphene is the best component to be used to obtain performant Si-anodes. Si/graphene nanocomposite has been extensively studied [91–96]. Gan et al. synthesized spherical graphite/Si/graphene (graphite/Si@reGO) composites through spray drying [97]. The surface of this composite was primarily composed of flake graphite sheets and the Si nanoparticles were uniformly distributed in the inner structure of the composite, with the entire structure encapsulated in reduced graphene oxide. A conductive and protective network with reduced graphene oxide (rGO) and CVD-implemented carbon was constructed using silicon nanoparticles embedded inside (Fig. 11.10) [98]. In addition to the wrapping of rGO on the Si nanoparticles, the additional carbon layer provided extra protection from fractures during cycling and contributed to the formation of a stable solid-electrolyte interface. This composite delivered a capacity of 1139 and 894 mAh/g at 0.1 A/g and 1C, respectively, and retained 94% of its initial capacity after 300 cycles at 1C. Luo et al. fabricated a freestanding composite anode using a one-step gel-coating-reduction approach enabling control over the uniform insertion of Si nanoparticles into the pores between graphene sheets. Over long-term cycling of 1300 cycles at 400 mA/g, a capacity decay as low as 0.06% per cycle and an average coulombic efficiency of 99.8% were achieved [99]. Used as an anode, a carbon-coated Si nanoparticles/reduced graphene oxide multilayer anchored to a nanostructured current collector delivered a capacity that remained above 800 mAh/g after 350 cycles at a current density of 2.0 A/g [100]. In general, the combination of graphene and a porous silicon structure leads to optimal results [101]. Microsized walnut-like porous silicon/reduced graphene oxide (P-Si/rGO) core-shell composites with an inner porous structure delivered a capacity of 1258 mAh·g^{-1} after 300 cycles at a current density of

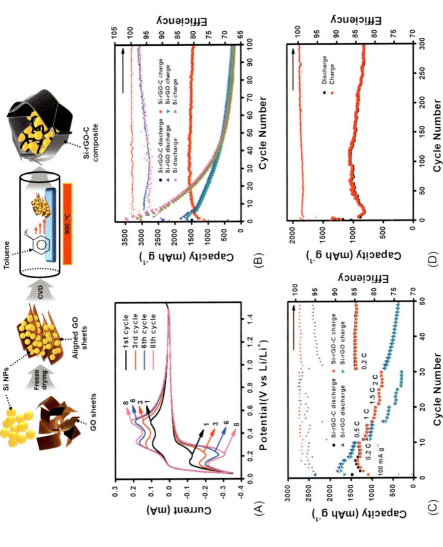

FIG. 11.10 Top: Schematic illustration of the synthesis procedure of Si-rGO-C composite. Bottom: (A) CV test of the Si-rGO-C composite with 1st, 3rd, 6th, and 8th cycles shown. (B) Cycling performance of Si-rGO-C, Si-rGO composites, and bare Si at 0.2C after activation. (C) Rate capability tests for Si-rGO-C and Si-rGO composites. (D) Long-term cycling performance of Si-rGO-C composite at 1C after activation. (Reproduced with permission from K. Feng, W. Ahn, G. Lui, H. W. Park, A. G. Kashkooli, G. Jiang, X. Wang, X. Xiao, Z. Chen. Implementing an in-situ carbon network in Si/reduced graphene oxide for high performance lithium-ion battery anodes. Nano Energy 19 (2016) 187–197. Copyright 2016 Elsevier.)

1 A/g [102]. Carbon-coated Si nanoparticles (50 nm thick) encapsulated in a 3D graphene network used as an anode with mass loading of 0.55 mg/cm delivered a capacity of 2883 mAh/g under a constant rate of 0.1 A/g. A capacity of 2041 mAh/g was retained over 100 cycles at 0.5C, and a capacity of ~752 mAh/g was achieved after 270 cycles at a rate of $5\,A\,g^{-1}$ [103]. This technology is sufficiently mature, such that this graphene/silicon composite anode material is already commercialized by Angstron Energy (AEC), a subsidiary of Global Graphene Group with a production capacity of 50 tons per year [104].

11.5 Nano-structured SiO_x and its use as a lithium-ion battery anode

SiO_x exhibits a high capacity and has the advantage of a small volume variation during cycling. However, its drawback is a large initial irreversible loss of capacity due to Li_2O and lithium silicates during lithiation [105]. Thus, the results are dependent on the value of x. As x increases, so does the formation of irreversible products serving as a buffer layer to alleviate the change in volume during cycling; however, the capacity decreases because the amount of active Si decreases. Therefore, a compromise has to be made. Takezawa et al. estimated that the best compositions are $x = 1.02$ and 1.34, where excellent cyclability was achieved [106]. However, Suh et al. estimated the optimum value at $x = 1.06$ for granulated SiO_x nanoparticles synthesized using thermal plasma and the water granulation process [107]. The corresponding anode delivered an initial capacity of 1196 mAh/g with a coulombic efficiency of 75%. Recent studies demonstrated that stable Li-ion batteries can be obtained with SiO_x-based anodes [108–115]. In particular, an anode of SiO_x encapsulated in graphene bubble film delivered 80% of the original capacity (780 mAh/g) after 1000 cycles at a current rate of 1 A/g [116].

Microscopic SiO_x/C particles were also considered [117, 118]. SiO particles were combined with graphite (G) to increase the conductivity, and CNTs to obtain a better dispersion of the particles [119]. Better results were obtained using porous materials. In particular, a micro-sized Si-C composite composed of interconnected Si and carbon nanoscale building blocks through disproportion of SiO followed by HF etching delivered a capacity of 1459 mAh/g after 200 cycles at 1 A/g with a capacity retention of 97.8% and had a high tap density of $0.78\,g\,cm^{-3}$ [72].

The use of nanometric wire structures (or nanowires) accommodates deformations in the radial direction of the fibers, avoiding the pulverization of the anode material and the loss of electrical contacts [27, 28]. In addition, a fine dispersion of silicon can be produced in an inactive matrix serving to relax the mechanical stresses and ensuring electrical continuity [28, 29].

11.5.1 SiO$_x$ nanowire synthesis

Hydro-Quebec developed a method to synthesize SiO$_x$ at high temperatures (1450–1600°C) to obtain nanostructured SiO$_x$ wires [120]. SiO$_x$ NWs are produced from the homogeneous nucleation of silicon monoxide in a carrier gas at atmospheric pressure. In a high-temperature induction reactor, gaseous SiO is formed from liquid silicon and solid silica. Contact with oxygen in the air should be avoided to avoid oxidation of the newly formed SiO; therefore, argon is used as a protective atmosphere. In addition, the use of argon as a carrier gas shifts the equilibrium of the reaction toward the production of SiO.

The temperature of the liquid silicon in the graphite crucible is increased to approximately 1500°C. After adding the silica sand, the injection of argon by the graphite cover of the crucible enables the oxygen present in the reactor to be driven out. When the oxygen is completely purged, the parasitic reaction for the production of silica smoke stops, and the reaction for the production of SiO$_x$ particles begins. Therefore, the color of the product changes from white (silica fume, SiO$_2$) to brown (SiO$_x$).

The material produced can be observed using SEM under high magnification (Fig. 11.11). The SiO$_x$ condensate produced has spherical agglomerates from 2 to 10 μm in diameter composed of a nanometric fibrous structure.

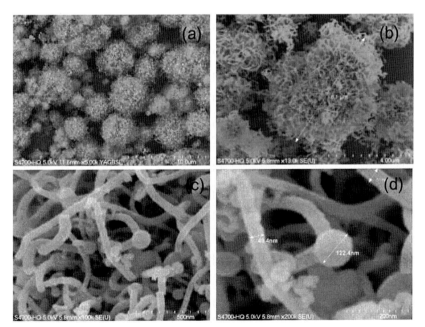

FIG. 11.11 Electron micrograph of SiO$_x$ particles synthesized in the high-temperature metallurgical process.

The nanofilaments have an approximate diameter of 50 nm and are linked together by spheres approximately 100–150 nm in diameter. The X-ray diffraction analysis indicates that the particles are composed of amorphous silica, crystalline silicon, and silicon carbide (β form).

11.5.2 Electrochemical evaluation of SiO$_x$ nanowires

A composite electrode was fabricated by mixing the active material produced (SiO$_x$) with 25% w/w of carbon black (Denka Black) and 25% w/w of binder (sodium alginate, Aldrich) in a solvent consisting of deionized water to obtain a homogeneous dispersion (slurry). The slurry was deposited on a copper current collector. The electrode was dried at 110°C for 20 h under vacuum. A CR2032 button cell battery was assembled in a glove box filled with helium. The electrolyte used was 1 mol L^{-1} LiPF$_6$ in a 3:7 (v/v) mixture of ethylene carbonate (EC) and diethyl carbonate (DEC) with 2% w/w of vinylene carbonate (VC) (Ube). The counter electrode was a thin film of lithium (200 μm). The electrochemical tests on the battery were conducted by discharge-charge cycling in the galvanostatic mode over a 0.005–2.5 V potential range with a current of C/24. The SiO$_x$ produced exhibited a higher initial capacity than graphite (950 vs. 372 mAh/g) and the initial irreversibility loss was 37%. Subsequently, the battery was cycled to measure its stability at a speed of C/6 (Fig. 11.12). After formation cycling, the reversible capacity at C/6 of 600 mAh/g increased continuously over 200 cycles, where it reached more than 900 mAh/g. This last observation was unusual; the most plausible assumption was that the specific surface of the material increased with the number of charge-discharge cycles. New surfaces may be produced by the cracking of materials without losing electrical contacts.

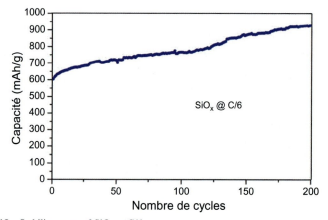

FIG. 11.12 Stability curve of SiO$_x$ at C/6.

Advanced Si-based electrodes for high-energy Li-ion batteries **Chapter | 11** **437**

The prelithiation of Si-materials was proposed [121–129], including that of SiO$_x$ [130], to overcome the irreversible loss of capacity in the first cycle. The prelithiation of SiO$_x$ based on electrical shorting with lithium metal foil enabled Kim et al. to obtain increased coulombic efficiencies in the first three cycles, reaching 94.9%, 95.7%, and 97.2%, respectively [128]. However, chemical lithiation is not envisioned for industrial applications, and it is hindered by the difficult synthesis of a highly active lithiation reagent. We further discuss silicon oxide as a coating layer in Section 11.7.2.

11.6 Binder selection

In Si anodes, the binder has an important function to keep the electrode film integrated during cycling. We tested four binders in coin half-cells with the same electrodes. Acrylic resin exhibited the highest first cycle efficiency (CE) (84.2% with a capacity of 2973 mAh g^{-1}), followed by acrylic resin (84.0%) plus additive (2% vinyl carbonate, VC), alginate (81.6%), and polyimide (72.8%); however, polyamide exhibited the highest reversible capacity (3080 mAh/g) during the first formation cycle.

Fig. 11.13 shows the cycle performance of Si/C powder with different binders in coin half-cells. Acrylic resin exhibited cycle retention of 2600 mAh/g after 60 cycles, while the lowest retention was observed for polyimide (2000 mAh/g with 31% capacity fade). Sodium alginate is often used in

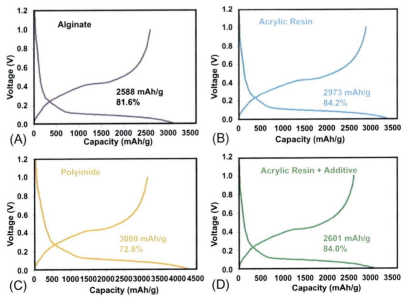

FIG. 11.13 Formation voltage profile of Si/C powder with different binders: (A) alginate, (B) acrylic resins, (C) polyimide, and (D) acrylic resin +2% VC additive.

particular for NWs because of weak interaction between sodium alginate and the electrolyte. In addition, this binder offers access to Li^+ ions to the surface of Si, and it is useful for stabilizing the SEI [131]. According to our results, shown in Fig. 11.13, the alginate binder exhibited a capacity fade of 25% with a capacity of 1900 mAh/g after 60 cycles. Based on these data, we selected the acrylic resin to fabricate the electrode for full-cell tests to evaluate the effect of different types of electrolytes. During the cycle test, full cells containing acrylic resin exhibited better cycle retention with better efficiency than the other electrolytes. However, the cycle life performance was not acceptable compared with that of commercial anodes. The loss of contact between silicon anode particles and the mechanical disintegration of the anode film were the primary causes of this capacity fade.

The efficiency of the binder in Si-based anodes is due to the hydrogen bonds constructed between the binder and Si powder for enhanced adhesion. Conductive polymers have this potential. However, the hydrogen bonding between the carboxyl group of the polymer and the hydroxyl group of the silicon surface is sufficient to increase the cycling ability in low loading scenarios [131–134]. However, this is not sufficient to maintain the integrity of the Si particles in high-areal-capacity Si anodes. Polyrotaxanes/polyacrylic acid (PAA) [135] and polyvinyl alcohol (PVA)/PAA [136] combine both the hydrogen bonding and mechanical properties, enabling a long cycle life with high areal capacity. Recently, efforts have been conducted to determine polymeric binders that result in high mass loading without the requirement of a significant hydrogen bonding. Table 11.3 shows such robust binders along with the mass loading of the cell tested with these binders and their performance [136, 138–144]. Among them, poly(3,4-ethylenedioxythiophene): poly(styrenesulfonate) (PEDT:PSS) by itself, with its linear structure, cannot effectively alleviate the volume expansion of Si, and it must be cross-linked with another polymer (Table 11.3). In addition, glycerol was introduced as a cross-linker to PEDOT:PSS [145]. The Si-nanoparticles anode with the glycerol-crosslinked binder exhibited a high reversible capacity of 1951 mAh/g after 200 cycles at 0.5 A/g and superior rate capability (804 mAh/g at a high current of 8.0 A/g). Liu et al. fabricated an effective 3D network with a high strength by interweaving the "hard" and "soft" polymer chains in the composite binder, namely a hard poly(furfuryl alcohol) as the skeleton and soft polyvinyl alcohol (PVA) as the filler. The resulting Si anode exhibited an areal capacity of >10 mAh/cm and enabled an energy density of >300 Wh kg^{-1} in a full lithium-ion battery (LIB) cell [137].

A self-healing conductive hydrogel binder (ESVCA) was prepared by the addition of gelation agents (ammonium persulfate (APS)) in the composite solution by complexing PEDOT: PSS with poly(vinyl alcohol) (PVA), followed by modification with 4-carbonxybenzaldehyde (CBA) [144]. The corresponding Si-electrode with a-Si loading mass varying from 2.00 to 2.48 mg/cm exhibited

TABLE 11.3 Polymeric binders used and tested as Si anodes for LIBs.

Type of polymeric binders	Areal mass loading (mg/cm)	Capacity normalized to active materials (mAh/g)	Capacity normalized to total electrode (mAh/g)	Highest areal capacity (mAh/cm)	Ref.
PFA-PVA	2–5	2200 @ 1.0A/g, 60th	1320	10 (initial) 6 (50th)	[137]
PEDT:PSS	0.4–1.5	1927@2A/g, 100th 2186@0.5A/g, 100th	1547 1749	3	[138]
PVDF-g-PtBA	2.0	1635 @ 5C, 30th	981	3.2	[139]
PAA-PVA	2.4	2283 @ 1C, 100th 1800 @ 4A/g, 50th	1370 1080	4.3	[136]
PFM+ calending	1.32–3.39	1000 @ 0.1C, 150th	900	3.5	[140]
PEDOT:PSS+ VAA	2.0	1660 @ 0.84A/g 700th	1500	5.13	[141]
PEDOT:PSS+ PEO+PEI	–	1840 @ 1.0A/g, 60th	1472	4	[142]
PAA-P(HEA-co-DMA)	2–6	2394 @ 1A/g, 200th 1855 @ 5A/g, 200th	2155	4	[143]

PFA=poly(furfuryl alcohol), PVA=polyvinyl alcohol, PVDF-g-PtBA=poly(vinylidene difluoride)-graft-poly(tert-butylacrylate), PAA=poly(acrylic acid), PFM=poly(9,9-dioctylfluorene-co-fluorenone-co-methyl benzoic ester), VAA=vinyl acetate-acrylic, PEDOT=poly (3,4-ethylenedioxythiophene), PSS=poly(styrenesulfonate), PEI=polyethylen-imine, PEO=poly (ethylene oxide), PAA-P(HEA-co-DMA)=poly(acrylic acid)-poly(2-hydro-xyethyl acrylate-co-dopamine methacrylate).

a capacity of 1786 mAh/g at 500 mA/g with a capacity retention of 71.3% after 200 cycles at ambient temperature. An n-type multi-block polymer PPy-b-PB of poly(1-pyrenemethyl methacrylate) (PPy) and polybutadiene (PB) was designed as an anode with a high reversible capacity (2274 mAh/g) and capacity retention (87.1%) at the rate of 0.2C (0.84 A g^{-1}) after 200 cycles [146], but the mass loading of Si of the entire anode was small (~0.24 mg/cm) compared with the other data reported in Table 11.3. Another polymeric binder (polyvinylidene fluoride vs. polyacrylic acid/carboxymethyl cellulose) was considered in a LIB using Li_6PS_5Cl-infiltrated Si electrodes to obtain an all-solid-state $LiCoO_2$/Si battery with an energy density of 338 Wh kg^{-1} [147].

11.7 Reducing surface reactivity of nano-silicon

11.7.1 Protection against water

The composite silicon electrode is created using an aqueous binder solution such as sodium alginate. Nanoscale silicon particles have a large specific surface and are reactive. Silicon reacts with water and releases gas, more probably hydrogen. The oxidation of silicon using water can change the surface chemistry of the particles by introducing an insulating oxidation film of varying thickness, and hydrogen poses a safety concern (pressurization of airtight containers with flammable gas). In addition, we experienced the generation of gas during mixing and coating of the composite anode, producing quality problems such as unstable loading level and inhomogeneous surface quality.

Thus, decreasing the reactivity of nano-silicon powders using a surface core-shell approach or surface modification is imperative. We coated nano-silicon powder with PAA using the spray drying method. For bare nano-silicon powder, we observed a strong release of gas in the suspension. However, the nano-silicon powder that was surface treated with PAA exhibited no gas evolution during the slurry mixing and coating process, even after aging the slurry overnight.

The spray-dry process enabled us to produce secondary particles of a nano-Si/C composite with 5–60 μm particle sizes. The cross-section of the particle obtained using an ion beam indicated that the secondary particles consisted of hollow nano-Si particles, with an interior that contained a mixture of polymer, carbon black, and a large fraction of void space. For the large-scale production using this technique (1–10 kg), we used an Agglomaster machine and piloting furnace. In the cycle performance of full-cells, cycle retention is always worse than that of half-cells because of the poor discharge/charge efficiency during cycling. In addition, the continuous formation and destruction of the unstable SEI increase Li$^+$ consumption during cycling. Consequently, the formation of a stable SEI is a significant factor in improving the performance of the cycle life.

11.7.2 Stabilization of the SEI

Another problem due to the large expansion of Si during lithiation is the difficulty to stabilize the SEI because the SEI passivation layer can be broken during the delithiation as the silicon shrinks. This re-exposes a fresh Si surface to the electrolyte such that an additional SEI is formed; further, the SEI becomes increasingly thicker with the cycling and the battery ages. Thus, the stabilization of the SEI is crucial. Therefore, much effort has been devoted to the surface and interface engineering of silicon-based anode materials [148].

One solution is the coating the Si structure with a protective material, with careful engineering to control the thickness of the shell, which must be sufficiently thick to protect but sufficiently thin to limit its weight. The beneficial effect can be evidenced by the increase in the first cycle coulombic efficiency to 83% by a 10 nm-thick carbon coating on Si NWs of 90 nm in diameter [149], and even 90.3% by replacing the carbon with 10 nm-thick Cu [150]. Carbon coating nano-Si increases the cycle life [151–153]. Si nanoparticles confined in an N, O-dual-doped mesoporous carbon with a void between the Si and the C delivered 1401 mAh/g (per gram of silicon plus carbon) after the 500th cycle [151]. In practice, the structure of the coating carbon is significant. An amorphous carbon (AC) coating exhibits a better conductivity, while an ordered carbon (OC) coating exhibits a good tolerance to volume variation. This led Fang et al. to realize an amorphous/ordered dual carbon (AC/OC) coating structure on silicon nanoparticles to obtain a synergetic effect of the two types of carbon [154]. As a result, this composite delivered a capacity of 1020 mAh/g at 200 cycles and 650 mAh/g after 500 cycles at a current density of 0.36 A/g. Chen et al. proposed a double carbon shell designed to perform dual functions of encapsulating the volume change of silicon and stabilizing the SEI layer [155].

A full cell with this anode and a $LiNi_{0.45}Co_{0.1}Mn_{1.45}O_4$ cathode exhibited an average discharge voltage of 4.2 V, a high energy density of 473.6 Wh kg^{-1}, and a good cycling performance. As an alternative, ultra-thin graphene and Li_4SiO_4-double-shell structures were proposed by Ai et al. to obtain a synergetic effect of an electron and ion conductor [156]. The corresponding anode delivered a capacity of 1370 mAh/g at a current rate of 0.5C, a good rate capability, and high capacity retention of 1105 mAh/g after 200 cycles. Cu-coating had a similar effect to Si thin films [157]. Coating Si NWs with a 100 nm-thick coat composed of Ag and a conductive polymer, poly(3,4-ethylenedioxythiophene) (PEDOT), increased the capacity retention to 80% after 100 cycles, compared with 30% before coating [158]. Song et al. reported an alloy-forming approach to convert a-Si-coated copper oxide (CuO) core-shell NWs into hollow and highly interconnected Si-Cu alloy (mixture) [159]. The corresponding anode, without the use of any binder or conductive agent, delivered a capacity of 1010 mAh/g (or 780 mAh/g) after 1000 cycles at 3.4 A/g (or 20 A/g), with a capacity retention rate of $\approx$84% ($\approx$88%). Positive results were also recorded with copper-silicon core-shell nanotube arrays [160]. Porous Si/Cu composite films were fabricated by applying the pre-deposited Cu-nanoparticle-assembled film as the growth direction template for the subsequent deposition of a-Si active layer using the cluster beam deposition technique [161]. This film delivered a capacity of 3124 mAh/g after 1000 cycles at 1 A/g, which was also evidence of the efficiency of porous nanostructured Si-based anodes.

Aluminum coating has a singular effect as it does not improve the coulombic efficiency, but it increases the cycling ability [162] because the Al-coat is stiff and efficiently decreases the mechanical stress in the Si core during cycling [163, 164]. Al_2O_3-coatings act differently. Contrary to the other coating materials we have mentioned, aluminum oxide is insulating. During the first lithiation, Al_2O_3 transforms into an Al-Li-O glass that is a good ionic conductor but remains an electrical insulator such that it has the attributes of an SEI substitute [165]. However, owing to its poor electrical conductivity, its thickness must be smaller. For instance, a 10 nm-thick Al_2O_3 coating obtained using atomic laser deposition was tested on Si thin films [166, 167] and Si NWs [168], for which the coating resulted in a 45% increase in the cycle life. Song et al. combined the advantages of both Al_2O_3 coating and Cu in an Al_2O_3 coated hierarchical copper-silicon (HCS) NW structure [169], which delivered a capacity of ∼830 mAh/g, 85% capacity retention after 800 cycles, and high areal capacity (∼3.5 mAh cm^{-2}) at a high Si mass loading (1.2 mg/cm), with a lifespan of more than 6000 cycles (Fig. 11.14). Jin et al. enclosed Si within a TiO_2 shell thinner than 15 nm to act as an artificial self-healing SEI. The full cell with this anode material, compressed to obtain a tap density of 1.4 g cm^{-3}, against a 3 mAh/cm $LiCoO_2$ cathode, achieved a stabilized areal capacity of 1.6 mAh/cm at the 100th cycle with a stabilized coulombic efficiency of 99.9% [170].

A core-shell Si-SiO_x anode material was also investigated. It can be formed simply by exposing Si to oxygen; this composite has a good cycling ability [19]. Si oxide-coated Si NWs exhibited a reversible capacity of 1503 mAh/g over 560 cycles [20]. Nevertheless, the Si@SiO_x anodes have a high initial irreversibility loss due to the insulating character of silicon oxide and the irreversible reaction between Li^+ and SiO_x [171]. Significantly better results were demonstrated by Chen et al. with porous Si NWs stabilized by a surface oxide layer fabricated using a low-cost metal-assisted chemical etching technique [172]. This Si@SiO_x anode delivered a reversible capacity of 1503 mAh/g at the 560th cycle at a current density of 600 mA/g, demonstrating an average of only 0.04% drops per cycle compared with its initial capacity. This result illustrated the importance of the porosity in the performance of Si-based nanoparticles as anodes. Another solution to limit the initial irreversible loss of capacity was recently proposed by Zhu et al. by preparing an in-situ-formed artificial SEI membrane (Li_2SiO_3) to protect the Si anode by selectively lithiating the Si@SiO_x coating layer [173]. As an anode, Si@Li_2SiO_3 exhibits a high ICE of 89.1%, an excellent rate capability of 959 mAh/g even at 30 A/g, and a long cycling lifespan of 1091 mAh/g at 3 A/g even after 1000 cycles.

The surface clamping of porous silicon structures was also efficient, specifically under the form of Si-C composites. Control of the SEI was obtained by coating double-walled porous nanotubes with rigid carbon [174]. The corresponding anode delivered a capacity of ∼600 mAh g^{-1} and ∼88% capacity retention over 6000 cycles at a 12-C rate. However, note that the loading was small. Si-C core-shell nanoparticles were designed with a void space between

Advanced Si-based electrodes for high-energy Li-ion batteries Chapter | 11 443

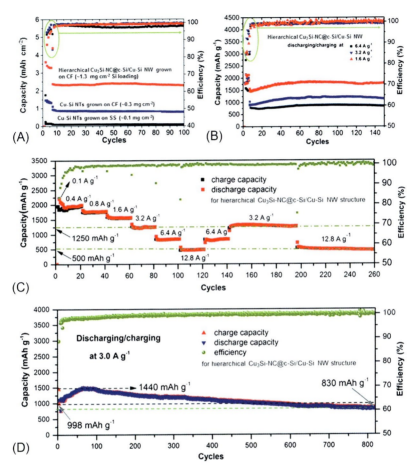

FIG. 11.14 Electrochemical performance of the HCS NW anode with a high Si loading of ~1.2 to 1.5 mg/cm. (A) Areal capacities and the corresponding coulombic efficiencies of the HCS NW structure grown overall on the Cu foam, a Cu-Si nanotube on the Cu foam, and a Cu/a-Si NW core-shell structure on the SS substrate. (B) Discharge capacities and corresponding coulombic efficiencies of the HCS NW anodes with a discharge-charge current of 1.6, 3.2, and 6.4 A/g over 150 cycles, respectively. (C) Rate performance from the 0.1–12.8 A/g discharge-charge current. (D) Long performance of the HCS NW hybrid anodes. *(Reproduced with permission from H. C. Song, S. Wang, X. Y. Song, H. F. Yang, G. H. Du, L. W. Yu, J. Xu, P. He, H. S. Zhou, K. J. Chen, A bottom-up synthetic hierarchical buffer structure of copper silicon nanowire hybrids as ultra-stable and high-rate lithium-ion battery anodes. J. Mater. Chem. A 6 (2018) 7877–7886. Copyright 2018 Royal Society of Chemistry.)*

the Si core and the carbon shell to accommodate the volume change of the Si particle without damaging the carbon shell. Liu et al. fabricated pomegranate-inspired Si/C clusters designed and fabricated through micro-emulsion and step-growth polymerization methods [69]. Here, the improvement in the cycling ability was due to the multifunctional carbon component that acted as an electrolyte-blocking layer contributing to stable SEI formation on the outer surface of the microparticles. However, the Si loading was small (0.2 mAh/cm), and small loading means a small volumetric density. Si@void@C was prepared by coating Si with SiO_2 by the hydrolysis of tetraethoxysilane (TEOS), followed by coating C with sucrose [175], glucose [176], polyacrylonitrile or polystyrene [177], polyvinylidene fluoride (PVDF) [178], polydopamine [179, 180], and resorcinol-formaldehyde (RF) resin as carbon sources [69]. After carbonization, the hard templates (SiO_2) were etched with an HF solution, leaving void spaces to form the yolk or shell structure. Recently, Huang et al. succeeded in avoiding the use of HF, which is a dangerous product. Instead, they used polystyrene (PS) and polyaniline (PANI) as the pore and the C-shell sources, respectively. This Si@C@void@C composite was cycled at 1000 mA/g, exhibiting a reversible capacity of approximately 630 mAh/g, and no capacity degradation was observed after 500 cycles [181].

A new yolk-shell structured Si/C anode used a rigid carbon-coated SiO_2 shell to better confine the incorporated multiple Si nanoparticles. In addition, Fe_2O_3 nanoparticles embedded in flexible CNT networks were filled into the YS-Si/C microspheres as a conductive "highway" to bridge the void between the inner Si-yolks and outer C/SiO_2 double-shell [182]. The average thicknesses of the outer C-shell and inner SiO_2-shell were 60 and 200 nm, respectively. The size distributions of CNTs and Fe_2O_3 nanoparticles were both between 50 and 60 nm. As an anode, this YS-Si/C structure delivered a capacity of 1700 mAh g^{-1} at 1 A g^{-1} and 1500 mAh g^{-1} at 2 A g^{-1}, with over 95% reversible capacity maintained after 450 cycles at 0.5 A g^{-1}. Recently, with the use of RF and TEOS as precursors, a Si/C yolk-shell composite was obtained, with Si nanoparticles (80–100 nm) well encapsulated in a ~12-nm thick hollow carbon shell and a 60 nm thick void space between the silicon core and outer carbon layer [183]. This composite delivered a capacity of 1807 mAh/g and retained 999 mAh/g at 100 mA/g beyond 100 cycles. Furthermore, a reversible capacity of 500 mAh/g after 2000 cycles at 1 A/g was demonstrated. A Si@NG@void@C-gel (SNGC-gel), where NG is N-doped graphene, was obtained by combining the conductive polymer pyrrole (py) and graphene. This structure delivered a stable reversible specific capacity of 1480.3 mAh/g at a current density of 2 A/g after 400 cycles. A long cycling life of more than 3000 cycles was demonstrated at high current densities (494.5 mAh/g at 10 A/g and 366.8 mAh/g at 20 A/g) [184].

11.8 Conclusion

Despite many advantages of silicon, including the existence of a well-developed industrial fabrication for electronics, cheap price, and a very high specific capacity, its use as an anode for Li-ion batteries has been impeded by several drawbacks, primarily the large increase in volume during lithiation, which results in cracking or even pulverization of the Si particles. To remedy this problem, many efforts have been conducted to prepare nano-sized Si particles because small particles are less prone to cracking. Considering nano-sized silicon-based anodes, we note that large amounts of binder and carbon conductive materials are necessary to obtain high specific capacities in nano-sized Si anodes, which means decreased volumetric energy density. Until recently, the current manufacturing processes for mixing binder and carbon conductive materials with the active materials could not effectively overcome the severe agglomeration of nano-sized Si. In addition, conductive polymers can adhere firmly to the Si-based composites and avoid the cracking of the solid-electrolyte interface (SEI) that would expose the composites to fresh surface contact and create a thick and resistive SEI. Significantly large amounts of binder and carbon conductive materials were necessary to obtain high specific capacities in nano-sized Si anodes, which means decreased volumetric energy densities.

However, progress in recent years with conductive polymer binders has enabled Si-based anodes without significant aggregation to be obtained. Graphite-based anodes can typically store a charge of 4 mAh/cm^2, as they use 50 μm-thick films on the current collector. Therefore, for a-Si capacity of 3800 mAh/g, an areal mass of 1 mg/cm^2 of Si is required to be competitive with graphite anodes. Some Si-based anodes have an areal mass twice larger using nano-sized Si and have been tested successfully over 1000 cycles. However, this performance can be only achieved by preparing the nano-Si with appropriate geometries and structures. Long cycle life and high rate performance require composites derived from nanostructured Si, with a core-shell or yolk-shell structure construction, and a combination with graphene in particular. Porosity is also a key parameter to alleviate the change in volume of Si during cycling. The best results have been obtained by controlling the porosity in the aforementioned composites.

Parallel to the construction of nano-sized Si-based anodes, the search for micro-sized particles has experienced a renewed vitality and, from a commercial perspective, has a practical interest. An improved cycling ability was achieved with structures that incorporate nano-sized Si onto a micrometer-sized host, such as graphite. The advantages of micro-sized Si over nano-sized Si are the higher capacity and Si-loading of the corresponding anode. The drawback is a lower C-rate. In addition, the synthesis process is more traditional and, thus, less expensive. In practice, combining a high-rate capability and high-energy density in a single battery is difficult, such that both micro- and nanostructures might have commercial opportunities, depending on which parameter should be favored according to the use envisioned for the cells.

References

[1] H. Wu, G. Zheng, N. Liu, T.J. Carney, Y. Yang, Y. Cui, Engineering empty space between Si nanoparticles for lithium-ion battery anodes, Nano Lett. 12 (2012) 904–909.

[2] K. Yoo, J. Kim, Y.S. Jung, K. Kang, Scalable fabrication of silicon nanotubes and their application to energy storage, Adv. Mater. 24 (2012) 5452–5456.

[3] J.R. Szczech, S. Jin, Nanostructured silicon for high capacity lithium battery anodes, Energy Environ. Sci. 4 (2011) 56–72.

[4] M.A. Rahman, G. Song, A.I. Bhatt, Y.C. Wong, C. Wen, Nanostructured silicon anodes for high-performance lithium-ion batteries, Adv. Funct. Mater. 26 (2016) 647–678.

[5] K.T. Lee, J. Cho, Roles of nanosize in lithium reactive nanomaterials for lithium ion batteries, Nano Today 6 (2011) 28–41.

[6] L. Su, Y. Jing, Z. Zhou, Li ion battery materials with core–shell nanostructures, Nanoscale 3 (2011) 3967–3983.

[7] R. Yi, M.L. Gordin, D. Wang, Integrating Si nanoscale building blocks into micro-sized materials to enable practical applications in lithium-ion batteries, Nanoscale 8 (2016) 1834–1848.

[8] M. Ashuri, Q.R. He, L.L. Shaw, Silicon as a potential anode material for Li-ion batteries: where size, geometry and structure matter, Nanoscale 8 (2016) 74–103.

[9] K. Feng, M. Li, W.W. Liu, A.G. Kashkooli, X.C. Xiao, M. Cai, Z.W. Chen, Silicon-based anodes for lithium-ion batteries: from fundamentals to practical applications, Small 14 (2018) 1702737.

[10] J. Ryu, T. Bok, S. Kim, S. Park, Fundamental understanding of nanostructured Si electrodes: preparation and characterization, Chem. Nano Mat. 4 (2018) 319–337.

[11] V. Dosaj, M. Kroupa, R. Bittar, Silicon and silicon alloys, chemical and metallurgical, in: Kirk-Othmer Encyclopedia of Chemical Technology, John Wiley & Sons Inc., 2000.

[12] W.R. Runyan, Silicon, in: Kirk-Othmer Encyclopedia of Chemical Technology, John Wiley & Sons Inc., 2000.

[13] A. Schei, J.K. Tuset, H. Tveit, Production of High Silicon Alloys, Akademika Publishing, 1998.

[14] J. Li, G. Chen, P. Zhang, W. Wang, J. Duan, Technical challenges and progress in fluidized bed chemical vapor deposition of polysilicon, Chin. J. Chem. Eng. 19 (2011) 747–753.

[15] W.O. Filtvedt, A. Holt, P.A. Ramachandran, M.C. Melaaen, Chemical vapor deposition of silicon from silane: review of growth mechanisms and modeling/scaleup of fluidized bed reactors, Sol. Energy Mater. Sol. Cells 107 (2012) 188–200.

[16] S. Masatomo, T. Akizuki, K. Itaka, M. Kubota, K. Tsubouchi, T. Ishigaki, H. Koinuma, Effect of hydrogen radical on decomposition of chlorosilane source gases, J. Phys. Conf. Ser. 441 (2013), 012003.

[17] Y. Jiao, A. Salce, W. Ben, F. Jiang, X. Ji, E. Morey, D. Lynch, Siemens and Siemens-like processes for producing photovoltaics: energy payback time and lifetime carbon emissions, JOM 63 (2011) 28–31.

[18] K. Yasuda, T. Okabe, Solar-grade silicon production by metallothermic reduction, JOM 62 (2010) 94–101.

[19] H. Li, X. Huang, L. Chen, Z. Wu, Y. Liang, A high capacity Nano Si composite anode material for lithium rechargeable batteries, Electrochem. Solid-State Lett. 2 (1999) 547–549.

[20] A.A. Griffith, VI. The phenomena of rupture and flow in solids, Phil. Trans. R. Soc. London 221 (1921) 163.

[21] J.P. Baïlon, J.M. Dorlot, Des matériaux, Presses Intern. Polytechnique, 2000.

[22] J.J. Gilman, Direct measurements of the surface energies of crystals, J. Appl. Phys. 31 (1960) 2208–2218.

[23] X.H. Liu, L. Zhong, S. Huang, S.X. Mao, T. Zhu, J.Y. Huang, Size-dependent fracture of silicon nanoparticles during lithiation, ACS Nano 6 (2012) 1522–1531.

[24] K. Zaghib, P. Charest, M. Dondigny, A. Guerfi, M. Lagacé, A. Mauger, M. Kopec, C.M. Julien, LiFePO$_4$: from molten ingot to nanoparticles with high-rate performance in Li-ion batteries, J. Power Sources 195 (2010) 8280–8288.
[25] B.A. Wills, J. Finch, Wills' Mineral Processing Technology: An Introduction to the Practical Aspects of Ore Treatment and Mineral Recovery, Elsevier Science, 2015.
[26] K. Zaghib, Electrode architecture-assembly of battery materials and electrodes, in: Advanced Battery Materials Research (BMR) Program. FY 2016—Q2, U.S. Department of Energy, 2015, pp. 6–8.
[27] K. Zaghib, Electrode architecture-assembly of battery materials and electrodes, in: Advanced Battery Materials Research (BMR) Program. FY 2016—Q3, U.S. Department of Energy, 2015, pp. 8–10.
[28] D. Leblanc, R. Dolbec, A. Guerfi, J. Guo, P. Hovington, M. Boulos, K. Zaghib, Silicon nanopowder synthesis by inductively coupled plasma as anode for high-energy Li-ion batteries: Arrays, functional materials, and industrial nanosilicon, in: Silicon nanomaterials Sourcebook, 2017, pp. 463–484.
[29] J. Guo, R. Dolbec, M. Boulos, D. Leblanc, A. Guerfi, K. Zaghib, Nanoparticles comprising a core covered with a passivation layer, process for manufacture and uses thereof, Word Patent WO2018,157,256, 7 September 2018.
[30] K.-S. So, H. Lee, T.H. Kim, S. Choi, D.W. Park, Synthesis of silicon nanopowder from silane gas by RF thermal plasma, Phys. Status Solidi A 211 (2014) 310–315.
[31] K. Sony, B.W. Sheldon, X. Xiao, A.F. Bower, M.W. Verbrugge, Diffusion mediated lithiation stresses in Si thin film electrode batteries and energy storage, J. Electrochem. Soc. 159 (2012) A1520–A1527.
[32] K. Zaghib, A. Guerfi, D. Leblanc, Particulate anode materials and methods for their preparation, US Patent 9,559,355, 31 January 2017.
[33] D. Leblanc, P. Hovington, C. Kim, A. Guerfi, D. Bélanger, K. Zaghib, Silicon as anode for high-energy lithium ion batteries: from molten ingot to nanoparticles, J. Power Sources 299 (2015) 529–536.
[34] T. Takamura, S. Ohara, M. Uehara, J. Suzuki, K. Sekine, A vacuum deposited Si film having a Li extraction capacity of over 2000 mAh g^{-1} with a long cycle life, J. Power Sources 129 (2004) 96–100.
[35] S. Ohara, J. Suzuki, K. Sekine, T. Takamura, Li insertion/extraction reaction at a Si film evaporated on a Ni foil, J. Power Sources 119–121 (2003) 591–596.
[36] L.B. Chen, J.Y. Xie, H.C. Yu, T.H. Wang, An amorphous Si thin film anode with high capacity and long cycling life for lithium ion batteries, J. Appl. Electrochem. 39 (8) (2009) 1157–1162.
[37] X. Zhu, J. Shen, X. Chen, Y. Li, W. Peng, G. Zhang, F. Zhang, X. Fan, Enhanced cycling performance of Si-MXene nanohybrids as anode for high performance lithium ion batteries, J. Chem. Eng. 378 (2019) 122212.
[38] S. Suresh, Z.P. Wu, S.F. Bartolucci, S. Basu, R. Mukherjee, T. Gupta, P. Hundekar, Y.F. Shi, T.M. Lu, N. Koratkar, Protecting silicon film anodes in lithium-ion batteries using an atomically thin graphene drape, ACS Nano 11 (2017) 5051–5061.
[39] H. Wu, Y. Cui, Designing nanostructured Si anodes for high energy lithium ion batteries, Nano Today 7 (2012) 414–429.
[40] H.T. Nguyen, F. Yao, M.R. Zamfir, C. Biswas, K.P. So, Y.H. Lee, S.M. Kim, S.N. Cha, J.M. Kim, D. Pribat, Highly interconnected Si nanowires for improved stability Li-ion battery anodes, Adv. Energy Mater. 1 (2011) 1154–1161.

[41] V. Schmidt, J.V. Vittemann, U. Gösele, Growth, thermodynamics and electrical properties of silicon nanowires, Chem. Rev. 110 (2010) 361–388.

[42] H. Kim, B. Han, J. Choo, J. Cho, Three-dimensional porous silicon particles for use in high-performance lithium secondary batteries, Angew. Chem. Int. Ed. 47 (2008) 10151–10154.

[43] Y. Yao, M.T. McDowell, I. Ryu, H. Wu, N. Liu, L. Hu, W.D. Nix, Y. Cui, Interconnected silicon hollow nanospheres for lithium-ion battery anodes with long cycle life, Nano Lett. 11 (2011) 2949–2954.

[44] X.L. Wang, W.Q. Han, Graphene enhances Li storage capacity of porous single-crystalline silicon nanowires, ACS Appl. Mater. Interfaces 2 (2010) 3709–3713.

[45] J.P. Rong, C. Masarapu, J. Ni, Z.J. Zhang, B.Q. Wei, Tandem structure of porous silicon film on single-walled carbon nanotube macrofilms for lithium-ion battery applications, ACS Nano 4 (2010) 4683–4690.

[46] J. Guo, A. Sun, C. Wang, A porous silicon-carbon anode with high overall capacity on carbon fiber current collector, Electrochem. Commun. 12 (2010) 981–984.

[47] X. Li, M. Gu, H. Shenyang, R. Kennard, P. Yan, X. Chen, C. Wang, M.J. Sailor, J.-G. Zhang, J. Liu, Mesoporous silicon sponge as an anti-pulverization structure for high-performance lithium-ion battery anodes, Nat. Commun. 5 (2014) 4105.

[48] B.M. Bang, H. Kim, H.K. Song, J. Cho, S. Park, Scalable approach to multi-dimensional bulk Si anodes via metal-assisted chemical etching, Energy Environ. Sci. 4 (2011) 5013–5019.

[49] M. Ge, X. Fang, J. Rong, C. Zhou, Review of porous silicon preparation and its application for lithium-ion battery anodes, Nanotechnology 24 (2013) 422001.

[50] M. Ge, J. Rong, X. Fang, A. Zhang, Y. Lu, C. Zhou, Scalable preparation of porous silicon nanoparticles and their application for lithium-ion battery anodes, Nano Res. 6 (2013) 174–181.

[51] K.Q. Peng, J.J. Hu, Y.J. Yan, Y. Wu, H. Fang, Y. Xu, S.T. Lee, J. Zhu, Fabrication of single-crystalline silico nanowires by scratching a silicon surface with catalytic metal particles, Adv. Funct. Mater. 16 (2006) 387–394.

[52] Y. Zhao, X.Z. Liu, H.Q. Li, T.Y. Zhai, H.S. Zhou, Hierarchical micro/nano porous silicon Li-ion battery anodes, Chem. Commun. 48 (2012) 5079–5081.

[53] B.M. Bang, J.I. Lee, H. Kim, J. Cho, S. Park, High-performance microporous bulk silicon anodes synthesized by template-free chemical etching, Adv. Energy Mater. 2 (2012) 878–883.

[54] X. Zhu, S.H. Cho, R. Tao, X. Jia, Y. Lu, Building high-rate silicon anodes based on hierarchical Si@C@CNT nanocomposite, J. Alloys Compd. 791 (2019) 1105–1113.

[55] X.Y. Yue, W. Sun, J. Zhang, F. Wang, K.N. Sun, Facile synthesis of 3D silicon/carbon nanotube capsule composites as anodes for high-performance lithium-ion batteries, J. Power Sources 329 (2016) 422–427.

[56] M. Thakur, R.B. Pernites, N. Nitta, M. Isaacson, S.L. Sinsabaugh, M.S. Wong, S.L. Biswal, Freestanding macroporous silicon and pyrolysed polyacylonitrile as a composite anode for lithium ion batteries, Chem. Mater. 24 (2012) 2998–3003.

[57] M. Thakur, M. Isaacson, S.L. Sinsabaugh, M.S. Wong, S.L. Biswal, Gold-coated porous silicon films as anodes for lithium-ion batteries, J. Power Sources 205 (2012) 426–432.

[58] H. Kim, J. Cho, Superior lithium electroactive mesoporous Si-carbon core-shell nanowires for lithium battery anode material, Nano Lett. 8 (2008) 3688–3691.

[59] J. Entwistle, A. Rennie, S. Patwardhan, A review of magnesiothermic reduction of silica to porous silicon for lithium-ion battery applications and beyond, J. Mater. Chem. A 38 (2018) 18344–18356.

[60] Y. Cen, Y. Fan, Q. Qin, R.D. Sisson, D. Peleian, J. Liang, Synthesis of Si anode with a microsized-branched structure from recovered Al scrap for use in Li-ion batteries, J. Power Sources 410–411 (2019) 31–37.

[61] D.A. Agyeman, K. Song, G.-H. Lee, M. Park, Y.-M. Kang, Carbon-coated Si nanoparticles anchored between reduced graphene oxides as an extremely reversible anode material for high energy-density Li-ion battery, Adv. Energy Mater. 6 (2016) 1600904.

[62] X. Feng, J. Yang, Y. Bie, J. Wang, Y. Nuli, W. Lu, Nano/micro-structured Si/CNT/C composite from nano-SiO$_2$ for high power lithium ion batteries, Nanoscale 6 (2014) 12532–12539.

[63] X. Gao, J. Li, Y. Xie, D. Guan, C. Yuan, A multilayered silicon-reduced graphene oxide electrode for high performance lithium-ion batteries, ACS Appl. Mater. Interfaces 7 (2015) 7855–7862.

[64] D.S. Jung, T.H. Hwang, S.B. Park, J.W. Choi, Spray drying method for large-scale and high-performance silicon negative electrodes in Li-ion batteries, Nano Lett. 13 (2013) 2092–2097.

[65] J.I. Lee, N.S. Choi, S. Park, Highly stable Si-based multicomponent anodes for practical use in lithium-ion batteries, Energy Environ. Sci. 5 (2012) 7878–7882.

[66] D.C. Lin, Z.D. Lu, P.C. Hsu, H.R. Lee, N. Liu, J. Zhao, H.T. Wang, C. Liu, Y. Cui, A high tap density secondary silicon particle anode fabricated by scalable mechanical pressing for lithium-ion batteries, Energy Environ. Sci. 8 (2015) 2371–2376.

[67] B. Liu, P. Soares, C. Checkles, Y. Zhao, G. Yu, Three-dimensional hierarchical trnary nanostructures for high-performance Li-ion battery anodes, Nano Lett. 13 (2013) 3414–3419.

[68] J. Liu, P. Kopold, P.A. van Aken, J. Maier, Y. Yu, Energy storage materials from nature through nanotechnology: a sustainable route from reed plants to a silicon anode for lithium-ion batteries, Angew. Chem. Int. Ed. 54 (2015) 9632–9636.

[69] N. Liu, Z. Lu, J. Zhao, M.T. McDowell, H.W. Lee, W. Zhao, Y. Cui, A pomegranate-inspired nanoscale design for large-volume-change lithium battery anodes, Nat. Nanotechnol. 9 (2014) 187–192.

[70] A. Magasinski, P. Dixon, B. Hertzberg, A. Kvit, J. Ayala, G. Yushin, High-performance lithium-ion anodes using a hierarchical bottom-up approach, Nat. Mater. 9 (2010) 353–358.

[71] C.M. Xiao, N. Du, X.X. Shi, H. Zhang, D.R. Yang, Large-scale synthesis of Si@C three-dimensional porous structures as high-performance anode materials for lithium-ion batteries, J. Mater. Chem. A 2 (2014) 20494–20499.

[72] R. Yi, F. Dai, M.L. Gordin, S.R. Chen, D.H. Wang, Micro-sized Si-C composite with interconnected nanoscale building blocks as high-performance anodes for practical application in lithium ion batteries, Adv. Energy Mater. 3 (3) (2013) 295–300.

[73] Q. Yun, X. Qin, Y.-B. He, W. Lv, Y.V. Kaneti, B. Li, Q.-H. Yang, F. Kang, Micron-sized spherical Si/C hybrids assembled via water/oil system for high-performance lithium ion battery, Electrochim. Acta 211 (2016) 982–988.

[74] Q.B. Yun, X.Y. Qin, W. Lv, Y.B. He, B.H. Li, F.Y. Kang, Q.H. Yang, "Concrete" inspired construction of a silicon/carbon hybrid electrode for high performance lithium ion battery, Carbon 93 (2015) 59–67.

[75] Z. Zhang, Y. Wang, W. Ren, Q. Tan, Y. Chen, H. Li, Z. Zhong, F. Su, Scalable synthesis of interconnected porous silicon/carbon composites by the Rochow reaction as high-performance anodes of lithium ion batteries, Angew. Chem. Int. Ed. 53 (20) (2014) 5165–5169.

[76] J.-Y. Li, Q. Xu, G. Li, Y.-X. Yin, L.-J. Wan, Y.-G. Guo, Research progress regarding Si-based anode materials towards practical application in high energy density Li-ion batteries, Mater. Chem. Front. 1 (2017) 1691–1708.

[77] S.Y. Kim, J. Lee, B.H. Kim, Y.J. Kim, K.S. Yang, M.S. Park, Facile synthesis of carbon-coated silicon/graphite spherical composites for high-performance lithium-ion batteries, ACS Appl. Mater. Interfaces 8 (2016) 12109–12117.

[78] M. Ko, S. Chae, J. Ma, N. Kim, H.-W. Lee, Y. Cui, J. Cho, Scalable synthesis of silicon-nanolayer-embedded graphite for high-energy lithium-ion batteries, Nat. Energy 1 (2016) 16113.

[79] F.F. Cao, J.W. Deng, S. Xin, H.X. Ji, O.G. Schmidt, L.J. Wan, Y.G. Guo, Cu-Si Nanocable arrays as high-rate anode materials for lithium-ion batteries, Adv. Mater. 23 (2011) 4415–4420.

[80] Y.C. Zhang, Y. You, S. Xin, Y.X. Yin, J. Zhang, P. Wang, X.S. Zheng, F.F. Cao, Y.G. Guo, Rice husk-derived hierarchical silicon/nitrogen-doped carbon/carbon nanotube spheres as low-cost and high-capacity anodes for lithium-ion batteries, Nano Energy 25 (2016) 120–127.

[81] Q. Xu, J.Y. Li, J.K. Sun, Y.X. Yin, L.J. Wan, Y.G. Guo, Watermelon-inspired Si/C microspheres with hierarchical buffer structures for densely compacted lithium-ion battery anodes, Adv. Energy Mater. 7 (2017) 1601481.

[82] X.S. Zhou, L.J. Wan, Y.G. Guo, Electrospun silicon nanoparticle/porous carbon hybrid nanofibers for lithium-ion batteries, Small 9 (2013) 2684–2688.

[83] X.S. Zhou, Y.G. Guo, A PEO-assisted electrospun silicon-graphene composite as an anode material for lithium-ion batteries, J. Mater. Chem. A 1 (2013) 9019–9023.

[84] X.S. Zhou, A.M. Cao, L.J. Wan, Y.G. Guo, Spin-coated silicon nanoparticle/graphene electrode as a binder-free anode for high-performance lithium-ion batteries, Nano Res. 5 (2012) 845–853.

[85] Q. Xu, J.Y. Li, Y.X. Yin, Y.M. Kong, Y.G. Guo, L.J. Wan, Nano/micro-structured Si/C anodes with high initial coulombic efficiency in Li-ion batteries, Asian J. Chem. 11 (2016) 1205–1209.

[86] S. Xin, Y.G. Guo, L.J. Wan, Nanocarbon networks for advanced rechargeable Lithium batteries, Acc. Chem. Res. 45 (2012) 1759–1769.

[87] X.L. Wu, Y.G. Guo, L.J. Wan, Rational design of anode materials based on group IVA elements (Si, Ge, and Sn) for Lithium-ion batteries, Chem. Asian J. 8 (2013) 1948–1958.

[88] F. Luo, B. Liu, J. Zheng, G. Chu, K. Zhong, H. Li, X. Huang, L. Chen, Review—nano-silicon/carbon composite anode materials towards practical application for next generation Li-ion batteries, J. Electrochem. Soc. 162 (2015) A2509–A2528.

[89] S. Chae, S.-H. Choi, N. Kim, J. Sung, J. Cho, Integration of graphite and silicon anodes for the commercialization of high-energy lithium-ion batteries, Angew. Chem. Int. Ed. 59 (2020) 110–115.

[90] T. Xu, D. Wang, P. Qiu, J. Zhang, Q. Wang, B.J. Xia, X.H. Xie, *In situ* synthesis of porous Si dispersed in carbon nanotube intertwined expanded graphite for high-energy lithium-ion batteries, Nanoscale 10 (2018) 16638–16644.

[91] H.-C. Tao, L.-Z. Fan, Y. Mei, X. Qu, Self-supporting Si/reduced graphene oxide nanocomposite films as anode for lithium ion batteries, Electrochem. Commun. 13 (2011) 1332–1335.

[92] X. Xin, X. Zhou, F. Wang, X. Yao, X. Xu, Y. Zhu, Z. Liu, A 3D porous architecture of Si/graphene nanocomposite as high-performance anode materials for Li-ion batteries, J. Mater. Chem. 22 (2012) 7724–7730.

[93] X. Zhao, C.M. Hayner, M.C. Kung, H.H. Kung, In-plane vacancy-enabled high-power Si-graphene composite electrode for lithium-ion batteries, Adv. Energy Mater. 1 (2011) 1079–1084.

[94] W. Sun, R. Hu, M. Zhang, J. Liu, M. Zhu, Binding of carbon coated nano-silicon in graphene sheets by wet ball-milling and pyrolysis as high performance anodes for lithium-ion batteries, J. Power Sources 318 (2016) 113–120.

[95] W. Sun, R. Hu, H. Zhang, Y. Wang, L. Yang, J. Liu, M. Zhu, A long-life nano-silicon anode for lithium ion batteries: supporting of graphene nanosheets exfoliated from expanded graphite by plasma-assisted milling, Electrochim. Acta 187 (2016) 1–10.

[96] S. Chen, P. Bao, X. Huang, B. Sun, G. Wang, Hierarchical 3D mesoporous silicon@graphene nanoarchitectures for lithium ion batteries with superior performance, Nano Res. 7 (2013) 85–94.

[97] L. Gan, H. Guo, Z. Wang, X. Li, W. Peng, J. Wang, S. Huang, M. Su, A facile synthesis of graphite/silicon/graphene spherical composite anode for lithium-ion batteries, Electrochim. Acta 104 (2013) 117–123.

[98] K. Feng, W. Ahn, G. Lui, H.W. Park, A.G. Kashkooli, G. Jiang, X. Wang, X. Xiao, Z. Chen, Implementing an in-situ carbon network in Si/reduced graphene oxide for high performance lithium-ion battery anodes, Nano Energy 19 (2016) 187–197.

[99] Z. Luo, Q. Xiao, G. Lei, Z. Li, C. Tang, Si nanoparticles/graphene composite membrane for high performance silicon anode in lithium ion batteries, Carbon 98 (2016) 373–380.

[100] Z.J. Liu, P.Q. Guo, B.L. Liu, W.H. Xie, D.Q. Liu, D.Y. He, Carbon-coated Si nanoparticles/reduced graphene oxide multilayer anchored to nanostructured current collector as lithium-ion battery anode, Appl. Surf. Sci. 396 (2017) 41–47.

[101] L.S. Jiao, J.Y. Liu, H.Y. Li, T.S. Wu, F.H. Li, H.Y. Wang, L. Niu, Facile synthesis of reduced graphene oxide-porous silicon composite as superior anode material for lithium-ion battery anodes, J. Power Sources 315 (2016) 9–15.

[102] W. Zhai, Q. Ai, L.N. Chen, S.Y. Wei, D.P. Li, L. Zhang, P.C. Si, J.K. Feng, L.J. Ci, Walnut-inspired microsized porous silicon/graphene core-shell composites for high-performance lithium-ion battery anodes, Nano Res. 10 (2017) 4274–4283.

[103] X. Yi, W.-J. Yu, M.A. Tsiamtsouri, F. Zhang, W. He, Q. Dai, S. Hu, H. Tong, J. Zheng, B. Zhang, J. Liao, Highly conductive C-Si@G nanocomposite as a high-performance anode material for Li-ion batteries, Electrochim. Acta 295 (2019) 719–725.

[104] https://www.graphene-info.com/global-graphene-group-launches-graphene-silicon-li-ion-battery-anode-material. On line 14 May 2019.

[105] Y. Hwa, C.-M. Park, H.-J. Sohn, Modified SiO as a high performance anode for Li-ion batteries, J. Power Sources 222 (2013) 129–134.

[106] H. Takezawa, K. Iwamoto, S. Ito, H. Yoshizawa, Electrochemical behaviors of nonstoichiometric silicon suboxides (SiO_x) film prepared by reactive evaporation for lithium rechargeable batteries, J. Power Sources 244 (2013) 149–157.

[107] S.S. Suh, W.Y. Yoon, D.H. Kim, S.U. Kwon, J.H. Kim, Y.U. Kim, C.U. Jeong, Y.Y. Chan, S.H. Kang, J.K. Lee, Electrochemical behavior of SiO_x anodes with variation of oxygen ratio for Li-ion batteries, Electrochim. Acta 148 (2014) 111–117.

[108] C. Huang, A. Kim, D.J. Chung, E. Park, N.P. Young, K. Jurkschat, H. Kim, P.S. Grant, Multiscale engineered Si/SiO_x nanocomposite electrodes for lithium-ion batteries using layer-by-layer spray deposition, ACS Appl. Mater. Interfaces 10 (2018) 15624–15633.

[109] X. Zhang, L. Huang, Q. Shen, X. Zhou, Y. Chen, Hollow boron-doped Si/SiO_x nanospheres embedded in the vanadium nitride/nanopore-assisted carbon conductive network for superior lithium storage, ACS Appl. Mater. Interfaces 11 (2019) 45612–45620.

[110] T. Kang, J. Chen, Y. Cui, Z. Wang, H. Xu, Z. Ma, X. Zuo, X. Xiao, J. Nan, Three-dimensional rigidity-reinforced SiOx anodes with stabilized performance using an aqueous multicomponent binder technology, ACS Appl. Mater. Interfaces 11 (29) (2019) 26038–26046.

[111] T. Wang, X. Guo, H. Duan, C. Chen, H. Pang, SiO-based ($0 < x \leq 2$) composites for lithium-ion batteries, Chin. Chem. Lett. 31 (2020) 654–666.

[112] X. Cai, W. Liu, S. Yang, S. Zhang, Q. Gao, X. Yu, J. Li, H. Wang, Y. Fang, Dual-confined SiO embedded in TiO$_2$ shell and 3D carbon nanofiber web as stable anode material for superior lithium storage, Adv. Mater. Interfaces 6 (2019) 1801800.

[113] W. Yang, H. Liu, Z. Ren, N. Jian, M. Gao, Y. Wu, Y. Liu, H. Pan, A novel multi-element, multiphase, and B-containing SiO$_x$ composite as a sable anode material for Li-ion batteries, Adv. Mater. Interfaces 6 (2019) 1801631.

[114] Z.L. Li, H.L. Zhao, P.P. Lv, Z.J. Zhang, Y. Zhang, Z.H. Du, Y.Q. Teng, L.N. Zhao, Z.M. Zhu, Watermelon-like structured SiO$_x$-TiO$_2$@C nanocomposite as a high-performance lithium-ion battery anode, Adv. Funct. Mater. 28 (2018) 1605711.

[115] Q. Yu, P.P. Ge, Z.H. Liu, M. Xu, W. Yang, L. Zhou, D.Y. Zhao, L.Q. Mai, Ultrafine SiO$_x$/C nanospheres and their pomegranate-like assemblies for high-performance lithium storage, J. Mater. Chem. A 6 (2018) 14903–14909.

[116] Q. Xu, J.K. Sun, Z.L. Yu, Y.X. Yin, S. Xin, S.H. Yu, Y.G. Guo, SiO$_x$ encapsulated in graphene bubble film: an ultrastable Li-ion battery anode, Adv. Mater. 30 (2018) 1707430.

[117] H. Zhao, Q. Yang, N. Yuca, M. Ling, K. Higa, V.S. Battaglia, D.Y. Parkinson, V. Srinivasan, G. Liu, A convenient and versatile method to control the electrode microstructure toward high-energy lithium-ion batteries, Nano Lett. 16 (2016) 4686–4690.

[118] X. Liu, H.L. Zhao, J.Y. Xie, P.P. Lv, K. Wang, J.J. Cui, SiO$_x$ (0 < x <= 2) Based anode materials for lithium-ion batteries, Prog. Chem. 27 (2015) 336–348.

[119] Y. Ren, J. Ding, N. Yuan, S. Jia, M. Qu, Z. Yu, Preparation and characterization of silicon monoxide/graphite/carbon nanotubes composite as anode for lithium-ion batteries, J. Solid State Electrochem. 16 (2012) 1453–1460.

[120] M. Mamiya, H. Takei, M. Kikuchi, C. Uyeda, Preparation of fine silicon particles from amorphous silicon monoxide by the disproportionation reaction, J. Cryst. Growth 229 (2001) 457–461.

[121] N. Liu, L. Hu, M.T. McDowell, A. Jackson, Y. Cui, Prelithiated silicon nanowires as an anode for lithium-ion batteries, ACS Nano 5 (2011) 6487–6493.

[122] M.W. Forney, M.J. Ganter, J.W. Staub, R.D. Ridgley, B.J. Landi, Prelithiation of silicon–carbon nanotube anodes for lithium-ion batteries by stabilized lithium metal powder (SLMP), Nano Lett. 13 (2013) 4158–4163.

[123] J. Zhao, Z.D. Lu, H.T. Wang, W. Liu, H.W. Lee, K. Yan, D. Zhuo, D.C. Lin, N. Liu, Y. Cui, Artificial solid electrolyte interphase-protected Li$_x$Si nanoparticles: an efficient and stable prelithiation reagent for lithium-ion batteries, J. Am. Chem. Soc. 137 (2015) 8372–8375.

[124] J. Zhao, H.W. Lee, J. Sun, K. Yan, Y.Y. Liu, W. Liu, Z.D. Lu, D.C. Lin, G.M. Zhou, Y. Cui, Metallurgically lithiated SiO$_x$ anode with high capacity and ambient air compatibility, Proc. Natl. Acad. Sci. U. S. A. 113 (2016) 7408–7413.

[125] Y. Zhang, C.Q. Zhang, S.M. Wu, X. Zhang, C.B. Li, C.L. Xue, B.W. Cheng, High-columbic-efficiency lithium battery based on silicon particle materials, Nanoscale Res. Lett. 10 (2015) 395.

[126] N.H. Yang, Y.S. Wu, J. Chou, H.C. Wu, N.L. Wu, Silicon oxide-on-graphite planar composite synthesized using a microwave-assisted coating method for use as a fast-charging lithium-ion battery anode, J. Power Sources 296 (2015) 314–317.

[127] M. Marinaro, M. Weinberger, M. Wohlfahrt-Mehrens, Toward pre-lithiatied high areal capacity silicon anodes for lithium-ion batteries, Electrochim. Acta 206 (2016) 99–107.

[128] H.J. Kim, S. Choi, S.J. Lee, M.W. Seo, J.G. Lee, E. Deniz, Y.J. Lee, E.K. Kim, J.W. Choi, Controlled prelithiation of silicon monoxide for high performance lithium-ion rechargeable full cells, Nano Lett. 16 (2015) 282–288.

[129] S. Chang, J. Moon, K. Cho, M. Cho, Multiscale analysis of prelithiated silicon nanowire for Li-ion battery, Comput. Mater. Sci. 98 (2015) 99–104.
[130] A. Guerfi, P. Charest, M. Dondigny, J. Trottier, M. Lagacé, P. Hovington, A. Vijh, K. Zaghib, SiO_x–graphite as negative for high energy Li-ion batteries, J. Power Sources 196 (2011) 5667–5673.
[131] I. Kovalenko, B. Zdyrko, A. Magasinski, B. Hertzberg, Z. Milicev, R. Burtovyy, I. Luzinov, G. Yushin, A major constituent of brown algae for use in high-capacity lithium-ion batteries, Science 334 (2011) 75–79.
[132] B. Koo, H. Kim, Y. Cho, K.T. Lee, N.S. Choi, J. Cho, A highly cross-linked polymeric binder for high-performance silicon negative electrodes in lithium-ion batteries, Angew. Chem. Int. Ed. 51 (2012) 8762–8767.
[133] A. Magasinski, B. Zdyrko, I. Kovalenko, B. Hertzberg, R. Burtovyy, C.F. Huebner, T.F. Fuller, I. Luzinov, G. Yushin, Toward efficient binders for Li-ion battery Si-based anodes: polyacrylic acid, ACS Appl. Mater. Interfaces 2 (2010) 3004–3010.
[134] X. Yu, H. Yang, H. Meng, Y. Sun, J. Zheng, D. Ma, X. Xu, Three-dimensional conductive gel network as an effective binder for high-performance Si electrodes in lithium-ion batteries, ACS Appl. Mater. Interfaces 7 (2015) 15961–15967.
[135] S. Choi, T.W. Kwon, A. Coskun, J.W. Choi, Highly elastic binders integrating polyrotaxanes for silicon microparticle anodes in lithium ion batteries, Science 357 (2017) 279–283.
[136] J. Song, M. Zhou, R. Yi, T. Xu, M.L. Gordin, D. Tang, Z. Yu, M. Regula, D. Wang, Interpenetrated gel polymer binder for high-performance silicon anodes in lithium-ion batteries, Adv. Funct. Mater. 24 (2014) 5904–5910.
[137] T. Liu, Q. Chu, C. Yan, S. Zhang, Z. Lin, J. Lu, Interweaving 3D network binder for high-areal-capacity Si anode through combined hard and soft polymers, Adv. Energy Mater. 9 (2019) 1802645.
[138] T.M. Higgins, S.H. Park, P.J. King, C.J. Zhang, N. McEvoy, N.C. Berner, D. Daly, A. Shmeliov, U. Khan, G. Duesberg, V. Nicolosi, J.N. Coleman, A commercial conducting polymer as both binder and conductive additive for silicon nanoparticle-based lithium-ion battery negative electrodes, ACS Nano 10 (2016) 3702–3713.
[139] J.I. Lee, H. Kang, K.H. Park, M. Shin, D. Hong, H.J. Cho, N.R. Kang, J. Lee, S.M. Lee, J.Y. Kim, C.K. Kim, H. Park, N.S. Choi, S. Park, C. Yang, Amphiphilic graft copolymers as a versatile binder for various electrodes of high-performance lithium-ion batteries, Small 12 (2016) 3119–3127.
[140] H. Zhao, N. Yuca, Z. Zheng, Y. Fu, V.S. Battaglia, G. Abdelbast, K. Zaghib, G. Liu, High capacity and high density functional conductive polymer and SiO anode for high-energy lithium-ion batteries, ACS Appl. Mater. Interfaces 7 (2015) 862–866.
[141] L. Wang, T. Liu, X. Peng, W. Zeng, Z. Jin, W. Tian, B. Gao, Y. Zhou, P.K. Chu, K. Huo, Highly stretchable conductive glue for high-performance silicon anodes in advanced lithium-ion batteries, Adv. Funct. Mater. 28 (2018) 1704858.
[142] W. Zeng, L. Wang, X. Peng, T. Liu, Y. Jiang, F. Qin, L. Hu, P.K. Chu, K. Huo, Y. Zhou, Enhanced ion conductivity in conducting polymer binder for high-performance silicon anodes in advanced lithium-ion batteries, Adv. Energy Mater. 8 (2018) 1702314.
[143] Z. Xu, J. Yang, T. Zhang, Y. Nuli, J. Wang, S.-I. Hirano, Silicon microparticle anodes with self-healing multiple network binder, Joule 2 (2018) 950–961.
[144] S. Hu, L. Wang, T. Huang, A. Yu, A conductive self-healing hydrogel binder for high-performance silicon anodes in lithium-ion batteries, J. Power Sources 449 (2020) 227472.

[145] X. Liu, J. Zai, A. Iqbal, M. Chan, N. Ali, R. Qi, R. Xiang, Glycerol-crosslinked PEDOT:PSS as bifunctional binder for Si anodes: improved interfacial compatibility and conductivity, J. Colloid Interface Sci. 565 (2020) 270–277.

[146] Q. Ye, P. Zhang, X. Ao, D. Yao, Z. Lei, Y. Deng, C. Wang, Novel multi-block conductive binder with polybutadiene for Si anodes in lithium-ion batteries, Electrochim. Acta 315 (2019) 58–66.

[147] D.H. Kim, H.A. Lee, Y.B. Song, J.W. Park, S.-M. Lee, Y.S. Jung, Sheet-type Li_6PS_5Cl-infiltrated Si anodes fabricated by solution process for all-solid-state lithium-ion batteries, J. Power Sources 426 (2019) 143–150.

[148] W. Luo, X. Chen, Y. Xia, M. Chen, L. Wang, Q. Wang, W. Li, J. Yang, Surface and interface engineering of silicon-based anode materials for lithium-ion batteries, Adv. Energy Mater. 7 (2017), 1701083.

[149] H.X. Chen, Z.X. Dong, Y.P. Fu, Y. Yang, Silicon nanowires with and without carbon coating as anode materials for lithium-ion batteries, J. Solid State Electrochem. 14 (2010) 1829–1834.

[150] H. Chen, Y. Xiao, L. Wang, Y. Yang, Silicon nanowires coated with copper layer as anode materials for lithium-ion batteries, J. Power Sources 196 (2011) 6657–6662.

[151] R.T. Xu, G. Wang, T.F. Zhou, Q. Zhang, H.P. Cong, S. Xin, J. Rao, C.F. Zhang, Y.K. Liu, Z.P. Guo, Rational design of Si@carbon with robust hierarchically porous custard-apple-like structure to boost lithium storage, Nano Energy 39 (2017) 253–261.

[152] W. Wang, L. Gu, H.L. Qian, M. Zhao, X. Ding, X.S. Peng, J. Sha, Y.W. Wang, Carbon-coated silicon nanotube arrays on carbon cloth as a hybrid anode for lithium-ion batteries, J. Power Sources 307 (2016) 410–415.

[153] S. Jeong, X. Li, J. Zheng, P. Yan, R. Cao, H.J. Jung, C. Wang, J. Liu, J.G. Zhang, Hard carbon coated nano-Si/graphite composite as a high-performance anode for Li-ion batteries, J. Power Sources 329 (2016) 323–329.

[154] G. Fang, X.L. Deng, J.Z. Zou, X.R. Zeng, Amorphous/ordered dual carbon coated silicon nanoparticles as anode to enhance cycle performance in lithium ion batteries, Electrochim. Acta 295 (2019) 498–506.

[155] S.Q. Chen, L.F. Shen, P.A. van Aken, J. Maier, Y. Yu, Dual-functionalized double carbon shells coated silicon nanoparticles for high performance lithium-ion batteries, Adv. Mater. 29 (2017) 1605650.

[156] Q. Ai, P. Zhou, W. Zhai, X.X. Ma, G.M. Hou, X.Y. Xu, L.N. Chen, D.P. Li, L. Chen, L. Zhang, Synergistic double-shell coating of graphene and Li_4SiO_4 on silicon for high performance lithium-ion battery application, Diam. Relat. Mater. 88 (2018) 60–66.

[157] V.A. Sethuraman, K. Kowolik, V. Srivinasan, Increased cycling efficiency and rate capability of copper-coated silicon anodes in lithium-ion batteries, J. Power Sources 196 (2011) 393–398.

[158] Y. Yao, N. Liu, M.T. McDowell, M. Pasta, Y. Cui, Improving the cycling stability of silicon nanowire anodes with conductive polymer coatings, Energy Environ. Sci. 5 (2012) 7927–7930.

[159] H. Song, H.X. Wang, Z. Lin, X. Jiang, L. Yu, J. Xu, Z. Yu, X. Zhang, Y. Liu, P. He, L. Pan, Y. Shi, H. Zhou, K. Chen, Highly connected silicon–copper alloy mixture nanotubes as high-rate and durable anode materials for lithium-ion batteries, Adv. Funct. Mater. 26 (2016) 524–531.

[160] L.M. Sun, X.H. Wang, R.A. Susantyoko, Q. Zhang, Copper-silicon core-shell nanotube arrays for free-standing lithium ion battery anodes, J. Mater. Chem. A 2 (2014) 15294–15297.

[161] L. Lin, Y.T. Ma, Q.S. Xie, L.S. Wang, Q.F. Zhang, D.L. Peng, Copper-nanoparticle-induced porous Si/Cu composite films as an anode for lithium ion batteries, ACS Nano 11 (2017) 6893–6903.

[162] E.L. Memarzadeh, W.P. Kalisvaart, A. Kohandehghan, B. Zahiri, C.M.B. Holt, D. Mitlin, Silicon nanowire core aluminum shell coaxial nanocomposites for lithium ion battery anodes grown with and without a TiN interlayer, J. Mater. Chem. 22 (2012) 6655–6668.

[163] I. Ryu, J.W. Choi, Y. Cui, W.D. Nix, Size-dependent fracture of Si-nanowire battery anodes, J. Mech. Phys. Solids 59 (2011) 1717–1730.

[164] X.H. Liu, H. Zheng, L. Zhong, S. Huang, K. Karki, L.Q. Zhang, Y. Liu, A. Kushima, W.T. Liang, J.W. Wang, J.H. Cho, E. Epstein, S.A. Dayeh, S.T. Picraux, T. Zhu, J. Li, J.P. Sullivan, J. Cumings, C. Wang, S.X. Mao, Z.Z. Ye, S. Shang, J.H. Huang, Anisotropic swelling and fracture of silicon nanowires during lithiation, Nano Lett. 11 (2011) 3312–3318.

[165] Y. Liu, N.S. Hudak, D.L. Huber, S.J. Limmer, J.P. Sullivan, J.Y. Huang, In situ transmission electron microscopy observation of pulverisation of aluminum nanowires and evolution of the thin surface Al_2O_3 layers during lithiation-delithiation cycles, Nano Lett. 11 (2011) 4188–4194.

[166] X. Xiao, P. Lu, J.R. Dahn, Ultrathin multifunctional oxide coatings for lithium ion batteries, Adv. Mater. 23 (2011) 3911–3915.

[167] Y. He, X. Yu, Y. Wang, H. Li, X. Huang, Alumina-coated patterned amorphous silicon as the anode for a lithium-ion battery with high coulombic efficiency, Adv. Mater. 23 (2011) 4938–4941.

[168] H.T. Nguyen, M.R. Zamfir, L.D. Duong, Y.H. Lee, P. Bondavalli, D. Pribat, Alumin-coated silicon-based nanowire arrays for high quality lithium-ion battery anodes, J. Mater. Chem. 22 (2012) 24618–24626.

[169] H.C. Song, S. Wang, X.Y. Song, H.F. Yang, G.H. Du, L.W. Yu, J. Xu, P. He, H.S. Zhou, K.J. Chen, A bottom-up synthetic hierarchical buffer structure of copper silicon nanowire hybrids as ultra-stable and high-rate lithium-ion battery anodes, J. Mater. Chem. A 6 (2018) 7877–7886.

[170] Y. Jin, S. Li, A. Kushima, X.Q. Zheng, Y.M. Sun, J. Xie, J. Sun, W.J. Xue, G.M. Zhou, J. Wu, Self-healing SEI enables full-cell cycling of a silicon-majority anode with a coulombic efficiency exceeding 99.9%, Energy Environ. Sci. 10 (2017) 580–592.

[171] P. Hovington, M. Dontigny, A. Guerfi, J. Trottier, M. Lagacé, A. Mauger, C.M. Julien, K. Zaghib, In situ scanning electron microscope study and microstructural evolution of nano silicon anode for high energy Li-ion batteries, J. Power Sources 248 (2014) 457–464.

[172] Y. Chen, L.F. Liu, J. Xiong, T.Z. Yang, Y. Qin, C.L. Yan, Porous Si nanowires from cheap metallurgical silicon stabilized by a surface oxide layer for lithium ion batteries, Adv. Funct. Mater. 25 (2015) 6701–6709.

[173] Y. Zhu, W. Hu, J. Zhou, W. Cai, Y. Lu, J. Liang, X. Li, S. Zhu, Q. Fu, Y. Qian, Prelithiated surface oxide layer enabled high-performance Si anode for Lithium storage, ACS Appl. Mater. Interfaces 11 (20) (2019) 18305–18312.

[174] H. Wu, G. Chan, J.W. Choi, I. Ryu, Y. Yao, M.T. McDowell, S.W. Lee, A. Jackson, Y. Yang, L.B. Hu, Y. Cui, Stable cycling of double-walled silicon nanotube battery anodes through solid-electrolyte interphase control, Nat. Nanotechnol. 7 (2012) 310–315.

[175] X.Y. Zhou, J.J. Tang, J. Yang, J. Xie, L.L. Ma, Silicon@carbon hollow core–shell heterostructures novel anode materials for lithium ion batteries, Electrochim. Acta 87 (2013) 663–668.

[176] Z. Sun, S.Y. Tao, X.F. Song, P. Zhang, L. Gao, A silicon/double-shelled carbon yolk-like nanostructure as high-performance anode materials for lithium-ion battery, J. Electrochem. Soc. 162 (2015) A1530–A1536.

[177] X.C. Xiao, W.D. Zhou, Y.N. Kim, I. Ryu, M. Gu, C.M. Wang, G. Liu, Z.Y. Liu, H.J. Gao, Regulated breathing effect of silicon negative electrode for dramatically enhanced performance of Li-ion battery, Adv. Funct. Mater. 25 (2015) 1426–1433.

[178] L.Y. Yang, H.Z. Li, J. Liu, Z.Q. Sun, S.S. Tang, M. Lei, Dual yolk-shell structure of carbon and silica-coated silicon for high-performance lithium-ion batteries, Sci. Rep. 5 (2015) 10908.

[179] N. Liu, H. Wu, M.T. McDowell, Y. Yao, C.M. Wang, Y. Cui, A yolk-shell design for stabilized and scalable Li-ion battery alloy anodes, Nano Lett. 12 (2012) 3315–3321.

[180] L. Pan, H.B. Wang, D.C. Gao, S.Y. Chen, L. Tan, L. Li, Facile synthesis of yolk–shell structured Si–C nanocomposites as anodes for lithium-ion batteries, Chem. Commun. 50 (2014) 5878–5880.

[181] X. Huang, X. Sui, H. Yang, R. Ren, Y. Wu, X. Guo, J. Chen, HF-free synthesis of Si/C yolk/shell (YS-Si/C) particles for lithium-ion batteries, J. Mater. Chem. A 6 (2018) 2593–2599.

[182] L. Zhang, C. Wang, Y. Dou, N. Cheng, D. Cui, Y. Du, P. Liu, M. Al-Mamun, S. Zhang, H. Zhao, A yolk–shell structured silicon anode with superior conductivity and high tap density for full lithium-ion batteries, Angew. Chem. Int. Ed. 58 (26) (2019) 8824–8828.

[183] W. Zhang, J. Li, P. Guan, C. Lv, C. Yang, N. Han, X. Wang, G. Song, Z. Peng, One-pot sol-gel synthesis of Si/C yolk-shell anodes for high performance lithium-ion batteries, J. Alloys Compd. 835 (2020) 155135. Available online 15 April 2020.

[184] J. Wu, J. Liu, Z. Wang, X. Gong, M. Qi, Y. Wang, N-doped gel-structures for construction of long cycling Si anodes at high current densities for high performance lithium-ion batteries, J. Mater. Chem. A 6 (2019) 11347–11354.

Chapter 12

Batteries for integrated power and CubeSats: Recent developments and future prospects

Aloysius F. Hepp[a], Prashant N. Kumta[b,c,d], Oleg I. Velikokhatnyi[c], and Ryne P. Raffaelle[e]

[a]*Nanotech Innovations, LLC, Oberlin, OH, United States,* [b]*Department of Chemical and Petroleum Engineering, University of Pittsburgh, Pittsburgh, PA, United States,* [c]*Department of Bioengineering, University of Pittsburgh, Pittsburgh, PA, United States,* [d]*Mechanical Engineering and Materials Science, University of Pittsburgh, Pittsburgh, PA, United States,* [e]*Department of Physics, Rochester Institute of Technology, Rochester, NY, United States*

12.1 Introduction

Rechargeable batteries are ideal for numerous applications: electronics, electric vehicles, stationary energy storage, and aerospace systems [1–3]. Lithium-ion batteries (LIBs) are prevalent among rechargeable options due to their high energy density, long cycle life, and environmental compatibility [4–8]. Graphite storage in the form of LiC_6 has been known to be a potential anode material for decades [9]. A recognition of severe shortcomings with graphite anodes by the 1990s resulted in an intense level of activity to explore multiple approaches for employing silicon-based materials as anodes in a variety of LIB structures [10–18]. However, LIBs are known to have potential safety issues [19,20], there are also issues related to Si anode irreversible swelling [21–23]. Recent reviews highlight attempts to address these issues [24–27]. Over the past several decades, our research groups have examined aspects of advanced (nano)materials, including Si alloys for aerospace power applications [2,15,28–32]. These applications present unique challenges such as temperature fluctuations, rapid gravitational fluctuations, high-energy particles and radiation environments, atomic oxygen, hard-ultraviolet light, thermal management and the necessity or weight- and space savings [29,33–36].

The first topic covered in this chapter is the Starshine 3 integrated (micro-) power source (system) I(M)PS experiment and successful in-space flight testing

[29,37–40]. Integrated power technologies that involve energy storage and energy conversion address several concerns about limitations imposed upon space flight hardware and systems related to power, space, and mass limitations. We continue with a summary of integrated power technology studies and reviews from the years 2005–2020 [29,41–44]. The final topic that we address in this chapter is the use of batteries for energy storage in challenging aerospace applications and environments [45–48], particularly CubeSats [49–51].

It is important to note at the outset that while a variety of battery technologies could be employed, Li-ion batteries with Si-containing anodes may certainly be the optimal choice in certain situations. The important take-home lesson of this chapter is the criticality of energy storage for future advanced (space) power systems, especially for small platforms, including potential future applications for advanced (integrated) technologies with energy conversion, other structures, and electronic devices.

12.2 Integrated micropower source in space: Starshine 3

By the late 1990s, it was clear that integrating power with other active devices was potentially enabling numerous applications such as biomedicine [52–54] advanced electronics for communications [55], and wearable devices [56]. In this section, we discuss the background to efforts at NASA Glenn Research Center in the late 1990s to develop integrated power technologies for small satellites. An integrated power device was successfully demonstrated in low-Earth orbit. In the next section, we address follow-on efforts by numerous other researchers to develop (Li–ion) batteries integrated with solar energy conversion and other technologies.

12.2.1 Integrated power technologies: Multi-junction devices and motivation

While developing a lower-temperature process to enable the growth of ternary chalcogenide solar cells on lightweight substrates such as Kapton® [57–59], we investigated the viability of multi-junction thin-film solar cells (Fig. 12.1) to potentially drive the AM0 efficiency of these devices towards 20% [59–62]. At the same time, we were investigating carbon nanotubes and Si-based materials for use as anodes in Li-ion batteries [2, 15, 28, 30, 31, 63]. We recognized that we could develop a related device structure of a solar cell integrated with a battery (Fig. 12.2); we subsequently initiated an (internally-funded) effort to develop an integrated (micro-)power source or system (IMPS or IPS) [28–30, 40, 41, 64–66].

Integrated power devices (or systems), combining three traditionally separate power system components (energy conversion and storage and control), would be ideal for scientific payloads, experiments, and smaller satellites. The volume formerly required by traditionally separately located chemical

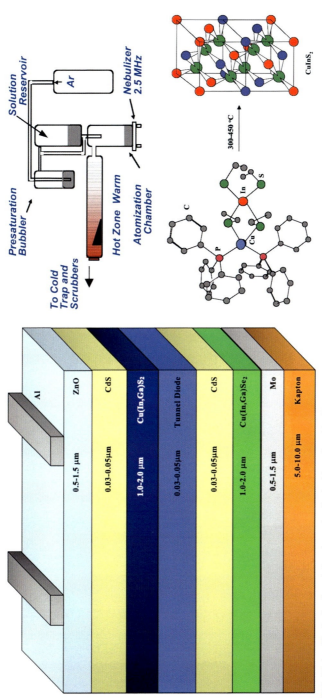

FIG. 12.1 Clockwise from left: schematic of a dual junction thin-film solar cell, diagram of atmosphere-assisted chemical spray pyrolysis or CVD system, diagram of single-source precursor decomposing into (ternary chalcopyrite) CuInS$_2$ solar cell material. (*Courtesy NASA.*)

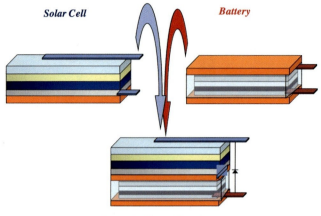

FIG. 12.2 Schematic of integrated power system that combines photovoltaic power generation with lithium ion battery storage in a single autonomous device. Provides continuous power under varying or cyclic illumination. Eliminates design constraints imposed by integrating all power requirements on a single centralized electrical bus. *(Courtesy NASA.)*

batteries frees up valuable space for other systems or an increased payload. Another benefit of using an I(M)PS as a distributed power system is a reduction in spacecraft complexity, especially with respect to power distribution wiring, simplifying spacecraft integration. Integrated power systems could also be used in specialized instances, serving as decentralized (or distributed) power sources or uninterruptible power supplies for discrete components. Of course, this requires components to be located such that they have a view of the Sun for at least during some portion of the orbit. The use of IMPS for multichip module sensors that may be placed wherever they are needed in a "postage stamp" fashion could have tremendous benefit in future nanosatellite design. Finally, with future improvement in thin-film power generation and energy storage (vide infra 12.2.3), IMPS should also find application as the main power system for upcoming missions using constellations of very small spacecraft: microsatellites, nanosatellites, picosatellites [49, 67, 68], or currently CubeSats [50, 67–70].

12.2.2 Starshine 3 flight opportunity: Integrated micro power system design

At the annual Small Satellite Conference at Utah State in 1999, we became acquainted with the team from Starshine 3, the third in a series of small satellite launches intended to engage students and amateur radio enthusiasts [71] and joined the team to include two NASA GRC power experiments [37–40]. The primary mission of Starshine 3 was to measure atmospheric density as a function of altitude. Starshine 3 was essentially a passive satellite. There was no use

Batteries for integrated power and CubeSats **Chapter | 12** 461

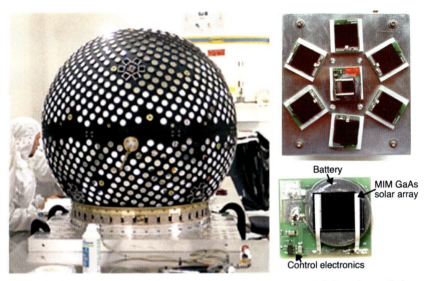

FIG. 12.3 Photos clockwise from left: Starshine 3 small satellite; expanded view of constellation (one of five on satellite) of a seven-junction, 1 cm^2 monolithically interconnected GaAs module (MIM) surrounding an integrated (micro-)power device; detail of an IPS showing a single MIM GaAs solar array, commercial Li-ion battery, and control electronics. *(Courtesy NASA.)*

of electrical power to assist the orbital tracking. This made Starshine 3 an excellent platform to test new power technologies without the burden of mission success depending on the power system. Traditionally, power systems are one of the most conservatively designed systems on a spacecraft. It is therefore especially difficult to introduce new technologies into the power system design due to the necessity that any new technology must be flight-proven before it can be flown. This reality of space technology development can greatly reduce the pace of innovation.

The Starshine 3 satellite (Fig. 12.3) was 1 m in diameter with a mass of 88 kg. Its surface was covered with 1000 student-polished mirrors, 31 laser retroreflectors, 48 triple-junction 2 × 2 cm commercial (EMCORE Corp.) solar cells and five IMPS assembled at NASA Glenn Research Center. The Starshine 3 IMPS consisted of a solar array, a rechargeable battery, and power management electronics all fitting on one square inch of a circuit board. This would be the first in-space demonstration of an IMPS, although given the short lead time, we chose low-risk and/or commercial off-the-shelf (COTS) technologies to successfully demonstrate proof-of-concept: an in-house produced gallium arsenide (GaAs) monolithically integrated module (MIM) solar cell array, a commercial Li-ion thick, coin battery and fool-proof (COTS) control electronics [37, 38, 40, 66].

The IMPS were designed to deliver a constant 20 μA through a 1000 Ω platinum temperature sensor. The solar array is one square centimeter, monolithically interconnected module (MIM) of seven GaAs solar cells connected in series. The array output was nearly 7 Volts and could deliver up to 3 mA of current to the load and/or charging of the battery. The IMPS energy storage was a high capacity 3-Volt Li-MnO$_2$ rechargeable (Panasonic ML2020 coin cell) battery. The battery had a diameter of 2.0 cm, a thickness of 2.0 mm, and a mass of 2.2 g with a nominal capacity of 45 mA-h rated for a continuous 100 μA load [37, 38, 66].

Ideally, the output of the high-voltage, small-area MIM would be designed to match the open-circuit voltage of the Li-ion battery. The MIM used in this flight demonstration experiment had more than enough voltage and current to both charge a Li-ion battery and power an equivalent load. The power management electronics consisted of a micro-power voltage regulator (MAXIM 1726EUK) and a blocking diode. The voltage regulator kept the battery from charging above 3 Volts; the blocking diode prevented current from flowing back through the array when it is in the dark. The load side included two, p-type MOSFETS that shut off the load from the IMPS below 2.3 V. All of the electronic components were selected to minimize parasitic losses in the circuit and avoid draining the battery [37, 38, 40]. Most of these electronics were necessary to avoid damage to the battery that would reduce its cycle life. Ideally, the solar array voltage and size could be matched to the battery voltage and charge current [37, 40]. A blocking diode could also have been integrated into the array eliminating the need for some of the electronics [41].

12.2.3 Starshine 3 integrated power system flight experiment and results

Starshine 3 was launched from a Lockheed Martin Athena I rocket from Kodiak, AK on September 29, 2001, at 6:40 P.M. Alaska Daylight Time (22:40 EDT); it was deployed to a circular orbit of 475 km (low-earth orbit (LEO)) at a 67 degrees inclination with a fixed rotational velocity of 5 degrees/s [39]. The Starshine 3 satellite was covered with mirrors; the spherical geometry allowed for easy modeling of the orbital decay. By observing the orbital decay, the mission allowed for a calculation of atmospheric density versus altitude [39, 71]. The storage capacity of each IMPS battery was 45 mAh, capable of powering the load for 90 days without recharging. When Starshine 3 launched, a month later than originally scheduled, it had been 60 days since the IMPS units were fully charged; thus, they had only one-third of their initial charge left [38, 40].

While on orbit, the five independent power supplies experienced quite adverse operating conditions: Starshine 3 was deployed in an orbit where the solar beta angle passed through ±90 degrees. This meant that there were periods when the solar array on an IMPS did not "see" the sun for days at a time. Any charging would have to come from light reflected off of the surface of the Earth (albedo). Furthermore, when illuminated by only the Earth's albedo, the operating temperature of the IMPS sank as low as −18°C; only 2°C above the

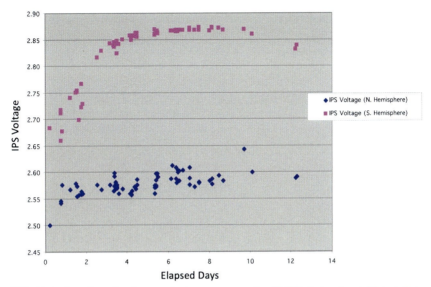

FIG. 12.4 Data from five integrated micro power supplies (IMPS) designed and fabricated at NASA GRC that operated on-board Starshine 3 from late-September to mid-October 2001. Four of the IMPS units had particularly challenging operating conditions due to a severe initial solar beta angle that provided mostly Earth albedo (light reflected off the Earth) charging and an operating temperature of −10°C. *(Courtesy NASA.)*

battery's operating range. The operating voltage of two IMPS units versus days in orbit is shown in Fig. 12.4. The IMPS marked "N. Hemisphere" was located 45 degrees above Starshine's equator as defined by the photograph in Fig. 12.3. The IMPS marked "S. Hemisphere" was located 45 degrees below the equator. When deployed, Starshine 3 had a solar beta angle of −61 degrees. A beta angle of at least −44 degrees is needed for any direct sunlight on the N. Hemisphere IMPS. That meant that the IMPS units in the upper hemisphere did not see the sun at all for the first 18 days, 5 days later than the last data point in Fig. 12.4 [38, 40].

Despite these adverse conditions, the IMPS on the upper hemisphere (Fig. 12.4) still managed to charge under albedo lighting only. This is evident by the slowly increasing voltage during the first 13 days. The net average current measured over the first 18 days was 26 μ-amps, enough to account for the increasing voltage during albedo illumination. The IMPS in the lower hemisphere (S. Hemisphere in Fig. 12.4) quickly charged to its maximum voltage of 2.9 V. The peak voltages reached by each IMPS are set by the voltage regulators and are not directly comparable. They reflect the set point of each IMPS voltage regulator. During the first 100 days, all five IMPS units maintained a voltage greater than 2.5 V. The voltage fluctuations shown in Fig. 12.4 are attributed to the changing solar exposure due to beta angle and eclipse time.

All communication from Starshine 3 ceased after 100 days. The satellite was intentionally de-orbited and burned up on January 21, 2003, after 7434 revolutions around Earth, after 479 days in orbit with one orbit lasting approximately 90 min. Starshine 3 marked the first deployment and proof-of-concept of IMPS technology via a successful space experiment. As discussed earlier, the IMPS units had particularly challenging operating conditions due to a severe initial solar beta angle that provided only Earth albedo (light reflected off the Earth) charging and an average operating temperature of $-10°C$ [38–40].

12.3 Integrated power technologies—2005–2020

It was our goal during our initial I(M)PS development to have all components seamlessly integrated onto a common substrate using thin-film batteries and thin-film solar cells. Scaling up the manufacturing methods should allow an IPS to deliver the highest specific power and energy for the lowest cost. This section covers follow-on efforts to Starshine 3 by many others and includes improvements in system simplicity as voltage matching and monolithic integration have been demonstrated. We review recent results and the state-of-the-art in integrated technologies, address potential applications and discuss advantages, challenges, and practical considerations.

12.3.1 Next steps after Starshine 3: Solar charging of electric vehicle batteries

Reducing power system mass, which is typically 20%–30% spacecraft dry mass, will help reduce launch mass, perhaps enough to enable a mission concept previously too heavy to fly, or allow the use of a smaller, cheaper launch vehicle. The Starshine 3 IMPS flight experiment is just the first step towards this goal and has provided valuable experience in the design and operation of IMPS in an actual space environment. As discussed earlier, the development of very small nano- or picosatellites (now CubeSats) [49, 67–70] has generated a need for smaller lightweight power systems. The advent of CubeSats (1U format = $(10\,\text{cm})^3$) [49, 67, 68] with current sizes of 1U, 2U, 3U, 6U, and 12U [49, 68, 69] provide many more opportunities to test novel flight hardware, scientific payloads and power concepts on-orbit or on missions that venture away from Earth [50, 67–70, 72–75]. CubeSats provide an excellent platform for providing proof-of-concept flight experiment and qualification opportunities for higher-risk and novel flight hardware; this topic will be covered in more detail in the next section.

Further technology developments for integrated power devices will likely include integration at the device level shown notionally in Fig. 12.2 as opposed to the system level (Fig. 12.3). An alternative concept for such an integrated power module was patented about 15 years ago [41]. The patented concept for this integrated power device includes the methodology for constructing a complete system including electronics. This represents an advance over our crude space-tested IMPS device. However, a simpler approach would eliminate control electronics and circuitry.

An example of a simplified systems approach was detailed in a 2010 GM study of direct solar PV charging of batteries for electrically powered vehicles [76]. Several scenarios were compared for the efficiency of charging electric vehicles via direct solar (100%), grid-electricity (95%), and hydrogen storage (30%). A demonstration of optimizing the efficiency of the charging of electric vehicles was accomplished with a commercial crystalline silicon solar array (Table 12.1) to charge a series of strings of 10-, 12-, 13-, 14-, 15-, and 16-cells of commercial $LiFePO_4$ cells (Table 12.2) [76].

The cells and modules were characterized by fully charging the cells to their maximum charge voltage of 3.8–4.2 V/cell using a conventional battery charger. The voltages of individual cells in a series string were balanced by using the charger and load bank to bring all the cells to the same state of charge (SOC) as shown by their voltage (discharged: 2.5 V/cell; fully charged: 4.0 V/cell). The highest solar energy to charge efficiency (14.5%) was achieved when the 50.2 V solar PV module was connected to a 15-cell battery module (Fig. 12.5).

TABLE 12.1 Electrical, solar to electrical efficiency, and temperature coefficient specifications modules under standard test conditions, STC (1000 W m^{-2}, 25°C, AM1.5 solar spectrum) and typical operating conditions (52°C).

Parameter	Module value (25°C, AM1.5)	Module value (52°C)
Maximum power P_{max}, watts	190	175
Maximum power point voltage, V_{mpp}, volts	54.8	50.2
Maximum power point current, I_{mpp}, amperes	3.47	3.49
Open circuit voltage, V_{oc}, volts	67.5	62.0
Short-circuit current, I_{sc}, amperes	3.75	3.77
Module efficiency, %	16.1	14.8
Solar cell efficiency, %	18.5	17.0

Adapted from T.L. Gibson, N.A. Kelly, Solar photovoltaic charging of lithium-ion batteries, J. Power Sources 195 (12) (June 2010) 3928–3932. Copyright Elsevier (June 2010).

TABLE 12.2 Battery cell specifications for high power lithium-ion battery cells.

Specification	Value
Cell type	Lithium iron phosphate (LiFePO$_4$)
Operational voltage	3.3 V
Nominal capacity	2.3 Ah
Cell dimensions	26 mm OD × 65.5 mm
Cell assembly mass	72 g
Max tab current	100 A continuous
Recom, charge voltage	3.6 V
Recom. float-charge voltage	3.45 V
Recom. cut-off current (100% SOC)	0.05 A
Max. recommended charge voltage	3.8 V
Max. allowable charge voltage	4.2 V
Max. continuous charge current (−20°C to 60°C)	10 A
Recommended discharge cut-off voltage	2.0 V
Max. discharge current	60 A
Max. recommended cell temperature	70°C
Max. allowable cell temperature	85°C

Reprinted with permission from T.L.Gibson, N.A. Kelly, Solar photovoltaic charging of lithium-ion batteries, J. Power Sources 195 (12) (June 2010) 3928–3932. Copyright Elsevier (June 2010).

The solar energy-to-battery charging efficiency η_s (efficiency of integrated device or system) can be determined using Eq. (12.1) [76]:

$$\eta_s = (V_b \cdot C_b)/(P \cdot A \cdot t \cdot 1/3600) \cdot 100 \qquad (12.1)$$

where V_b, C_b, P, A, and t are the battery average voltage (V), battery charge increase (Ah), solar irradiance (W m^{-2}), PV area (m^2), and the charging time (s), respectively.

The GM study provided a proof-of-concept for a direct connection (no intervening electronics), optimized, and self-regulating PV solar battery charging system for providing energy for battery-powered (or hybrid) vehicles. The coupling of the PV and battery systems was optimized by matching the maximum power point voltage output of the PV system (at PV operating conditions) to the

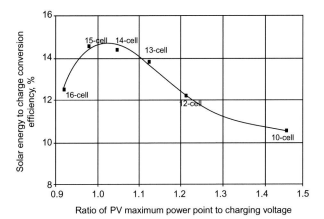

FIG. 12.5 Solar energy to battery charge conversion efficiency comparison for 10-, 12-, 13-, 14-, 15-, and 16-cell modules, rate = 1.5C, full charge at 40 min. The optimum efficiency was approximately 14.5% when the battery charge voltage was equal to the PV MPP ($V_{mpp}/V_{battery}$ charging near unity). *(Reprinted with permission from T.L. Gibson, N.A. Kelly, Solar photovoltaic charging of lithium-ion batteries, J. Power Sources 195 (12) (June 15, 2010) 3928–3932 (Copyright Elsevier, June 2010).)*

maximum charge voltage of the battery. The optimized solar charging system efficiency reached 14.5%, by combining a 15% PV system solar to electrical efficiency and a nearly 100% electrical to battery charge efficiency. The rapid drop in power from the PV system as the battery voltage passed the PV maximum power point provided a self-regulating feature to the system, preventing overcharging of the batteries and potential risks of excessive heat and battery damage. This system is a more economical approach than our one-of-a-kind rapid turnaround system [29,37,38,40] and a notional integrated patented system [41] in that it relies on voltage matching and does not include electronics.

12.3.2 Monolithic integration and further examples of integrated power technologies

The primary benefit resulting from combining two (lightweight) devices providing separate functions into an integrated power device or system is a less complex, reduced volume, lightweight unit or system providing an integrated function. Thin-film batteries and solar cells are ideally suited to such applications [12, 28, 60–62]; as discussed previously, a thin-film IPS could be used for commercial as well as aerospace applications [2, 29, 41–44]. The next generation of advancement would not only eliminate electronics as the GM systems study successfully demonstrates [76] but would go one step further and utilize monolithic integration as well (Fig. 12.6) [43], as we had originally envisioned (Fig. 12.2) for space applications [29,30,40].

468 PART | V Applications and future directions

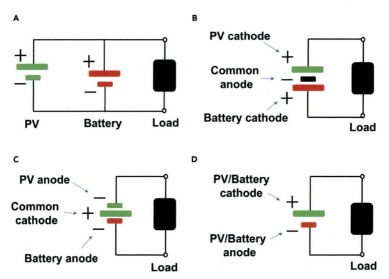

FIG. 12.6 Circuit Representation of PV-Battery Systems: (A) Conventional discrete charging; (B–D) Integrated charging: (B) Three-electrode configuration with common anode; (C) three-electrode configuration with common cathode; (D) two-electrode configuration. *(Reprinted with permission from A. Gurung, Q. Qiao, Solar charging batteries: advances, challenges, and opportunities, Joule 2 (7) (July 2018) 1217–1230 (Copyright Elsevier July 2018).)*

A monolithically integrated device was fabricated and studied in 2016 [77]. A triple-junction thin-film silicon solar cell was deposited on a glass substrate with a suitable transparent front contact and a battery deposited directly on the back contact of the solar cell (Fig. 12.7). Further details of the films that comprise the solar cell [78–82], Li-ion battery [83–89], and contacts can be found in Table 12.3.

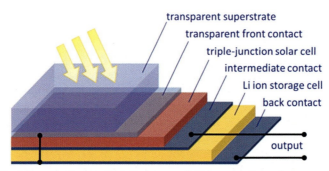

FIG. 12.7 Schematic of a monolithically-integrated PV-battery cell-to-cell concept. *(Reprinted with permission from S.N. Agbo, T. Merdzhanova, S. Yu, H. Tempel, H. Kungl, R.-A. Eichel, U. Rau, O. Astakhov, Development towards cell-to-cell monolithic integration of a thin-film solar cell and lithium-ion accumulator, J. Power Sources 327 (September 2016) 340–344 (Copyright Elsevier September 2016).)*

TABLE 12.3 Details of materials in the integrated power device shown in Fig. 12.7.

Layer	Material(s) (and process)	Device	Performance
Superstrate	Glass		
Transparent conductor	SnO$_2$:F (FTO)[a]	Triple junction TF silicon Solar cell[b]	$V_{oc} = 2.09$ V $FF = 76.1\%$ $J_{sc} = 2$ mA/cm^2 $P_{mpp} = 0.314$ mW $\eta = 8.6\%$ (atten. illumination of 37.4 mW/cm^2)
p-type Si	a-Si:H (B) RF-CVD @ 145°C[c]		
Intrinsic Si	a-Si:H by RF-CVD @ 180°C[c]		
n-type Si	mc-Si:H (P) RF-CVD @ 145°C[c]		
Contact	ZnO:Al/Ag/ZnO:Al		
Anode	Li$_4$Ti$_5$O$_{12}$ (LTO)	LTO/LPC Li–ion Battery[d]	$V = 1.9$ V (charge) $V = 1.85$ (discharge) ACap $= 1.45$ mAh/cm^2 Ch/Dc $= 0.07$ mA ACap $= 0.56$ mAh/cm^2 Ch/Dc $= 0.6$ mA
Separator	Glass Fiber Membrane soaked with LP30 (diphenylmethane diisocyanate)		
Electrolyte[e]	(DMC—O=C(OCH$_3$)$_2$: EC—(CH$_2$O)$_2$CO; 1 M LiPF$_6$)		
Cathode	LiFePO$_4$/C (LPC)		
Back contact	ZnO:Al/Ag/ZnO:Al		

[a]Fluorine-doped tin oxide.
[b]Photovoltaic metrics as defined earlier in text and Table 12.1.
[c]a-Si:H (D), hydrogenated amorphous silicon, p- (B) or n- (P) doped; RF-CVD, radio-frequency enhanced chemical vapor deposition.
[d]Battery metrics as defined earlier in text and in Table 12.2, ACap, areal capacity, Ch/Dc, charge/discharge.
[e]DMC, dimethyl carbonate, EC, ethylene carbonate.
Compiled from information included in S.N. Agbo, T. Merdzhanova, S. Yu, H. Tempel, H. Kungl, R.-A. Eichel, U. Rau, O. Astakhov, Development towards cell-to-cell monolithic integration of a thin-film solar cell and lithium-ion accumulator, J. Power Sources 327 (September 2016) 340–344.

A practical method to express the performance of the integrated device, the real-world energy gain of the system, is determined by subtracting the energy of the battery dissipated through the solar cell in the dark from the energy supplied to the battery during illumination. This power gain can be determined by Eq. (12.2) [77]:

$$P_g = d \cdot P_i - (1 - d \cdot P_d) \quad (12.2)$$

where P_g, P_i, P_d, and d are the average power gain, power produced by the solar cell under illumination, power lost during the discharge of the battery, and the duty factor, respectively. The actual use of an integrated PV-battery device involves alternating periods of illumination and darkness. The duty factor (d) is simply a ratio (Eq. 12.3) or fraction of time under illumination (t_i) to total time, t_i plus time in darkness t_d; $d = 0$ for total and continual darkness and $d = 1$ under constant illumination.

$$d = t_i/(t_i + t_d) \tag{12.3}$$

Fig. 12.8 plots the power gain (P_g) calculated using $J \cdot V$ under a range of duty factors; the optimal operating voltage (V_{opt}) of the PV-battery device at a given duty factor results in a maximum of P_g (V). It is apparent (Fig. 12.8) that $V_{opt} = 1.8\,V$ at the V_{mpp} for $d = 1$ and shifts to lower voltages as the duty factor decreases. Also, the plateau voltage ($V_{plateau} \approx 1.9\,V$) of the battery used by Agbo et al. [77] is outside the range of optimum charging voltages in Fig. 12.8. In an optimized case the PV-battery combination can adequately sustain the energy stored in the battery even without blocking diode; the PV-battery combined device would need to match V_{opt} and $V_{plateau}$. For example, during the summer, where $d \approx 0.6$, the triple junction thin-film solar cell used in [77] has a $V_{opt} = 1.7\,V$; this is 0.2 V less than $V_{plateau}$.

Eq. (12.1) can be used to determine the solar energy-to-battery charging efficiency; results (vs. charging time) are presented in Fig. 12.9. The solar

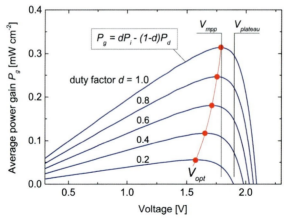

FIG. 12.8 Power gain of the PV-battery cell versus operating voltage calculated using the attenuated JV and dark JV for different duty factors. The power gain takes into account the different possible illumination time scenario defined by the duty factor (***d***). The $V_{plateau}$ is a feature of the battery and is independent of ***d*** while the optimum voltage for battery charging V_{opt} decreases with decreasing ***d***. *(Reprinted with permission from S.N. Agbo, T. Merdzhanova, S. Yu, H. Tempel, H. Kungl, R.-A. Eichel, U. Rau, O. Astakhov, Development towards cell-to-cell monolithic integration of a thin-film solar cell and lithium-ion accumulator, J. Power Sources 327 (September 2016) 340–344 (Copyright Elsevier September 2016).)*

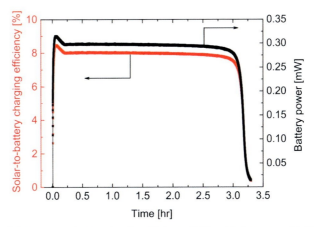

FIG. 12.9 Solar energy-to-battery charging efficiency (left) and battery power (right) calculated over the entire charging cycle. The solar-to-battery charging efficiency of 8.5% is maximum at the maximum power point and is nearly equal to the standalone efficiency of the solar cell. *(Reprinted with permission from S.N. Agbo, T. Merdzhanova, S. Yu, H. Tempel, H. Kungl, R.-A. Eichel, U. Rau, O. Astakhov, Development towards cell-to-cell monolithic integration of a thin-film solar cell and lithium-ion accumulator, J. Power Sources 327 (September 2016) 340–344 (Copyright Elsevier September 2016).)*

energy-to-battery charging efficiency and the power delivered to the battery vary as the operating point of the solar cell moves along the power curve; the conversion efficiency of the integrated device is dependent on the state of charge. The maximum solar energy-to-battery charging efficiency of 8.5% implies that virtually all of the available power generated by the solar cell can be transmitted to the battery; the charging efficiency quickly drops to 8% up to full charge at 3 h. As discussed earlier, it is clear that the solar cell-battery device can be optimized further by increasing the power point voltage of the solar cell to match the plateau voltage of the battery, this would improve solar-to-battery efficiency and suppress current discharge in the case of insufficient illumination.

Several key advances have been described for the integration of energy conversion and storage technologies: systems-level integration of energy conversion and storage with the aid of control circuitry was successfully demonstrated in LEO in 2001 [38,40], a device was patented that integrates the control circuitry [41] in 2006, a simpler system that efficiently charges a battery string of LiPO$_4$ cells that is voltage matched to a commercial silicon module with relevance to charging an electric vehicle was demonstrated in 2010 [76]; these three examples are variations of a discrete IPS as shown in Fig. 12.6A. In addition, the feasibility of monolithic integration (Fig. 12.6B–D) of a triple junction silicon solar cell and a lithium battery without control electronics, demonstrated in 2016 [77] is discussed earlier.

Several recently-published reviews [42–44,90,91] highlight the development of many integrated PV-battery (and supercapacitor [91]) technologies utilizing different combinations of solar cells and energy storage devices, in one case, comparing their efficiencies [43] with the current world-record solar cell efficiencies [62]. Within the past several years, numerous reports have appeared that highlight new results for IPS's with various combinations of solar cell and battery technologies, these are listed alongside the system (or device) structure (Fig. 12.6) with the overall system efficiency (Table 12.4). A more detailed schematic and description of the circuit representation of a three-electrode integrated device is shown in Fig. 12.10 [44].

TABLE 12.4 Comparison of integrated power technologies from literature 2010–2020.

Year	Solar cell technology[a]	Device structure[b]	Battery tech.[c]	System η_s percent[d]	Top solar cell η_{sc} Percent[e]	Ref.
2010	c-Si	Discrete	LIB	14.5	22	76
2012	DSSC	Int: 3-electrode	LIB	0.8	10	95
2013	DSSC	Int: Redox Flow	RFB	0.05	10	92
	a-Si	Int: 3-electrode	SSLB	4	12	96
2015	PSC	Discrete	LIB	7.4	20	97
2016	p-i-n Si	Int: 3-electrode	LIB	8.5	12	77
	mc-Silicon	Int: Redox Flow	RFB	3.2	20	93
	mc-Silicon	Int: Redox Flow	RFB	1.7	20	94
2017	PSC	Discrete	LIB	9.4	22	98
	mc-Silicon	Int: 3-electrode	LIB	7.6	20	99
	p-i-n Si	Int: 3-electrode	SSLB	8.1	12	100
	DSSC	Int: 2-electrode	LMB	0.08	10	101
2018	2J-III-V	Int: Redox Flow	RFB	14.1	35	102

TABLE 12.4 Comparison of integrated power technologies from literature 2010–2020—cont'd

Year	Solar cell technology	Device structure	Battery tech.	System η_s percent	Top solar cell η_{sc} Percent	Ref.
2019	PSC	Int: 3-electrode	AIB	12	25	103
	c-Si	Int: Redox Flow	RFB	5.4	27	104
	DSSC	Int: Redox Flow	RFB	0.1	10	105
2020	PSC	Discrete	ALIB	9.3	25	106
	PSC	Discrete	LIB	9.9	25	107
	PSC	Discrete	LIB	7.3	25	108
	p-i-n Si	Int: Redox Flow	RFB	4.9	14	109

[a]*Solar cell technology—crystalline silicon (c-Si); dye-sensitized solar cell (DSSC); amorphous (thin-film) silicon (a-Si); perovskite solar cell (PSC); multi-junction thin-film silicon (p-i-n Si); multi-crystalline silicon (mc-Si); tandem III–V solar cells (2J III–V).*
[b]*Device structures as shown in Fig. 12.11.*
[c]*Battery technology—Li-ion battery (LIB), redox flow battery (RFB), solid-state thin-film lithium battery (SSLB), liquid metal battery (LMB), aluminum iron battery (AIB), aqueous lithium ion battery (ALIB).*
[d]*Working efficiency (η) of overall system.*
[e]*Best efficiency of solar cell technology in year study published [62].*
Compiled from recent literature reviews [42–44, 90, 91].

An alternative organization of devices is employed by Lennon et al. [44]. The integrated redox-flow structure including redox flow batteries (RFB's) is alternatively considered a monolithic four-electrode device for energy storage [44] and can include photo-electrochemical cells for solar energy harvesting [92, 93]; an electrically-connected four-electrode device is essentially a discrete device as pictured in Fig. 12.6A. Table 12.4 only includes reports [76, 77, 92–109] that mention system efficiency gathered by consulting previously compiled literature [42–44, 90, 91] and the most recent literature (post-2019) results. Finally, an IP fiber device, composed of a DSSC and LIB with a system efficiency of 5.7% [110], has been included in Table 12.5, a compilation of reports [111–113] of integrated fiber-format LIBs with potential applications mentioned in several reviews published over the past 5 years [44, 90, 114].

Several trends from the IPS literature summary in Table 12.4 are readily apparent. Discrete device structures average much higher (~10%) efficiency, either voltage matched or performance enhanced by electronics, thus are much

474 PART | V Applications and future directions

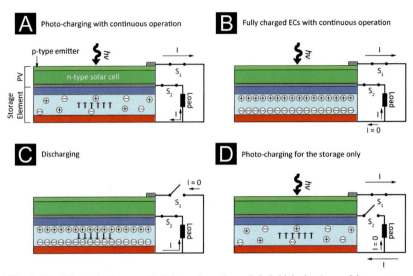

FIG. 12.10 Schematic diagrams depicting a three-electrode hybrid device (comprising an n-type solar cell with a p-type emitter) operating under different modes by closing or opening a series of switches. When the device is solar charging (A), current flows through both the resistive load and the energy storage circuit element in a parallel configuration positioned on the rear of electrode of the solar cell. With electrical connection to the common electrode, energy can be continuously provided to the resistive load even when the energy storage electrode is fully charged (B). When discharging in (C), current flows through the inner circuit with switch S2 closed and switch S1 opened to prevent discharging through the solar cell. (D) Depicts the case when fast charging of the energy storage electrode is required (i.e., switch S2 is opened to prevent the solar-generated current from passing through the load). The symbols $\oplus$ and $\ominus$ represent cations and anions in the electrolyte, respectively, and the arrows show the direction of positive ion flow in the electrolyte. *(Reprinted with permission from D. Lau, N. Song, C. Hall, Y. Jiang, S. Lim, I. Perez-Wurfl, Z. Ouyang, A. Lennon, Hybrid solar energy harvesting and storage devices: The promises and challenges, Mater. Today Energy 13 (September 2019) 22–44 Copyright Elsevier (September 2019).)*

more readily optimized for practical applications. Proven solar cell technologies including III-V (Starshine 3 or RFB), various silicon, and perovskite devices, particularly when combined with LIB's or AIB have very good IPS performance. Integrated devices or systems that are (monolithically) integrated only have demonstrated efficiency near or above 5% when utilizing proven solar cell technologies. While the list of IPS's with calculated system efficiencies of battery/solar cell devices is relatively limited, there have been numerous integrated power devices and systems reported; these are summarized and details provided in the referenced literature reviews [42–44, 90,91]. The review by Vega-Garita et al. discusses photovoltaic-supercapacitor (PSC) integrated devices as well [91].

TABLE 12.5 Summary of flexible-fiber integrated power device and Li-ion batteries.

Composition and structure	Device	Performance metrics (capacity)	Further comments	Ref.
DSSC/LiMn$_2$O$_4$// Li$_4$Ti$_5$O$_{12}$ battery	Integrated Power	18.4 mAh	5.7% system efficiency	[110]
Nickel/tin-coated copper//LiCoO$_2$	Full-cell	1 mAh cm^{-1} at 0.1C	High mechanical stability	[111]
CNT/Si fiber//CNT/ LiMn$_2$O$_4$ yarn	Full-cell	106.5 mAh g^{-1} at 1C	Can be woven into a textile	[112]
CNT/Li$_4$Ti$_5$O$_{12}$ yarn// CNT/LiMn$_2$O$_4$ yarn	Full-cell	138 mAh g^{-1} at 0.01 mA	High stretch-ability	[113]

Selected and compiled from [44, 90, 114].

12.3.3 Integrated power technologies: Applications, challenges, and practical concerns

We have fully addressed several specific examples of IPS's earlier. As discussed in detail in recent reviews [43,44,91,115], there are numerous applications which may benefit from combining or more closely coupling solar energy conversion and storage, including: aerospace [29,38–41,64–71], electric vehicle recharging [76], low-power electronics [116–119], sensors [120–122], electrochromic smart-windows [123,124], lighting products [125, 126], and self-charging wearable electronic devices [127–134]. Pictures of a wearable device (Fig. 12.11), schematic and film details (Fig. 12.12), and electrical characteristics of a flexible PV module charging a flexible battery (Fig. 12.13) from Arias et al. [130] are included below for illustrative purposes. These applications exploit one or more advantages enabled by the IPS paradigm including a (more) flexible form factor, reduced cost, reducing power usage and/or loss from reduced external wiring, and reduced device volume where surface area, volume, and weight are all potential limitations as in aerospace applications, discussed in detail for the Starshine 3 IMPS experiment mentioned earlier [38–40,64–66].

Some performance details for various applications are outlined in several reviews [44, 90]; a convenient summary of illustrative IPS applications [38, 40, 76, 119, 121, 125–127, 130] with relevant details are included in Table 12.6.

476 PART | V Applications and future directions

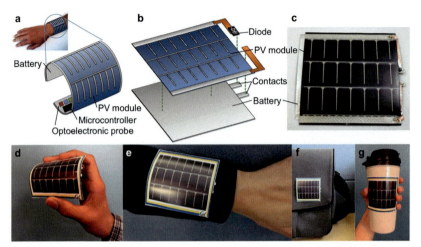

FIG. 12.11 (A) Illustration of activity-tracking wristband concept containing flexible battery, PV energy harvesting module, and pulse oximeter components. (B) Diagram and (C) photograph of a flexible energy harvesting and storage system comprising PV module, battery, and surface-mount Schottky diode, showing the components and attachment points. The diode is included to prevent discharge of the battery into the PV module in low-light conditions. (D–G) Photographs of the device being flexed in the hand (D) and on various flexible and curved surfaces: jacket sleeve (E), bag (F), and travel mug (G). *(Reprinted from A.E. Ostfeld, A.M. Gaikwad, Y. Khan, A.C. Arias, High-performance flexible energy storage and harvesting system for wearable electronics, Sci. Rep. 6 (May 2016) 26122, https://doi.org/10.1038/srep26122, Creative Commons 4.0 International (CC BY 4.0) https://creativecommons.org/licenses/by/4.0/.)*

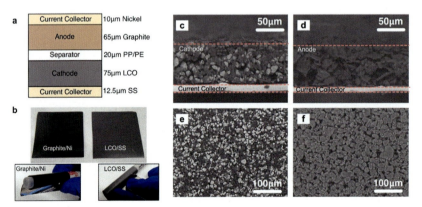

FIG. 12.12 (A) Cross-sectional schematic of the lithium ion battery. (B) Optical image of graphite electrode on nickel foil and LCO electrode on stainless steel, flat (top) and flexed over a pen with diameter of 10 mm (bottom). Cross-sectional SEM micrographs of LCO (C), and graphite (D) electrodes, respectively. Topographical SEM micrographs of LCO (E), and graphite (F) electrodes, respectively. *(Reprinted from A.E. Ostfeld, A.M. Gaikwad, Y. Khan, A.C. Arias, High-performance flexible energy storage and harvesting system for wearable electronics, Sci. Rep. 6 (May 2016) 26122, https://doi.org/10.1038/srep26122, Creative Commons 4.0 International (CC BY 4.0) https://creativecommons.org/licenses/by/4.0/.)*

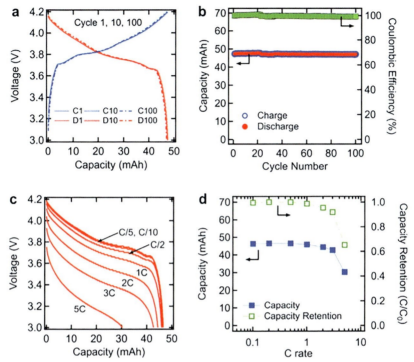

FIG. 12.13 Electrical characteristics of flexible PV module charging the flexible battery: (A) current-voltage characteristics of the PV module; (B) battery voltage over time as it is charged by the PV module, under different illumination conditions; (C) battery voltage profiles over 10 charge/discharge cycles, charging is performed using the PV module under 100 mW/cm² irradiance and discharging is performed at a constant current of 20 mA. (D) Time for the battery to charge to 4.2 V and discharge to 3.6 V over the same 10 cycles as in (C). *(Reprinted from A.E. Ostfeld, A.M. Gaikwad, Y. Khan, A.C. Arias, High-performance flexible energy storage and harvesting system for wearable electronics, Sci. Rep. 6 (May 2016) 26122, https://doi.org/10.1038/srep26122, Creative Commons 4.0 International (CC BY 4.0) https://creativecommons.org/licenses/by/4.0/.)*

For classification purposes, IPS's can be divided into two broad categories: high-power and low-power devices [44]. Devices with a PV generation-rated power less than 10-watt peak (W_p) can be considered "low-power," this would include every example from Table 12.6, except the GM electric vehicle charging system [76]. Devices able to deliver more than 10 W_p can be classified as high power systems [135]. To put this value into perspective, charging a cell phone requires 4 ± 2 W.

Thus far, we have emphasized the advantages that can be realized when (more closely) integrating energy conversion and storage technologies via several device designs, with and without electronics. There are numerous complicating issues related to designing, fabricating, and using integrated power

TABLE 12.6 Illustrative examples of IPS applications from our work and the literature.

Application	Power details	Performance metrics and/or further comments	Ref.
Starshine 3: 1st LEO Flight Demonstration of IMPS	7 V MIM-GaAs Array, COTS control electronics and commercial 45 mAh LIB	Maintained 2.5 V or more for 5 IMPS for 100 days in LEO	[38] [39] [40]
Solar Recharging of Electric Vehicle Batteries	Commercial c-Si solar cell array charging a string of LiFePO$_4$ (3.3 V; 2.3 Ah) batteries	Voltage-matched 15% array (175 watts) recharged string of 14 batteries with 14.5% system efficiency; full-charge (1.5C) after 40 min	[76]
Bluetooth Low Energy Beacons	Li rechargeable coin cell (3.6 V) can be re-charged by 88 cm^2 solar array (Si)	Transmission power of 1 mwatt, advertising interval of 800 ms	[119]
Temperature Sensor	Thin-film Li battery (12 µAh); 2 1 mm^2 solar cells (η-5.48%); power management (dc/dc converter)	System consumes 7.7 µW; power stored in battery can power sensor for ~5 years	[121]
Solar Lamp	Flexible polymer solar cells; Li polymer battery; white LED lamp	12.5 × 8.8 × 2.4 cm; mass = 50 g	[125]
WakaWaka Light (Commercial Product)	Integrated lithium polymer (Li-Po) battery of 500 mAh, c-Si PV cell ($\eta = 18$%) S.A. = 11 × 6 cm^2	LED lights up to 25 lm	[126]
Wearable Textile Solar Batteries	Six polymer solar cells in series connected to textile battery: Li$_4$Ti$_5$O$_{12}$ Anode/ LiFePO$_4$ cathode	Supplied LED load consumed 4.2 mW	[127]
Wearable Health Monitoring Device (Pulse Oximeter)	a-Si solar cell (~230 mW); graphite anode and LiCoO$_2$ cathode, capacity = 47.5 mAh	With appropriate load duty cycle, average load current can be matched to solar module current; battery maintains constant state of charge	[130]

Selected and compiled from [38–40, 44, 76, 91].

devices. These have been addressed in our earlier work for aerospace applications [29, 37, 42, 64–66], and in some detail, including in-depth analysis in several review articles [43, 44, 91, 115]. A summary of critical issues and concerns [136–141] and relevance to potential applications is provided in Table 12.7.

Challenges exist concerning durability, material compatibility, electrode arrangements, form factors, and operating modes, see references in Table 12.7. Device architecture and material choices need to be carefully selected according to the specific intended application to ensure adequate durability to enable practical outcomes. For example, the advantage of a three-electrode device (depicted in Fig. 12.10) over two-electrode architecture is illustrated in Fig. 12.14 for continual operation of a load connected to the c-Si solar cell-powered hybrid device, note that this is a low-power device. Further innovations and improvements in fabrication techniques are required to develop and demonstrate new hybrid devices that present real operating or cost advantages with quantifiable metrics over alternative solutions comprising individual solar harvesting and energy storage devices [43, 44, 91, 115]. However, at present, integrated devices employing three-electrode monolithically or discrete/four-electrode monolithically integrated architectures appear to offer the most flexibility for practical applications for the foreseeable future [43, 44, 91].

12.4 Energy storage options for CubeSats

As discussed earlier, CubeSats are a recently-developed platform for testing and/or demonstrating numerous novel or high(er)-risk space technologies in low-Earth orbit and beyond [49, 50, 67–70, 72–75, 142–146]. CubeSats require far fewer solar cells than high-altitude platforms such as airships and solar electric aircraft (or high-altitude pseudo-satellites (HAPs)) [46], therefore more expensive multi-junction III-V solar cells with efficiencies in the range of 25%–30% are typically utilized [50, 68, 143]. Despite the greater efficiency and power generated, as with HAPS [46], the total power required for spacecraft operations is not always available (or sufficient) from solar cells or arrays depending upon the orbit, mission duration, distance from the Sun, or peak loads. This will therefore necessitate stored, onboard energy, mainly from batteries.

Primary batteries are not rechargeable and thus are used only for short mission durations (typically a day, up to 1 week) [68]. Therefore, one of the numerous rechargeable or secondary energy storage options must be utilized [46, 147–149]. This section of the chapter will provide a further introduction to CubeSats, present challenges, novel battery designs, and intriguing options for space exploration missions.

TABLE 12.7 Key issues, practical concerns, and IPS applications from the literature.

Broad area of concern or issue	Details and practical considerations	Relevant application(s) and/or further comments	Ref.
Durability and autonomous operations	Individual solar charging and energy storage components need to be durable over time	Applies to all applications; short lifetimes offer few benefits when compared with devices powered by low-cost rechargeable batteries; cost not an issue for aerospace or off-grid	[29, 44, 91, 121]
Device design for specific application	Device configuration, materials and component technologies, and power requirements	All applications; consider impact of switch from solar-powered and dark operation and unique constraints imposed by operating conditions	[43, 44, 91, 115]
Material choices for components of hybrid devices	Health and safety issues, mainly toxicity and long-term reactions that cause harmful by-products; leak or puncture could cause leak of harmful materials	Medical implants or wearable devices, contact with tissue(s) may occur; Si solar cells though limited by lower V_{oc}, may be more suitable for such applications as the material is nonhazardous	[44, 91, 127, 130, 136]
Corrosion of energy storage component	Volatile solvents pose flammability as well as toxicity concerns, use of liquid electrolytes raises device leakage concerns	Impact on performance of two and three-electrode systems; expected to be more problematic for devices employing aqueous electrolytes; solid electrolytes may be preferred	[43, 44, 137, 138]
Device architecture: two-electrode	Once device is fully charged, current ceases to flow through circuit; power cannot be provided to load; current must flow through solar cell on discharge	Two-electrode device essentially a solar-charged energy storage device rather than a hybrid device providing a continuous source of power to drive a load; discharge likely to result in high equivalent series resistance (ESR)	[43, 44, 91, 102, 139]
Device architecture: three-electrode	Preferred over two-electrode configurations: removes need to discharge through the solar cell	Requiring switches to control current flow during charging and discharging; autonomous switching can be achieved by using micro-controllers	[43, 44, 91, 126, 140]
Device architecture: four-electrode	More accurately referred to as system vs. device; RFB-based IPS has potential advantages of cyclability and scalability due to higher energy capacity	Enables numerous applications, allows for use of sensors and other devices and higher power operations, trade-offs: increased complexity and space utilization; higher mass and cost, adds issues (i.e., integration and thermal)	[43, 44, 76, 91, 141]

Adapted and compiled from [43, 44, 91].

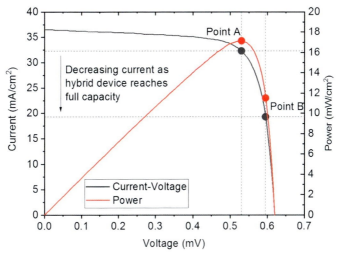

FIG. 12.14 Graph showing an *I-V* curve (black) and power-*V* (*red*) curve of a c-Si p-type commercial solar cell. When the storage electrode is incorporated at the rear of the solar cell (i.e., via a common electrode), the charging current will reduce from Point A to B as storage electrode approaches its maximum capacity. Once the potential difference across the common electrode and the counter electrode exceed the V_{oc} of the solar cell then current will cease to flow across the storage electrodes. If electrical switches are used as shown in Fig. 12.10, then the solar cell can continue to provide power to an external load. *(Reprinted with permission from D. Lau, N. Song, C. Hall, Y. Jiang, S. Lim, I. Perez-Wurfl, Z. Ouyang, A. Lennon, Hybrid solar energy harvesting and storage devices: The promises and challenges, Mater. Today Energy 13 (September 2019) 22–44. Copyright Elsevier (September 2019).)*

12.4.1 CubeSats: Background, present issues, and future missions

Fig. 12.15 correlates CubeSat form factor and small satellite size designation [143]. The IMPS test on Starshine 3 [38–40] is an example of a successful flight demonstration during the beginning of the CubeSat era when the first picosatellite mission, the Orbiting Picosatellite Activated Launcher (OPAL), was launched in 2000 [142, 146]; the first actual CubeSat (2U) was launched from the Space Shuttle Endeavor (STS-113) on December 2, 2002 [144]. It is difficult to precisely determine the eventual outcome of some CubeSats due to issues with launches, deployments, and operations [68, 69, 142–146]. A reasonable assessment is a CubeSat mission success rate of ∼60%, the confirmed CubeSat performance failure rate is ∼20%, with various other mishaps, including launch failure at 20% [145]; it should be noted that the success rate is increasing over time [69, 145, 146]. The type of CubeSat (form factor) launched is broadening and including a higher proportion of larger sizes. As of mid-2018: 3U CubeSats accounted for 64%, followed by 1U (19%), 2U (8%), 1.5U (5%), and 6U (4%) of launches [145]. Looking forward, including CubeSats yet to be launched, the mix of types (by form factor) changes to 3U (44%), 1U (14%), 6U (13%), 8U (8%). 2U (7%), 12U and 1.5U (3%), and other sizes (8%) [69].

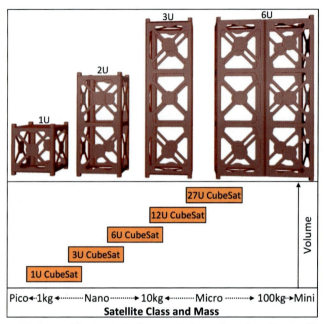

FIG. 12.15 CubeSat specifications in the framework of overall small satellite classifications. The volume of a 1U unit is equal to $10 \times 10 \times 10$ cm (or 1000 cm^3). *(Reprinted with permission from A. Poghosyan, A. Golkar, CubeSat evolution: Analyzing CubeSat capabilities for conducting science missions, Progr. Aerosp. Sci. 88 (January 2017) 59–83. Copyright Elsevier (January 2017).)*

Given the >1200 CubeSats launched to date (~45% in constellations) by industry (45%), academia, (30%), and government entities (15%) [69, 145, 146], this subsection will focus on the notable aspect(s) of energy storage (especially battery) technology relevant to CubeSats, as they are primarily dependent upon solar cells and arrays for power generation. A recent assessment is that approximately 85% of all nanosatellite spacecraft were equipped with solar panels and rechargeable batteries [68]. A general breakdown of primary mission types is educational, communications, earth imaging, military, science, and technology demonstration [146]; this is roughly increasing in form factor from 1U to 6U and beyond. The significant fraction of educational institution involvement is a consequence of the lower cost and launch opportunities afforded to student groups [68].

A visual summary of major systems for a notional 6U CubeSat is shown in Fig. 12.16 [150]. Of course, it is certainly feasible to accomplish multiple missions with one CubeSat. Fig. 12.17 is a picture of RainCube, a 6U CubeSat currently in operation, that is primarily a technology demonstration mission that is also earth imaging [151]. Fig. 12.18 is a schematic of Biosentinel, a 6U CubeSat that will accomplish both science and technology demonstration missions. Fig. 12.19 is a montage of pictures, a schematic and an artist's rendering of NEA Scout, another dual mission (science and technology demonstration) 6U

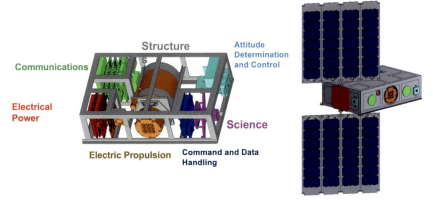

FIG. 12.16 Subsystem schematic and technical drawing of a 6U CubeSat proposed for an asteroid fly-by mission as described in: G.A. Landis, S. Oleson, M. McGuire, A. Hepp, J. Stegman, M. Bur, L. Burke, M. Martini, J. Fittje, L. Kohout, J. Fincannon, T. Packard, A Cubesat Asteroid Mission: Design Study and Trade-offs, Paper IAC-14,B4,8,9,x26216, 65th International Astronautical Congress, Toronto, Canada, September 29–October 3, 2014, 6 pp. *(Courtesy NASA.)*

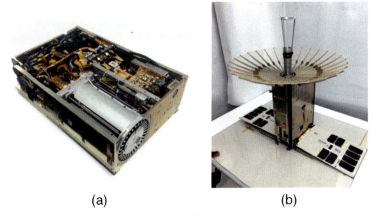

(a)   (b)

FIG. 12.17 Pictures of the NASA/JPL RainCube (Radar in a CubeSat) launched on May 21, 2018 from Wallops Island (NASA GSFC) and deployed from the ISS on June 25, 2018: (A) The integrated radar payload and flight avionics in the 6-U bus chassis; (B) The fully integrated RainCube satellite including the solar panels and the deployed radar antenna. It is the first operating radar instrument on a CubeSat (6U) designed to track large storms. Mission information: https://www.jpl.nasa.gov/cubesat/missions/raincube.php. *(Courtesy NASA.)*

CubeSat developed to visit a Near-Earth asteroid. Biosentinel and NEA Scout are to be deployed from Artemis-1, scheduled to be launched in late 2021 [152].

As briefly discussed earlier, there are numerous rechargeable or secondary energy storage options with various performance metrics [46, 147–149]. Secondary-type batteries include chemistries such as nickel-cadmium (NiCd), nickel-hydrogen (NiH$_2$), nickel-metal hydride (Ni-MH), lithium-ion (LIB),

484 PART | V Applications and future directions

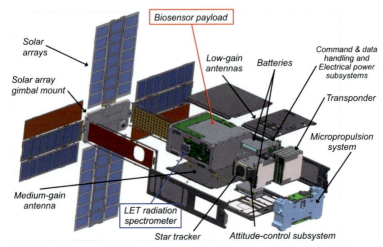

FIG. 12.18 NASA BioSentinel 6U CubeSat General System Configuration—Exploded View. Mission information: https://www.nasa.gov/ames/biosentinel, Accessed August 21, 2020. *(Courtesy NASA.)*

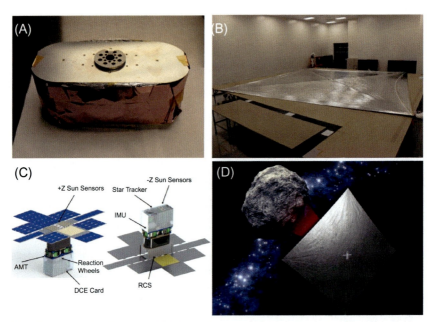

FIG. 12.19 NASA NEA Scout 6U CubeSat (A) solar sail flight unit, spooled; (B) solar sail subsystem flight unit, deployed; (C) CubeSat 6U bus; (D) artist's rendition of NEA Scout surveying asteroid **1991 VG**. *(Mission information: https://www.nasa.gov/content/nea-scout, Accessed August 21, 2020. (A) and (B) Courtesy NASA, (C) and (D) Reprinted from J. Pezent, R. Sood, A. Heaton, High-fidelity contingency trajectory design and analysis for NASA's near-earth asteroid (NEA) Scout solar sail Mission, Acta Astronaut. 159 (June 2019) 385–396. Copyright Elsevier (June 2019).)*

lithium sulfide (Li-S), and lithium polymer (LiPo), many have been used extensively in the past on small spacecraft or suborbital HAPS. Each battery type is associated with certain applications that depend on performance parameters, including energy density, cycle life, and reliability [68, 147, 148]. As other chapters in this book attest, lithium-based secondary batteries (3.6 V, >150 Wh/kg) are commonly used in portable electronic devices because of their rechargeability, low weight, and high energy [4–8]; this technology is commonly employed for spacecraft missions. A comparison of battery energy densities for key commercial manufacturers is shown in Fig. 12.20 [68, 153–156]. Table 12.8 lists leading COTS secondary battery components for CubeSats, including vendor, mass-specific energy, cell types, technology readiness level (TRL) [157] and product reference(s) [68, 154, 156, 158–161].

12.4.2 Potential failure modes for commercial Li-ion batteries in low-earth orbit

A very recent study examined failure modes of three commercial LIB cathode technologies with the same graphite anode material by testing in a simulated LEO environment [162]: lithium nickel cobalt aluminum oxide (NCA) [163], lithium nickel manganese cobalt oxide (NMC) [164], and lithium iron phosphate (LFP) [165]. Cells were grouped three-in-parallel (3P—producing a minimum, median, and maximum value of each cell group for any metric of interest) for each cathode type for simulated low-Earth orbit (LEO) CubeSat cycling, representing a typical 2U sized CubeSat battery pack. Parallel groups of each cell type were used: one group was cycled at atmospheric pressure

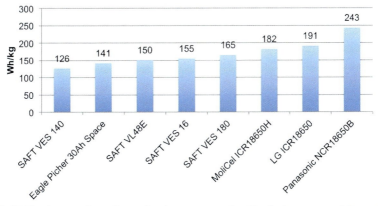

FIG. 12.20 A comparison of secondary battery energy densities for key commercial manufacturers. *(Reprinted from E. Agasid, R. Burton, R. Carlino, G. Defouw, A.D. Perez, A.G. Karacalıoğlu, B. Klamm, A. Rademacher, J. Schalkwyk, R. Shimmin, J. Tilles, S. Weston, State of the Art Small Spacecraft Technology, NASA/TP—2018–220027, December 2018, 207 pp. (Courtesy NASA).)*

TABLE 12.8 Commercial-off-the-shelf (COTS) secondary battery components for CubeSats including vendor, specific energy, cell technology, technology readiness level [157], and product reference(s).

Product	Manufacturer	Specific energy (Wh/kg)	Cell technology	TRL [157]	Reference
40W-hr CubeSat battery	AAC-Clyde	119	Clyde Li-Polymer	9	[158]
BAT-110	Berlin Space Technologies	58	Lithium-Ferrite	9	[159]
BP-930S	Canon	132	Four 18,650 Li-ion cells	9	[68, 156]
COTS 18650 Li-ion battery	Enersys	90–243	Sony, LG, Molicell, Sanyo, Samsung	8	[156, 160]
NanoPower BP4	GomSpace	143	GomSpace NanoPower LIB	9	[161]
NanoPower BPX	GomSpace	154	GomSpace NanoPower LIB	9	[161]
Rechargeable space battery (NPD-002271)	EaglePicher	154	EaglePicher LIB	7	[154]

Adapted from E. Agasid, R. Burton, R. Carlino, G. Defouw, A.D. Perez, A.G. Karacalıoğlu, B. Klamm, A. Rademacher, J. Schalkwyk, R. Shimmin, J. Tilles, S. Weston, State of the art small spacecraft technology, NASA/TP—2018-220027, December 2018, 207 pp. Available at: https://www.nasa.gov/sites/default/files/atoms/files/soa2018_final_doc-6.pdf, to be updated Fall 2020, Accessed 30 July 2020.

(101 kPa), the other group in vacuum (0.2 kPa), with both groups at 10°C, this temperature coincides with the observed average temperature in LEO [166].

A typical LEO (orbit time varies with altitude and velocity) requires ~90 min, resulting in 16 orbits per day period; a 55 min (3300 s) solar charging time is followed by an eclipse time of 35 min (2100 s) [167]. The LEO cycle in this study was accelerated by a factor of 3, with 700 s discharge periods and 1100 s charge periods for a total "orbit period" of 30 min. The cells were subject to a representative LEO power cycle consisting of constant power (CP) discharge, and sinusoidal power charge. Reference cycles (operate over the full (100%) change in the state of charge (ΔSoC) range to identify capacity degradation) completed at 25°C after every 480 "LEO cycles" (equivalent to 10 days) were required to obtain remaining capacity, energy efficiency, and IR measurements, as the accelerated LEO cycle produced only a partial state of charge. After each reference cycle, another 10 days of accelerated LEO cycles were initiated. Table 12.9 summarizes all of the cell reference data and key experimental results.

From Table 12.9, it is clear that NCA and NMC cells have much greater energy densities than LFP, thus both materials systems have a narrower ΔSoC operating window [162]. These superior attributes combined with their ability to degrade substantially in capacity while still completing discharge-charge cycles might lead CubeSat power system designers to overlook LFP technology. The important result of this study is that LFP batteries have a much lower degradation rate compared to the other chemistries, especially at lower temperatures, and outperform them in both application cycle life and energy efficiency (Fig. 12.21).

In summary, the primary failure modes determined were: current interrupt device (3P to 2P) for NCA and significant internal resistance increase for NMC cells; the reader is encouraged to consult the study by Cook et al. for further details of the quite in-depth failure analysis [162]. A practical outcome is that the low degradation rate of LFP cells suggests that CubeSats employing this cell type would not be required to oversize the initial capacity by a substantial amount, unlike NCA and NMC batteries. As discussed earlier, NCA or NMC cathode or primary batteries may be suitable for shorter-term missions [68, 162]. In future, it would be quite useful to develop correlations between values given in manufacturer specification sheets [153–161, 163–165] and the performance that CubeSat battery system designers predict in LEO [68, 162] or deep space missions [50, 51, 152].

12.4.3 Integrating batteries into structural elements to enhance performance

As documented earlier and discussed in detail in the following section, CubeSats with a form factor of 6U are being developed more frequently in recent years; the larger form factor(s) and extra volume can accommodate propulsion technology for longer, more complex missions [67–69, 143–146, 150–152]. However, to enhance the available energy storage capacity in a small volume, one approach is to integrate a LIB into a (lightweight) structural element; this would be particularly valuable for a small form factor (i.e., 1U) CubeSat. A recent study demonstrates the integration of a pouch-free full cell LIB into a carbon fiber-containing composite matrix to produce a high-performance multifunctional structural battery [70]. Such a strategy provides a clear system-level performance advantage as inactive materials for the LIB serve a dual purpose as the active materials for the carbon fiber epoxy composite matrix.

TABLE 12.9 Commercial-off-the-shelf (COTS) single cell 18–65 cylindrical secondary battery cathode material datasheet specifications, experimental measurements, and results summary.

Cell parameters [Ref.]	NCA [163]	NMC [164]	LFP [165]
Mass (g)	45.5	43.7	39.2
Jelly roll length (mm)	610	660	840
Cathode elemental composition	LiNi$_{0.83}$Co$_{0.14}$Al$_{0.03}$O$_2$	LiNi$_{0.5}$Mn$_{0.3}$Co$_{0.2}$O$_2$	LiFePO$_4$
Anode, average particle diameter	Graphite, 20 µm	Graphite, 30 µm	Graphite, 2 µm
Low voltage limit (V)	2.5	2.75	2.0
High voltage limit (V)	4.2	4.2	3.6
Nominal voltage (V)	3.6	3.6	3.3
Max continuous discharge (C rate)	2.0	2.0	27.3
Max continuous charge current (C rate)	0.5	1.0	3.6
Initial coulombic capacity (Ah)	3.15	2.40	1.15
Initial energy capacity (Wh)	10.7	8.40	3.70
Initial IR (mΩ)	38.5	45.7	12.2
Energy density (Wh/L)	647	507	223
Specific energy (Wh/kg)	235	192	94
Power density (W/L)	1436	1132	5991
Discharge temperature range (°C)	−20 to +60	−20 to +60	−30 to +55
Charge temperature range (°C)	0 to +45	0 to +45	0 to +55

TABLE 12.9 Commercial-off-the-shelf (COTS) single cell 18–65 cylindrical secondary battery cathode material datasheet specifications, experimental measurements, and results summary—cont'd

Cell parameters [Ref.]	NCA [163]	NMC [164]	LFP [165]
Cycle life to 80% capacity	250	300	4000
LEO cycling (#) condition status 0.2 kPa[a]/10°C	Failed #4518	Failed #1406	Operational
LEO cycling (#) condition status 101 kPa[b]/10°C	Failed #1161	Failed #854	Operational
Normalized coulombic discharge capacity 10°C (%)[c]	37/71	43/66	79/81
Discharge energy throughput[c]	10,800/2700	3000/2000	10,380[d]
Equivalent cycles of initial measured coulombic capacity[c]	916/247	380/240	2841/2770
100% ΔSoC energy efficiency[c] (%)	79/82	81/83	95/96

[a] Laboratory vacuum (LV).
[b] Atmosphereic pressure (AP).
[c] (LV/AP).
[d] Throughput is the same for both LV and AP.
Adapted from R.W. Cook, L.G. Swan, K.P. Plucknett, Failure mode analysis of lithium ion batteries operated for low Earth orbit CubeSat applications, J. Energy Storage 31 (October 2020) 101561. Copyright Elsevier (October 2020).

Pint and coworkers utilized carbon fiber materials as the current collector via graphite/carbon fiber anodes and LFP/carbon fiber cathodes; these were directly integrated into the carbon fiber panels via a traditional composite layup process, see Fig. 12.22 [70]. A total energy density above 35 Wh/kg was demonstrated relative to all active and composite packaging materials. In Fig. 12.23, this pouch-free battery composite material can be used to fabricate 1U CubeSat structural walls to absorb the electrical energy storage capability into the

490 PART | V Applications and future directions

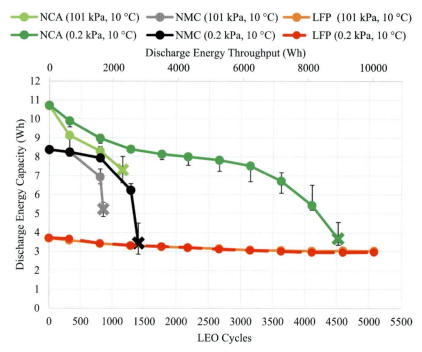

FIG. 12.21 Discharge energy capacity measured at reference cycles *vs.* completed LEO cycles and discharge energy throughput for each cell type (NCA, NMC, LFP) and LEO cycling pressure condition (101 kPa, 0.2 kPa). *(Reprinted from R.W. Cook, L.G. Swan, K.P. Plucknett, Failure mode analysis of lithium ion batteries operated for low Earth orbit CubeSat applications, J. Energy Storage 31 (October 2020) 101561. Copyright Elsevier (October 2020).)*

CubeSat structure, thereby increasing usable interior CubeSat volume for scientific instruments or other mission-critical components.

12.4.4 Use of commercial batteries on a successful inter-planetary CubeSat mission

The Mars Cube One twin (MarCO-A and B, nicknamed Eve and Wall-E, respectively) 6U microsatellites (10–100 kg) are the first CubeSats to complete an interplanetary mission (Fig. 12.24) [50, 168, 169]. The NASA Interior Exploration using Seismic Investigations, Geodesy and Heat Transport (InSight) mission landed on Elysium Planitia, an equatorial site on Mars, on November 26, 2018; the InSight lander's goal is to study the crust-mantle and core of Mars. InSight (with Eve and Wall-E) was launched on May 5, 2018, after a 26-month delay due to a test anomaly of one of InSight's scientific instruments. One of several unique features of the InSight mission was the addition of the redundant MarCO 6U CubeSats. Although launched together with

Batteries for integrated power and CubeSats **Chapter | 12** **491**

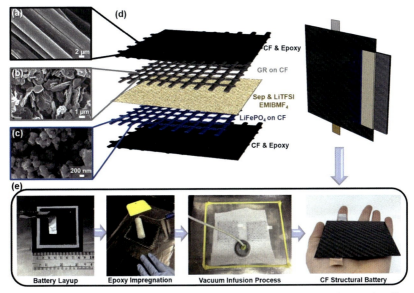

FIG. 12.22 Carbon fiber battery composite fabrication as shown by SEMs of (A) carbon fiber, (B) graphite, and (C) lithium iron phosphate, (D) a scheme showing the stacking of the individual layers of the composite battery along a picture of these layers cured into a composite material and (E) composite layup process along with a picture of a carbon fiber composite structural battery panel being held. *(Reprinted with permission from: K. Moyer, C. Meng, B. Marshall, O. Assal, J. Eaves, D. Perez, R. Karkkainen, L. Roberson, C.L.Pint, Carbon fiber reinforced structural lithium-ion battery composite: Multifunctional power integration for CubeSats, Energy Storage Mater. 24 (January 2020) 676–681. Copyright Elsevier (January 2020).)*

InSight, the two MarCO spacecraft separated soon after launch to fly their own trajectories to Mars to test their endurance and ability to navigate in deep space, thus also becoming the first CubeSats to fly in deep space. The two spacecraft were kept approximately 10,000 km (6200 mi) away from InSight at either flank during flight to protect the main spacecraft from potential mishap; this distance was reduced to 3500 km as the three spacecraft approached Mars.

The primary objectives of MarCO were to demonstrate new miniaturized communication and navigation technologies in the CubeSat form factor in a deep space environment, and more importantly, provide a real-time communications relay during the InSight lander's entry, descent, and landing (EDL) phase, the most critical and uncertain phase of any surface mission. Though only lasting 7–10 min, the EDL phase involves activating several pyrotechnic devices for critical operations to decelerate a lander or rover, including opening parachutes, firing retro-rockets, etc. [170]. The MarCO-A and B CubeSats were intended to provide real-time data relay, with minimal cost to the overall mission (Fig. 12.24); otherwise, InSight would have relayed flight information to the Mars Reconnaissance Orbiter (MRO) which does not have that capability.

492 PART | V Applications and future directions

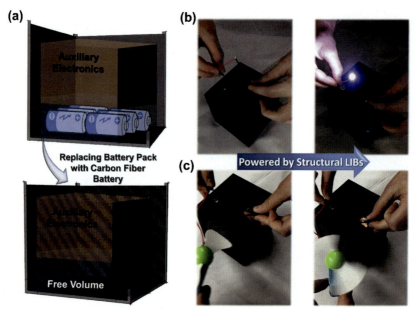

FIG. 12.23 Replacing interior external battery pack with structural battery creates free volume within the CubeSat chassis; (A). Electrochemical performance of 4 composite structural battery panels in series in a 1U prototype CubeSat frame, (B) lighting a LED and (C) operating a fan. *(Reprinted with permission from: K. Moyer, C. Meng, B. Marshall, O. Assal, J. Eaves, D. Perez, R. Karkkainen, L. Roberson, C.L.Pint, Carbon fiber reinforced structural lithium-ion battery composite: Multifunctional power integration for CubeSats, Energy Storage Mater. 24 (January 2020) 676–681. Copyright Elsevier (January 2020).)*

An additional objective of MarCO was to test new miniaturized communication and navigation technologies in a Deep Space environment; see Fig. 12.25 for a photograph of solar array testing of a MarCO CubeSat at JPL; a listing of key components [50, 163, 168, 169, 171–177] of a MarCO CubeSat is included in Table 12.10.

Of interest and relevance to the topic of this chapter is the successful demonstration of COTS batteries (Table 12.10 and Fig. 12.26) for an inter-planetary (Mars) mission. NASA has examined several factors for the selection of LIB's for Mars exploration [178]. Also, based on performance characteristics of the Panasonic NCR18650B cells [163], this chemistry was initially investigated for NASA's upcoming Europa Clipper mission but did not meet that mission's low-temperature performance requirement [50]. As part of this evaluation for Europa Clipper, a detailed assessment of this technology and related cathode chemistries was performed under various operating conditions, including radiation exposure [179]; all three chemistries performed well after radiation with minimal loss in performance [50].

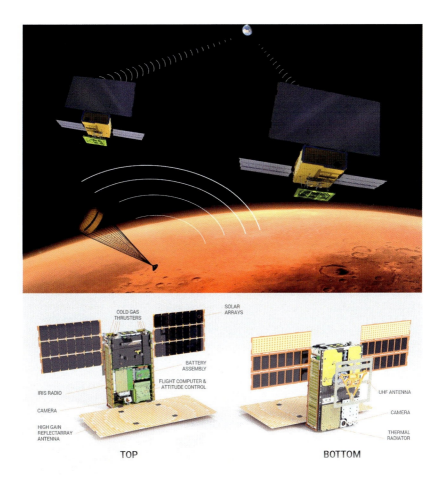

FIG. 12.24 Top: MarCO A and B monitoring the InSight landing (artist's concept). Bottom: illustration of the Mars Cube One spacecraft, indicating location of the battery assembly. *(Reprinted with permission from F.C. Krause, J.A. Loveland, M.C. Smart, E.J. Brandon, R.V. Bugga, Implementation of commercial Li-ion cells on the MarCO deep space CubeSats, J. Power Sources 449 (February 2020) 227544. Copyright Elsevier (February 2020).)*

While simulated LEO testing [162] of COTS batteries for Earth-orbiting CubeSats discovered shortcomings for NCA chemistry (Panasonic NCR18650B [163]), this technology proved to be successful for the MarCO mission [50]. A critical difference between LEO simulation and the MarCO mission conditions is the absence of significant solar array shadowing during the transit to and short-term mission performance around Mars. The flight batteries (Fig. 12.26) were kept at 50% charge (A @ 5°C) and (B @ 20°C) during the 26-month launch delay. Both spacecraft remained warm from launch, with onboard internal temperatures around 18°C; the batteries were nearly fully

FIG. 12.25 A Jet Propulsion Laboratory (JPL) Engineer uses sunlight to test the solar arrays on one of the Mars Cube One (MarCO) spacecraft at NASA's JPL. The MarCOs were the first CubeSats flown into deep space. They accompanied NASA's InSight lander on its cruise to Mars. After making the journey to Mars, they relayed data about InSight's entry, descent and landing back to Earth. Though InSight's mission did not depend on the success of the MarCOs, they were a successful test of CubeSat performance in deep spaced; see: https://www.jpl.nasa.gov/cubesat/missions/marco.php, accessed December 11, 2020. *(Photo courtesy NASA/Caltech Jet Propulsion Laboratory.)*

charged at deployment. During the seven-month journey to Mars, there were one telecom pass and two desaturation maneuvers in a day, averaging 1.5 cycles per day over 7 months for a total of ~300 cycles. The maximum depth of discharge for the 300 cycles is <50%. Based on the anticipated mission conditions and usage, and the known performance loss during storage and cycling of Panasonic NCR-B cells, an overall loss of 18% in performance from the initial cell level values to what would be available for MarCO spacecraft after completing its cruise to Mars was estimated [50]. The estimated performance of the MarCO battery after Mars flyby as a function of discharge current at the allowable flight temperature (AFT) is illustrated in Fig. 12.27.

The MarCO 6 month-mission was completed on November 26, 2018. Overall, MarCO-A successfully transmitted 93% of the InSight EDL data and MarCO-B sent 97%. The first image InSight took of Mars was relayed to Earth by the MarCO CubeSats. Over the following days, MarCO-B transmitted several images of Mars taken by its wide field-of-view camera, including the landing site; the poles of Mars, white from the ice caps, several volcanoes; even two small dim images of Phobos; and finally, a 'farewell' image of Mars (Fig. 12.28). Due to the loss of communication, the primary MarCO mission was declared complete on February 2nd, 2019. Based on trajectory calculations, MarCO-B

TABLE 12.10 Components of the MarCO 6U CubeSat.[a]

Component	Purpose	Details	Reference
UHF[b] antenna	Receiving data from InSight lander	Deployable loop antenna; InSight sent the data to the two CubeSats via the UHF radio channel (401 MHz); used DSN[c]	[171]
X-band antennas (three)	Transmitting/receiving data with earth	1 kg FPR[d]; 8.425-GHz Mars-to-Earth communication	[172]
Small radio	Communicate with ground using X-band; receive data from InSight using UHF; collect tracking measurements for navigation	Iris V2 radio; included a UHF receiver; software-defined radio has up to 4 W RF output at X-band frequencies, fully DSN[c] compatible, with 4 receive and 4 transmit ports	[173]
Star tracker	Attitude determination and control	XACT (fleXible ADCS Cubesat Technology) attitude control unit: star tracker, gyro, coarse sun sensors, and 3-axis reaction wheels	[174]
(Cold-gas) propulsion system	Attitude control and trajectory correction maneuvers	Vacco propulsion system with R-236FA[e] propellant	[175]
(Miniature) wide-angle and narrow-angle cameras	Verify deployments; capture outreach images	Color wide-field engineering camera with 138° diagonal field of view and color narrow-field camera with a 6.8° diagonal field of view-both cameras can produce images 752 by 480 pixels in resolution	[176]
Deployable solar panels (two)	Solar power generation	Three-panel solar arrays 21 3J-GaAs solar cells Panels provided: 35 W near Earth and 17 W near Mars	[177]
Li-ion batteries	Energy storage	Panasonic NCR18650B capacity=3.2 Ah; spec. Energy=~240Wh/kg; energy density>700Wh/L	[163]

[a]Stowed size of 14.4 in. (36.6 cm) by 9.5 in. (24.3 cm) by 4.6 in. (11.8 cm), 13.5 kg (30 lb).
[b]Ultra-high frequency.
[c]Deep space network.
[d]Folded-panel reflectarray.
[e]1,1,1,3,3,3-Hexafluoropropane, a fire suppressant.
Compiled from [50, 163, 168, 169, 171–177].

496 PART | V Applications and future directions

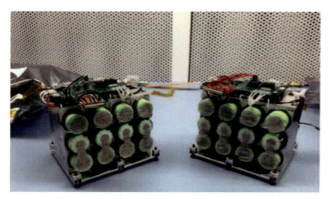

FIG. 12.26 Picture of the final assembled MarCO batteries. The bus voltage for the MarCO power system was 12 V, and based on the energy requirement, the MarCO battery pack was designed in a 3S4P configuration, i.e., four parallel strings of three cell in series per string. *(Reprinted with permission from F.C. Krause, J.A. Loveland, M.C. Smart, E.J. Brandon, R.V. Bugga, Implementation of commercial Li-ion cells on the MarCO deep space CubeSats, J. Power Sources 449 (February 2020) 227544. Copyright Elsevier (February 2020).)*

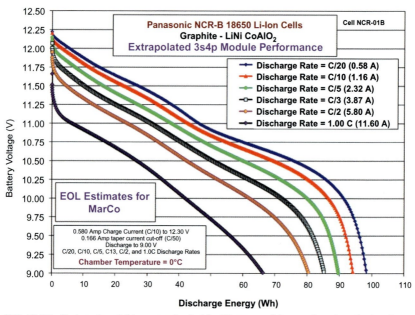

FIG. 12.27 Projected available energy for the MarCO spacecraft battery (based on single cell test data) upon reaching Mars. Discharging was performed between C/20 and C rate, at the lowest Allowable Flight Temperature of 0°C. *(Reprinted with permission from F.C. Krause, J.A. Loveland, M.C. Smart, E.J. Brandon, R.V. Bugga, Implementation of commercial Li-ion cells on the MarCO deep space CubeSats, Journal of Power Sources 449 (February 2020) 227544. Copyright Elsevier (February 2020).)*

Batteries for integrated power and CubeSats **Chapter | 12 497**

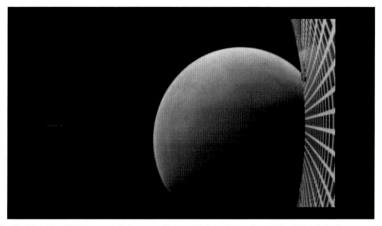

FIG. 12.28 MarCO-B, one of the experimental Mars Cube One (MarCO) CubeSats, took this image of Mars from about 4700 miles (7600 km) away during its flyby of the Red Planet on November 26, 2018. *(This image was taken at about 3:10 p.m. EST while MarCO-B was flying away from the planet after InSight had landed; see: https://www.jpl.nasa.gov/cubesat/missions/marco.php (Accessed 11 December 2020). Photo courtesy NASA/Caltech Jet Propulsion Laboratory.)*

is currently more than 1 million miles (1.6 million kilometers) past Mars; MarCO-A is almost 2 million miles (3.2 million kilometers) past Mars [176].

In conclusion, battery technologies for CubeSats, mainly for LEO missions, have been discussed earlier, and local (Earth-Moon system) missions are cursorily addressed in several above-referenced reports [36, 68]. For readers interested in further information on deep space exploration missions, beyond the few CubeSat missions discussed in this chapter, a recent NASA publication provides an excellent historical overview [180]; future missions can be found on an agency website [181]. More in-depth discussions of relevant battery technologies have been addressed in recent literature [182–185] and will be reviewed in a soon-to-be-published book chapter [186].

12.5 Conclusions

Rechargeable batteries are ideal for numerous applications: electronics, electric vehicles, stationary energy storage, and aerospace systems. Over the past several decades, our research groups have examined the aspects of advanced (nano) materials, including Si alloys for aerospace power applications. These applications present unique challenges such as temperature fluctuations, rapid gravitational fluctuations, high-energy particles and radiation environments, atomic oxygen, hard-ultraviolet light, thermal management and the necessity or weight- and space savings.

Researchers at NASA Glenn Research Center with different research partners developed integrated power technology concepts suitable for small satellites. Integrated power technologies that involve energy storage and energy conversion address several concerns about limitations imposed upon space flight hardware and systems related to power, space, and mass limitations. An integrated power device was demonstrated in low-Earth orbit on the Starshine 3 microsatellite in late 2001 or early 2002.

We reviewed efforts by numerous other researchers to develop (Li-ion) batteries integrated with solar cells and other technologies. Improvements in system simplicity (via voltage matching) and monolithic integration have been demonstrated over the past decade. We outlined the basic types of integrated power devices (or systems), addressed potential applications and discussed advantages, challenges, and practical considerations. It is important to note that while a variety of battery technologies could be employed, Li-ion batteries with Si-containing anodes may certainly be the optical choice in certain situations. Scaling up the manufacturing methods should allow an IPS to deliver the highest specific power and energy for the lowest cost.

We reviewed a variety of battery technologies (with a focus on LIBs) for small satellites, especially CubeSats. Commercial LIBs have been successfully demonstrated for numerous exploration missions to several classes of solar system destinations (Earth-Moon system, Mars, Jupiter, and small bodies) over the past 20 years, including the remarkable combined lander/CubeSat mission to Mars in the past several years. The next 20 years will most certainly see further developments in integrated power technologies and space exploration by small satellites enabled by advanced (lithium-ion) batteries.

Acknowledgments

We dedicate this chapter in part to Philip P. Jenkins, our long-time colleague and collaborator at the U.S. Naval Research Laboratory who passed away at the end of 2019. We thank the numerous graduate students, postdoctoral associates, and professionals at numerous academic, government, and industrial laboratories who contributed to various research studies and projects referenced throughout the chapter. Prashant N. Kumta would also like to acknowledge the Edward R. Weidlein Endowed Chair Professorship funds at the University of Pittsburgh for providing resources to support this effort. Finally, we acknowledge the vision and commitment to America's space program of our colleagues at the NASA Glenn Research Center and the Jet Propulsion Laboratory.

References

[1] J.N. Mrgudich, P.J. Bramhall, J.J. Finnegan, Thin-film rechargeable solid-electrolyte batteries, IEEE Trans. Aerosp. Electron. Syst. AES-1 (3) (1965) 290–296.
[2] M.D. Kankam, V.J. Lyons, M.A. Hoberecht, R.R. Tacina, A.F. Hepp, Recent GRC aerospace technologies applicable to terrestrial energy systems, J. Propuls. Power 18 (2) (2002) 481–488.
[3] F. Cheng, J. Liang, Z. Tao, J. Chen, Functional materials for rechargeable batteries, Adv. Mater. 23 (15) (2011) 1695–1715.

[4] S. Megahed, B. Scrosati, Lithium-ion rechargeable batteries, J. Power Sources 51 (1–2) (1994) 79–104.
[5] R.A. Huggins, Lithium alloy negative electrodes, J. Power Sources 81–82 (1999) 13–19.
[6] P. Poizot, S. Laruelle, S. Grugeon, L. Dupont, J.-M. Tarascon, Nano-sized transition-metal oxides as negative-electrode materials for lithium-ion batteries, Nature 407 (6803) (2000) 496–499.
[7] J.-M. Tarascon, M. Armand, Issues and challenges facing rechargeable lithium batteries, Nature 414 (2001) 359–367.
[8] M.W. Verbrugge, P. Liu, Electrochemical characterization of high-power lithium ion batteries using triangular voltage and current excitation sources, J. Power Sources 174 (1) (2007) 2–8.
[9] R. Yazami, P. Touzain, A reversible graphite-lithium negative electrode for electrochemical generators, J. Power Sources 9 (1983) 365–371.
[10] S. Bourderau, T. Brousse, D.M. Schleich, Amorphous silicon as a possible anode material for Li-ion batteries, J. Power Sources 81–82 (1999) 233–236.
[11] H. Li, X. Huang, L. Chen, Z. Wu, Y. Liang, High capacity nano-Si composite anode material for lithium rechargeable batteries, Electrochem. Solid-State Lett. 2 (11) (1999) 547–549.
[12] J.B. Bates, N.J. Dudney, B. Neudecker, A. Ueda, C.D. Evans, Thin-film lithium and lithium-ion batteries, Solid State Ionics 135 (1–4) (2000) 33–45.
[13] M. Yoshio, H. Wang, K. Fukuda, T. Umeno, N. Dimov, Z. Ogumi, Carbon-coated Si as a lithium-ion battery anode material, J. Electrochem. Soc. 149 (12) (2002) A1598–A1603.
[14] Z.S. Wen, J. Yang, B.F. Wang, K. Wang, Y. Liu, High capacity silicon/carbon composite anode materials for lithium ion batteries, Electrochem. Commun. 5 (2) (2003) 165–168.
[15] J.P. Maranchi, A.F. Hepp, P.N. Kumta, High-capacity, reversible silicon thin-film anodes for lithium ion batteries, Electrochem. Solid-State Lett. 6 (9) (2003) A198–A201.
[16] H. Jung, M. Park, Y.-G. Yoon, G.-B. Kim, S.-K. Joo, Amorphous silicon anode for lithium-ion rechargeable batteries, J. Power Sources 115 (2) (2003) 346–351.
[17] N. Dimov, S. Kugino, M. Yoshio, Carbon-coated silicon as anode material for lithium ion batteries: advantages and limitations, Electrochim. Acta 48 (11) (2003) 1579–1587.
[18] X.-W. Zhang, P.K. Patil, C. Wang, A.J. Appleby, F.E. Little, D.L. Cocke, Electrochemical performance of lithium ion battery, nano-silicon-based, disordered carbon composite anodes with different microstructures, J. Power Sources 125 (2) (2004) 206–213.
[19] C. Mikolajczak, M. Kahn, K. White, R.T. Long, Lithium-Ion Batteries Hazard and Use Assessment, Exponent Failure Analysis Associates, Inc., Menlo Park, CA, July 2011 94025, prepared for Fire Protection Research Foundation, Quincy, MA 02169-7471, Doc. no. 1100034.000 AOFO 0711 CMOl.
[20] X.-B. Cheng, R. Zhang, C.-Z. Zhao, Q. Zhang, Toward safe lithium metal anode in rechargeable batteries: a review, Chem. Rev. 117 (15) (2017) 10403–10473.
[21] J.B. Goodenough, Y. Kim, Challenges for rechargeable Li batteries, Chem. Mater. 22 (3) (2010) 587–603.
[22] B. Xu, D. Qian, Z. Wang, Y.S. Meng, Recent progress in cathode materials research for advanced lithium ion batteries, Mater. Sci. Eng. R 73 (5–6) (2012) 51–65.
[23] N. Nitta, F. Wu, J.T. Lee, G. Yushin, Li-ion battery materials: present and future, Mater. Today 18 (5) (2015) 252–264.
[24] A. Casimir, H. Zhang, O. Ogoke, J.C. Amine, J. Lu, G. Wu, Silicon-based anodes for lithium-ion batteries: effectiveness of materials synthesis and electrode preparation, Nano Energy 27 (2016) 359–376.

[25] Y. Sun, K. Liu, Y. Zhu, Recent progress in synthesis and application of low-dimensional silicon based anode material for lithium ion battery, J. Nanomater. 2017 (2017) 4780905, 15 p.

[26] P. Li, G. Zhao, X. Zheng, X. Xu, C. Yao, W. Sun, S.X. Dou, Recent progress on silicon-based anode materials for practical lithium-ion battery applications, Energy Storage Mater. 15 (2018) 422–446.

[27] H.-P. Feng, L. Tang, G.-M. Zeng, Y. Zhou, Y.-C. Deng, X. Ren, B. Song, C. Liang, M.-Y. Wei, J.-F. Yu, Core-shell nanomaterials: applications in energy storage and conversion, Adv. Colloid Interf. Sci. 267 (2019) 26–46.

[28] R.P. Raffaelle, J.D. Harris, D. Hehemann, D. Scheiman, G. Rybicki, A.F. Hepp, A facile route to thin-film solid state lithium microelectronic batteries, J. Power Sources 89 (1) (2000) 52–55.

[29] R.P. Raffaelle, A.F. Hepp, G.A. Landis, D.J. Hoffman, Mission applicability assessment of integrated power components and systems, Prog. Photovolt. Res. Appl. 10 (6) (2002) 391–397.

[30] J.P. Maranchi, I.S. Kim, A.F. Hepp, P. Kumta, Nanostructured electrochemically active materials: opportunities for aerospace power applications, in: AIAA Proceedings of the 1st International Energy Conversion Engineering Conference, Portsmouth, VA, USA, 2003, AIAA-2003-5952, 14 pp.

[31] R.P. Raffaelle, B.J. Landi, J.D. Harris, S.G. Bailey, A.F. Hepp, Carbon nanotubes for power applications, Mater. Sci. Eng. B 116 (3) (2005) 233–243.

[32] J.J. Wu, W.R. Bennett, Fundamental investigation of Si anode in Li-ion cells, in: 2012 IEEE Energytech, Cleveland, OH, USA, 2012, pp. 1–5.

[33] J.A. Caffrey, D.M., Hamby, a review of instruments and methods for dosimetry in space, Adv. Space Res. 47 (4) (2011) 563–574.

[34] J. Gonzalo, D. Domínguez, D. López, On the challenge of a century lifespan satellite, Prog. Aerosp. Sci. 70 (2014) 28–41.

[35] NASA, Space technology roadmaps and priorities: restoring NASA's technological edge and paving the way for a new era in space, Steering Committee for NASA Technology Roadmaps, Aeronautics and Space Engineering Board, Division on Engineering and Physical Sciences, National Research Council, National Academy Press, Washington, DC, 2012, 376 pp.

[36] Energy Storage Technologies for Future Planetary Science Missions, Jet Propulsion Laboratory, prepared for Planetary Science Division, Science Mission Directorate, NASA HQ, December 2017, JPL D-101146.

[37] P. Jenkins, D. Scheiman, D. Wilt, R. Raffaelle, R. Button, M. Smith, T. Kerslake, T. Miller, Advance power technology experiment for the Starshine 3 satellite, in: Proceedings of the 15th Annual/USU Conference on Small Satellites, Logan, UT, USA, August 2001, paper SSC01-VI-8, https://digitalcommons.usu.edu/smallsat/, (Accessed 30 July 2020).

[38] P. Jenkins, D. Scheiman, D. Wilt, R. Raffaelle, R. Button, T. Kerslake, M. Batchelder, D. Lefevre, R.G. Moore, Results from the advance power technology experiment on the Starshine 3 satellite, in: Proceedings of the 16th Annual/USU Conference on Small Satellites, Logan, UT, USA, August 2002, paper SSC02-X-3, https://digitalcommons.usu.edu/smallsat/, (Accessed 30 July 2020).

[39] http://www.azinet.com/starshine/, February 2003, (Accessed 30 July 2020).

[40] D.M. Wilt, A.F. Hepp, M. Moran, P.P. Jenkins, D.A. Scheiman, R.P. Raffaelle, Integrated micropower system (IMPS) development at NASA Glenn research center, in: Electrochemical Society Proceedings 2002-25, 2003, pp. 194–202.

[41] E.J. Simburger, J.H. Matsumoto, P.A. Gierow, A.F. Hepp, Integrated Thin Film Battery and Circuit Module, U.S. Patent 7,045,246, (May 16, 2006).

[42] Q. Li, Y. Liu, S. Guo, H. Zhou, Solar energy storage in the rechargeable batteries, Nano Today 16 (2017) 46–60.
[43] A. Gurung, Q. Qiao, Solar charging batteries: advances, challenges, and opportunities, Joule 2 (7) (2018) 1217–1230.
[44] D. Lau, N. Song, C. Hall, Y. Jiang, S. Lim, I. Perez-Wurfl, Z. Ouyang, A. Lennon, Hybrid solar energy harvesting and storage devices: the promises and challenges, Mater. Today Energy 13 (2019) 22–44.
[45] X. Zhu, Z. Guo, Z. Hou, Solar-powered airplanes: a historical perspective and future challenges, Prog. Aerosp. Sci. 71 (2014) 36–53.
[46] J. Gonzalo, D. López, D. Domínguez, A. García, A. Escapa, On the capabilities and limitations of high altitude pseudo-satellites, Prog. Aerosp. Sci. 98 (2018) 37–56.
[47] Y. Xu, W. Zhu, J. Li, L. Zhang, Improvement of endurance performance for high-altitude solar-powered airships: a review, Acta Astronaut. 167 (2020) 245–259.
[48] https://spinoff.nasa.gov/Spinoff2019/t_1.html, (Accessed 31 July 2020).
[49] https://www.nasa.gov/content/what-are-smallsats-and-cubesats, (Accessed 30 July 2020).
[50] F.C. Krause, J.A. Loveland, M.C. Smart, E.J. Brandon, R.V. Bugga, Implementation of commercial Li-ion cells on the MarCO deep space CubeSats, J. Power Sources 449 (2020) 227544.
[51] A. Probst, R. Förstner, Spacecraft design of a multiple asteroid orbiter with re-docking lander, Adv. Space Res. 62 (8) (2018) 2125–2140.
[52] A.J. Appleby, D.Y.C. Ng, S.K. Wolfson Jr., H. Weinstein, Implantable biological fuel cell with an air-breathing cathode, in: Proc. of 4th Intersociety Energy Conversion Engineering Conference, Washington, DC, USA; September 22–26, 1969.
[53] B.Y.C. Wan, A.C.C. Tseung, Some studies related to electricity generation from biological fuel cells and galvanic cells, in vitro and in vivo, Med. Biol. Eng. 12 (1) (1974) 14–28.
[54] D.C. Jeutter, A transcutaneous implanted battery recharging and biotelemeter power switching system, IEEE Trans. Biomed. Eng. BME-29 (5) (May 1982) 314–321.
[55] M. Maeda, M. Nishijima, H. Takehara, C. Adachi, H. Fujimoto, O. Ishikawa, A 3.5 V, 1.3 W GaAs power multi-chip IC for cellular phones, IEEE J. Solid State Circuits 29 (10) (1994) 1250–1256.
[56] R. Hahn, H. Reichl, Batteries and power supplies for wearable and ubiquitous computing, in: Digest of papers, Third International Symposium on Wearable Computers, San Francisco, CA, USA, October 18–19, 1999, pp. 168–169.
[57] J.A. Hollingsworth, A.F. Hepp, W.E. Buhro, Spray CVD of copper indium sulfide films: control of microstructure and crystallographic orientation, Chem. Vap. Depos. 3 (1999) 105–108.
[58] K.K. Banger, J.A. Hollingsworth, J.D. Harris, J. Cowen, W.E. Buhro, A.F. Hepp, Ternary single-source precursors for polycrystalline thin-film solar cells, Appl. Organomet. Chem. 16 (2002) 617–627.
[59] A.F. Hepp, R.P. Raffaelle, K.K. Banger, M.H. Jin, J.E. Lau, J.D. Harris, J.E. Cowen, S.A. Duraj, Chemical vapor deposition for ultralightweight thin-film solar arrays, in: Proceedings of the 37th Intersociety Energy Conversion Engineering Conference, Washington, DC, USA, 2002, pp. 198–203.
[60] A.F. Hepp, M.A. Smith, J.H. Scofield, J.E. Dickman, G.B. Lush, D.L. Morel, C. Ferekides, N.G. Dhere, Multi-Junction Thin-Film Solar Cells on Flexible Substrates for Space Power, NASA, October 2002, 12, TM/2002-211834.
[61] A.F. Hepp, J.S. McNatt, S.G. Bailey, R.P. Raffaelle, B.J. Landi, S.-S. Sun, C.E. Bonner, K.K. Banger, D. Rauh, Ultra-lightweight space power from hybrid thin-film solar cells, IEEE Aerosp. Electron. Syst. Mag. 23 (9) (2008) 31–41.

[62] https://www.nrel.gov/pv/cell-efficiency.html, (Accessed 30 July 2020).
[63] J.D. Harris, R.P. Raffaelle, T. Gennett, B.J. Landi, A.F. Hepp, Growth of multi-walled carbon nanotubes by injection CVD using cyclopentadienyliron dicarbonyl dimer and cyclooctatetraene iron tricarbonyl, Mater. Sci. Eng. B 116 (2005) 369–374.
[64] R.P. Raffaelle, J. Underwood, D. Scheiman, J. Cowen, P. Jenkins, A.F. Hepp, J. Harris, D.M. Wilt, Integrated solar power systems, in: Conference Record of the 28th IEEE Photovoltaic Specialists Conference, Anchorage, AK, USA, 2000, pp. 1370–1373.
[65] R.P. Raffaelle, J. Underwood, J. Maranchi, P. Kumta, O.P. Khan, J. Harris, C.R. Clark, D. Scheiman, P. Jenkins, M.A. Smith, D.M. Wilt, R.M. Button, A.F. Hepp, W.F. Maurer, Integrated microelectronic power supply (IMPS), in: Proceedings of the 36th Intersociety Energy Conversion Engineering Conference vol. 1, Savannah, GA, USA, 2001, pp. 239–242.
[66] R.P. Raffaelle, P. Jenkins, D. Scheiman, M.A. Smith, D. Wilt, T. Kerslake, A.F. Hepp, Microsat power supplies, in: 40th AIAA Aerospace Sciences Meeting & Exhibit, Reno, NV, USA, 2002, p. 4, AIAA-2002-0719.
[67] J. Bouwmeester, J. Guo, Survey of worldwide pico- and nanosatellite missions, distributions and subsystem technology, Acta Astronaut. 67 (7–8) (2010) 854–862.
[68] E. Agasid, R. Burton, R. Carlino, G. Defouw, A.D. Perez, A.G. Karacalıoğlu, B. Klamm, A. Rademacher, J. Schalkwyk, R. Shimmin, J. Tilles, S. Weston, State of the Art Small Spacecraft Technology, NASA/TP—2018–220027, (December 2018) 207. Available at: https://www.nasa.gov/sites/default/files/atoms/files/soa2018_final_doc-6.pdf, to be updated Fall 2020, Accessed July 30, 2020.
[69] https://www.nanosats.eu/#page-top, (Accessed 30 July 2020).
[70] K. Moyer, C. Meng, B. Marshall, O. Assal, J. Eaves, D. Perez, R. Karkkainen, L. Roberson, C.L. Pint, Carbon fiber reinforced structural lithium-ion battery composite: multifunctional power integration for CubeSats, Energy Storage Mater. 24 (2020) 676–681.
[71] B. Braun, C. Butkiewicz, I. Sokolsky, J. Vasquez, G. Moore, The Starshine satellite: from concept to delivery in four months, in: Proceedings of the 13th Annual/USU Conference on Small Satellites, Logan, UT, USA, August 1999, paper SSC99-I-7, https://digitalcommons.usu.edu/smallsat/, (Accessed 30 July 2020).
[72] S. Schwartz, R.T. Nallapu, P. Gankidi, G. Dektor, J. Thangavelautham, Navigating to small-bodies using small satellites, in: 2018 IEEE/ION Position, Location and Navigation Symposium (PLANS), Monterey, CA, USA, 2018, pp. 1277–1285.
[73] J. Thangavelautham, E. Asphaug, S. Schwartz, An on-orbit cubesat centrifuge for asteroid science and exploration, in: 2019 IEEE Aerospace Conference, Big Sky, MT, USA, 2019, pp. 1–11.
[74] H. Chen, T.S. du Jonchay, L. Hou, K. Ho, Integrated in-situ resource utilization system design and logistics for Mars exploration, Acta Astronaut. 170 (2020) 80–92.
[75] M.F. Palos, P. Serra, S. Fereres, K. Stephenson, R. González-Cinca, Lunar ISRU energy storage and electricity generation, Acta Astronaut. 170 (2020) 412–420.
[76] T.L. Gibson, N.A. Kelly, Solar photovoltaic charging of lithium-ion batteries, J. Power Sources 195 (12) (2010) 3928–3932.
[77] S.N. Agbo, T. Merdzhanova, S. Yu, H. Tempel, H. Kungl, R.-A. Eichel, U. Rau, O. Astakhov, Development towards cell-to-cell monolithic integration of a thin-film solar cell and lithium-ion accumulator, J. Power Sources 327 (2016) 340–344.
[78] R. Rüther, G. Kleiss, K. Reiche, Spectral effects on amorphous silicon solar module fill factors, Sol. Energy Mater. Sol. Cells 71 (2002) 375–385.
[79] F. Urbain, V. Smirnov, J.P. Becker, U. Rau, J. Ziegler, F. Yang, B. Kaiser, W. Jaegermann, S. Hoch, M. Blug, F. Finger, Solar water splitting with earth-abundant materials using

amorphous silicon photocathodes and Al/Ni contacts as hydrogen evolution catalyst, Chem. Phys. Lett. 638 (2015) 25–30.
[80] Y. Li, N.J. Grabham, S.P. Beeby, M.J. Tudor, The effect of the type of illumination on the energy harvesting performance of solar cells, Sol. Energy 111 (2015) 21–29.
[81] F. Urbain, V. Smirnov, J.-P. Becker, A. Lambertz, U. Rau, F. Finger, Light-induced degradation of adapted quadruple junction thin film silicon solar cells for photoelectrochemical water splitting, Sol. Energy Mater. Sol. Cell 145 (2016) 142–147.
[82] S.N. Agbo, T. Merdzhanova, S. Yu, H. Tempel, H. Kungl, R.-A. Eichel, U. Rau, O. Astakhov, Development towards cell-to-cell monolithic integration of a thin-film solar cell and lithium-ion accumulator, J. Power Sources 327 (2016) 340–344.
[83] L.Q. Sun, R.H. Cui, A.F. Jalbout, M.J. Li, X.M. Pan, R.S. Wang, H.M. Xie, LiFePO$_4$ as an optimum power cell material, J. Power Sources 189 (2009) 522–526.
[84] P. Reale, A. Fernicola, B. Scrosati, Compatibility of the Py$_{24}$TFSI-LiTFSI ionic liquid solution with Li$_4$Ti$_5$O$_{12}$ and LiFePO$_4$ lithium ion battery, J. Power Sources 194 (2009) 182–189.
[85] S.W. Oh, S.-T. Myung, S.-M. Oh, K.H. Oh, K. Amine, B. Scrosati, Y.-K. Sun, Double carbon coating of LiFePO4 as high rate electrode for rechargeable lithium batteries, Adv. Mater. 22 (2010) 4842–4845.
[86] M.R. Hill, G.J. Wilson, L. Bourgeios, A.G. Pandolfo, High surface area templated LiFePO$_4$ from a single source precursor molecule, Energy Environ. Sci. 4 (2011) 965–972.
[87] T.F. Yi, S.Y. Yang, Y. Xie, Recent advances of Li$_4$Ti$_5$O$_{12}$ as a promising next generation anode material for high power lithium-ion batteries, J. Mater. Chem. A 3 (2015) 5750–5777.
[88] Y. Honda, S. Muto, K. Tatsumi, H. Kondo, K. Horibuchi, T. Kobayashi, T. Sasaki, Microscopic mechanism of path-dependence on charge–discharge history in lithium iron phosphate cathode analysis using scanning transmission electron microscopy and electron energy-loss spectroscopy spectral imaging, J. Power Sources 291 (2015) 85–94.
[89] L. Ruiyi, J. Yuanyuan, Z. Xiaoyan, L. Zaijun, G. Zhiguo, W. Guangli, L. Junkang, Significantly enhanced electrochemical performance of lithium titanate anode for lithium ion battery by the hybrid of nitrogen and sulfur co-doped graphene quantum dots, Electrochim. Acta 178 (2015) 303–311.
[90] D. Schmidt, M.D. Hager, U.S. Schubert, Photo-rechargeable electric energy storage systems, Adv. Energy Mater. 6 (1) (2016) 1500369.
[91] V. Vega-Garita, L. Ramirez-Elizondo, N. Narayan, P. Bauer, Integrating a photovoltaic storage system in one device: a critical review, Prog. Photovolt. 27 (4) (2019) 346–370.
[92] P. Liu, Y.L. Cao, G.R. Li, X.P. Gao, X.P. Ai, H.X. Yang, A solar rechargeable flow battery based on photoregeneration of two soluble redox couples, ChemSusChem 6 (2013) 802–806.
[93] S. Liao, X. Zong, B. Seger, T. Pedersen, T. Yao, C. Ding, J. Shi, J. Chen, C. Li, Integrating a dual-silicon photoelectrochemical cell into a redox flow battery for unassisted photocharging, Nat. Commun. 7 (2016) 11474.
[94] W. Li, H.-C. Fu, L. Li, M. Cabán-Acevedo, J.-H. He, S. Jin, Integrated photoelectrochemical solar energy conversion and organic redox flow battery devices, Angew. Chem. Int. Ed. 55 (2016) 13104–13108.
[95] W. Guo, X. Xue, S. Wang, C. Lin, Z.L. Wang, An integrated power pack of dye-sensitized solar cell and Li battery based on double-sided TiO$_2$ nanotube arrays, Nano Lett. 12 (2012) 2520–2523.
[96] R.B. Ye, K. Yoshida, K. Ohta, M. Baba, Integrated thin-film rechargeable battery on α-Si thin-film solar cell, Adv. Mater. Res. 788 (2013) 685–688.
[97] J. Xu, Y. Chen, L. Dai, Efficiently photo-charging lithium-ion battery by perovskite solar cell, Nat. Commun. 6 (2015) 8103.

[98] A. Gurung, K. Chen, R. Khan, S.S. Abdulkarim, G. Varnekar, R. Pathak, R. Naderi, Q. Qiao, Highly efficient perovskite solar cell photocharging of lithium ion battery using DC-DC booster, Adv. Energy Mater. 7 (2017) 1602105.

[99] H.-D. Um, K.-H. Choi, I. Hwang, S.-H. Kim, K. Seo, S.-Y. Lee, Monolithically integrated, photo-rechargeable portable power sources based on miniaturized Si solar cells and printed solid-state lithium-ion batteries, Energy Environ. Sci. 10 (4) (2017) 931–940.

[100] F. Sandbaumhüter, S.N. Agbo, C.-L. Tsai, O. Astakhov, S. Uhlenbruck, U. Rau, T. Merdzhanova, Compatibility study towards monolithic self-charging power unit based on all-solid thin-film solar module and battery, J. Power Sources 365 (2017) 303–307.

[101] A. Paolella, C. Faure, G. Bertoni, S. Marras, A. Guerfi, A. Darwiche, P. Hovington, B. Commarieu, Z. Wang, M. Prato, Light-assisted delithiation of lithium iron phosphate nanocrystals towards photo-rechargeable lithium ion batteries, Nat. Commun. 8 (2017) 14643.

[102] W. Li, H.-C. Fu, Y. Zhao, J.-H. He, S. Jin, 14.1% efficient monolithically integrated solar flow battery, Chem 4 (11) (2018) 2644–2657.

[103] Y. Hu, Y. Bai, B. Luo, S. Wang, H. Hu, P. Chen, M. Lyu, J. Shapter, A. Rowan, L. Wang, A portable and efficient solar-rechargeable battery with ultrafast photo-charge/discharge rate, Adv. Energy Mater. 9 (28) (2019) 1900872.

[104] W. Li, E. Kerr, M.-A. Goulet, H.-C. Fu, Y. Zhao, Y. Yang, A. Veyssal, J.-H. He, R.G. Gordon, M.J. Aziz, S. Jin, A long lifetime aqueous organic solar flow battery, Adv. Energy Mater. 9 (31) (2019) 1900918.

[105] A. Khataee, J. Azevedo, P. Dias, D. Ivanou, E. Dražević, A. Bentien, A. Mendes, Integrated design of hematite and dye-sensitized solar cell for unbiased solar charging of an organic-inorganic redox flow battery, Nano Energy 62 (2019) 832–843.

[106] G.-M. Weng, J. Kong, H. Wang, C. Karpovich, J. Lipton, F. Antonio, Z.S. Fishman, H. Wang, W. Yuan, A.D. Taylor, A highly efficient perovskite photovoltaic-aqueous Li/Na-ion battery system, Energy Storage Mater. 24 (2020) 557–564.

[107] L.-C. Kin, Z. Liu, O. Astakhov, S.N. Agbo, H. Tempel, S. Yu, H. Kungl, R.-A. Eichel, U. Rau, T. Kirchartz, T. Merdzhanova, Efficient area matched converter aided solar charging of lithium ion batteries using high voltage perovskite solar cells, ACS Appl. Energy Mater. 3 (1) (2020) 431–439.

[108] A. Gurung, K.M. Reza, S. Mabrouk, B. Bahrami, R. Pathak, B.S. Lamsal, S.I. Rahman, N. Ghimire, R.S. Bobba, K. Chen, J. Pokharel, A. Baniya, M.A.R. Laskar, M. Liang, W. Zhang, W.-H. Zhang, S. Yang, K. Xu, Q. Qiao, Rear-illuminated perovskite photorechargeable lithium battery. Adv. Funct. Mater. (2020) https://doi.org/10.1002/adfm.202001865, pre-print.

[109] M. Liu, M. Du, G. Long, H. Wang, W. Qin, D. Zhang, S. Ye, S. Liu, J. Shi, Z. Liang, C. Li, Iron/Quinone-based all-in-one solar rechargeable flow cell for highly efficient solar energy conversion and storage, Nano Energy 76 (2020) 104907.

[110] H. Sun, Y. Jiang, S. Xie, Y. Zhang, J. Ren, A. Ali, S.-G. Doo, I.H. Son, X. Huang, H. Peng, Integrating photovoltaic conversion and lithium ion storage into a flexible fiber, J. Mater. Chem. A 4 (20) (2016) 7601–7605.

[111] Y.H. Kwon, S.-W. Woo, H.-R. Jung, H.K. Yu, K. Kim, B.H. Oh, S. Ahn, S.-Y. Lee, S.-W. Song, J. Cho, H.-C. Shin, J.Y. Kim, Cable-type flexible lithium ion battery based on hollow multi-helix electrodes, Adv. Mater. 24 (38) (2012) 5192–5197.

[112] W. Weng, Q. Sun, Y. Zhang, H. Lin, J. Ren, X. Lu, M. Wang, H. Peng, Winding aligned carbon nanotube composite yarns into coaxial fiber full batteries with high performance, Nano Lett. 14 (6) (2014) 3432–3438.

[113] J. Ren, Y. Zhang, W. Bai, X. Chen, Z. Zhang, X. Fang, W. Weng, Y. Wang, H. Peng, Elastic and wearable wire-shaped lithium-ion battery with high electrochemical performance, Angew. Chem. Int. Ed. 53 (30) (2014) 7864–7869.

[114] X. Wang, K. Jiang, G. Shen, Flexible fiber energy storage and integrated devices: recent progress and perspectives, Mater. Today 18 (5) (2015) 265–272, (April 2019) 346–370.

[115] A. Vlad, N. Singh, C. Galande, P.M. Ajayan, Design considerations for unconventional electrochemical energy storage architectures, Adv. Energy Mater. 5 (19) (2015) 1402115.

[116] G. Dennler, S. Bereznev, D. Fichou, K. Holl, D. Ilic, R. Koeppe, M. Krebs, A. Labouret, C. Lungenschmied, A. Marchenko, D. Meissner, E. Mellikov, J. Méot, A. Meyer, T. Meyer, H. Neugebauer, A. Öpik, N.S. Sariciftci, S. Taillemite, T. Wöhrle, A self-rechargeable and flexible polymer solar battery. Sol. Energy 81 (8) (2007) 947–957, https://doi.org/10.1016/j.solener.2007.02.008.

[117] F.C. Krebs, J. Fyenbo, M. Jørgensen, Product integration of compact roll-to-roll processed polymer solar cell modules: methods and manufacture using flexographic printing, slot-die coating and rotary screen printing. J. Mater. Chem. 20 (41) (2010) 8994–9001, https://doi.org/10.1039/c0jm01178a.

[118] Z. Gao, C. Bumgardner, N. Song, Y. Zhang, J. Li, X. Li, Cotton-textile-enabled flexible self-sustaining power packs via roll-to-roll fabrication. Nat. Commun. 7 (2016) 11586, https://doi.org/10.1038/ncomms11586.

[119] K.E. Jeon, T. Tong, J. She, Preliminary design for sustainable BLE beacons powered by solar panels. in: 2016 IEEE Conference on Computer Communications Workshops (INFOCOM WKSHPS), San Francisco, CA, 2016, pp. 103–109, https://doi.org/10.1109/INFOCOMW.2016.7562054.

[120] Y. Wang, L. Zhang, K. Cui, C. Xu, H. Li, H. Liu, J. Yu, Solar driven electrochromic photoelectrochemical fuel cells for simultaneous energy conversion, storage and self-powered sensing. Nanoscale 10 (2018) 3421–3428, https://doi.org/10.1039/c7nr09275j.

[121] M. Fojtik, D. Kim, G. Chen, Y.-S. Lin, D. Fick, J. Park, M. Seok, M.-T. Chen, Z. Foo, D. Blaauw, D. Sylvester, A millimeter-scale energy-autonomous sensor system with stacked battery and solar cells. IEEE J. Solid State Circuits 48 (3) (2013) 801–813, https://doi.org/10.1109/JSSC.2012.2233352.

[122] K.S. Adu-Manu, N. Adam, C. Tapparello, H. Ayatollahi, W. Heinzelman, Energy-harvesting wireless sensor networks (EH-WSNs): a review, ACM Trans. Sens. Netw. 14 (2) (2018) 1–50.

[123] G.A. Niklasson, C.G. Granqvist, Electrochromics for smart windows: thin films of tungsten oxide and nickel oxide, and devices based on these, J. Mater. Chem. 17 (2) (2007) 127–156.

[124] Z. Xie, X. Jin, G. Chen, J. Xu, D. Chen, G. Shen, Integrated smart electrochromic windows for energy saving and storage applications. Chem. Commun. 2014 (5) (2014) 608–610, https://doi.org/10.1039/c3cc47950a.

[125] F.C. Krebs, T.D. Nielsen, J. Fyenbo, M. Wadstrom, M.S. Pedersen, Manufacture, integration and demonstration of polymer solar cells in a lamp for the "lighting Africa" initiative, Energy Environ. Sci. 3 (5) (2010) 512–525.

[126] Waka Waka, Datasheet. Waka Waka Light—solar-powered LED flashlight, https://wakawaka.com/en/manuals/, 2014, (Accessed 24 July 2020).

[127] Y.H. Lee, J.S. Kim, J. Noh, I. Lee, H.J. Kim, S. Choi, J. Seo, S. Jeon, T.S. Kim, J.Y. Lee, J.W. Choi, Wearable textile battery rechargeable by solar energy. Nano Lett. 13 (11) (2013) 5753–5761, https://doi.org/10.1021/nl403860k.

[128] K. Jost, G. Dion, Y. Gogotsi, Textile energy storage in perspective. J. Mater. Chem. A 2 (28) (2014) 10776–10787, https://doi.org/10.1039/c4ta00203b.

[129] S. Lemey, F. Declercq, H. Rogier, Textile antennas as hybrid energy-harvesting platforms. Proc. IEEE 102 (11) (2014) 1833–1857, https://doi.org/10.1109/JPROC.2014.2355872.

[130] A.E. Ostfeld, A.M. Gaikwad, Y. Khan, A.C. Arias, High-performance flexible energy storage and harvesting system for wearable electronics, Sci. Rep. 6 (2016) 26122.

[131] C. Li, M.M. Islam, J. Moore, J. Sleppy, C. Morrison, K. Konstantinov, S.X. Dou, C. Renduchintala, J. Thomas, Wearable energy-smart ribbons for synchronous energy harvest and storage. Nat. Commun. 7 (2016) https://doi.org/10.1038/ncomms13319.

[132] W. Liu, M.S. Song, B. Kong, Y. Cui, Flexible and stretchable energy storage: recent advances and future perspectives. Adv. Mater. 28 (1) (2017) 1603436, https://doi.org/10.1002/adma.201603436.

[133] X. Pu, W. Hu, Z.L. Wang, Toward wearable self-charging power systems: the integration of energy-harvesting and storage devices. Small 14 (1) (2018) 1702817, https://doi.org/10.1002/smll.201702817.

[134] T. Hughes-Riley, T. Dias, C. Cork, A historical review of the development of electronic textiles, Fibers 6 (2) (2018) 34.

[135] G. Apostolou, A.H.M.E. Reinders, Overview of design issues in product-integrated photovoltaics, Energy Technol. 2 (3) (2014) 229–242.

[136] N.P. Lebedeva, L. Boon-Brett, Considerations on the chemical toxicity of contemporary Li-ion battery electrolytes and their components, J. Electrochem. Soc. 163 (6) (2016) A821–A830.

[137] A. Manthiram, X. Yu, S. Wang, Lithium battery chemistries enabled by solid-state electrolytes, Nat. Rev. Mater. 2 (2017) 16103.

[138] J.Y. Luo, W.J. Cui, P. He, Y.Y. Xia, Raising the cycling stability of aqueous lithium-ion batteries by eliminating oxygen in the electrolyte, Nat. Chem. 2 (2010) 760–765.

[139] T. Miyasaka, T.N. Murakami, The photocapacitor: an efficient self-charging capacitor for direct storage of solar energy, Appl. Phys. Lett. 85 (17) (2004) 3932–3934.

[140] T.N. Murakami, N. Kawashima, T. Miyasaka, A high-voltage dye-sensitized photocapacitor of a three-electrode system, Chem. Commun. 2005 (26) (2005) 3346–3348.

[141] T. Chen, L. Qiu, Z. Yang, Z. Cai, J. Ren, H. Li, H. Lin, X. Sun, H. Peng, An integrated "energy wire" for both photoelectric conversion and energy storage, Angew. Chem. Int. Ed. 51 (48) (2012) 11977–11980.

[142] D.J. Barnhart, T. Vladimirova, A.M. Baker, M.N. Sweeting, A low-cost femtosatellite to enable distributed space missions, Acta Astronaut. 64 (11−12) (2009) 1123–1143.

[143] A. Poghosyan, A. Golkar, CubeSat evolution: analyzing CubeSat capabilities for conducting science missions, Prog. Aerosp. Sci. 88 (2017) 59–83.

[144] K. Lemmer, Propulsion for CubeSats, Acta Astronaut. 134 (2017) 231–243.

[145] T. Villela, C.A. Costa, A.M. Brandão, F.T. Bueno, R. Leonardi, Towards the thousandth CubeSat: a statistical overview, Int. J. Aerosp. Eng. 2019 (2019) 5063145, (13 pp.), https://doi.org/10.1155/2019/5063145.

[146] https://sites.google.com/a/slu.edu/swartwout/home/cubesat-database, (Accessed 17 August 2020).

[147] G. Zubi, R. Dufo-López, M. Carvalho, G. Pasaoglu, The lithium-ion battery: state of the art and future perspectives, Renew. Sust. Energ. Rev. 89 (2018) 292–308.

[148] D. Pande, D. Verstraete, Impact of solar cell characteristics and operating conditions on the sizing of a solar powered nonrigid airship, Aerosp. Sci. Technol. 72 (2018) 353–363.

[149] S. Koohi-Fayegh, M.A. Rosen, A review of energy storage types, applications and recent developments, J. Energy Storage 27 (2020) 101047.

[150] G.A. Landis, S. Oleson, M. McGuire, A. Hepp, J. Stegman, M. Bur, L. Burke, M. Martini, J. Fittje, L. Kohout, J. Fincannon, T. Packard, A Cubesat asteroid mission: design study and trade-offs, in: Paper IAC-14,B4,8,9,x26216, 65th International Astronautical Congress, Toronto, Canada, September 29–October 3, 2014, p. 6. Available at: https://ntrs.nasa.gov/citations/20150002091, (Accessed 20 August 2020).
[151] https://www.jpl.nasa.gov/cubesat/missions/raincube.php, (Accessed 20 August 2020).
[152] D.M. McIntosh, J.D. Baker, J.A. Matus, The NASA Cubesat missions flying on artemis-1, in: Proceedings of the 34th Annual AIAA/USU Conference on Small Satellites, Logan, UT, USA, July 2020, p. 11. paper SSC20-WKVII-02, https://digitalcommons.usu.edu/smallsat/2020/all2020/44/, (Accessed 20 August 2020).
[153] https://www.saftbatteries.com/products-solutions/products, (Accessed 21 August 2020).
[154] https://www.eaglepicher.com/markets/space/, (Accessed 21 August 2020).
[155] http://www.molicel.com/products-applications/, (Accessed 21 August 2020).
[156] https://www.batteryspace.com/18650seriesli-ioncells.aspx, (Accessed 21 August 2020).
[157] https://www.nasa.gov/topics/aeronautics/features/trl_demystified.html, (Accessed 17 August 2020).
[158] https://www.aac-clyde.space/satellite-bits/batteries, (Accessed 22 August 2020).
[159] https://www.berlin-space-tech.com/portfolio/battery-bat-110/, (Accessed 22 August 2020).
[160] https://www.enersys.com/, (Accessed 22 August 2020).
[161] https://gomspace.com/shop/subsystems/power/default.aspx, (Accessed 22 August 2020).
[162] R.W. Cook, L.G. Swan, K.P. Plucknett, Failure mode analysis of lithium ion batteries operated for low Earth orbit CubeSat applications, J. Energy Storage 31 (2020) 101561.
[163] Panasonic, NCR18650B Data Sheet, https://www.batteryspace.com/prod-specs/NCR18650B.pdf, (Accessed 23 August 2020).
[164] J.J. Kim, S.H. Ahn, LG Chem ICR18650B4 Data Sheet, vol. 21 (1), (2020) https://www.batteryspace.com/prod-specs/5457_B4.pdf, (Accessed 23 August 2020).
[165] LithiumWerks, APR18650M1B Data Sheet, https://www.batteryspace.com/prod-specs/6612-APR18650M1B.pdf, (Accessed 23 August 2020).
[166] G.A. Harvey, W.H. Kinard, MISSE 1 & 2 Tray Temperature Measurements, MISSE Post-Retrieval Conference Orlando FL, Available from: https://ntrs.nasa.gov/citations/20060020702, June 2006, (Accessed 23 August 2020).
[167] K.K. de Groh, B.A. Banks, D.C. Smith, Environmental Durability Issues for Solar Power Systems in Low Earth Orbit, NASA TM 106775, Available from:https://ntrs.nasa.gov/citations/19950009355, March 1995, (Accessed 23 August 2020).
[168] https://solarsystem.nasa.gov/missions/mars-cube-one/in-depth/, (Accessed 24 August 2020).
[169] T.K. Andrew, J. Baker, J. Krajewski, MarCO: flight review and lessons learned, in: Proceedings of the 33rd Annual AIAA/USU Conference on Small Satellites, Logan, UT, USA paper: SSC19-III-04, August 2019, p. 6. https://digitalcommons.usu.edu/cgi/viewcontent.cgi?article=4575&context=smallsat, (Accessed 24 August 2020).
[170] https://www.jpl.nasa.gov/infographics/infographic.view.php?id=10776, (Accessed 24 August 2020).
[171] E. Decrossas, N. Chahat, P.E. Walkemeyer, B.S. Velasco, Deployable circularly polarized UHF printed loop antenna for Mars Cube One (MarCO) CubeSat, in: 2019 IEEE International Symposium on Antennas and Propagation and USNC-URSI Radio Science Meeting, Atlanta, GA, USA, 2019, pp. 1719–1720.
[172] R.E. Hodges, N. Chahat, D.J. Hoppe, J.D. Vacchione, A deployable high-gain antenna bound for mars: developing a new folded-panel reflectarray for the first CubeSat mission to Mars.

IEEE Antennas Propag. Mag. 59 (2) (2017) 39–49, https://doi.org/10.1109/MAP.2017.2655561.
[173] C.B. Duncan, A.E. Smith, F.H. Aguirre, Iris transponder—communications and navigation for deep space, in: Proceedings of the 28th Annual AIAA/USU Conference on Small Satellites, Logan, UT, USA, paper: SSC14-IX-3, August 2014, 10 pp.
[174] https://bluecanyontech.com/components, (Accessed 24 August 2020).
[175] https://www.cubesat-propulsion.com/cusp-propulsion-system/, (Accessed 24 August 2020).
[176] https://directory.eoportal.org/web/eoportal/satellite-missions/m/marco#references, (Accessed 24 August 2020).
[177] https://mmadesignllc.com/products/solar-arrays/, (Accessed 24 August 2020).
[178] B.V. Ratnakumar, M.C. Smart, C.K. Huang, D. Perrone, S. Surampudi, S.G. Greenbaum, Lithium ion batteries for Mars exploration missions. Electrochim. Acta 45 (8–9) (2000) 1513–1517, https://doi.org/10.1016/S0013-4686(99)00367-9.
[179] B.V. Ratnakumar, M.C. Smart, L.D. Whitcanack, E.D. Davies, K.B. Chin, F. Deligiannis, S. Surampudi, Behavior of Li-ion cells in high-intensity radiation environments, J. Electrochem. Soc. 151 (4) (2004) A652–A659.
[180] A.A. Siddiqi, Beyond Earth: A Chronicle of Deep Space Exploration, 1958–2016, second ed., National Aeronautics and Space Administration, Office of Communications, NASA History Division, Washington, DC, 2018. NASA SP2018-4041, The NASA History Series, https://www.nasa.gov/connect/ebooks/beyond_earth_detail.html, (Accessed 27 August 2020).
[181] https://nssdc.gsfc.nasa.gov/planetary/upcoming.html, (Accessed 31 August 2020).
[182] B.H. Foing, G. Racca, A. Marini, D. Koschny, D. Frew, B. Grieger, O. Camino-Ramos, J.L. Josset, M. Grande, SMART-1 science and technology working team, SMART-1 technology, scientific results and heritage for future space missions, Planet. Space Sci. 151 (2018) 141–148.
[183] G. Johnson, Memories and safety lessons learned of an Apollo electrical engineer, J. Space Saf. Eng. 7 (1) (2020) 18–26.
[184] T. Hoshino, S. Wakabayashi, M. Ohtake, Y. Karouji, T. Hayashi, H. Morimoto, H. Shiraishi, T. Shimada, T. Hashimoto, H. Inoue, R. Hirasawa, Y. Shirasawa, H. Mizuno, H. Kanamori, Lunar polar exploration mission for water prospection—JAXA's current status of joint study with ISRO, Acta Astronaut. 176 (2020) 52–58.
[185] M.E. Evans, L.D. Graham, A flexible lunar architecture for exploration (FLARE) supporting NASA's Artemis program, Acta Astronaut. 177 (2020) 351–372.
[186] A.F. Hepp, P.N. Kumta, O. Velikokhatnyi, M.K. Datta, Batteries for aeronautics and space exploration: recent developments and future prospects, in: A.F. Hepp, P.N. Kumta, O. Velikokhatnyi, M.K. Datta (Eds.), Lithium-Sulfur Batteries: Advances in High-Energy Density Batteries, Elsevier, New York, 2021.

Index

Note: Page numbers followed by *f* indicate figures, and *t* indicate tables.

A

Ab initio molecular dynamics (AIMD), 217–218
Acrylic resin, 437–438, 437*f*
Active and inactive phases, of Si-based materials
  carbon coating, 268, 269*f*
    graphitization, 268–271
  lithiation process, 265–268
  nonstoichiometric SiO_x particles, 265–268
  organic matrix, 271
  organic precursors, 268, 270*f*
  petroleum pitch, 268–271
  silicon dioxide (SiO_2) surface layer, 265
Alginate, 24–25
Alloy-forming approach, 20
Ambient temperature, carbon-coated 24 h-milled silicon, 397–398, 398*f*
Ammonium persulfate (APS), 360–362, 438–440
Amorphous carbon (AC) coating, 34, 341–342, 441
Amorphous Si (a-Si), 425–426
  cracking of, 128
  delamination, 165–166, 167*f*
  effect of
    buffer layer stiffness, 166–172
    interface properties, 172–174, 173*f*
  lithiation process, 125–127, 127*f*
  square-shaped, 164–165
Anodization method, 426–427
Atomic force microscopy (AFM) technique, 12–13, 212–214
Atomic layer deposition (ALD), 158, 343–344
Attenuated total reflection Fourier transform infrared spectroscopy (ATR-FTIR), 10

B

Berendsen barostat, 102
Berendsen thermostat, 102
BET surface area, Si particles, 138, 141*t*
Binder
  conventional, 222–224
  polymeric, 224–247
  pulverization and delamination, 21–28
  selection, 437–440
  self-healing polymers, 245–246
  Si electrodes, 132–138, 272–277
Biomass, 69–71
Boron-doping, 426–427
"Bottom-up" mechanism, 191
Buffer layer stiffness, 166–172
Butler-Volmer equation, 161–162, 161*t*, 301–303

C

Cahn-Hilliard formulation, 301–302, 312–314, 319
Carbon-coated Si nanoparticles, 432–434
Carbon nanofiber (CNF), 349–350
Carbon nanotubes (CNTs), 429–430, 432, 458
Carbon-silicon hybrid materials, 49–57
4-Carbonxybenzaldehyde (CBA), 438–440
Carbothermal process, metallurgical silicon, 414
Carboxymethyl cellulose (CMC), 23, 132–133, 136*f*, 272
Catechol, 26
Catechol-functionalized chitosan cross-linked by glutaraldehyde (CS-CG+GA), 137–138
Cellulose fiber, 175–176
Charge-discharge curves, 132–133, 132*f*
  etched and coated silicon anode, 391–393, 392*f*
  Si-Ti thin films, 150, 150*f*
  Si *vs.* SiO electrodes, 143–144, 143*f*
Charge distribution, lithiated Si, 104, 105*f*
Charge rate (C-rate), 164, 176–177*t*
Chevrier-Dahn Delithiation, 100, 100*f*
Chevrier-Dahn Lithiation, 97, 97*f*
Chromatography, 288–289
Clays, 72
Commercial off-the-shelf (COTS), 461, 486*t*, 488–489*t*
Computational method, 164
Conducting polymers, 21, 277
Coulombic efficiency (CE), 34, 143–144, 143*f*, 219–220, 222–245, 283, 337–339, 341–342, 347–349
  etched and coated silicon anode, 393, 394*f*
  SiO@C electrode, 144–145, 146*f*
  Si powder, wet-milled, 420–422, 422*f*

509

Coulombic efficiency (CE) *(Continued)*
　Si thin film, 148, 149*f*
　Si-Ti thin films, 150–152, 151*f*
Crack formation, 125–130, 128–129*f*
Crystalline silicon (c-Si), 373–374
　anisotropy, 125
　morphology changes, 125, 126*f*
　solar cell-powered hybrid device, 479, 481*f*
　volume expansion, 125
CubeSats, 464, 482–484*f*
　carbon fiber battery composite fabrication, 489–490, 491*f*
　commercial batteries, 490–497
　current issues, 481–485
　energy storage options for, 479–497
　failure modes for commercial Li-ion batteries in low-earth orbit, 485–487
　future missions, 481–485
　integrating batteries into structural elements, 487–490
　MarCO 6U, 491–492, 495*t*
　pouch-free battery composite, 489–490, 492*f*
Cu-foil, 53–55, 55*f*
CuO nanobelt, 65–66, 66*f*
Cyanoresin organogel, 285
Cyclic voltammetry, 377, 388, 389*f*
Cycling performance of cells
　carbon-coated and uncoated milled silicon electrodes, 395, 395*f*
　full cell with NCM cathode and carbon-coated 24 h-milled silicon, 399–400, 401*f*
　of Si/C powder with different binders, 437–438, 437*f*
　of Si/Li coin cells, 419–420, 419*f*

D

De-agglomeration, 376–377, 385–386
Degradation phenomena, in silicon anodes
　Cahn-Hilliard formulation, 312–314, 319
　concentration and damage profiles, in silicon particle with surface film, 317–321
　Fick's law of diffusion, 312–314, 319
　fracture phase map and capacity fade correlation, 322–326
　lithium ion transport kinetics, 312
　mechano-electrochemical stochastics, 312–316
　Poisson's ratio, 315–317
　surface film on damage, effect of, 321–322
　Young's modulus, 315–317, 321–324

Delithiation process, MD simulation, 100–102, 101–102*f*
　amorphous $Li_{15}Si_4$ nanosheet, 101–102
　Chevrier-Dahn Delithiation, 100, 100*f*
Density functional theory (DFT), 217–219
Density functional theory and Green's function (DFT-GF) approach, 208
"Depth of discharge" test, 133, 133*f*
Diamond wire cutting technology, 69
Diatomite, 75, 77*f*
Diethyl carbonate (DEC), 418–419, 436
Dilatometer, 131, 131*f*
Dimethyl carbonate (DMC), 193–195, 202–203, 278–279
1,3-Dioxolane (DOL), 279
Discharge capacity, 133–135, 134*f*
Discharge test, 133, 133*f*
Dry grinding, 415, 418
d-SiO/C@NCNBs, 66–67, 67*f*
　preparation process, 66–67, 67*f*
　SEM images, 67–68, 68*f*

E

EIS. *See* Electrochemical impedance spectroscopy (EIS)
Elasticity, 416
Elastic strain energy, 163, 169–170, 174*f*
Electric vehicle batteries, solar charging of, 464–467
Electric vehicles (EVs), 331, 374, 411–412
Electrochemical impedance spectroscopy (EIS), 377–379
　carbon-coated 24 h-milled silicon, 399, 400*t*
　etched Si anode, 390–391, 390*f*
　milled silicon and carbon-coated milled silicon electrodes, 396, 396*t*
Electrochemical reduction of SiO_2 ($ERSiO_2$), 80, 81*f*
Electrochemistry transport dynamics, in silicon anodes
　nanorods, 304–306
　nanospheres, 302–304
　radial diffusion, 301
　single-particle model, 301
　single phase diffusion, 301–302
Electrode resistance, etched silicon electrode, 390–391, 391*t*
Electroless etching, 426–427
Electrolytes
　and additives
　　charge/discharge cycles, 281

cost-efficient approach, 281
ethylene carbonate derivative additives, 281–282
reactive organic molecules, 283
salt additives, 282–283
SEI film, 281
silicon anode, for lithium ion batteries, 277–278
solid electrolyte interface (SEI), 277–278
analytical methods, of SEIs, 287–289
carbonate electrolytes and lithium salts, 278–281, 279f
ionic liquid (IL) electrolytes, 286–287
polymer electrolytes, 283–285
Electromotive force (EMF), 104–106
Energy Materials Industrial Research Initiative (EMIRI), 36–37
Ethylene carbonate (EC), 183–184, 191, 193–195, 202–212, 217–219, 278–279, 418–419, 436
Ethyl methyl carbonate (EMC), 278–279
Expanded graphite (EG), 432

F

Faraday's constant, 303
FFA. *See* Furfuryl alcohol (FFA)
Fick's law of diffusion, 312–314, 319
Field emission scanning electron microscopy, 376–377
Finite element-based computational code, 164
Flexible-fiber integrated power device, 473, 475t
Fluoroethylene carbonate (FEC), 193–207, 218–219, 281–282, 418–419
Fourier transform infrared spectroscopy (FTIR), 288
Furfuryl alcohol (FFA), 376–378, 385–386

G

Galvanostatic lithiation, 165
Gas chromatography mass spectrometry (GCMS), 287–288
Gel polymer electrolytes (GPE), 285
Gibbs free energy, 332–333
Glass derived Si (gSi), 69
Graphite, 4, 6t, 119, 412–413
Graphite/Si/graphene (graphite/Si@reGO) composites, 432–434
Graphite/Si-porous carbon composite, 49–50, 50f
Green-Lagrange strain tensor, 314–315

Griffith's theory, 417
Guar gum, 136, 137f

H

Half-cell discharge process, 162–164, 162f
Hardening modulus, 166t
Hierarchical copper-silicon (HCS) NW anode, 441, 443f
High-altitude pseudo-satellites (HAPs), 479
Highest occupied molecular orbital (HOMO), 189–190
High-purity silicon production, via Siemens process, 414–415
Hollow Si nanotubes (h-SiNTs), 51–52
Hybrid and alloy-based silicon materials, 49
carbon-silicon hybrid materials, 49–57
oxide-containing anodes, 61–68, 64f, 68f
silicon-metal alloy anodes, 58–61, 63t
Hybrid anodes, processing, 58, 59t
Hydrostatic stress, 381–382

I

IMPS. *See* Integrated micro power system (IMPS)
Inductively coupled plasma (ICP), 422
In-situ dilatometer, 131, 131f, 135f
SiSPC electrode, 141, 143f
Si thin film, 148–150, 149f
Si-Ti thin films, 152, 152f
In situ TEM lithiation, of silicon nanospheres, 424–425
In situ lithiation, 97, 98f
Integrated micro power system (IMPS), 458–463, 463f
Integrated power device, 464–465, 469t, 471–472, 474f
Integrated power sources (IPS), 472–473t
applications, 475–479, 478t, 480t
challenges, 475–479
issues, 477–479, 480t
monolithic integration, 467–474, 468f
multijunction devices and motivation, 458–460, 459–460f
practical concerns, 475–479, 480t
solar charging of electric vehicle batteries, 464–467, 465–466t, 467f
in space, 458–464
Starshine 3, 460–464
Interface fracture strength, 166t
Interface fracture toughness, 166t

Interface properties, on a-Si pattern anode stability, 172–174, 173f
Internal stress, 316, 334–335, 356, 416
Ionic liquid (IL) electrolytes, 286–287
IPS. *See* Integrated power sources (IPS)
Irreversible capacity loss (ICL), 419–420
Isopropyl alcohol (IPA), 420

J
Jet milling, 418

L
Large-scale Atomic/Molecular Massively Parallel Simulator (LAMMPS), 97
LEO. *See* Low-earth orbit (LEO)
Linear polyamine, polyethyleneimine (LPEI), 73, 74f
Liquid chromatography mass spectrometry (LCMS), 287–288
Liquid silicon, 414
Li-Si system
  binary phase diagram, 412, 412f
  crystal structure data of, 412–413, 413t
  redox potentials, 412, 412t
Lithiated silicon
  anode, 21
  charge distribution, 104, 105f
  electromotive force, 104–106
  structure of, 120–121, 121f, 122–123t
  volume change, 102, 103f
  Young's modulus, 103, 104t
Lithiation process, MD simulation, 97–100, 99f
  amorphous and crystalline Si, 99f
  Chevrier-Dahn Lithiation, 97, 97f
  in situ lithiation, 97, 98f
Lithium bis (oxalato)borate (LiBOB), 281
Lithium bis (fluorosulfonyl)imide (LiFSI), 281
Lithium cobalt oxide (LiCoO$_2$), 4
Lithium ethylene dicarbonate (LEDC), 278–279
Lithium ethylene mono-carbonate (LEMC), 278–279
Lithium hexafluorophosphate (LiPF$_6$), 281
Lithium-ion batteries (LIBs), 3, 119, 411–412, 457, 466t, 483–485
Lithium iron phosphate (LFP), 485–486
Lithium nickel cobalt aluminum oxide (NCA), 485–486
Lithium nickel manganese cobalt oxide (NMC), 485–486
Lithium perchlorate (LiClO$_4$), 4
Lithium polymer (LiPo), 483–485
Lithium sulfide (Li-S), 483–485
Lithium, transport equation, 160–161, 161t
Low-earth orbit (LEO), 458, 462, 479, 485–487, 493–494
Lowest unoccupied molecular orbital (LUMO), 189–190
Low-pressure thermal chemical vapor deposition (LPCVD), 51–52

M
Magnesiothermic reduction, 73–74, 74f, 76t
Magnetron sputtering, 425–426
MarCO spacecraft, 490–492, 493f
Mass capacity, Si nanosheet, 112, 113f, 115f
Matrix-assisted laser desorption/ionization (MALDI) mass spectrometry, 287–288
Mechanical attrition process, 417–418
Mechanical integrity, 157, 165–167
Metal-assisted wet chemical etching, 426–427
Metallic melt processing, 73–74, 75f
Metallurgical-grade silicon (MG-Si), 415
  carbothermal process, 414
  lumps, 417–418
Methylene ethylene carbonate (MEC), 282
Micron-sized Si (m-Si), 138, 140f
Micro-patterned layered anode, 159
Modulus mismatch, 159
Molecular dynamics (MD) simulation, 96–97
  delithiation process of Si, 100–102, 101–102f
  lithiation process of Si, 97–100, 99f
  for porous Si, 112, 113f
Monolithically interconnected module (MIM), 461–462
Monolithic integration, integrated power sources (IPS), 467–474, 468f
Multi-junction thin-film solar cells, 458, 459f

N
Nano-Si/graphene composite, 432–434
Nano-Si/graphite microparticles, 430–432
Nano-silicon
  diatomite-derived, 75, 77f
  surface reactivity reduction, 440–444
    protection against water, 440
    stabilization of SEI, 440–444
  synthesis
    chemical process, 423–424
    physical process, 423
Nano-Si/pyrolysis carbon microparticles, 429–430
Nanostructured 3D electrode architectures, 331–332
  mitigation strategies, 335–340
  pulverization, 340

Index 513

silicon anodes, mechanism of, 334–335
structure and electrochemistry, of silicon, 332–334
three carbon-based scaffold, 352–360
three porous conductive polymer framework, 360–363
three porous metallic current collector, as scaffold, 346–352
three porous silicon-carbon composite, 340–346
Nanostructured electrodes, 14
Nanowires (NWs), 426
NCM cathodes, 399, 401f
Neo-Hookean constitutive model, 165
Neutron reflectometry, 29, 30–31f
Newton-Raphson scheme-based linearization, 164
Nexelion anode, 5
Nickel-cadmium (NiCd), 483–485
Nickel-hydrogen (NiH$_2$), 483–485
Nickel-metal hydride (Ni-MH), 483–485
N-methyl-2-pyrrolidone (NMP), 23
Nuclear magnetic resonance (NMR), 287–288

O

One-electron reduction mechanism, 191, 193–195
One-step gel-coating-reduction approach, 432–434
Open circuit potential (OCP), 303
Operando color microscopy, 214–217
Orbiting Picosatellite Activated Launcher (OPAL), 481
Oxide-containing anodes, 61–68, 64f, 68f

P

Particle size, 138–141
Particle size distribution (PSD), 418
Patterned anode, 159, 160f
Pectin, 25
Pentafluorophenyl isocyanate (PFPI), 283
Performance modeling, in silicon anodes
  crack coalescence, 299
  mechano-electrochemical interaction dynamics, 299–300
  morphology performance interactions
    charge transfer kinetics, 300
    electrochemistry transport dynamics, 301–306
    nanospheres vs. nanorods with same volume and diameter, 306–307
    nanospheres vs. nanorods with same volume and varying diameter, 308–312
    radial diffusion, 306

silicon nanospheres, 300
next-generation anode, for LIBs, 299
volumetric fluctuations, 299
Periodic boundary conditions (PBCs), 96–97
Phenolic resins, 341–342
Phosphorus-doped Si, 128–130
Physical vapor deposition (PVD), 425–426
Piola-Kirchoff stress tensor, 314
Plasma-enhanced chemical vapor deposition (PECVD), 343–344
Poisson's ratio, 166t, 315–317
Polyacrylate binders, 23–24
Polyacrylic acid (PAA), 136, 137f, 225–245
Polyacrylonitrile (PAN), 428–429
Polyaniline (PANI), 22
Polybutadiene (PB), 438–440
Polydopamine, 52–53
Poly (3,4-ethylenedioxythiophene) (PEDOT), 441
Polyethylene glycol dilaurate (PEGDL), 285
Polyethyleneoxide (PEO), 275
Polyimide, as binder, 141
Polymeric binders, 22, 438, 439t
Polypyrrole (PPy), 22
Polyrotaxanes/polyacrylic acid (PAA), 438
Polysaccharide binders, 25
Polyvinyl alcohol (PVA), 225–245, 438–440
Polyvinylidene difluoride (PVDF), 224
Polyvinylidene fluoride (PVDF), 23, 272
Porous silicon, 426–429
Pristine Si, Young's modulus, 103, 104t
Propylene carbonate (PC), 3, 193–195, 202–203, 278–279
PV-battery systems
  circuit representation of, 468f
  current-voltage characteristics, 475, 477f
  monolithically-integrated, 468, 468f
  power gain of, 470, 470f

Q

Quartz, carbothermal reduction of, 414

R

RainCube, 482–483, 483f
Raman spectra, 376–377, 387–388
Raw materials, 68–80
  from biomass and clays, 69–72
  silicon-containing industrial sources, recycling of, 69, 70f
Reactive force field (ReaxFF), 97, 104, 105f
Recycling, of silicon-containing industrial sources, 69, 70f
Redox flow battery (RFB), 473

Reduced graphene oxide (rGO), 432–434
Representative volume element (RVE), 96–97
Residual stress, 375–376, 379–385
Resorcinol-formaldehyde (RF) resin, 52–53
Rice husk, 69–71

S

Scanning electron microscopy (SEM), 420, 421*f*
SCB. *See* Superconducting carbon black (SCB)
SEI. *See* Solid-electrolyte-interphase (SEI)
Self-healing polymers (SHPs), 138, 139*f*, 245–246
Shear lag zone (SLZ), 214–217
SHP-Si anode, 138, 139*f*
Si@C@CNT electrode, 427–428, 428*f*
Si/C composites
    bottom-up assembly, 18, 19*f*
    electrochemical characteristics, 55–56, 56*f*, 57*t*
    SEM and TEM profiles, 50–51, 51*f*
    yolk-shell structure, 52–53, 54*f*
Si/C microspheres, spray drying process, 430–432, 431*f*
Siemens process, high-purity silicon production via, 414–415
SiGe alloy, 20–21
Si/graphite@carbon, 72, 73*f*
Si/graphite/pyrolytic carbon (SiGC) composite, 50*f*
    charge/discharge profiles, 52, 53*f*
    fabrication process, 77, 78*f*
Silanes, 33
Si-Li alloy
    electrode perspective, 11, 12*f*
        pulverization and delamination, 21–28
        silicon/electrolyte interphase, 28–34, 30–31*f*
        volume expansion and material pulverization, 12–21
    material perspective, 8–11
    physical parameters, 121, 124*t*
Silicon (Si), 80
    electrochemical properties, 5, 6*t*
    high-purity, Siemens process, 414–415
    as low-cost anode material, 414–415
    3D nano-porous structure, 16
    volume expansion, 121, 124*f*, 373–374
Silicon-aluminum (Si-Al) film, 347–349
Silicon dioxide, 414
Silicon electrode binders
    adhesion, 272–275
    bottom-up functional binder design, for Si materials, 277
    carboxymethyl cellulose (CMC), 272
    electrical conductivity, 276
    electrochemical stability, 272
    functional conductive polymer adhesive binders, 272
    ionic conductivity, 275
    lithium polyacrylate (LiPAA), 272
    methacrylate radical polymerization processes, 277
    side chain-conducting polymer approach, 277
Silicon/electrolyte interphase, 28–34, 30–31*f*
Silicon/graphene/carbon nano fiber (FSiGCNFs) 3D hybrid electrode architecture, 359–360
Silicon majority anode (SIMA), 245–246
Silicon-metal alloy anodes, 58–61, 63*t*
Silicon microparticles (SiMPs), 219, 225–246
Silicon monoxide (SiO), 36, 141–142
    charge-discharge performance, 143*f*, 144*t*
    thickness change, 144, 145*f*
Silicon nanoparticles (SiNPs), 219–222, 225–246
Silicon nanopowders, mechanically milled
    decrease in mechanical size of silicon, 417–418
    fracture mechanics, 416–417, 416–417*f*
    size effects on electrochemical properties, 418–420, 419*f*
    wet grinding optimization, 420–422, 421–422*f*
Silicon nanospheres, using induced plasma, 422–434
    chemical process, 423–424
    in situ TEM lithiation, 424–425
    nano-Si/graphene composite, 432–434
    nano-Si/graphite microparticles, 430–432
    nano-Si/pyrolysis carbon microparticles, 429–430
    physical process, 423
    porous silicon, 426–429
    Si nanowires, 426
    Si thin films, 425–426
Silicon nanowires (Si NWs), 14, 426
Silicon powder
    etched anode from
        cell assembly, 377
        characterizations, 376–377
        coating on hierarchical nanostructures, 385–388, 386–387*f*

electrochemical characterizations, 377, 388–394, 389–390f, 391t, 392f, 394f
electrode preparation, 377
synthesis, 376–377
unetched anode from
ambient temperature on electrode performance, 397–399
cell assembly, 378–379
characterizations, 378
charging current on electrode performance, 397
coating, 394–397
electrochemical characterizations, 378–379
electrode preparation, 378–379
full cell, 399–400
synthesis, 378
Siloxanes, 33
Silver-coated 3D macro-porous silicon anode, 20
Si/metal oxide composites, 21
Si-metal/silicide composite, 61, 63t
Si nanosheets, as anode, 106
lithiation extents of, 110–111, 110f
modeling and computation, 107–110, 108f
Si/N-doped carbon/CNT (SNCC) nano/micro spheres, 71–72, 72f
Single-crystal silicon (c-Si) anode
preparation, 375–376
Raman spectra, 379, 380f
residual stress, 375–376, 379–385, 381t
surface morphology, 383–385, 384f
SiO@C electrode
cycle performance and Coulombic efficiency, 144–145, 146f
postmortem SEM images, 146–148, 147f
voltage-time and thickness-time profiles, 144–145, 147f
SiO_2 nanobelts, 65–66, 65f
SiO-PAA composite electrode, 24
SiO_x nanowires
electrochemical evaluation of, 436–437
electron micrograph of, 435f
stability curve, 436, 436f
synthesis, 435–436
Si/rGO/CNT (SiGC) electrode, 77, 79f
Si-secondary particle cluster (SiSPC)
in-situ dilatometer, 141, 143f
SEM images, 138, 140f
Si@SiO_x/GH composites, 49–50, 50f
Si/Sn composite, 58–61, 60f, 63t
Si thin films, 425–426

Si/TiO_2 nanowire array (TNA), 61–65, 64f
Si-Ti thin films
charge-discharge curves, 150, 150f
cycle performance and Coulombic efficiency, 150–152, 151f
in-situ dilatometry profiles, 152, 152f
Sodium alginate, 136, 137f, 437–438
Solar charging, of electric vehicle batteries, 464–467
Solar energy, battery charge conversion efficiency, 465–466, 467f
Solid electrolyte interphase (SEI), 373–374, 388, 390–391
binders
conventional binders, 222–224
polymeric binders, 224–247, 226–243t
self-healing polymers (SHPs), 245–246
chemical/electrochemical reactions, 183–184
compositions, electrolyte and salt decomposition, 191–195, 196–201t
electrolyte additives and their roles, 195–207
electrolyte formulation, solvents used for, 184–187
electron tunneling effect, 183–184
energetics of, 189–190
fluoroethylene carbonate (FEC) additive, 203–207
formation
coating strategies, 219–222
computational work on, 217–219
core-shell/yolk-shell morphology, 219–222
growth model, 212
mechanical deformation and associated strain of, 212–217
nano-silicon, surface reactivity reduction, 440–444
negative electrodes, for Li-ion batteries, 184–187, 185–186f
nonaqueous electrolytes, electrochemical stability of, 183–184
properties, 208
Si/SiO_x electrodes, formation on, 208–212
surface passivation, of lithium metal surface, 184–187
vinylene carbonate (VC), 202–203
voltage-dependent phenomenon, 187–188
Space Shuttle Endeavor (STS-113), 481
Specific surface area (SSA), of spherical graphite, 430–432
Spray drying process, 430–432, 431f
graphite/Si/graphene (graphite/Si@reGO) composites, 432–434

Spray drying process *(Continued)*
  nano-silicon, surface reactivity reduction, 440
  Si/C microspheres, 430–432, 431*f*
Stability improvement, Si electrodes
  active matrix with O, 141–148
  binder, 132–138
  inactive matrix with Ti, 150–152
  particle size, 138–141
  thin film, 148–150
Starshine 3, 460–464, 461*f*
Strategic Energy Technology (SET) plan, 36–37
Stress relaxation mechanism, 417
Styrene butadiene rubber (SBR), 23
Succinic anhydride (SA), 283
Superconducting carbon black (SCB), 386–387, 387*f*
Surface clamping, of porous silicon, 442–444
Surface engineering, 158–159
Surface state of charge (SOC), 303

T

Tap density, Si particles, 138, 141*t*
Thermal chemical vapor deposition (TCVD), 356
Thermal decomposition, 415
Thermal plasma synthesis, 423
Thermostat, Berendsen, 102
Thin-film batteries, 148–150, 149*f*
3D carbon-based scaffold electrode, 80
Time-of-flight secondary ion mass spectrometry (TOF-SIMS) method, 272–275
TiO_2 coated core-shell Si particles, 21
Titanium dioxide (TiO_2), 61
Transmission electron microscope (TEM), in situ lithiation of silicon nanospheres, 424–425
Transport equation, lithium, 160–161, 161*t*
Trichlorosilane (HSiCl3), 415

Triethylene glycol methyl ether methacrylate (TEGMA) additive, 288–289
Trimethoxy methyl silane (TMMS), 283
Triphenyl phosphate (TPP), 283
Tris(pentafluorophenyl) borane (TPFPB), 283

U

Ultrathin-graphite foam (UGF), 353

V

Vinylene carbonate (VC), 193–203, 281–282, 436
Volume change, 6*t*, 26
  lithiated Si, 102, 103*f*
  measurement of, 130–132
Volume expansion, 125–130
Volumetric capacity, porous Si, 112, 113*f*, 115*f*

W

Wearable device, 475, 476*f*
Wet grinding, 415, 418, 420–422

X

X-ray diffraction (XRD), 377, 386–387
X-ray photoelectron spectroscopy (XPS), 287–288
X-ray reflectivity (XRR), 208–212

Y

Young's modulus, 315–317, 321–324
  amorphous Si (a-Si), 166*t*
  lithiated Si, 103, 104*t*

Z

Zero-dimensional (0D) nest-like Si nanospheres, 14

Printed in the United States
by Baker & Taylor Publisher Services